EXAM 1 Ch. 4, 6 ✓ W9-AXO-678

11. 1, 2, 3, 6

5. 1, 2, 3, 4, 5, 6, 7, 9

6: 1, 2, 3, 4, 5,

4: 1, 2, 3, 4, 5, 6, 7, 8, 17

Questions:

T/F has history?

Third Edition

Air Pollution Control

Third Edition

Air Pollution Control

A Design Approach

C. David Cooper
Professor
University of Central Florida

F. C. Alley
Professor Emeritus
Clemson University

WAVELAND

PRESS, INC.

Long Grove, Illinois

For information about this book, contact:
  Waveland Press, Inc.
  4180 IL Route 83, Suite 101
  Long Grove, IL 60047-9580
  (847) 634-0081
  info@waveland.com
  www.waveland.com

This book is dedicated to our students,
past, present, and future

CONTENTS

viii   Contents

8 AUXILIARY EQUIPMENT: HOODS, DUCTS, FANS, AND COOLERS 239

9 A PARTICULATE CONTROL PROBLEM 291

10   PROPERTIES OF GASES AND VAPORS   307

11   VOC INCINERATORS   321

PREFACE TO THE THIRD EDITION

Engineers working in air pollution control have many responsibilities. One of the most demanding, yet satisfying, of these is the design of air pollution control systems. This textbook describes the philosophy and procedures for the design of such systems, and will help young engineers prepare for the challenges and rewards awaiting them as designers. In addition, this text presents numerous chapters on specialized control equipment that contain the necessary equations and data to design and specify new systems and/or to analyze existing systems. Therefore, it also serves as a good source of information about air pollution control for more experienced engineers.

Our text has two main objectives. The first is to present information about the general topic of air pollution and its control. The second, perhaps more important, objective is to aid in the formal design training and instruction of engineering students. Design of equipment and systems has often been underrepresented in air pollution books; however, it is a key function of engineers, and should be emphasized in engineering curricula.

Engineering textbooks generally must be updated on a regular basis because of ongoing discoveries, innovations, and developments in all the engineering fields. In this Third Edition, two new chapters have been included—one on biofiltration as a control technology for VOCs and odors, and one on indoor air quality. The other chapters have been updated extensively. Reference to the World Wide Web as a data source is demonstrated frequently, and more emphasis has been placed on PC-based spreadsheet applications in solving problems throughout the book. We have included an appendix containing several practice P.E. Exam problems and step-by-step solutions. Hopefully, students will find these extra problems helpful in learning to master the problem-solving techniques needed for this exam.

Regarding **global climate change,** our thoughts have crystallized and our beliefs have strengthened with each passing year. Please read the preface to the Second Edition, written almost ten years ago. Fortunately, more and more governments now realize the importance of slowing emissions of carbon dioxide and other "greenhouse gases." The Kyoto Protocol has started many nations on the path toward stopping the growth of CO_2 emissions, and eventually reducing those emissions. Phillip Abelson, editor of *Science*, has endorsed the idea of removing large quantities of CO_2 from fossil fuel combustion gases, and then sequestering it underground in abandoned oil fields and coal mines (*Science*, *289*, August 25, 2000). Undoubtedly, in the years to come, many of today's students will be engaged in finding solutions to this global problem.

The authors would like to express our acknowledgments to Ms. Kimberly Giramma and her student assistants at UCF who typed numerous drafts and had to deal with typing engineering equations. We also want to acknowledge the helpful and professional work of Laurie Prossnitz, our excellent editor at Waveland Press.

C. D. Cooper
F. C. Alley

PREFACE TO THE SECOND EDITION

Since the publication of the first edition, there have been a number of developments in the field of air pollution control, including increased awareness of the major impacts of mobile sources on urban air quality, the new Clean Air Act Amendments in the United States, and a rapidly growing concern for air quality throughout the world. Although the primary strength of this book is its emphasis on design, we have added new material in several other areas, including updates on air quality trends, a new section on photochemical smog formation, and an entire chapter on mobile sources.

The authors believe it necessary to offer a few words about **global climate change** that may not be appropriate for inclusion in the body of this text. Most scientists and engineers are trained to state conclusions only with sufficient data to be certain of those conclusions. However, the issues involved in global climate change are of such importance, the time lags in the system are so long, and the possible consequences of inaction are so great that the authors feel we must take a stand now, even without complete certainty. Therefore, the next few sentences represent our beliefs.

We believe that global warming has already started and that events are already in motion that will result in significant changes in large-scale weather patterns by the middle of the twenty-first century, if not sooner—changes that many will not welcome. A number of political and scientific leaders in the world want to begin reducing carbon emissions now. However, some politicians and special interest groups in the United States and elsewhere refuse to act, partly because of substantial economic impacts on particular industries or segments of the economy.

In our view, this attitude is shortsighted, self-serving, and dangerous. The United States needs to be at the forefront of these issues. As engineers, we must pursue up-to-date information, inform ourselves of the facts, and act in the policy-making debates. As individuals, we must do all we can to try to reduce our emissions of carbon dioxide and trace gases that are contributing to global climate change. As the popular phrase goes—*think globally, act locally*.

C. D. Cooper
F. C. Alley

 CHAPTER 1

AN OVERVIEW

*As I was walking in your Majesties Palace at Whitehall . . . a pre-
sumptuous Smoake . . . did so invade the Court . . . [that] men could
hardly discern one another for the Clowd. . . . And what is all this, but
that Hellish and dismall Clowd of SEA-COALE . . . [an] . . . impure
and thick Mist, accompanied with a fuliginous and filthy vapour. . . .*

John Evelyn, 1661

♦ 1.1 INTRODUCTION ♦

The preceding quote was chosen to herald the start of this chapter
and this book because it clearly demonstrates that air pollution is not a
new phenomenon, but was a problem in some local areas centuries ago.
In fact, according to Te Brake (1975), the smoke from the burning of
"sea-coale" in lime kilns in London was a serious problem as early as
A.D. 1285. The air pollution situation in London persisted, and, in
1307, King Edward I banned the burning of sea coal in lime kilns (Te
Brake 1975). By the last quarter of the fourteenth century, the problem
diminished, only to reappear by the middle of the sixteenth century.

According to Te Brake (1975), the periods of peak air pollution
problems in preindustrial London corresponded roughly to periods of
population expansion and fuel "crises"; that is, sea coal (the less-desir-
able and more-polluting fuel) was burned when wood (the preferred
fuel) went into periods of short supply and/or high prices. British
woodlands were subjected to many population-related pressures
including the need for arable land, the need for building materials,
and the need for fuel. The sudden switch in the fifteenth century from
the use of the polluting sea coal to clean-burning wood may have been
the result of the sudden drastic decline in London's population caused
by the Black Death (plague).

Today much of our air pollution is directly related to the combustion of fuels for industrial production, for transportation, and for pro-

1

duction of electricity for domestic use. Although isolated air pollution problems were of local significance centuries ago, air pollution did not become a global concern until the advent of the industrial revolution. Fuel combustion in a country is directly related to the number of people and their standard of living (that is, energy consumption). Since the early 1800s, world population has increased about one order of magnitude, and per-capita energy consumption has increased about two orders of magnitude. Of course, much greater than average increases have occurred in the industrialized urban centers of the world. Therefore, it is not surprising that air pollution has become an international concern in the recent past.

However, if Te Brake's arguments and conclusions are accepted, then some of the reasons for the air pollution problems of today's society are similar to those that existed 700 years ago. Let us hope that our present environmental problems are resolved by our technological abilities and not by a drastic global calamity.

◆ 1.2 DEFINITIONS AND TYPES OF POLLUTANTS ◆

According to one dictionary, *pollution* is a synonym for contamination. Therefore, air pollutants are things that contaminate the air in some manner. The federal government, as well as each state, has incorporated into law a more precise definition of air pollution. The legal definition in the state of Florida (all such definitions are similar) is as follows:

> *Air pollution is the presence in the outdoor atmosphere . . . of any one or more substances or pollutants in quantities which are or may be harmful or injurious to human health or welfare, animal or plant life, or property, or unreasonably interfere with the enjoyment of life or property, including outdoor recreation.*

> *(Florida Administrative Code 1982)*

By the preceding definition, any solid, liquid, or gas that is present in the air in a concentration that causes some deleterious effect is considered an air pollutant. However, there are several substances that, by virtue of their massive rates of emission and harmful effects, are considered the most significant pollutants.

National Ambient Air Quality Standards (NAAQSs) have been established for six **criteria air pollutants**—five **primary** (meaning emitted directly) and one **secondary pollutant** (so-called because it is formed in the lower atmosphere by chemical reactions among primary pollutants). The term **criteria pollutant** comes from the fact that health-based criteria were used to establish the NAAQSs for these pollutants.

The five primary criteria pollutants are *particulate matter less than 10 µm* in diameter (PM-10), *sulfur dioxide* (SO_2), *nitrogen dioxide*

(NO_2) *carbon monoxide* (CO), and *particulate lead*; the secondary criteria pollutant is *ozone* (O_3). Of these, the first four are emitted in the United States (and other large industrialized countries) in quantities measured in millions of metric tons per year and are sometimes called *major* primary pollutants. When speaking of emissions of the criteria pollutants, one often sees the data presented in classifications of particulate matter (PM), sulfur oxides (SO_x), and nitrogen oxides (NO_x). Another class of compounds—*volatile organic compounds* (VOCs)—though not a criteria pollutant, is recognized as a major primary pollutant because of its large emissions and its importance in the reactions that form ground-level *ozone*.

Some clarification is needed in regard to VOCs. As used in this text, VOCs include all organic compounds with appreciable vapor pressures. Some VOCs (for example, propylene) are reactive in the atmosphere, whereas others (for example, methane) are inert. Some VOCs are hydrocarbons (contain only hydrogen and carbon), but others may be aldehydes, ketones, chlorinated solvents, and so on.

In the United States, emissions of the primary air pollutants increased rapidly following the end of the Great Depression. As the population increased and became more mobile and as industrial production soared, it was inevitable that air pollution would increase. However—perhaps because of the development of legislative and technical controls—the emissions of all the major pollutants peaked in the early 1970s and then began to decline. Unfortunately, recent data suggest that for some of the pollutants, this declining trend has slowed and might be reversing (U.S. Environmental Protection Agency 2000). Figure 1.1 illustrates the trends in U.S. emissions for particulates, sulfur oxides, nitrogen oxides, and VOCs. Carbon monoxide and lead data are presented separately in Figure 1.2 because different scales are required.

It is important to note that air pollution is not just a problem of the United States, but rather it is an international problem. While much of the data in this text comes from the American experience, industrialized and developing nations throughout the world have experienced and are continuing to experience severe air pollution problems. According to one author (Nadakavukaren 2000), the less-developed countries (LDCs) have the most serious air pollution problems in the world today. The largest, most crowded cities are particularly impacted; examples are Jakarta, Indonesia; São Paulo, Brazil; Cairo, Egypt; Warsaw, Poland; Beijing, China; Santiago, Chile; Taipei, Taiwan; and of course, the world's largest (and most polluted) urban area, Mexico City, Mexico.

Furthermore, as the populations of the LDCs grow and their economies expand (both at rapid rates), their contributions to global air pollution will become a greater part of the whole. In the early 1970s, it was thought that the United States contributed about 30–50% of world air pollution emissions, depending upon the specific pollutant. How-

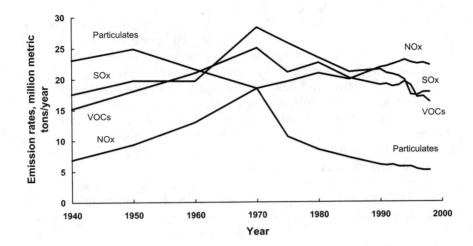

Figure 1.1
Trends in U.S. annual emission rates for SO$_x$, particulates, VOCs, and NO$_x$.
(Adapted from U.S. Environmental Protection Agency, EPA-454/R-00-003, 2000.)

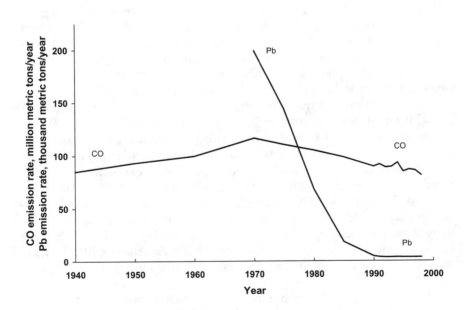

Figure 1.2
Trends in U.S. annual emission rates for CO and lead.
Note: Lead data are unavailable prior to 1970.
(Adapted from U.S. Environmental Protection Agency, EPA-454/R-00-003, 2000.)

ever, owing to vigorous and effective air pollution control efforts over the past thirty years, the United States has reduced its total emissions, whereas in many other countries, emissions have risen rapidly. By 1987, it was estimated that the United States emitted from 12% (for particulate matter) to 40% (for hydrocarbons) of total world emissions (Faiz et al. 1992). By the late 1990s, due more to growth in the LDCs than to improved pollution control in the United States, U.S. emissions accounted for less than 10% of the world's particulate emissions, and no more than about one-third of worldwide emissions of any other pollutant. Over the last ten years, leaders in the World Bank and other organizations, and the individual governments of these countries have recognized these crucial air quality problems. It is encouraging to note that they have now started to address their air pollution problems through technical and regulatory approaches.

The only secondary pollutant for which there is an AAQS and which is also of major concern in urban centers throughout the world is ozone (or more generally, *photochemical oxidants*). Oxidants are secondary pollutants because they are not emitted directly; rather, they are formed in the lower atmosphere by chemical reactions involving sunlight, hydrocarbons, and nitrogen oxides. It is important to distinguish between ozone near the ground (the pollutant) and ozone in the upper atmosphere (which helps protect us from ultraviolet radiation).

The pollutants mentioned above mostly impact people and the environment on a local or urban scale. However, a pollutant with serious regional-scale impact is acid rain (more correctly, *acidic deposition*). Acid precursors, such as sulfur oxides and nitrogen oxides, react with oxygen and water in the atmosphere to form acids that can then precipitate with rain, snow, sleet, or dry particulates. The two most important constituents of acid deposition are HNO_3 and H_2SO_4, which contribute about 98% of the free acidity found in acid rain (Likens 1976). In the 1970s, over two-thirds of the acidity in rainfall was sulfur based and one-third was nitrogen based. During the past twenty years, there has been relatively more success in the United States with controlling SO_2 emissions than with NO_x emissions, and now about 60% of the acidity in rainfall derives from SO_2 and 40% from NO_x (Lear 2000). More information about acid deposition—including maps of the average pH of rainfall throughout the country—can be obtained from the Web site: http://nadp.sws.uiuc.edu.

Monitoring in the eastern United States and in Scandinavia showed a marked decrease in the pH of rainfall from the mid-1950s through the mid-1970s. Rainfall pHs were measured in the range from less than 2 to 5.5 (U.S. Environmental Protection Agency 1979). In many lakes with little natural buffering capacity, lake water pH dropped rapidly. At pHs less than about 4.5, fish die because of the acidity itself and because of the acid leaching of toxic metals from nearby soils into the water (U.S. Environmental Protection Agency 1979). Other effects include disruption of terrestrial ecosystems, corrosion of steel structures, damage to historical artifacts, and possible direct or indirect effects on human health (Bubenick, Record, and Kindya 1981). An aggressive program to reduce emissions of SO_x from power plants over the last twenty years has helped stop this trend in increasing acidity. Current pH monitoring data can be found at http://nadp.sws.uiuc.edu.

◆ 1.3 POLLUTANTS OF GLOBAL CONCERN ◆

For the first time since humans began roaming the earth, we (in the twenty-first century) have the numbers and the power to change our environment on a global scale. Two air pollution problems that confront modern society fall into the category of pollutants of global concern.

OZONE DEPLETION

In the 1930s, chemists invented a "miracle" chemical. It was extremely stable, nontoxic, nonflammable, and could be used in many commercial applications. This chemical and its derivatives that followed are called chlorofluorocarbons (CFCs). CFCs (also known as freons) came to be used throughout the world (as refrigerants, aerosol propellants, foam-blowing agents, cleaning solvents, air conditioning gases, and other substances).

For a long time, no one suspected any adverse consequences to the use of CFCs. However, in 1974, the theory was put forward that CFCs—which are stable in the lower atmosphere—break down in the stratosphere, releasing chlorine atoms (Molina and Rowland 1974). Chlorine atoms and other radicals remove stratospheric ozone very effectively through a set of catalytic reactions that regenerate the chlorine atom or radical; thus, tens of thousands of ozone molecules can be destroyed by one chlorine atom before it is removed from the stratosphere. Stratospheric ozone is a key factor in protecting all life on earth, because it absorbs almost all of the ultraviolet (UV) radiation coming into the earth's atmosphere, preventing the UV radiation from reaching ground level.

In 1985, the dramatic discovery of a huge ozone "hole" over Antarctica proved the theory of ozone depletion. The hole (as big as the United States) showed as much as 50% reduction in the protective ozone layer in that region during the winter months. Since that time, ozone depletion also has been observed over the northern latitudes, including parts of Canada, the United States, Europe, and Russia. Such ozone depletion has likely already accounted for millions of cases of skin cancers and cataracts among humans, similar effects among livestock and wild animals, perhaps billions of dollars of damages in reduced crop yields, and degradation of plastics due to the increased UV radiation reaching the earth's surface.

The discovery of the ozone hole was dramatic; it shocked the world into action. In 1987, 46 countries manufacturing CFCs developed a treaty (the Montreal Protocol) to reduce CFC production and use on a scheduled basis, and by 1989, 39 countries had ratified it. In 1992, the U.S. Congress voted to accelerate the phaseout of CFCs. However, no provisions have been made for recovering and destroying the millions of tons of CFCs that still exist in items such as old refrigerators and old cars. Because CFCs released in the past are still working their way up to the stratosphere, ozone depletion will be a concern for many years to come. For more information, a good Web site to visit is www.epa.gov/docs/ozone/index.html.

GLOBAL CLIMATE CHANGE

As severe a problem as ozone depletion is, at least the world has undertaken steps to solve it. There is, however, another air pollution problem that far overshadows ozone depletion, and that is **global climate change (GCC)**. This problem may be the most significant and the most difficult problem ever faced by humankind. Global climate change (also called global warming or the greenhouse effect) is a complex issue, and can only be addressed briefly in this text. Nevertheless, its importance must not be underestimated. Engineers, scientists, and political leaders throughout the world must constantly be aware of this problem if we are to mitigate its potentially devastating effects.

The term *greenhouse effect* is popular but a bit misleading. The name refers to the retention of infrared (IR) radiation (heat) by certain gases in the atmosphere before that heat is lost to space. The first reason that the popular name is misleading is that atmospheric heat retention is essential to life on Earth. Our comfortably warm climate is only possible because of this heat-trapping ability of (primarily) carbon dioxide and water vapor. Without this natural "greenhouse effect" in our atmosphere, Earth would be approximately 33 degrees C (60 degrees F) colder than it is right now. So, when people talk of the greenhouse effect, they really refer to the recent, rapid, unwanted *increases* in the atmosphere's heat-retention ability.

An interesting example of the natural greenhouse effect is obtained by comparing the so-called *radiation temperatures* (the calculated average temperatures that would be obtained if there were no absorption of IR energy by the planetary atmospheres) of Mars, Earth, and Venus. Mars has essentially no atmosphere, while Venus has a very dense atmosphere that is 97% carbon dioxide. The results of these calculations produce the following conclusions (U.S. Environmental Protection Agency 1983): Mars exhibits a 3-degree C greenhouse effect, Earth shows a 33-degree effect, but Venus obtains a huge temperature boost of 468 degrees C!

Another reason that the term greenhouse effect is misleading is that it conveys a vision of a mild warming of the earth. However, a change in the average temperature of the world of even a few degrees is certain to result in severe changes in large-scale regional weather patterns. Therefore, it is not just the warming that is of concern.

Shifting rainfall patterns and ocean currents combined with more energy retention in the atmosphere likely will result in big increases in the frequency and intensity of the *extremes* of weather—such as hurricanes, tornados, heat waves, droughts, and floods. Major shifts in rainfall might well result in massive crop failures in some areas, much like what is thought to have happened in the fertile crescent of the Middle East 3000 years ago. Another likely major impact is a rise in average sea level, by as much as one meter or more, over the next hundred years (Singer 1989). Even a rise of half this much will flood a large percent of coastal wetlands and severely impact low-lying countries like Bangladesh, Egypt, or the Netherlands. For all these reasons, we believe that *global climate change* is a more descriptive phrase than greenhouse effect.

An indicator of global warming is the earth's **average global temperature (AGT)**. The AGT for the earth in the twentieth century was about 15 C, but increased significantly during the last 25 years. Furthermore, from the depths of the last ice age 18,000 years ago (when AGT was about 9 C) to the present time, average global temperature has increased by 6 degrees C (Barnola et al. 1987). This is a drastic change for AGT, the reasons for which are not completely understood. Therefore, any examination of recent temperature records, especially over short periods of time, and especially in trying to determine cause and effect, must be made with extreme caution.

The AGT is not the best measure of climate change. For one thing, it is a single number being used to measure a massive and complicated system. Also, averaging temperatures near the poles with other temperatures near the equator does not allow tracking of any regional trends. A more precise way to track the changes over time is to measure the *temperature deviation* or *temperature anomaly* at each station. The temperature anomalies can be tracked and averaged with more justification since each tracks the temperature change over time at just one place. Temperature anomaly is a "noisy" variable (meaning that the deviations from year to year can be large compared to the rate of change of the long-term average), nevertheless, it is being reported and used today by many groups.

The average global temperature anomaly (AGTA) for the past 120 years is plotted in Figure 1.3. The average of the years 1951 through 1980 is the base period from which the temperature deviations are calculated. Examination of Figure 1.3 reveals an interesting and sobering trend. From 1880 to 1950 (the beginning of the relatively stable base period), the AGTA increased by about 0.4 degrees C. In the 20 years since the end of the base period, the AGTA has increased by about another 0.3 degrees C, for a total change of about 0.7 degrees C within the last 120 years. This is a huge change over such a short period of time. It seems an almost inescapable conclusion that such a change must be due in large part to anthropogenic emissions, but that conclusion is still considered premature by some people.

As can be seen in Figure 1.3, the bulk of the warming has occurred in two periods—from 1880 through 1940 and from 1976 to the present. Part of the argument against global climate change in the past has been that climate models have been unable to properly reproduce the temperature deviations (both AGTA and those at various locations throughout the world). This is especially true of the earlier period of temperature rise (Zwiers and Weaver 2000). Thus, it was argued that natural forcings (such as changes in the sun's intensity and volcanic eruptions) are the predominant cause of global warming. However, in a comprehensive modeling study using state-of-the-art climate models, Stott and associates have produced excellent agreement of their model predictions with both land and sea temperature changes over the entire period from 1860 through the present (Stott et al. 2000). They included both natural and anthropogenic forcings, and showed that both must be considered. They further extended their model to the year 2100 with a standard set of assumptions for continued emissions. The resulting prediction was that average temperatures rise by 2.5 C over the next 100 years (as compared with an average rise of 0.6 C over the past 100 years).

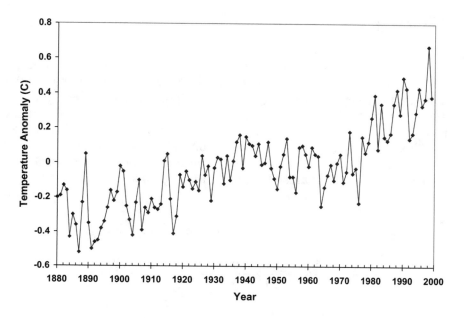

Figure 1.3
Recent behavior of average global temperature anomaly.
(http://www.giss.nasa.gov/data/update/gistemp/graphs)

SOURCES OF GLOBAL CLIMATE CHANGE

The main contributor to global climate change over the past century has been carbon dioxide; however, three other gases are now significant. First, we will discuss the effects of carbon dioxide, then we will return to the importance of these other gases.

The idea of a greenhouse effect—that is, that the burning of fossil fuels puts more carbon dioxide into the air, which in turn warms the earth—is not a new idea. It was first published in a scientific journal by a Swedish chemist, Svante Arrhenius, back in 1896. Arrhenius had noticed that people of his day were burning more and more wood and coal, and he put two and two together. According to Weiner (1990), Arrhenius summed it all up as follows:

> *We are evaporating our coal mines into the air. . . . [which must eventually cause] a change in the transparency of the atmosphere.*

The main source of excess carbon dioxide emissions into the atmosphere is the burning of fossil fuels—coal, oil, and gas. Worldwide, coal is the biggest source of energy for electricity and is the biggest contributor to carbon emissions. Liquid petroleum is the second largest source of carbon emissions. Worldwide, energy consumed by burning liquid petroleum fuels (mainly for transportation) is actually greater than the energy consumed by coal, but accounts for slightly less carbon emissions. Natural gas combustion is the third largest source of carbon dioxide.

▸ ▸ ▸ ▸ ▸ ▸ ▸ ▸ ▸

Example 1.1

Assume an average car in the United States gets 20 miles per gallon of gasoline, is driven 12,000 miles per year, and weighs 3500 pounds. Further assume that gasoline weighs 5.9 pounds per gallon and contains 85% carbon by weight. Is there any truth to the statement that each car emits its own weight in carbon dioxide each year!?

Next, given that there are about 600 million vehicles worldwide, estimate the annual global carbon emissions from motor vehicles. Give your answer in Teragrams (1 Tg = 1 trillion grams = 1 million metric tons) per year.

Solution

The carbon contained in the gasoline burned annually is

$$\frac{12{,}000 \text{ mi}}{\text{year}} \times \frac{1 \text{ gal}}{20 \text{ mi}} \times \frac{5.9 \text{ lbs}}{\text{gal}} \times 0.85 = \frac{3010 \text{ lbs C}}{\text{year}}$$

The carbon dioxide emitted is

$$3010 \text{ lbs C/yr} \times 44 \text{ lb } CO_2/12 \text{ lb C} = 11{,}040 \text{ lbs } CO_2/\text{yr}$$

So the average U.S. car emits *much more* than its own weight in carbon dioxide each year!

To estimate worldwide emissions from vehicles, we must make a number of gross assumptions as to the average vehicle in the world (including cars, trucks, buses, mopeds, and so on). Let us assume that the average vehicle in the world travels 24,000 km per year, gets 9 km per liter, and burns fuel with a density of 0.75 kg/L and with a carbon content of 87%. With these assumptions, annual carbon emissions from vehicles are:

$$\frac{24{,}000 \text{ km}}{\text{year}} \times \frac{1 \text{ L}}{9 \text{ km}} \times \frac{0.75 \text{ kg}}{\text{L}} \times 0.87 \times 600(10)^6 \text{ veh} \times \frac{1 \text{ Tg}}{10^9 \text{ kg}} =$$

$$\frac{1044 \text{ Tg}}{\text{year}} \text{ (1044 million metric tons/yr)}$$

It is interesting to note that the answer from this simple approach compares favorably with published estimates.

◀ ◀ ◀ ◀ ◀ ◀ ◀ ◀ ◀

Altogether, it is estimated that about 5.5 billion metric tons *per year* of carbon emissions (about 20 billion metric tons per year of carbon dioxide) come from the combustion of fossil fuels worldwide. The burning of fossil fuels is the major contributor to the increase in atmospheric carbon dioxide, but it is not the only contributor—deforestation is a major factor, as is decomposition of solid waste. (In fact, landfills not only emit CO_2, they also release methane, one of the three other significant greenhouse gases.)

Deforestation hurts in two ways. First, when large tracts of forest lands are cleared and burned (almost all the wood is burned within months of clearing)—and, by the way, they are being cleared at an incredible rate worldwide—a lot of tree-stored carbon gets put back into the atmosphere. Second, deforestation decreases that part of the earth's biomass that removes CO_2 from the air. Trees are especially good at storing carbon and keeping it out of the air for years.

The earth's biomass (plants and animals) used to be essentially balanced with regard to uptake and release of carbon dioxide. However, because of huge increases in human population and huge decreases in forestlands over the past 300 years, the biomass is now out of balance, with a result of net emissions of CO_2 into the air. Reasonable estimates of net biomass additions of carbon emissions to the air range between 0.5 and 1.6 billion metric tons per year (Leggett 1990). Taking 1.0 billion tons annually as the average thus increases our estimate of net global carbon emissions up to 6.5 billion metric tons per year.

Other processes remove carbon dioxide from the air besides the photosynthesis activity of trees. The oceans are the major sink for

CO_2, holding far more CO_2 than the atmosphere. Plankton and other plants in the water remove CO_2 through photosynthesis like plants on land. Also, CO_2 dissolves directly from the air into the water. However, the oceans also release CO_2 back into the air, and right now the oceans and the atmosphere are pretty much in balance.

The fact remains that about 6.5 billion tons of carbon (24 billion tons of carbon dioxide) continue to be added to our air every year by burning fossil fuels, by deforestation, and by several other pathways. Over the last hundred years or so, about half of the new emissions have remained airborne (Masters 1991), steadily increasing the CO_2 content of the atmosphere. Based on measurements of air bubbles trapped in glacial ice corings, it is widely accepted that, prior to the industrial revolution, the carbon dioxide content of the atmosphere was fairly stable at 280 parts per million (ppm). [*Parts per million* is a common unit of measure for gas concentrations and will be defined later in Eq. (1.4).] By 1900, the level had reached about 300 ppm, reflecting the net increase in global emissions of carbon dioxide.

In 1958, the first accurate and precise measurements of atmospheric CO_2 concentrations were begun by Charles D. Keeling at the Mauna Loa observatory in Hawaii (see Figure 1.4). His now-classic work showed that the 1958 concentration of CO_2 was 315 ppm (Bacastow, Keeling, and Whorf 1982). Compared with the CO_2 level of 200 years earlier, the 1958 level of 315 ppm was a 12.5% increase (giving an average annual rate of increase over two centuries of 0.0625% per year). By 1980, the CO_2 level was 340 ppm, and by 1999 it had reached 370 ppm—a 17% increase from its 1958 value (giving an average rate over 41 years of 0.41% per year, almost a sevenfold increase in the *rate of growth*).

The other three gases in the atmosphere that are responsible for the recent increases in the heat retention capability of the atmosphere are methane, nitrous oxide, and CFCs. All have increased rapidly. Methane grew from 1.48 ppm in 1978 to 1.69 ppm in 1988, a 14% increase in 10 years. Nitrous oxide (N_2O) grew from 296 parts per billion (ppb) in 1978 to 307 ppb in 1989, a 4% increase. CFC-11 concentration in the atmosphere grew from about 157 parts per trillion (ppt) in 1978 to 232 ppt in 1987 (Studt 1991), a 48% increase in nine years.

There have been a number of estimates made about how much of a contribution each of these gases makes to the overall warming effect. Each gas absorbs infrared differently, with some CFCs being as much as 15,000 times as powerful as carbon dioxide (molecule for molecule) in terms of heat retention. Estimates (Flavin 1989) put the relative contributions as follows: carbon dioxide, 57%; CFCs, 25%; methane, 12%; and nitrous oxide, 6%, based on their current concentrations in the air. However, based on their recent rates of increases mentioned above, the effects of the trace gases are even more important.

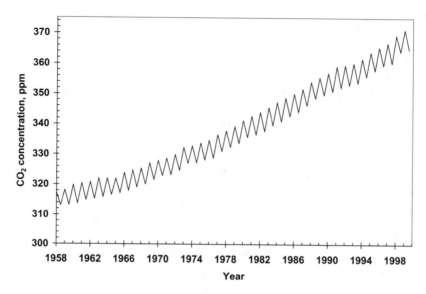

Figure 1.4
Growth in concentration of atmospheric CO_2 as measured at
Mauna Loa, Hawaii.
(Keeling and Whorf, 2000: http://cdiac.esd.ornl.gov/trends/co2/sio-mlo.html)

There is no doubt that greenhouse gases are increasing rapidly in our atmosphere. AGT appears to be increasing as well, but what evidence is there that global climate change is occurring? Other evidence of global warming is more anecdotal, partly because "the weather" is subject to large fluctuations from season to season and from year to year, and partly because the earth is so huge that it takes a long time for "real changes" to show up. "Real changes" are defined here as those that are large enough to say "without a doubt" that they are not part of the "noise" (normal random fluctuations about the mean), and that they are caused by global warming. However, there are several examples of anecdotes that seem particularly compelling. The U.S. Weather Service keeps temperature records that show that 8 of the 10 hottest years in the previous century occurred in the ten years from 1990 through 1999. The year 2000 was the warmest since such records have been kept. There have been widespread incidents of droughts and fires throughout the United States during the 1990s. The Inuit Indians in northwestern Canada have stated in research interviews that there are fewer seals and polar bears to hunt due to thinning sea-ice, and warmer weather has brought more mosquitoes that stay longer (*Orlando Sentinel* 2000). Biologists have noted that the range where certain butterflies live has crept northward by more than 100 miles during the past 25 years. It was noticed in the year 2000 (for the first time since such observations have been recorded) that there was no solid ice—only open water—at the north pole.

Because there are so many apparently random influences on local weather, and so much natural variation from place to place and from year to year, most people cannot grasp the concept of climate changes over a 100- to 200-year period. They want hard scientific evidence of cause and effect in weather differences over, say, a five-year period. So far, the changes in climate due to GCC have been small enough to be largely masked by the "noise" in the system. Despite the many pieces of anecdotal evidence, there are still many skeptics. A common technical criticism is that local surface temperature measurements may be contaminated by local urban heat island effects. Temperature records at rural sites, and records of sea surface temperatures, do not show the same rapidly increasing trends that have been observed at urban sites. For more data and discussions of some of the technical issues involved with GCC, visit the following Web sites:

http://www.giss.nasa.gov/data/update/gistemp/graphs
http://www.epa.gov/globalwarming/climate/trends/temperature.html
http://www.microtech.com.au/daly/graytemp/surftemp/html

Everyone agrees that steps to mitigate GCC are very expensive, and may require substantial sacrifices by many people. It has been argued that to undertake such steps prematurely would be foolish. However, to wait much longer may be even more foolish. The authors' views on global climate change were stated in 1994 in the Preface to the second edition of this text. Our global climate models, while good, are far from perfect. Yet, these models do predict that major effects will occur within the next several decades. The sad fact is that the inertia of the earth-atmosphere system is so great that, once begun, these changes cannot easily nor quickly be reversed.

WHAT CAN WE DO?

Anecdotal evidence suggests that GCC has already begun. World political leaders created the United Nations Framework Convention on Climate Change (UNFCCC) and opened it for signatures in Rio de Janeiro, Brazil, in June 1992. By June 1993, there were 66 country signatories to the "Rio Accord," and now many more have joined. This was a very general treaty, but established three very important agreements. First, it acknowledged that there was a problem, that the problem likely was of human making, and that countries should begin taking steps to address the problem. Second, it strongly supported the concept of sustainable development for future growth. Third, it laid the groundwork for future, more specific agreements. For more information, visit the Web site http://www.unfccc.de/resource/convkp.html.

The Kyoto Protocol (visit http://www.cnie.org/nle/clim-3.html) was one subsequent product of the UNFCCC. This agreement was adopted in Kyoto, Japan, in December 1997, and as of September 2000 had 84 signatories. The Kyoto Protocol commits the industrial countries to reducing greenhouse gas emissions to 7% below their 1990 levels by 2008–2012. Assuming normal economic growth, this is projected to be about 20–30% below the level of emissions that otherwise would be occurring at that time. Many actions to achieve the reductions, including emissions trading, are allowed. The United States and a number of other countries have not ratified this treaty; the U.S. has stated that meaningful commitments to reducing greenhouse gas emissions must be made by developing countries as well.

In our view, the United States needs to be a leader in adopting an energy policy that stresses *reducing carbon emissions*. A strong commitment to *energy conservation* and to the rapid development of commercial *solar power* should be the cornerstones of our national policy. In addition, it might prove necessary to capture carbon dioxide from the stack gases of fossil fuel-fired power plants, and then sequester it underground (Abelson 2000). While expensive, this would be a direct and effective way to reduce emissions to the atmosphere. A strong international commitment is also needed to control emissions of trace gases. We *as a nation* should strive to halt deforestation and encourage reforestation in countries around the world. Finally, we as a nation should work toward *world population control*.

In the interim, what can be done by citizens? Obviously, we as *individuals* can reduce our own energy consumption, we can reduce waste, we can plant trees locally, and, by our purchasing practices, we can encourage recycling and product substitution in all areas that involve the greenhouse gases. As *environmental engineers*, we should each maintain a keen awareness of this issue, and be prepared to discuss it publicly. We should try to get involved politically and to educate our political and industrial leaders. And of course, in the application of our profession, we must continue to look for ways to improve technology to reduce emissions and/or find alternative sources for energy and certain products. We must all do all we can to try to mitigate global climate change. Let us follow the popular advice—think globally, act locally.

Unfortunately, the discussion in this section has been an all-too-brief review of the complex issues of global climate change. However, as stated earlier, the focus of this book is on the engineering design and analysis of air pollution control systems. These systems typically are built to control local or regional air pollution problems, and thus, our attention will focus on the major pollutants as defined earlier. However, before we discuss the causes, sources, and effects of these pollutants, we will briefly review the legislative and regulatory history of air pollution control. If we are familiar with the events of the past, we may be better able to understand and deal with the present, and possibly the future.

◆ 1.4 LEGISLATIVE AND REGULATORY ◆ TRENDS IN THE UNITED STATES

Chicago and Cincinnati passed smoke control ordinances in 1881, and most other large American cities followed by 1912 (Council on Environmental Quality 1970). Smoke pollution in large industrial cities reached its peak in the 1930s to 1950s, and was brought under control not only by better control devices but also by the switch from coal to cleaner-burning oil and gas. However, these early local efforts did not address the (at the time unsuspected) more complex problems of air pollution (such as photochemical smog).

The first concerted efforts to do something about these more complex problems came in California in the late 1940s and early 1950s. Research by Professor A. J. Haagen-Smit in the late 1940s indicated that a photochemical reaction was responsible for the harmful pollutants in Los Angeles smog, rather than direct emissions of those particular pollutants. His findings were attacked by several groups (Krier and Ursin 1977), but were later proved to be true. Los Angeles County established an Air Pollution Control District (APCD) in 1946, and state and local governments began to exert more and more control over air pollution emissions. An excellent review of the efforts in California and at the federal level from 1940 to 1975 is presented by Krier and Ursin (1977). In this text, we will briefly mention some of the federal efforts.

During this same period of awakening interest in California (mid-1940s to early 1950s), two air pollution episodes occurred elsewhere that dramatically focused attention on the emerging air pollution problem. In 1948 in Donora, Pennsylvania (an industrialized town in a small valley), a four-day smog caused illness in about 7000 people, including the deaths of 20 people. In 1952, when a three-day smog blanketed London, England, the results were even more disastrous. There were an estimated 4000 deaths attributed to that air pollution episode. The deaths occurred primarily among those people who had histories of bronchitis, emphysema, or heart trouble (Wark and Warner 1981).

FEDERAL LEGISLATION

The history of federal legislative efforts begins with the **Air Pollution Control Act of 1955**. This act provided funds only for federal research and technical assistance (and not control), and was to be in effect only for five years. At the time the Air Pollution Control Act was proposed, there was a great deal of concern over federal "intervention" into what was thought to be a state and local problem. In fact, one of the early supporters of federal air pollution legislation, Senator Thomas Kuchel of California, "had assured the Senate that his proposal would not bring the federal government into control activities" (Krier and Ursin 1977).

As we know today, the intent of Congress has changed significantly from 1955, and rightfully so. Air pollution is a problem of international significance. Air movement does not recognize political boundaries. Although local and state efforts are extremely important to the successful development and implementation of any air pollution control program, overall management responsibility for our air resources should reside at the federal level.

The **Clean Air Act of 1963**, which was specifically requested by the president, replaced the 1955 act. Not only did it provide money for federal research, but also provided for grants to outside research agencies. This act also provided, for the first time, federal authority to address interstate air pollution problems (Council on Environmental Quality 1970).

The next piece of legislation was the **Motor Vehicle Air Pollution Control Act of 1965**. It established, for the first time, a federal program for the regulation of emissions from new motor vehicles. However, this was actually a benefit to the automotive industry because it helped the industry avoid having to make cars to meet possibly fifty different sets of emission regulations (for the various states). Although the emissions limits were first set for 1968 model cars based on technology existing in 1965, the law recognized the need to tighten controls as new technology became available.

The **Air Quality Act of 1967** was the next of the federal laws dealing with air pollution control. This piece of legislation not only extended the role of the federal government in research and development, but also articulated the new idea that research was not a substitute for regulation and that the federal government did in fact have the right and duty to enforce the use of control equipment. This act required the secretary of the Department of Health, Education, and Welfare (HEW) to designate air quality control regions to facilitate regional planning and control efforts. It also required the secretary of HEW to

> promulgate air quality criteria which, based on scientific studies, describe the harmful effects of an air pollutant on health, vegetation, and materials . . . [and to] . . . issue control technology documents showing availability, costs, and effectiveness of prevention and control techniques.

> *(Council on Environmental Quality 1970)*

During the Senate hearings for this legislation, a novel approach was introduced—namely, that regulatory standards could precede existing technology. Thus, industry could be forced to develop new technology to meet the standards by a certain deadline. This is known as *technology-forcing legislation*. Although not written into the 1967 act, this concept was used in the major 1970 amendments to the Clean Air Act.

The **Clean Air Act Amendments (CAAA) of 1970** is widely recognized as a powerful and important piece of environmental legislation. Along with the **National Environmental Policy Act,** which provided the authority to create the **Environmental Protection Agency (EPA),** it put some teeth into air pollution control enforcement. One of the major objectives of the CAAA of 1970 was to attain clean air by 1975. To meet this objective, new standards and new timetables for achieving them were established.

There were several parts of the CAAA of 1970 that were of major significance. We will mention some of the more important items. The act required the EPA to establish National Ambient Air Quality Standards (NAAQSs)—both primary standards (to protect public health) and stricter secondary standards (to protect public welfare). The CAAA of 1970 also required the various states to submit State Implementation Plans (SIPs) for attaining and maintaining the national primary standards within three years.

In light of the major contribution of mobile source emissions to the overall air pollution problem, automobile emissions were arbitrarily set at a 90% reduction from the 1970 (for CO and hydrocarbons) or 1971 (for NO_x) model-year emissions, to be achieved by 1975 (1976 for NO_x). This was a prime example of technology-forcing legislation as there was no proven way to achieve these goals when this law was enacted. In retrospect, these goals proved to be unattainable, and this part of the law has been amended several times.

Tough standards were to be written by the EPA for certain new industrial plants. These New Source Performance Standards (NSPSs) were national standards and were to be implemented and enforced by each state. This would prevent the possibility of weakened state air pollution standards resulting from the competition between states in trying to attract new industry. Industries were required to monitor their own emissions and to make the data available to the EPA and state pollution control agencies. Large fines and criminal penalties were authorized ($25,000/day and/or 1 year in prison) for willful violations of this act.

The **Clean Air Act Amendments of 1977** was a major piece of legislation (185 pages, single spaced) incorporating many modifications and additions to the Clean Air Act. However, the 1977 amendments retained the basic philosophy of federal management with state implementation. We will now discuss the highlights of the 1977 amendments.

The EPA was required to review and update, as necessary, air quality criteria and regulations as of January 1, 1980, and on five-year intervals thereafter. The EPA was required to continue the process (begun under the CAAA of 1970) of issuing performance standards for new and existing industrial sources.

A separate subpart was included for "prevention of significant deterioration" (PSD) of air quality in regions cleaner than the NAAQSs. Before the 1977 amendments, it was theoretically possible to pollute clean air up to the limits of the ambient standards by simply locating more industry in clean air regions. However, the PSD subpart defined Class I (pristine) areas, Class II (almost all other) areas, and Class III (industrialized) areas. Under the PSD subpart of the 1977 amendments, the ambient concentrations of pollutants would be allowed to rise by almost nothing in Class I areas, by specified amounts in Class II areas, and by larger amounts in Class III areas.

Another subpart of the 1977 amendments designated regulations for *nonattainment* areas, which are areas of the country that were already in violation of one or more of the NAAQSs. A policy known as *emissions offset* was adopted that allowed a new source to be constructed in a nonattainment area provided its emissions were *offset* by simultaneous reductions in emissions from existing sources. Hence, a company wishing to build a new plant in a nonattainment area could choose to install pollution control devices on one of its older plants, or could help pay for such devices on another source. Emissions offsets could be "banked" with the state for later use by a company, or could be sold or traded among companies. In addition, stricter emissions limits could be imposed on sources proposed for nonattainment areas than for sources in attainment areas. Thus, pollution abatement equipment for plants in nonattainment areas must be designed and operated to very strict standards.

The 1977 amendments allowed the attainment of the emissions standards for automobiles to be further delayed, and the standard for NO_x was permanently relaxed from the original goals of the CAAA of 1970. The carbon monoxide (CO) and hydrocarbon (HC) standards were set at a 90% reduction from the 1970 model year (to 3.4 g/mile for CO and 0.41 g/mile for HCs) to be achieved by the 1981 model year. The required NO_x standard was set at 1.0 g/mile by the 1982 model year (versus about 5.5 g/mile in 1970).

In addition, standards were proposed for heavy duty vehicles (HDVs) such as trucks and buses. The standards for HDVs were generally much less restrictive than for automobiles, and concerns were voiced about emissions from HDVs (Brown and Cooper 1978).

By 1990, Congress recognized that, despite previous efforts, there were still about 100 million Americans living in urban areas that did not meet EPA standards for healthful air. The **Clean Air Act Amendments of 1990** introduced sweeping changes to the Clean Air Act, including, for the first time, specific provisions addressing global air pollution problems. This comprehensive legislation (over 750 pages) consists of 11 separate titles that address several key issues: urban air pollution (particularly ozone, carbon monoxide, and PM-10), mobile sources, air toxics, acid deposition, and stratospheric ozone protection. These issues will be discussed briefly in the next few paragraphs.

Title I of the amendments applies to **urban areas** that are in **nonattainment** of one or more of the **NAAQSs**. The pollutants addressed specifically are ozone, carbon monoxide, and PM-10. The new law categorized American cities with regard to the level of *ozone* pollution (five categories: *marginal, moderate, serious, severe, and extreme*), and mandated various control measures (depending on category) to be implemented over differing time periods to bring these urban areas into attainment status. *Moderate, serious, severe*, and *extreme* areas must achieve a 15% reduction in volatile organic compounds within 6 years. For *serious* areas and above, a continued 3% per year reduction in VOCs is required until attainment is achieved. In an *extreme* area, such things as gasoline-powered lawnmowers and charcoal lighter fluids may well be restricted. Major NO_x sources must meet the same requirements as major VOC sources for ozone control unless the EPA finds no benefit.

For *carbon monoxide*, two categories (*moderate* and *serious*) were defined, and measures such as enhanced inspection/maintenance programs and required use of oxygenated fuels in winter were prescribed for certain CO areas. This title also will have a major impact on controlling the emissions of *PM-10*.

Because mobile sources emit such a large percent of total urban emissions, **Title II** of the amendments deals specifically with **mobile sources**. Revised emission standards are prescribed in two tiers—Tier I (by 1994 to 1996) is 0.25, 3.4, and 0.4 g/mile for non-methane hydrocarbons, CO, and NO_x, respectively. Cold temperature idle emissions are restricted, clean fuels are required for certain areas, gasolines must have lower volatility, diesel fuels are limited to 0.05% sulfur content, and as of January 1, 1996, lead was banned from use in motor vehicle fuel.

Title III of the amendments deals with **air toxics**. It lists 189 hazardous air pollutants (HAPs, also called air toxics), and requires the EPA to list source categories that emit one or more of those air toxics and to publish a schedule for regulation of those source categories. In addition to major sources like steel plants or petrochemical facilities, small area sources like dry cleaners, gas stations, and printers will be regulated.

Title IV addresses **acid deposition**, calling for a nationwide SO_2 emissions reduction of 10 million tons per year from 1980 levels, primarily from electric utilities. It promotes energy conservation and clean coal technologies, and utilizes an innovative market-based system for the SO_2 reductions, in which utility allowances for SO_2 emissions can be bought and sold. In addition, the EPA is required to develop a revised NO_x NSPS for utility boilers.

Title V establishes a new federal **operating permit** program to be administered by the states. States have three years to implement the permit program, and the program is subject to EPA approval. Permit fees will be assessed to pay for the program. Permit programs must include at least all Clean Air Act requirements applicable to each source, a schedule for compliance, and requirements for monitoring and reporting.

Title VI deals with **stratospheric ozone protection**, establishing a phase-out schedule for CFCs, halons, and carbon tetrachloride similar to the one initiated by the Montreal Protocol (which required use of the above pollutants to be eliminated by 2000). CFCs are being replaced by hydrochlorofluorocarbons (HCFCs), which are more degradable in the lower atmosphere, but are to be phased out by 2030. Mandatory recycling of CFCs from mobile source air conditioners was required by 1992.

Title VII makes the Clean Air Act more easily **enforceable**, and extends the range of civil and criminal penalties that may be sought for violators. Administrative fines of up to $200,000 can be assessed by the EPA. Other titles in the amendments deal with clean air research, disadvantaged businesses, employment transition assistance, and miscellaneous provisions.

The Clean Air Act Amendments of 1990 is a complex piece of legislation. Full compliance will take decades, and will be costly. Many industrial operations will be affected greatly, in addition to everyday activities like driving a car, mowing grass, getting laundry dry-cleaned, and even cooking outdoors on a charcoal grill.

FEDERAL REGULATIONS AND STANDARDS

Specific regulations are required to implement the intentions of the various federal and state laws regarding air pollution. It is usually desirable to write laws generally and then to allow governmental regulatory agencies to do the necessary research and to write the specific detailed regulations. This approach has been followed by Congress and the EPA.

The two types of standards are Ambient Air Quality Standards (AAQSs), which are those that deal with concentrations of pollutants in the outdoor atmosphere, and Source Performance Standards (SPSs), which are those that apply to emissions of pollutants from specific sources. AAQSs are always written in terms of concentration ($\mu g/m^3$ or ppm), whereas SPSs are written in terms of mass emissions per unit of time or unit of production (g/min or kg of pollutant per metric ton of product produced).

Under current federal law, AAQSs have been established for six pollutants: particulate matter (PM-10), sulfur dioxide, nitrogen dioxide, carbon monoxide, ozone, and lead (40 CFR 50). National AAQSs were set by the EPA for these pollutants based on two criteria: The *primary standards* were established to protect the public health, whereas the *secondary standards* were established to protect the public well-being. The difference is that public well-being can be related to nonhealth effects such as visibility reduction (aesthetic) or crop damage (economic). These standards are presented in Table 1.1. Note that some states (for example, California) have set their own standards, which are stricter than those listed in this table. There are many other

Table 1.1   National Ambient Air Quality Standards (NAAQSs)[a]

Pollutant	Averaging Time	Standard Value and Type[b]
PM-10	Annual arithmetic mean	$50\ \mu g/m^3$ (P&S)
	24-hour average	$150\ \mu g/m^3$ (P&S)
PM-2.5[c]	Annual arithmetic mean	$15\ \mu g/m^3$ (P&S)
	24-hour average	$65\ \mu g/m^3$ (P&S)
CO	1-hour average	35 ppm (P)
	8-hour average	9 ppm (P)
SO_2	Annual arithmetic mean	0.03 ppm (P)
	24-hour average	0.14 ppm (P)
	3-hour average	0.50 ppm (S)
NO_2	Annual arithmetic mean	0.053 ppm (P&S)
O_3[d]	1-hour average	0.12 ppm (P&S)
O_3[c, d]	8-hour maximum	0.08 ppm (P&S)

[a] Standards (other than those based on annual mean) are not to be exceeded more than once per year.
[b] Type of standard: P = primary, S = secondary.
[c] As of late 2001, the PM-2.5 and the 8-hour ozone standards are not in effect. A 1999 federal court ruling blocked the implementation of these standards. EPA has asked the U.S. Supreme Court to reconsider that decision.
[d] The standard is considered to be exceeded when the 3-year average of the annual fourth highest daily maximum exceeds the standard.

Adapted from U.S. EPA Web site (www.epa.gov/airs/criteria.html), Dec. 2000.

compounds that have been considered air pollutants, some of which may be extremely important in certain regions. The omission of specific listings in this text is not intended to downplay their importance, but is simply a matter of space limitation.

A number of other countries in the world (including developing nations like Brazil, Mexico, and Indonesia) have established AAQSs. Also, the World Health Organization (WHO) has published guidelines for air quality standards. In many cases, the standards for some of these other countries and/or the WHO guidelines are stricter than United States standards. For example, the 1-hour ozone standard in Brazil is $160\ \mu g/m^3$, while the WHO guidelines range from 150 to 200 $\mu g/m^3$ (Faiz et al. 1992).

National New Source Performance Standards (NSPSs) are written as mass emissions rates for specific pollutants from specific sources. NSPSs are derived from actual field tests at a number of industrial plants, and are numerous because of the variety of sources (40 CFR 50). NSPSs apply to new sources (those constructed after the rule is promulgated); existing sources often require special handling. Some examples are given in Table 1.2.

Table 1.2   Selected Examples of National New Source Performance
Standards (NSPSs)

1. Steam electric power plants (built after September 18, 1978):
   A. Particulates: 0.03 lb/million Btu of heat input (13 g/million kJ)
   B. NO_x: 0.20 lb/million Btu (86 g/million kJ) for gaseous fuel
      0.30 lb/million Btu (130 g/million kJ) for liquid fuel
      0.60 lb/million Btu (260 g/million kJ) for anthracite or bituminous coal
   C. SO_2: 0.20 lb/million Btu (86 g/million kJ) for gas or liquid fuel. For coal-fired
      plants, the SO_2 standard requires a scrubber that maintains at least a 70%
      SO_2 removal efficiency up to an emission rate of 0.60 lb/million Btu, but
      may require a scrubber that is more than 90% efficient depending on the
      percent of sulfur in the coal and the coal's heating value. Furthermore, the
      maximum permissible emission rate is 1.2 lb SO_2 per million Btu of heat
      input, but the permissible emission rate may be less depending on the
      coal's percent of sulfur and the required scrubber efficiency. [A convenient
      way to apply this regulation is to use Figure 1.5 on page 28 (developed by
      Molburg in 1980), in which the required removal efficiency is the ordinate
      and the allowable SO_2 emission rate is the abscissa. To find the precise
      operating point for any given coal, draw a straight line from the origin of the
      graph to the point within the curved lines that best represents the proper-
      ties of the coal. The point where the line intersects the shaded *admissible
      region* defines the operating point, which consists of a required efficiency
      and an allowable emission rate.]
2. Solid-waste incinerators: The particulate emission standard is a maximum 3-hr
   average concentration of 0.18 g/dscm* corrected to 12% CO_2.
3. Nitric acid plants: The standard is a maximum 3-hr average NO_x emission of
   1.5 kg/metric ton of 100% acid produced. All NO_x emissions are to be
   expressed as 100% NO_2. Also, the stack gases must meet 10% opacity (where
   0% opacity represents perfectly clear stack gas, and 100% opacity means
   completely opaque).
4. Sulfuric acid plants: The standard is a maximum 3-hr average emission of SO_2
   of 2 kg/metric ton of 100% acid produced. An acid mist standard is a maximum
   3-hr emission of 0.075 kg SO_2 per metric ton of acid produced. Also, the stack
   gases must meet 10% opacity.
5. Primary copper smelters: The particulate emission standard is 50 mg/dscm,
   the SO_2 standard is 0.065% by volume, and the opacity is limited to 20%.
6. Wet-process phosphoric acid plants: The total fluorides emission standard is
   10.0 g/metric ton of P_2O_5 feed.
7. Iron and steel plants: Particulate discharges may not exceed 50 mg/dscm, and
   the opacity must be 10% or less except for 2 minutes in any hour.
8. Sewage sludge incinerators: The particulate emission standard is 0.65 g/kg
   sludge input (dry basis). The opacity standard is 20%.

*dscm means dry standard cubic meter.
Adapted from 40 CFR (Code of Federal Regulations) 60, 1991.

A separate category of standards for emissions from point sources
has been created for hazardous air pollutants. These National Emission Standards for Hazardous Air Pollutants (NESHAPs) apply to

those substances that do not have AAQSs but that may result in "an increase in serious irreversible, or incapacitating, reversible illness." As of this writing, NESHAPs have been promulgated by the EPA for only a few sources of pollutants, but activity in this area has increased greatly as a result of the Clean Air Act Amendments of 1990. The NESHAPs are very specific as to sources and types of control methods. A brief summary is presented in Table 1.3.

In the CAAA of 1990, 189 hazardous air pollutants (HAPs) were identified for potential regulation. Industrial and commercial waste incinerators, industrial boilers and process heaters, and other combustion sources were suspected of emitting large quantities of many of these HAPs. Some HAPs are components of the waste and/or fuel, and others are formed during the combustion process.

During the last half of the 1990s, the U.S. EPA initiated a massive effort to develop maximum achievable control technology (MACT) regulations for HAPs from industrial and commercial combustion sources. The EPA's initiative involved government, industry, and environmental advocacy groups, and was called the *industrial combustion coordi-*

Table 1.3 Summary of National Emission Standards for Hazardous Air Pollutants (NESHAPs)

1. Beryllium:	The emissions from all point sources are limited to 10 grams of beryllium per 24 hours. If the EPA approves, the source owner/operator may substitute the requirement to meet an ambient air quality standard of 0.01 $\mu g/m^3$ averaged over a 30-day period. Separate standards are listed for rocket motor testing using a beryllium-containing propellant.
2. Mercury:	The emissions from mercury ore processing facilities and mercury cell chlor-alkali plants shall not exceed 2300 grams of mercury per 24 hours. Emissions from sludge incinerators or dryers shall not exceed 3200 g/24 hours.
3. Vinyl chloride:	The standards are listed for specific equipment and processes in ethylene dichloride plants, vinyl chloride plants, and PVC plants. In general, the standard is 10 ppm of vinyl chloride in any exhaust gases.
4. Benzene:	The standard is very specific and basically applies to plants and equipment within plants that handle benzene. The standards are designed to prevent or minimize leakage of benzene into the atmosphere.
5. Asbestos:	The standards apply to asbestos mills, eleven manufacturing operations using commercial asbestos, demolition and renovation of facilities containing asbestos, and other processes. Basically, the standard requires that any air exhausts must contain no visible emissions.

Adapted from 40 CFR (Code of Federal Regulations) 61, 1991.

nated rule-making (ICCR) process. Although the process at first worked well to help EPA gather data from industry, it was difficult to get all the groups to work together to establish guidelines for new regulations, and the process moved very slowly. After about two years, EPA stopped the ICCR in 1998, and went back to rule making as usual.

This major rule-making effort was not finished as of the writing of this edition; however, some of the HAPs that had been identified for possible regulation include HCl; metals such as mercury, lead, cadmium, nickel, and chromium; and organic HAPs such as benzene, 1,3-butadiene, chloroform (and dozens of other chlorinated organic compounds), chlorinated dioxins, formaldehyde, acrolein, and others. The reader should be aware that MACT regulations dealing with HAP emissions from industrial and commercial waste incinerators, industrial boilers, process heaters, large internal combustion engines, and gas turbines are either on the books as of 2001, or will be promulgated within the next two to three years.

Emissions standards are primarily used to regulate emissions from industrial sources, but can also be used to estimate the maximum rate of emissions from a proposed new source. Thus, we can use NSPSs or other standards to estimate control equipment performance requirements for design purposes. As an example, if a new solid-waste incinerator must meet the standard listed in Table 1.2, and the uncontrolled emissions are estimated at, say, 10 g/dscm, then a control device must be provided that is 98.2% efficient [(10 − 0.18)/10]. The uncontrolled emissions rates for a variety of processes can be estimated from data available in the EPA publication popularly known as AP-42 (*Compilation of Air Pollutant Emission Factors*, U.S. Environmental Protection Agency 1977).

In recent years, EPA has made all this information about emission factors available on its Web site. The Clearinghouse for Inventories and Emissions Factors (CHIEF) is searchable online and is updated regularly. The Web address is http://www.epa.gov/ttn/chief/ap42/index.html.

▶ ▶ ▶ ▶ ▶ ▶ ▶ ▶

Example 1.2

Estimate the collection efficiency required for a fluoride scrubber at a wet-process phosphoric acid plant that must meet the standard listed in Table 1.2. Assume that uncontrolled emissions would be about 120 grams of fluorides per metric ton of P_2O_5 feed.

Solution

$$\text{collection efficiency} = \frac{\text{amount collected}}{\text{amount input}} = \frac{\text{input} - \text{output}}{\text{input}}$$

$$= \frac{(120 - 10)}{120} \, 100\%$$

$$= 91.7\%$$

◀ ◀ ◀ ◀ ◀ ◀ ◀ ◀ ◀

New source performance standards can be used as emissions factors to estimate emissions from plants for regional airshed loading estimates. Another use is to estimate emissions from sources that are difficult to measure (such as the thousands of cars in a city). The next two example problems illustrate the use of NSPSs in estimating emissions.

Example 1.3

Calculate the daily SO_2 emissions from a new 200-ton-per-day sulfuric acid plant that will emit at the maximum allowable rate.

Solution

The standard is a maximum of 2 kg SO_2 per metric ton (MT) of sulfuric acid.

$$\frac{2 \text{ kg } SO_2}{\text{MT acid}} \times \frac{1 \text{ MT}}{1.102 \text{ T}} \times 200 \text{ tons/day} = 363 \text{ kg/day of } SO_2$$

Example 1.4

Calculate the daily emissions of particulates and SO_2 from a new 1000 megawatt (MW) coal-fired power plant that must meet the performance standards listed in Table 1.2. Assume the coal has a heating value of 12,000 Btu/lb and contains 3.0% sulfur. The plant is 39% efficient.

Solution

First, calculate the heat input rate for a 39% efficient plant.

$$E_{\text{in}} = \frac{1000 \text{ MW}}{0.39} \times \frac{1000 \text{ kW}}{\text{MW}} \times \frac{24 \text{ hr}}{\text{day}}$$

$$\times \frac{3412 \text{ Btu}}{\text{kWh}} = 2.10(10)^{11} \text{ Btu/day}$$

$$\text{Particulates emitted} = \frac{0.03 \text{ lb particulates}}{10^6 \text{ Btu}} \times \frac{2.10(10)^{11} \text{ Btu}}{\text{day}}$$

$$\times \frac{1 \text{ ton}}{2000 \text{ lb}} = 3.2 \text{ tons/day}$$

From Figure 1.5 as shown by the example line, we find an emission rate of exactly 0.60 lb SO_2 per million Btu heat input (at approximately 87% scrubber efficiency).

$$\text{SO}_2 \text{ emitted} = \frac{0.6 \text{ lb SO}_2}{10^6 \text{ Btu}} \times \frac{2.10(10)^{11} \text{ Btu}}{\text{day}} \times \frac{1 \text{ ton}}{2000 \text{ lb}} = 63 \text{ tons/day}$$

◀ ◀ ◀ ◀ ◀ ◀ ◀ ◀ ◀

THE PERMITTING PROCESS

Federal laws governing air pollution are designed to protect ambient air quality and public health. The U.S. Environmental Protection Agency, acting with the authority of such laws, establishes federal regulations that limit the emissions allowed from various sources. The specific regulations for each industrial category are documented in the Code of Federal Regulations (e.g., 40 CFR Part 60) and include emission limits, performance standards, permitting procedures, testing and monitoring requirements, record keeping, and reporting. Most states adopt the federal emission limits and standards by reference within their state laws. The states may create standards that are more stringent than the federal standards, if they so desire, but may not make them less stringent.

States have established their own air pollution permitting and compliance programs. Once a state plan is approved by the U.S. EPA, the state is delegated the authority to ensure clean and healthy air within its borders. The permitting process is the vehicle through which the states track and enforce their regulations. The state regulations set certain emission limits for various pollutants, above which a source may not emit without first obtaining a permit. Any source that fails to obtain a proper permit is subject to a significant fine (up to $10,000 per day).

One purpose of the state permit program is to allow the state to keep track of major point sources of air pollution. Permits provide a means to ensure that sources use appropriate technology to control pollution, and that they demonstrate ongoing compliance with emissions standards through routine testing, monitoring, record keeping, and reporting. A company applies for a construction permit first, which is reviewed for appropriate equipment selection and design. After construction of the plant in accordance with the controls specified in the permit, a temporary operating permit must be obtained. A regular operating permit is usually withheld until satisfactory performance is demonstrated in a plant test.

As mentioned previously, PSD (prevention of significant deterioration) increments were established to help prevent ambient concentration of pollutants in relatively clean air areas from increasing up to the NAAQSs. PSD considerations may be an important part of any permitting process, but are usually applicable to large projects. Federal PSD increments are presented in Table 1.4.

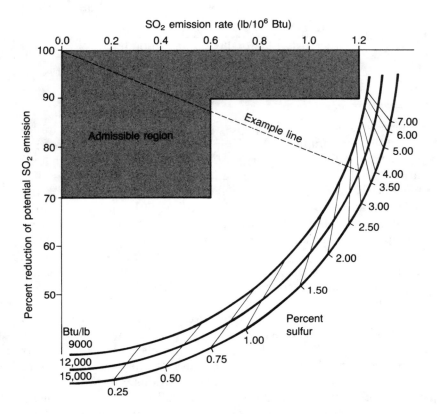

Figure 1.5
Graphical interpretation of the 1980 NSPS for SO_2 emissions from coal-fired power plants.
(Adapted from Molburg, 1980.)

Table 1.4   Federal PSD Allowable Increments

	Allowable Increment, $\mu g/m^3$		
Pollutant	**Class I**	**Class II**	**Class III**
Sulfur Dioxide			
Annual Average	2	20	40
24-Hour Maximum*	5	91	182
3-Hour Maximum*	25	512	700
Particulates			
Annual Geometric Mean	5	19	37
24-Hour Maximum*	10	37	75

*Maximums not to be exceeded more than once per year.
Adapted from 40 CFR (Code of Federal Regulations) 52, 1991.

◆ 1.5 THE IDEAL GAS LAW AND ◆ CONCENTRATION MEASUREMENTS IN GASES

Under normal conditions, dry ambient air contains approximately 78.08% nitrogen, 20.94% oxygen, 0.93% argon, 0.04% carbon dioxide, and traces of other gases (Tebbens 1968). For gases, percentages are usually expressed as percent by volume. For an ideal gas (ambient air approximates an ideal gas), volume percent is the same as mole percent. Recall that an ideal gas is one that always satisfies the ideal gas law

$$PV = nRT \tag{1.1}$$

where

P = absolute pressure

V = volume

n = number of moles

R = ideal gas law constant

T = absolute temperature

The units of all terms must be consistent. A common set of units is P in atm, V in liters, n in gmol, and T in degrees Kelvin. For this set of units, R has the value 0.08206 L-atm/gmol-K.

The ideal gas law is very important to air pollution engineers because it is well understood and is quite accurate at normal temperatures and pressures. The ideal gas law can also be written as

$$PV = \frac{M}{MW} RT \tag{1.2}$$

where

M = mass of the sample

MW = molecular weight of the gas

Equation (1.2) can be rearranged to give the mass density of an ideal gas as

$$\rho = \frac{M}{V} = \frac{(P)\, MW}{RT} \tag{1.3}$$

Using the set of units mentioned previously, and units of g/gmol for the molecular weight, the density ρ has units of g/liter.

Note that the ambient standards in Table 1.1 are given in units of parts per million (ppm) or micrograms per cubic meter ($\mu g/m^3$). These are common units of concentration measurement in air pollution work, and for gaseous pollutants they are related to each other through the ideal gas law, as will be shown later in this section. First, let us briefly review the meaning of ppm. The concentration measure *ppm* is simply the mole fraction or volume fraction of the pollutant in the gas mixture multiplied by a factor of 1,000,000; that is,

$$\text{ppm} = \frac{\text{volume of pollutant gas}}{\text{total volume of gas mixture}}\left(10^6\right) \qquad (1.4)$$

Note that the denominator in Eq. (1.4) is the total volume of gas and not just the volume of air. The results are similar at low concentrations but are not exactly the same, as shown in the following example problem.

▶ ▶ ▶ ▶ ▶ ▶ ▶ ▶ ▶

Example 1.5

(a) Exactly 500.0 mL of carbon monoxide (CO) is mixed with 999,500 mL of air. Calculate the resulting concentration of CO in ppm.

Solution

$$\frac{500 \text{ mL}}{999,500 \text{ mL} + 500 \text{ mL}}\left(10^6\right) = 500.0 \text{ ppm}$$

(b) Exactly 500.0 mL of CO is mixed with 1.000 m^3 (1,000,000 mL) of air. Calculate the resulting concentration of CO in ppm.

Solution

$$\frac{500}{1,000,000 + 500}\left(10^6\right) = 499.75 \text{ ppm}$$

◀ ◀ ◀ ◀ ◀ ◀ ◀ ◀ ◀

For engineering purposes, the results in Example 1.5 might well be considered identical, but at concentrations ranging upward from a few percent (2.0% = 20,000 ppm), a misunderstanding of the definition will result in significantly wrong answers.

Recall that the two common measures of concentration in air pollution work are ppm and μg/m^3. By using the ideal gas law, it is relatively straightforward to convert ppm to μg/m^3. However, it is necessary to specify a temperature and a pressure because a gas may occupy different volumes depending on these parameters. In air pollution work, the reference conditions are usually chosen as 25 C and 1.0 atm.

Solving the ideal gas law for the volume of the pollutant gas, and dividing by the volume of the total gas, the pollutant **volume fraction** in the gas can be written as:

$$\frac{V_p}{V_t} = \frac{n_p RT/P}{V_t} \tag{1.5}$$

where the subscripts p and t indicate pollutant gas and total gas, respectively. Note that a consistent set of units for Eq. (1.5) is as follows:

$V = L$

$P = atm$

$n = gmol$

$T = K$

$R = 0.082$ L-atm/(gmol-K)

Considering the left side of Eq. (1.5), recall that the concentration (in units of ppm) of one particular gas in an ideal gas mixture is equal to its volume fraction times one million. So multiply both sides of Eq. (1.5) by 10^6. Also, in the right side of Eq. (1.5), substitute for n_p the mass of the pollutant gas divided by molecular weight. The result of these two operations is Eq. (1.6):

$$C_{\text{ppm}} = 10^6 \frac{V_p}{V_t} = 10^6 \left(\frac{M_p / MW_p}{V_t} \right)(RT/P) \tag{1.6}$$

where

M_p = mass of pollutant gas, g

MW_p = molecular weight of pollutant gas, g/gmol

Equation (1.6) can be rearranged algebraically to isolate the concentration in units of mass per volume (in units of g/L):

$$M_p/V_t = \frac{C_{\text{ppm}} \times MW_p \times 10^{-6}}{RT/P} \tag{1.7}$$

From the ideal gas law, RT/P (the denominator of Eq. 1.7) is identically equal to V/n, the volume occupied by one mole of an ideal gas. Making the replacement, Eq. (1.7) becomes:

$$M_p/V_t = \frac{C_{\text{ppm}} \times MW_p \times 10^{-6}}{V/n} \tag{1.8}$$

The units of the left side of Eq. (1.8) are in grams/liter, which are not the most convenient for reporting pollutant concentrations in air. To convert to units of $\mu g/m^3$ (the more usual units), the left side of Eq. (1.8) must be multiplied by a factor of 10^9 (the product of 10^6 µg/g and 10^3 L/m^3). Of course, to maintain the equality of Eq. (1.8), the right side of the equation must be multiplied by 10^9 also.

Recall from high school chemistry that the volume per mole of an ideal gas (the V/n term) has the value 22.4 L/gmol at $T = 0$ C (273 K) and $P = 1$ atm. In air pollution work, it is typical to define standard temperature as 25 C (298 K). At this temperature, the value of V/n is 24.45 L/gmol. After multiplying both sides of Eq. (1.8) by 10^9, and after substituting for V/n, the final conversion equation is:

$$C_{mass} = \frac{1000\, C_{ppm} MW_p}{24.45} \qquad (1.9)$$

where

C_{mass} = mass concentration, $\mu g/m^3$

C_{ppm} = volume or molar concentration, ppm

Note that Eq. (1.9) applies only at 25 C and 1 atm. For any other temperature and pressure, an equation similar to Eq. (1.9) can be obtained by simply calculating an appropriate value of V/n (= RT/P) for the denominator (but remember that the units must be in L/gmol).

▶ ▶ ▶ ▶ ▶ ▶ ▶ ▶ ▶

Example 1.6

(a) What is the 1-hr AAQS for ozone, expressed in $\mu g/m^3$? (b) What is the 1-hr AAQS for sulfur dioxide, expressed in $\mu g/m^3$? (c) Assume that a sample of air at 25 C and containing SO_2 gas at a concentration equal to the AAQS is raised to 150 C. What is its SO_2 concentration at 150 C, in ppm and in $\mu g/m^3$?

Solution

(a) The ozone standard is 0.12 ppm. Using Eq. (1.9), with the MW of ozone (48),

$$C_{mass} = \frac{1000 \times 0.12 \times 48}{24.45}$$
$$= 236\ \mu g/m^3$$

(b) Similarly, using Eq. (1.9) with the MW of SO_2, we convert 0.14 ppm to units of $\mu g/m^3$,

$$C_{mass} = \frac{1000 \times 0.14 \times 64.06}{24.25}$$
$$= 367\ \mu g/m^3$$

(c) In units of ppm, which is a relative measure of the volume ratio of gases, the concentration remains the same, so the first answer is 0.14 ppm at 150 C. But to convert from ppm to $\mu g/m^3$ at this new temperature, first derive a new denominator for Eq. (1.9),

$$V/n = RT/P = \frac{0.08206 \times 423 \text{ K}}{1 \text{ atm}}$$

$$= 34.71 \text{ L/gmol}$$

Now, substitute into Eq (1.9),

$$C_{\text{mass}} = \frac{1000 \times 0.14 \times 64.06}{34.71}$$

$$= 258 \text{ }\mu\text{g/m}^3$$

♦ 1.6 OTHER APPLICATIONS OF ♦ THE IDEAL GAS LAW

As mentioned in the previous section, the ideal gas law is extremely important to air pollution engineers. Compliance with federal and state laws requires not only proper environmental engineering design and operation of pollution abatement equipment, but also careful analysis and accurate measurements of specified pollutants and environmental quality parameters. In the design process and in other types of calculations, engineers who have complete understanding of the ideal gas law will achieve greater accuracy with a smaller expenditure of their time. In this section, through the use of example problems, we will review various uses of the ideal gas law.

Emission sampling is another important area for which a good understanding of the ideal gas law and the equipment used to measure gas flow is important. Many poorly performing systems are due not to bad design practice, but to poor sampling technique. Some important considerations in gas sampling are equipment calibration, leaks, sample handling (condensation, adsorption), process operation fluctuations, and representative sampling of the process stream. We will address some of these considerations in the following examples.

When Eq. (1.1) was first presented, one widely used value of R was given for a specific set of units. There are many numerical values of R depending on the units chosen for P, T, V, and n. Table 1.5 lists several possible values of R, and the derivation of one such value is illustrated in the following example problem.

▶ ▶ ▶ ▶ ▶ ▶ ▶ ▶ ▶

Example 1.7

Given that the molar volume of an ideal gas is 22.4136 liters/gmol at 273.15 K and 1.000 atm, calculate the value of R in units of mm Hg-ft^3/lbmol-R.

Table 1.5   Values of the Ideal Gas Law Constant R [$R = (PV/nT)$]

R	Units (Pressure-Volume)/(Matter-Temperature)
0.08206	atm-L/gmol-K
82.06	atm-cm^3/gmol-K
62.36	mm Hg-L/gmol-K
1.314	atm-ft^3/lbmol-K
0.08314	bar-L/gmol-K
998.9	mm Hg-ft^3/lbmol-K
0.7302	atm-ft^3/lbmol-R
21.85	in. Hg-ft^3/lbmol-R
555.0	mm Hg-ft^3/lbmol-R
10.73	psia-ft^3/lbmol-R
1545.	psfa-ft^3/lbmol-R
8.314	Pa-m^3/gmol-K
R	**(Energy)/(Matter-Temperature)**
1.987	cal/gmol-K
8314.	J/kgmol-K

Note: 1 pascal (Pa) = 1 N/m^2; 1 N-m = 1 joule (J)

Solution

$$R = \frac{PV}{nT} = \left(\frac{V}{n}\right)\frac{P}{T}$$

$$R = \frac{22.4136 \text{ L}}{\text{gmol}} \times \frac{1.000 \text{ atm}}{273.15 \text{ K}} \times \frac{1 \text{ ft}^3}{28.316 \text{ L}} \times \frac{453.6 \text{ gmol}}{1 \text{ lbmol}}$$

$$\times \frac{760.0 \text{ mm Hg}}{1.000 \text{ atm}} \times \frac{1.000 \text{ K}°}{1.800 \text{ R}°}$$

$$= 555.0 \text{ mm Hg-ft}^3/\text{lbmol-R}$$

◀ ◀ ◀ ◀ ◀ ◀ ◀ ◀ ◀

In this example problem, more significant figures were used than are typically used in engineering calculations. Engineers often are satisfied with only three- or four-digit significance. Indeed, the data derived from field measurements are often accurate only to two significant figures. In an age of pocket calculators and personal computers when it has become easy to obtain numerical answers to eight decimal places, it is important to understand the significance of the numbers we work with. When high accuracy and precision are needed and justified, then many significant figures should be used. On the other hand, in many design problems the equations are modified by empirical relationships and must be considered to be approximate.

The weighted-average molecular weight of a mixture of ideal gases is calculated using mole fractions or volume fractions (not mass fractions) as the weighting factors. This is demonstrated in the following example problem.

▶ ▶ ▶ ▶ ▶ ▶ ▶ ▶ ▶

Example 1.8

Dry air can be considered to be 78.0% nitrogen ($MW = 28.0$), 21.0% oxygen ($MW = 32.0$), and 1.00% argon ($MW = 40.0$). Calculate the average molecular weight of air.

Solution

$$MW_{avg} = 0.78(28) + 0.21(32) + 0.01(40) = 28.96$$

Considering significant figures, $MW_{avg} = 29.0$.

◀ ◀ ◀ ◀ ◀ ◀ ◀ ◀ ◀

The density of an ideal gas mixture is a molar average of the densities of the pure components. The density can also be calculated by using the molar-average molecular weight of the gas mixture in Eq. (1.3), as shown in the next example. Note that the density of a gas depends on the pressure and temperature.

▶ ▶ ▶ ▶ ▶ ▶ ▶ ▶ ▶

Example 1.9

Calculate the density of air at standard conditions ($T = 25$ C and $P = 1$ atm).

Solution

$$\rho_{N_2} = \frac{(1.00)(28.0)}{(0.08206)(298)} = 1.145 \text{ g/L}$$

$$\rho_{O_2} = \frac{(1.00)(32.0)}{(0.08206)(298)} = 1.309 \text{ g/L}$$

$$\rho_{Ar} = \frac{(1.00)(40.0)}{(0.08206)(298)} = 1.636 \text{ g/L}$$

$$\rho_{Air} = 0.78(1.145) + 0.21(1.309) + 0.01(1.636)$$
$$= 1.184 \text{ g/L}$$

Alternatively,

$$\rho_{Air} = \frac{(1.00)(28.96)}{(0.08206)(298)} = 1.184 \text{ g/L}$$

◀ ◀ ◀ ◀ ◀ ◀ ◀ ◀ ◀

♦ 1.7 GAS FLOW MEASUREMENT ♦

A fixed number of moles of gas at a certain temperature and pressure occupies a certain volume. If the pressure and temperature of this fixed number of moles of gas are changed, then the new volume can be easily calculated. Equation (1.1) can be written as follows:

$$\frac{PV}{T} = nR \tag{1.10}$$

Since n is constant in this case, applying Eq. (1.10) to the two sets of conditions yields

$$\frac{P_{old}V_{old}}{T_{old}} = \frac{P_{new}V_{new}}{T_{new}} \tag{1.11}$$

or

$$V_{new} = V_{old}\,\frac{P_{old}}{P_{new}}\,\frac{T_{new}}{T_{old}} \tag{1.12}$$

Equation (1.12) can be applied to a static system or to a flow system as long as the number of moles under consideration remains constant.

▶ ▶ ▶ ▶ ▶ ▶ ▶ ▶ ▶

Example 1.10

A sample stream of dry gas is being withdrawn from a stack. The stack gases are at 200 C and 730 mm Hg. The stream flows through a heated filter, a set of cooled impingers, a small air pump, and then through a flow meter, as shown in Figure 1.6. The rate of flow is determined to be 30.0 liters/minute at 20 C and 790 mm Hg.

(a) Calculate the actual volumetric flow rate through the filter (at $T = 200$ C and $P = 730$ mm Hg).

(b) If 1.42 mg of solid particles are collected on the filter in 30 minutes, calculate the concentration of particles in the stack gas (in $\mu g/m^3$).

Solution

(a) $Q_{filter} = (30.0)\left(\dfrac{790}{730}\right)\left(\dfrac{473}{293}\right) = 52.4$ liters/min

(b) Total volume of gas sampled (at stack conditions):

$$52.4 \frac{L}{min} \times \frac{1 \, m^3}{1000 \, L} \times 30 \, min = 1.572 \, m^3$$

$$C_{part.} = \frac{1.42 \, mg}{1.572 \, m^3} \times \frac{10^3 \, \mu g}{1 \, mg} = 903 \, \mu g/m^3$$

◄ ◄ ◄ ◄ ◄ ◄ ◄ ◄ ◄

In the preceding example problem, if a concentration of, say, 500 ppm of NO_2 were measured in the gases exiting the flow meter, then we could state that the concentration in the stack was also 500 ppm. The reason for this is that ppm is a measure of *relative* concentration and as the total volume of the gas sample changes, so does the volume of NO_2. The *relative* volume of NO_2 (relative to the total) does not change. However, if some moles of a material that is gaseous in the stack are removed from the gas sample stream (such as water that is condensed in the impingers), we must account for the change in relative gas concentrations at the two places (upstream versus downstream of the impingers). This is shown in the following example problem.

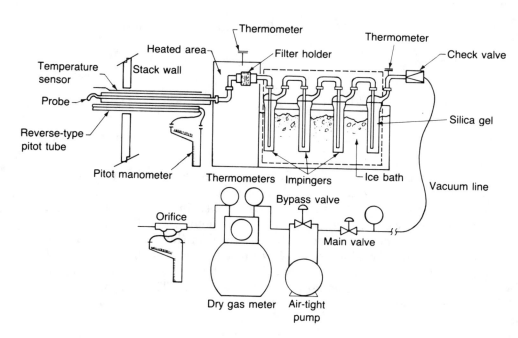

Figure 1.6
EPA Method 5—particulate sampling apparatus.

▶ ▶ ▶ ▶ ▶ ▶ ▶ ▶ ▶

Example 1.11

The same sampling tràin is employed as in Example 1.10, except that the stack gases now have a significant water vapor content. Over a 15-minute sampling time, 50.0 grams of water are condensed and collected in the impingers. The dry gas volumetric flow rate is measured as 30.0 L/min at 20 C and 790 mm Hg. An NO_2 concentration of 800 ppm is measured downstream of the flow meter. Calculate the NO_2 concentration in the stack, in which the humid gases are at $T = 200$ C and $P = 730$ mm Hg. Assume all the water is removed in the impingers.

Solution

The molar flow rate of dry gas is

$$n_d = \frac{(790)(30.0)}{(62.36)(293)} = 1.297 \text{ gmol/min}$$

The molar flow rate of water is

$$n_w = \frac{50.0 \text{ g}}{15 \text{ min}} \times \frac{1 \text{ gmol}}{18 \text{ g}} = 0.185 \text{ gmol/min}$$

and

$$n_{NO_2} = 800(10)^{-6} \times 1.297 = 0.001038 \text{ gmol/min}$$

The concentration of NO_2 in the stack gases is

$$C_{NO_2} = \frac{0.001038}{1.297 + 0.185}(10^6) = 700 \text{ ppm}$$

◀ ◀ ◀ ◀ ◀ ◀ ◀ ◀ ◀

The measurement of gas flow rates is important in air pollution work. Many different types of devices exist for measuring gas flow rates. Some commonly used devices are venturi meters, orifice meters, rotameters, wet test meters, and dry gas meters. The Bernoulli equation governs the operation of the first three of these widely used flow meters. However, since the density of gas varies with pressure and temperature, the ideal gas law also must be considered in deriving and using calibration equations for these meters. The Bernoulli equation can be written as

$$\frac{dP}{\rho} + g \, dz + \tfrac{1}{2} d\left(u^2\right) = 0 \qquad (1.13)$$

where

P = absolute pressure

ρ = density

g = gravitational acceleration

z = elevation

u = velocity

When dealing with the flow of gases, if the pressure drop is small compared to the upstream static pressure, the gas behaves essentially as if it is incompressible. Thus, equations for incompressible fluids give good results. For an incompressible fluid, Eq. (1.13) applied to two definite locations becomes

$$\frac{P_1}{\rho_1} + gz_1 + \frac{u_1^2}{2} = \frac{P_2}{\rho_2} + gz_2 + \frac{u_2^2}{2} \qquad (1.14)$$

Equation (1.14) ignores friction effects.

As the pressure drop increases, the effects of gas compressibility become significant. Finally, for large pressure drops, the gas velocity at the throat approaches sonic velocity. When this occurs in an orifice, the rate of flow becomes constant. These sonic or critical orifices are used widely because of their ability to control the flow rate to a known constant value without continual monitoring. [Well before reaching critical flow, however, the calibration equations based on Eq. (1.14) become invalid because of the assumption of incompressibility.]

For a horizontal venturi meter (and a constant density fluid), Eq. (1.14) becomes

$$u_2^2 - u_1^2 = \frac{2(P_1 - P_2)}{\rho_1} \qquad (1.15)$$

With the assumption of constant density, and through the use of the continuity equation, u_2 can be related to u_1 as follows:

$$u_1 = B^2 u_2 \qquad (1.16)$$

where $B = D_2/D_1$, the ratio of diameters at points 2 and 1; that is, B is the ratio of the smaller throat diameter to the larger upstream diameter

Substituting for u_1 in Eq. (1.15) results in

$$u_2 = \frac{1}{\sqrt{1 - B^4}} \sqrt{\frac{2(P_1 - P_2)}{\rho_1}} \qquad (1.17)$$

In practice, venturis are not perfectly frictionless, so an empirical venturi coefficient is introduced into Eq. (1.17)—that is,

$$u_2 = \frac{C_v}{\sqrt{1-B^4}} \sqrt{\frac{2(P_1-P_2)}{\rho_1}} \qquad (1.18)$$

For well-designed venturis, C_v is on the order of 0.98 or 0.99. Also, when D_2 is less than $D_1/4$, the term $1-B^4$ is approximately equal to 1.0. Finally, we note that in English units a g_c term must be included in the equations. [It would appear as a multiplicative factor in the numerator of the right side of Eq. (1.15) and in appropriate places in the equations following.]

The linear velocity through the throat is not the most convenient form of flow measurement. For liquids or for gases with "small" pressure drops, and for fully turbulent flows, the volumetric flow rate, Q, is obtained by multiplying the linear velocity by the throat area, A:

$$Q = \frac{C_v A_2 \sqrt{\frac{2(P_1-P_2)}{\rho_1}}}{\sqrt{1-B^4}} \qquad (1.19)$$

For gases, the pressure drop can be considered "small" if (P_1-P_2) is less than 10% of P_1 (Crawford 1976). If it is larger, compressibility is significant, and Eq. (1.19) should not be used. Thus, for ambient air monitoring, the pressure drop must be less than about 1.5 psi.

If a constant K is defined as the venturi coefficient, C_v, over the denominator of Eq. (1.19) and if ideal gas behavior is assumed, then Eq. (1.19) can be written as

$$Q = KA_2 \sqrt{\frac{2(P_1-P_2)RT_1}{P_1(MW)}} \qquad (1.20)$$

where

MW = gas molecular weight

R = ideal gas law constant

T_1 = absolute temperature at point 1

K = a constant $(C_v/\sqrt{1-B^4})$

Keep in mind that Eq. (1.20) implies that C_v is constant over all ranges of hydraulic conditions experienced by the meter.

If the pressure difference, P_1-P_2, is measured with a manometer, then Eq. (1.20) can be written as:

$$Q = KA_2 \sqrt{\frac{2\rho_l ghRT_1}{P_1 (MW)}} \tag{1.21}$$

where

ρ_l = density of the manometer fluid

h = height difference of the manometer fluid

For ordinary industrial applications and many laboratory applications, the venturi has certain practical disadvantages. It is expensive, relatively inflexible for changing flow rates, and it takes up a large space. An orifice meter overcomes these disadvantages, but at the cost of permanent pressure loss (power consumption). For design purposes, Eq. (1.18) is used, but with C_v (now called C_o) on the order of 0.6 rather than 0.98. Thus, the basic equation for an orifice or venturi meter is

$$Q = k\sqrt{\Delta P} \sqrt{\frac{T_c}{P_c MW_c}} \tag{1.22}$$

where

Q = volumetric flow rate

k = a calibration constant

ΔP = pressure drop

T_c, P_c = absolute temperature and absolute pressure of calibration

MW_c = molecular weight of the calibration gas

Equation (1.22) requires turbulent flow of a "constant-density" fluid through the meter. For a rotameter, a similar equation applies:

$$Q = k'RR \sqrt{\frac{T_c}{P_c MW_c}} \tag{1.23}$$

where RR = rotameter reading

Equation (1.23) is subject to assumptions and restrictions similar to those for Eq. (1.22).

As can be seen from the above equations, the gas flow rate varies as the square root of pressure drop for a venturi or an orifice meter, and varies linearly with the scale reading for a rotameter. However, both equations also have terms that are proportional to the square root of $T_c/(P_c MW_c)$. This means that if any of these meters is used at conditions different from those of calibration, the meter calibration equation must be corrected for this difference; that is, the *indicated* flow rate obtained directly from the meter reading and the old calibration equation is not correct if the meter is being operated at a different pressure or temperature or with a different gas than was used to calibrate the meter. The *actual* flow rate is then given by

$$Q_a = Q_i \sqrt{\frac{T_a}{P_a MW_a} \frac{P_c MW_c}{T_c}} \qquad (1.24)$$

where

Q_i = indicated flow rate

subscript a = actual conditions

subscript c = calibration conditions

Once the *corrected* actual flow rate is obtained, it can be *converted* to an equivalent flow rate at any temperature and pressure by using the ideal gas law. These correction and conversion procedures are demonstrated with the following problems.

Example 1.12

You have an orifice meter that was calibrated for air at 70 F and 1 atm. You are using it to measure an air flow, and you get an indicated reading of 25.0 ft^3/min at 100 F and 1.1 atm. What is the true gas flow rate in standard ft^3/min (scfm)?

Solution

Substituting into Eq. (1.24) to correct the indicated flow rate to an actual flow rate, and then Eq. (1.12) to convert the actual flow rate to an equivalent standard flow rate, we get

$$Q_a = 25.0 \text{ cfm } [(560 \text{ R/1.1 atm})(1.0 \text{ atm/530 R})]^{0.5}$$
$$= 24.5 \text{ cfm}$$
$$Q_s = 24.5 \text{ cfm} \times (537 \text{ R/560 R}) \times (1.1 \text{ atm/1.0 atm})$$
$$= 25.8 \text{ scfm}$$

Example 1.13

You have a venturi meter that was calibrated with pure nitrogen at 80 C and 1.5 atm. You wish to use it to measure the flow of air at 20 C and 0.9 atm. An old nitrogen calibration curve is presented on the next page. With air at the new conditions, you obtain a ΔP reading of 8 in. H$_2$O; what is the true air flow rate in standard m^3/min?

Flow meter calibration curve (N$_2$ gas at 80 C and 1.5 atm).

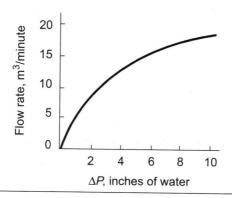

Solution

From the calibration curve, we obtain an indicated flow rate of 16.0 m^3/min. First we correct to an actual flow rate:

$$Q_a = 16.0 \text{ m}^3/\text{min} \sqrt{\frac{293}{(0.9)(29.0)} \times \frac{(1.5)(28.0)}{353}} = 18.5 \text{ m}^3/\text{min}$$

Next, convert this flow rate to standard conditions:

$$Q_s = 18.5 \text{ m}^3/\text{min} \left(\frac{298}{293}\right)\left(\frac{0.9}{1.0}\right) = 16.9 \text{ std m}^3/\text{min}$$

◀ ◀ ◀ ◀ ◀ ◀ ◀ ◀

For large streams of gas (as in a stack from an industrial furnace), it is impossible to measure the volumetric flow rate directly. However, one or more point measurements of the linear gas velocity can be made with a device called a **pitot tube.** A standard type of pitot tube has two concentric tubes, with the center tube open to the gas flow at its tip. If it is pointed directly into the flow, the center tube can be used to measure the total pressure of the flowing gas. This total pressure is the sum of the static pressure, P_s, and the dynamic pressure, $\frac{1}{2}\rho u^2$. The outer tube has holes around its circumference to measure the static pressure only. The difference between the two pressures, ΔP, is the dynamic pressure and is due to the velocity of the flow. A measurement of ΔP can be used to obtain the velocity:

$$u = k\sqrt{\frac{2\Delta P}{\rho}} \tag{1.25}$$

$$u = k\sqrt{\frac{2\rho_l g h}{\rho}} \qquad\qquad (1.26)$$

where k = calibration factor

The calibration factor for a well-constructed standard pitot tube is usually close to one. The lowest practical velocity that can be measured is about 2.5 m/s.

Because of its small openings, a standard pitot tube can quickly clog in a dusty or wet flow. Therefore a different type of pitot tube, known as a Stausscheibe or S-type pitot tube, has been developed for stack sampling. The S-type pitot tubes have openings on their ends that face in opposite directions. The tube is placed in a flow, with one opening pointing upstream and the other downstream. The tube facing upstream receives the total pressure, while the tube facing downstream receives a value less than the true static pressure owing to a low-pressure region in the wake of the sampler. The velocity is still obtained by Eq. (1.25) or (1.26), but the calibration factor must be found by field calibration of the individual S-type tube. Typical factors for commercial tubes are in a range of 0.8–0.9.

To measure the total volumetric flow in a stack or a duct, a cross-section of the duct is divided into equal area portions, and the velocity in each portion is measured with a pitot tube. The total flow is the sum of each velocity times its corresponding area in each portion. The next example illustrates this procedure.

▶ ▶ ▶ ▶ ▶ ▶ ▶ ▶ ▶

Example 1.14

A 2 m × 2 m square duct (shown below) was sampled at four positions with a pitot tube, with the following results:

Sampling points:

Point	ΔP (in. H_2O)
1	0.60
2	0.65
3	0.62
4	0.66

```
┌─────┬─────┐
│ 1   │ 2   │
│ .   │ .   │
├─────┼─────┤
│ 3   │ 4   │
│ .   │ .   │
└─────┴─────┘
```

The stack conditions are 279 F, 29.6 in. Hg static pressure, 28.1 average molecular weight. The correction factor for the S-type pitot tube is 0.87.

(a) Compute the linear velocity at each point (m/s).

(b) Compute the volumetric gas flow in the duct (m^3/s).

(c) Compute the mass flow of gas in the duct (kg/s).

Solution

$$u = k\sqrt{\frac{2\Delta P}{\rho}} = k\sqrt{\frac{2\rho_l g}{\rho}}\sqrt{h}$$

But

$$\rho = \frac{P\ MW}{RT} = \frac{(29.6\text{ in. Hg})(28.1\text{ lb/lbmol})}{\left(21.85\dfrac{\text{in. Hg ft}^3}{\text{lbmol R}}\right)(739\text{ R})} = 0.05151\text{ lb}_m/\text{ft}^3$$

$$\rho = \frac{0.05151\text{ lb}_m}{\text{ft}^3} \times \frac{35.3\text{ ft}^3}{\text{m}^3} \times \frac{\text{kg}}{2.205\text{ lb}_m} = 0.8246\text{ kg/m}^3$$

So

$$u = 0.87\sqrt{\frac{(2)\left(\dfrac{1000\text{ kg}}{\text{m}^3}\right)\left(\dfrac{9.8\text{ m}}{\text{s}^2}\right)}{0.8246\text{ kg/m}^3}}\sqrt{h,\text{ m}}$$

or

$$u = 134.1\sqrt{h}$$

(a)

Point	h, in. H_2O	h, m	$\sqrt{h}$	u, m/s
1	0.60	0.01524	0.1234	16.5
2	0.65	0.01651	0.1285	17.2
3	0.62	0.01575	0.1255	16.8
4	0.66	0.01676	0.1295	17.4

(b) $Q = \sum A_i u_i = 68\text{ m}^3/\text{s}$

(c) $\dot{M} = Q\rho = 68(0.8246) = 56.1\text{ kg/s}$

◀ ◀ ◀ ◀ ◀ ◀ ◀ ◀ ◀

◆ 1.8 CAUSES, SOURCES, AND EFFECTS ◆

In this section, we will briefly discuss some of the major pollutants with respect to their causes, sources, and effects. A **cause** is distinguished from a **source** in that a cause is fundamental, whereas a source is locational; that is, a cause explains why or how a pollutant is formed, whereas a source identifies what type of process, industry, or device discharges a particular pollutant. For reference in the following discussion, Table 1.6 presents data on emission rates from the major sources for the five major primary pollutants in the United States in 1989 and in 1998.

As the reader will see, the main focus of this text is on the control of air pollution from stationary sources. Relatively less consideration is given to the control of air pollution from motor vehicles. This is not to suggest that highway vehicle pollution is unimportant; on the contrary, from Table 1.6, we can see that emissions from automobiles, trucks, buses, and other vehicles account for approximately one-half of all the NO_x and VOC emissions and three-fourths of all the CO emissions in the United States. Internationally, the significance of mobile source pollution is growing at an even faster rate than industrial pollution. Faiz et al. (1992) point out that the growth rate of motorization in economically developing countries, particularly in Asia and South America, is even faster than their general population growth. Hence, the air pollution problem from mobile sources is growing more severe each year. In various countries around the world, motor vehicles account for some 10–60% of total air pollution emissions. In urban centers, their percentage contribution is much higher.

Despite our focus on the design of industrial-scale air pollution control equipment, in Chapter 18 we will investigate the topic of motor vehicle pollution and its control in more depth. We recognize that the *design* of vehicular pollution control equipment is a specialized field, limited essentially to the vehicle manufacturing companies, hence our decision to restrict the scope of our coverage of mobile sources in this text.

PARTICULATE MATTER

Particulate matter (or particulates) are very-small-diameter solids or liquids that remain suspended in exhaust gases and can be discharged into the atmosphere. They are *caused by* one of three fundamental processes. Materials-handling processes, such as crushing or grinding ores or loading dry materials in bulk, can result in the creation of fine dusts. Combustion processes can emit small particles of noncombustible ash or incompletely burned soot. Particles can also be formed by gas conversion reactions in the atmosphere between certain pollutant gases that were emitted previously. Major *sources* of particles include industrial processes, coal- and oil-burning electric power plants, residential fuel combustion, and highway vehicles.

Particulate effects include reductions in visibility such as smog or haze, soiling of buildings and other materials, corrosive and erosive damage of materials, and alteration of local weather. Also, particulates can damage human and animal health, and retard plant growth.

An object is visible to the eye because it contrasts with its background. As the distance between an object and an observer increases,

Table 1.6 National U.S. Emission Estimates, 1989 and 1998 (millions of tons per year)

Source Category	CO 1989	CO 1998	NOx 1989	NOx 1998	VOCs 1989	VOCs 1998	PM-10 1989	PM-10 1998	SO2 1989	SO2 1998
Transportation Sources										
On-Road Vehicles	66.05	50.39	7.68	7.77	7.19	5.33	0.37	0.26	0.57	0.33
Non-Road Sources*	17.78	19.91	4.53	5.28	2.55	2.46	0.48	0.46	0.78	1.08
SUBTOTAL	83.83	70.30	12.21	13.05	9.74	7.79	0.85	0.72	1.35	1.41
Stationary Sources										
Fuel Combustion										
Electric Utilities	0.32	0.42	6.59	6.10	0.04	0.05	0.27	0.30	16.22	13.22
Industrial Furnaces	0.67	1.11	3.21	2.97	0.13	0.16	0.24	0.25	3.09	2.90
Residential and Other	6.45	3.84	0.74	1.12	1.20	0.68	0.87	0.54	0.63	0.61
SUBTOTAL	7.44	5.37	10.54	10.19	1.37	0.89	1.38	1.09	19.94	16.73
Industrial Processes										
Chemicals and Petroleum	2.36	1.50	0.37	0.29	1.62	0.89	0.12	0.10	0.87	0.64
Metals Processing	2.13	1.50	0.08	0.09	0.07	0.08	0.21	0.17	0.70	0.44
Other	0.77	0.71	0.32	0.42	8.12	7.05	0.70	0.44	0.41	0.37
SUBTOTAL	5.26	3.71	0.77	0.80	9.81	8.02	1.03	0.71	1.98	1.45
Waste Disposal	1.75	1.15	0.08	0.10	0.94	0.43	0.25	0.31	0.04	0.04
Subtotal Stationary Sources	14.45	10.23	11.39	11.09	12.12	9.34	2.66	2.11	21.96	18.22
Miscellaneous & Area Sources	8.15	8.92	0.29	0.33	0.64	0.79	37.46	26.60	0.01	0.01
TOTAL ALL SOURCES	**106.40**	**89.50**	**23.90**	**24.50**	**22.50**	**17.90**	**41.00**	**29.40**	**23.30**	**19.60**

*Non-road includes boats, planes, trains, and construction equipment.
Notes: Miscellaneous CO is almost all from fires.
Miscellaneous PM-10 is almost all from fugitive (paved and unpaved roads, and construction) and agricultural sources.

Compiled from data in EPA 454/R-00-003 (March, 2000).

the apparent contrast decreases. Particulates in large concentrations can decrease visibility by scattering and absorbing light. As stated by Robinson (1968), the apparent contrast at a distance d is given by

$$C = C_0 e^{-\sigma d} \qquad (1.27)$$

where

C = apparent contrast at distance d

C_0 = actual contrast at zero distance

σ = extinction coefficient (the sum of a scattering coefficient σ_s and an absorption coefficient σ_a)

Note that σ must have units reciprocal to those of d. For several industrial hazes in England, the scattering and absorption coefficients were found to be about equal in magnitude (Middleton 1952).

The limit of visibility d_v is the distance where the limiting ratio of daytime visual contrast is reached. At that point, an observer can no longer distinguish an object from its background. At the limiting ratio of C/C_0 (usually taken to be 0.02), Eq. (1.27) can be solved for d_v.

$$d_v = \frac{3.91}{\sigma} \qquad (1.28)$$

Scattering of light owing to particles is a strong function of particle size. The most effective scatterers are particles that are about the same size as the wavelength of visible light—that is, those in the range from 0.1 to 1.0 microns (Friedlander 1977). High humidity increases scattering because tiny particles tend to act as condensation nuclei, forming droplets that increase the total number of effective particles.

Several authors have found correlations between the light scattering coefficient σ_s and the mass concentration of total suspended particulates (TSP) in the air (National Research Council 1979). Higher correlations can be expected when fine particulates dominate the mass concentration of the total distribution. A typical equation is as follows:

$$\sigma_s = a + b(C) \qquad (1.29)$$

where

σ_s = scattering coefficient, km^{-1}

C = TSP mass concentration, $\mu g/m^3$

a, b = regression constants

Values for a for Eq. (1.29) have been reported in the range from $-1.5(10)^{-2}$ to $-6.1(10)^{-2}$ km^{-1}, and for b in the range from $2.0(10)^{-3}$ to $3.6(10)^{-3}$ m^3/km-μg (National Research Council 1979). To use Eq. (1.29) to predict visibility, several assumptions must be made, as shown in the following example problem.

Example 1.15

Assume that Eq. (1.29) is valid for ambient particulate concentrations in a certain urban area, with $a = -2.0(10)^{-2}$ and $b = 3.0(10)^{-3}$. Also assume that σ_a is due to particulates alone and is equal to σ_s. Estimate the limit of visibility (km) in air that contains 150 µg/m^3 of TSP.

Solution

$$\sigma_s = -2.0(10)^{-2} + 3.0(10)^{-3}(150)$$
$$\sigma_s = 0.43 \text{ km}^{-1}$$

Since $\sigma = \sigma_a + \sigma_s$ and $\sigma_a = \sigma_s$,

$$\sigma = 0.86 \text{ km}^{-1}$$

From Eq. (1.28),

$$d_v = \frac{3.91}{0.86} = 4.5 \text{ km}$$

Fine particulates are responsible for the sometimes severe visibility reduction observed in many areas of the country, including our western national parks. Recent work on visibility modeling has produced well-defined relationships between particle size and light extinction coefficients. However, the relationships between gaseous pollutant emissions and subsequent secondary aerosol formation and resulting concentrations at downwind receptors remain elusive (Mathai 1990).

Particles in polluted urban air have been shown to increase fog formation and persistence (Robinson 1968). On a larger scale, increased particulate concentrations result in increased scattering and outright reflection of solar energy before it can reach the earth's surface. Increased rain and snow are also likely effects of particulates.

Deposition of particles can reduce the aesthetic appeal of structures and monuments. Structures and materials may need more frequent cleaning and/or painting. Particulates can intensify the chemical effects of other pollutants, especially corrosion due to acid gases. Deposition of cement kiln dust on plants has been shown to cause leaf damage, especially in moist environments (National Research Council 1979).

Airborne particles can affect human health in many ways. Certain pollutants may be toxic or carcinogenic (such as pesticides, lead, or arsenic). Particles may adsorb certain chemicals and intensify their effects by holding them in the lungs for longer periods of time. Of special interest are the physical effects of particles on the normal functioning of the respiratory system.

To cause lung damage, particles must penetrate the human respiratory system. Particles larger than about 10 microns generally do not penetrate deep into the lungs. They are intercepted by nasal hairs or settle onto the mucous membranes in the nasal or oral passages or trachea. Once captured on these membranes, insoluble particles are rapidly carried to the larynx through the normal combined action of ciliated and mucus-secreting cells. From there, particles can be swallowed or expectorated.

Very small particles (less than 0.1 microns) tend to be deposited in the tracheobronchial tree by diffusion. These are removed in the same manner as the large particles. Particles in the size range from 0.1 to 10 microns can penetrate deep into the lungs where they are then deposited in the respiratory bronchioles or alveolar sacs. This is the size range that has the most serious health effects, and is the range intended for control by the PM-2.5 standard. The progression of the ambient standards for PM has moved from total suspended particulate (TSP) to PM-10 to PM-2.5 over the years. This movement has occurred as our knowledge of the health effects of fine particles has increased, and as our abilities to sample and measure these small particles has improved. A very interesting animation of particulate matter deposition in the respiratory system is available at http://aerosol.ees.ufl.edu.

Ultimate effects of particles on human health are very difficult to quantify. Many epidemiological studies have been performed, but in all of the studies other pollutants were present with particulates. In the presence of high SO_2 concentrations, there is a direct relationship between TSP concentrations and hospital visits for bronchitis, asthma, emphysema, pneumonia, and cardiac disease (National Research Council 1979). Also, elderly persons who suffer from a respiratory or cardiac disease have a significantly higher-than-average risk of death when particulate concentrations are high for several days. More research is needed to quantify the effects of particulates on human health.

It is interesting to note that a detailed study of PM-10 in the Los Angeles area found that over 80% of the annual average PM-10 mass collected in samplers could be accounted for by five aerosol components (carbonaceous materials, nitrates, sulfates, ammonium, and soil-related material). This result indicates that much of the PM-10 is formed by atmospheric conversion reactions of pollutant gases (like NO or SO_2) or derives from fugitive sources (like windblown soil, road dust, or tire dust). Control of PM-10 thus may be very difficult.

SULFUR DIOXIDE

Sulfur oxides (SO_x) are *caused by* burning sulfur or any material containing sulfur. As can be seen from Table 1.6, the main *source*, by far, is fossil-fuel combustion for electric power generation, although certain industrial processes such as petroleum refining and nonferrous metal smelting can be important sources in specific locations. The primary sulfur oxide is SO_2, but some SO_3 is also formed in furnaces. In the atmosphere, SO_2 is slowly oxidized to SO_3. Furthermore, SO_2 and SO_3 can form acids when they hydrolyze with water, and the acids can then have detrimental effects on the environment, as discussed previously. In addition, SO_2 has been associated with human health problems, damage to plants and animals, smog and haze through the formation of acid mists, and corrosion of materials.

The early history of sulfur dioxide pollution is associated with extensive damage to vegetation (Brandt and Heck 1968). One of the major effects on green plants is chlorosis, or the loss of chlorophyll. Another is plasmolysis, or tissue collapse of many of the leaf cells. With SO_2, both effects can occur with either short exposures to high concentrations or long exposures to lower concentrations. More recently, studies of the effects of SO_2 on plants have focused on combined effects of acid deposition (most acid deposition is sulfur based) and ozone. Jensen and Dochinger (1989) found significant reductions in growth of six species of hardwood trees caused by increasing acidity of rainfall and by increasing ozone levels. Gaseous SO_2 (at typical ambient air levels) had no apparent effect.

The effects of short-term intermittent exposures to SO_2 on animals are similar to those on humans except that animals are much less sensitive. SO_2 is soluble and is readily absorbed in the upper respiratory tract. In humans, the threshold levels for taste and odor are 0.3 ppm and 0.5 ppm, respectively (Stokinger and Coffin 1968). At concentrations above 1 ppm, some bronchoconstriction occurs; above 10 ppm, eye, nose, and throat irritation is observed. SO_2 also stimulates mucus secretion, a characteristic of chronic bronchitis (Goldsmith 1968).

Sulfur dioxide effects are intensified by the presence of other pollutants, especially particulates. In Donora, Pennsylvania, in 1948, in London in 1952, and in other well-known air pollution episodes, both SO_2 and particulates were present in high concentrations simultaneously. Particulate sulfates or inert particles with adsorbed SO_2 can penetrate deep into the lungs and induce severe effects. Some data are presented in Table 1.7 (National Research Council 1979).

NITROGEN OXIDES

Nitrogen oxides (NO_x) are *formed* whenever any fuel is burned in air. At high temperature, N_2 and O_2 in the air combine to form NO and NO_2. Also, organically bound nitrogen atoms present in some fuels can contribute substantially to NO_x emissions. Total U.S. emissions of NO_x are almost equally distributed between *mobile sources* and *stationary combustion sources*. Nitrogen oxides contribute to smog, are injurious to plants and animals, and can affect human health. NO_x contributes to acidic deposition. Furthermore, NO_x reacts with reactive VOCs in the presence of sunlight to form photochemical oxidants.

Table 1.7 Health Effects and Dose Response for SO_2 and TSP

Location	Particles, $\mu g/m^3$	SO_2, $\mu g/m^3$	Measurement Averaging Time	Effect
London	2000	1040	24 hr	Mortality
	750	710	24 hr	Mortality
	500	500	24 hr	Exacerbation of bronchitis
New York City	145	286	24 hr	Increased prevalence of respiratory symptoms
Birmingham, AL	200	26	24 hr	Increased prevalence of respiratory symptoms
London	200	400	1 week	Increased incidence of respiratory illness
Britain	200	200	6 months	Bronchitis, sickness, absence from work
Britain	70	90	1 year	Lower respiratory infections

Adapted from National Research Council, 1979.

Excess NO_x concentrations in the air result in a brownish color because the gas strongly absorbs in the blue-green area of the visible spectrum (Robinson 1968). This discoloration is aesthetically displeasing and can reduce visibility. In Eq. (1.27), an extinction coefficient σ was defined as the sum of a scattering coefficient and an absorption coefficient. The absorption coefficient has both a particulate and an NO_2 component as follows:

$$\sigma_a = \sigma_{ap} + \sigma_{NO_2} C_{NO_2} \qquad (1.30)$$

where

σ_{ap} = absorption coefficient owing to particles

σ_{NO_2} = absorption coefficient owing to NO_2

C_{NO_2} = NO_2 concentration

The absorption due to NO_2 is a function of wavelength. Some data are presented in Table 1.8.

Wavelength, nanometers	σ_{NO_2}, ppm^{-1}mile^{-1}	Table 1.8
400	2.60	Absorption
450	2.07	Coefficient of NO_2
500	1.05	
550	0.47	
600	0.18	
650	0.062	

Adapted from Robinson, 1968.

Broad-leaved plants show necrosis at 2 to 10 ppm of NO_2, and retardation of growth can occur at about 0.5 ppm (Brandt and Heck 1968). The effects of NO_2 on people include nose and eye irritation, pulmonary edema (swelling), bronchitis, and pneumonia. The NO_2 concentrations usually encountered in polluted urban atmospheres are below the levels required to initiate such acute effects. Usually, concentrations in the 10–30 ppm range are necessary before irritation or pulmonary discomfort is experienced. Long-term exposure to NO_2 concentrations below this range can contribute to pulmonary fibrosis and emphysema (Stokinger and Coffin 1968).

More recent research has proved that NO_2 health impacts occur at much lower concentrations than previously thought. Hasselblad, Eddy, and Kotchmar (1992) used a technique known as meta-analysis to analyze the results of eleven previously published studies aimed at correlating increased levels of NO_2 and increased frequency of lower respiratory tract illness (LRI) in children. The previous studies (which had reported NO_2 concentrations between 20 and 200 µg/m^3) were, individually, inconclusive or contradictory. However, by analyzing the combined data, Hasselblad and associates were able to show that prolonged exposure to NO_2 levels that were elevated by only 30 µg/m^3 increased the odds of a LRI by about 20%.

The oxides of nitrogen react with certain VOCs (sometimes called *reactive hydrocarbons*) in the presence of sunlight to form photochemical oxidants, including ozone; this is the main reason for setting the NAAQS standard for NO_2 at 100 µg/m^3. These numerous reactions are complex, and a brief discussion of the reaction pathways is presented in Chapter 19. Here we will briefly review the effects of photochemical oxidants in the following section.

PHOTOCHEMICAL OXIDANTS AND VOCS

Photochemical oxidants are *caused by* a complex network of chemical reactions that occur in the ambient atmosphere. These reactions involve VOCs and NO_x, and are initiated by absorbing ultraviolet energy from sunlight. One of the main products of these reactions is

ozone, which is a very reactive oxidizing gas. (Note that ozone—a federal criteria pollutant—is not emitted directly, and so has no *source* per se.) Because VOCs and NO_x are both emitted in large quantities by motor vehicles and because of the importance of sunlight, photochemical oxidants are usually more prevalent in large, sunny urban areas with heavy traffic. Other major sources of VOCs are industrial processes like petrochemical processing, surface coating, printing, and other large operations involving organic solvents.

Ozone and other oxidants are severe eye, nose, and throat irritants; eye irritation occurs at 100 ppb, and severe coughing occurs at 2.0 ppm. Although the NAAQS for ozone is based on a 1-hour maximum, Lippmann (1989) points out that reductions in lung functions can occur in as little as five minutes of exposure to concentrations in the range of 20 to 150 ppb. Furthermore, these acute (short-term) effects worsen as the dose (concentration × exposure time) increases.

Lippmann (1989) also raises the concern of chronic (long-term) effects. For example, people exposed to seasonally elevated concentrations of ozone for years may experience irreversible, accelerated lung aging. EPA has proposed an 8-hour standard of 0.08 ppm, but perhaps a future ozone standard should include a seasonal or annual average maximum as well.

Other effects of oxidants include severe cracking of synthetic rubber and deterioration of textiles, paints, and other materials. Oxidants cause extensive damage to plants, including leaf discoloration and cell collapse (Brandt and Heck 1968), with effects starting at concentrations as low as 50 ppb. Ozone damage to crops in the United States is substantial, probably in the range of one billion dollars annually at current ambient levels (Adams et al. 1985). Ozone also has been shown to be a significant factor in forest losses in the United States and Europe (McLaughlin 1985).

Few VOCs have any known direct adverse effects on plants, animals, or materials. An exception is ethylene, which causes chlorosis and necrosis in broad-leaved plants (Brandt and Heck 1968). General retardation of growth for several kinds of plants has also been reported. Other than the displeasure experienced when exposed to certain odoriferous VOCs, no acute (short time frame) effects have been observed on humans at typical air pollution concentrations. Other VOCs are carcinogenic, and many VOCs are reactive in the atmosphere (forming oxidants).

CARBON MONOXIDE

Carbon monoxide (CO) is a colorless, odorless, tasteless gas that is *caused by* the incomplete combustion of any carbonaceous fuel. Power plants and other large furnaces are usually designed and operated carefully enough to ensure nearly complete combustion and do not emit much CO. Thus, the major *source* is the transportation sector. However, residential heating accounts for a significant fraction of total national CO emissions as do certain industrial processes.

Carbon monoxide is essentially inert to plants or materials but can have significant effects on human health. CO reacts with the hemoglobin in blood to prevent oxygen transfer. Depending on the concentration of CO and the time of exposure, effects on humans range from slight headaches to nausea to death.

The toxic effects of carbon monoxide on humans are due solely to the interactions of CO with blood hemoglobin (Stokinger and Coffin 1968). When a mixture of air and CO is breathed, both oxygen and CO are transferred through the lungs to the blood. Both adsorb onto hemoglobin, but the equilibrium coefficient for CO is approximately 210 times as great as that for oxygen. Thus, the equilibrium ratio of carboxyhemoglobin (HbCO) to oxyhemoglobin (HbO$_2$) is given by the following equation:

$$\frac{HbCO}{HbO_2} = (210)\frac{\overline{P}_{CO}}{\overline{P}_{O_2}} \tag{1.31}$$

where $\overline{P}_{CO}, \overline{P}_{O_2}$ = partial pressures of CO and O$_2$, respectively.

As more CO is breathed in and more HbCO is formed, the ability of the lungs and blood to supply oxygen to the rest of the body is decreased. Not only is the HbO$_2$ decreased, but that which remains tends to be more firmly bound because of the presence of HbCO (Goldsmith 1968). Equilibrium is not attained instantly. Typically, about 6 to 8 hours are required to reach equilibrium blood saturation when CO levels are approximately 50 ppm. The process is reversible, and once CO is removed from the air, HbCO will slowly break down, allowing CO to be expelled by the lungs.

▶ ▶ ▶ ▶ ▶ ▶ ▶ ▶ ▶

Example 1.16

Estimate the percentage of HbCO in the blood of a traffic officer exposed to 40 ppm CO for several hours. Assume the HbCO content reaches 60% of its equilibrium saturation value.

Solution

Air is approximately 20.9% oxygen, so at saturation

$$\frac{HbCO}{HbO_2} = (210)\frac{40}{209,000} = 0.040$$

At 60% saturation,

$$\frac{HbCO}{HbO_2} = 0.024$$

To find the fraction of HbCO in the blood, we must realize that
HbCO + HbO$_2$ = 1.0

$$HbCO + \frac{HbCO}{0.024} = 1.0$$

Thus, HbCO = 0.0234 and percent HbCO$_{actual}$ = 2.3%

◀ ◀ ◀ ◀ ◀ ◀ ◀ ◀ ◀

The most serious effects of atmospheric CO are expected for individuals already vulnerable to oxygen deficiencies. People with anemia, those with chronic heart or lung disease, and those living at high altitudes are more at risk than others. Levels of HbCO in the blood as low as 2% to 5% have been shown to induce some effects.

Cigarette smokers knowingly and voluntarily pollute their lungs with particulates, VOCs, and CO. Cigarette smoke typically contains 400–450 ppm CO (Wark and Warner 1981). The blood of most smokers averages between 5% and 10% HbCO, whereas nonsmokers typically have 2% or so (Goldsmith 1968). (A small amount of CO is produced by body metabolism.) A heavy smoker caught in a traffic jam in a high-altitude city such as Denver could easily exceed 10% to 20% HbCO, which could result in headaches, fatigue, and impaired driving ability.

In addition to CO, cigarette smoke contains an appreciable amount of formaldehyde (HCHO)—a HAP and a fairly stable product of incomplete combustion. In fact, a single cigarette emits about 1.4 mg of HCHO (Masters 1991). In many bars (where smokers seem to congregate and breathe each other's used smoke) it is not uncommon for the formaldehyde concentration in the air to exceed levels that can cause eye, nose, and throat irritation.

AN AIR POLLUTION INDEX

Based on an overall appraisal of the effects of various pollutants, a Pollutants Standards Index (PSI) has been adopted by the EPA (40 CFR 58). The PSI is reported and used by many U.S. cities to indicate their day-to-day air quality. The PSI assigns a numerical rating to air quality, which corresponds with a particular description of the air quality as shown in Table 1.9.

There are six subindices that contribute to the PSI—one each for TSP, SO$_2$, CO, O$_3$, and NO$_2$, and one for the product of the TSP and

SO_2 concentrations. Each subindex ranges from 0 to 500, with 100 corresponding to the primary NAAQS and 500 corresponding to significant harmful effects. The six subindices and their corresponding concentrations are listed in Table 1.10.

PSI Value	Air Quality Descriptor
0–50	Good
51–100	Moderate
101–199	Unhealthful
200–299	Very Unhealthful
≥300	Hazardous

Table 1.9 PSI Values and Air Quality Descriptors

Table 1.10 Individual PSI Subindex Breakpoints*

I Value	24-hr TSP $\mu g/m^3$	24-hr SO_2 $\mu g/m^3$	TSP × SO_2 $(\mu g/m^3)^2$	8-hr CO mg/m^3	1-hr O_3 $\mu g/m^3$	1-hr NO_2 $\mu g/m^3$
0	0	0	—	0	0	—
50	75	80	—†	5	118	—
100	260	365	—†	10	235	—†
200	375	800	65×10^3	17	400	1130
300	625	1600	261×10^3	34	800	2260
400	875	2100	393×10^3	46	1000	3000
500	1000	2620	490×10^3	57.5	1200	3750

*At 25 C and 760 mm Hg.
† No index values are reported at these concentration levels because there is no short-term NAAQS.

Adapted from 40 CFR (Code of Federal Regulations) 58, 1982.

The calculation of the PSI begins with first calculating the individual subindex for each of the six categories. Each subindex is a continuous but not smooth function of the pollution concentration. In fact, each subindex is defined to be a segmented linear function (40 CFR 58). This simply means that the discrete points defined in Table 1.10 are connected by straight lines. For any pollutant i at the tabular point j, the coordinates of the jth point are (X_{ij}, I_{ij}), where X_{ij} is the concentration of the ith pollutant at the jth point, and I_{ij} is the corresponding subindex value. For any observed concentration X_i, the value of the subindex I_i is then given by

$$I_i = \frac{I_{ij+1} - I_{ij}}{X_{ij+1} - X_{ij}}\left(X_i - X_{ij}\right) + I_{ij} \tag{1.32}$$

for $X_{ij} \leq X_i \leq X_{ij+1}$. Once a subindex is obtained for each pollutant, the overall PSI value is taken simply as the maximum of all the I_i values. Obviously, such indices are general guidelines only, and not totally definitive.

▶ ▶ ▶ ▶ ▶ ▶ ▶ ▶ ▶

Example 1.17

Calculate the PSI and give a verbal description of air that contains 7 mg/m^3 CO (8-hour average), 300 µg/m^3 TSP (24-hour average), and 300 µg/m^3 SO$_2$ (24-hour average).

Solution

By inspection of Table 1.10, the TSP concentration corresponds to a subindex value between 100 and 200. Also, by inspection, the CO concentration results in an I value between 50 and 100, as does the SO$_2$ concentration. CO and SO$_2$ can be eliminated from further consideration because only the maximum I value is needed. However, the product of TSP $\times$ SO$_2$ is $90(10)^3$, which has an I value between 200 and 300. Therefore, TSP $\times$ SO$_2$ is the maximum index. To calculate the actual I value, we use Eq. (1.32) as

$$I = \frac{300 - 200}{261(10^3) - 65(10^3)}\left[90(10)^3 - 65(10)^3\right] + 200$$

$$\text{PSI} = I = 213$$

The air quality description for this PSI value is "very unhealthful."

◀ ◀ ◀ ◀ ◀ ◀ ◀ ◀ ◀

◆ 1.9 NATIONAL AIR QUALITY TRENDS ◆

In a previous section, we presented and discussed trends in the national emissions of the primary pollutants. The changes in emissions in the first 20 years after the CAAA of 1970 are graphically summarized in Figure 1.7. It is important to monitor emissions, but the main objective of emission control programs is to improve ambient air quality. The EPA systematically compiles and publishes air monitoring data from several hundred sites across the country (U.S. Environmental Protection Agency 1991). In this section, we briefly present some of these data and findings. The ambient concentrations of total particulates (TP) as measured at nearly 2000 sites maintained a steady average of about 60 µg/m^3 through the last half of the 1970s, then declined to about 50 µg/m^3 in the 1980s. That concentrations of particulate matter did not decline with the decrease in PM emissions may be due in part to high background levels of naturally occurring PM.

The annual average SO$_2$ concentrations, as measured at 457 sites with continuous monitors, decreased by about two-thirds from 1975 to 1998 as shown in Figure 1.8 (U.S. Environmental Protection Agency 2000).

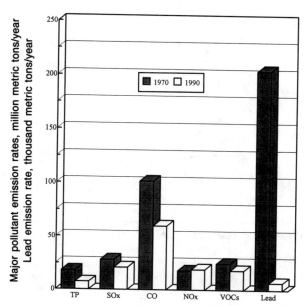

Figure 1.7
Pollutant emission rates, 1970 levels versus 1990 levels.
(Adapted from U.S. Environmental Protection Agency, EPA-450/4-91-023, 1991.)

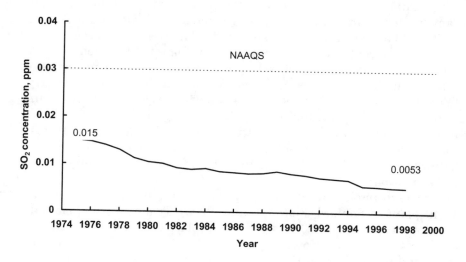

Figure 1.8
Trend in U.S. average annual SO_2 concentration, 1975–1998.
(Adapted from U.S. Environmental Protection Agency, EPA-454/R-00-003, 2000.)

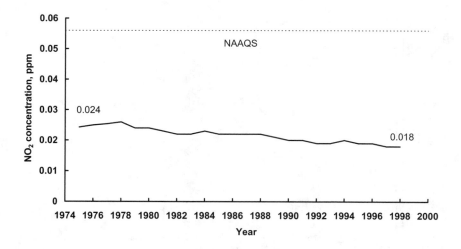

Figure 1.9
Trend in U.S. average annual NO₂ concentration, 1975–1998.
(Adapted from U.S. Environmental Protection Agency, EPA-454/R-00-003, 2000).

During this same period, emissions of SO_2 declined about 29% owing to better controls on combustion processes. Concentration data for TSP and SO_2 for several American cities in 1981 are presented in Table 1.11.

Annual average NO_2 concentrations and emissions followed the same pattern during this period. Both increased slightly from 1975 to 1978 and have declined slowly since then. The trend in average ambient NO_2 concentrations is presented in Figure 1.9. Even further reductions in both SO_2 and NO_x emissions are foreseen in the near future due to a very recent trend in the electric power industry towards "re-powering" projects. A re-powering project typically occurs at a large power plant and involves converting old boilers burning a heavy fuel oil to modern combustion turbines burning natural gas. This switch increases the efficiency of the plant and greatly reduces both SO_x (if the oil being burned has significant sulfur content) and NO_x emissions.

As an example, consider the re-powering of the Florida Power and Light plant in Sanford, Florida (Zahm 2000). Two heavy fuel oil boilers (total capacity of 870 MW) will be replaced by eight natural gas combustion turbines (total capacity of 2200 MW). Even with the increased power output, the annual emissions of SO_x will drop from more than 56,000 tons per year (tpy) to about 560 tpy, and NO_x emissions will drop from about 10,000 to about 5800 tpy. Even more impressive are the projected emission rates per unit of power gener-

Table 1.11 Annual Average Particulate and SO_2 Concentrations in Several American Cities in 1981

Location	SMSA* Population Range-millions	TSP $\mu g/m^3$	SO_2 ppm
New York, NY	>2	68	0.025
Los Angeles, CA	>2	121	0.011
Chicago, IL	>2	111	0.015
Washington, DC	>2	65	0.017
Houston, TX	>2	151	0.005
St. Louis, MO	>2	190	0.022
Pittsburgh, PA	>2	100	0.045
Atlanta, GA	>2	79	0.009
Miami, FL	1–2	97	0.003
Denver, CO	1–2	183	0.013
Phoenix, AZ	1–2	178	0.006
Cincinnati, OH	1–2	84	0.014
Portland, OR	1–2	114	0.012
New Orleans, LA	1–2	82	ND
Memphis, TN	0.5–1	74	0.018
Birmingham, AL	0.5–1	111	0.007
Honolulu, HI	0.5–1	51	0.007
Orlando, FL	0.5–1	67	0.006
Greenville-Spartanburg, SC	0.5–1	63	0.003
Fresno, CA	0.5–1	109	0.003

*Standard Metropolitan Statistical Area
Adapted from U.S. Environmental Protection Agency, 1983.

ated. Assuming 80% availability, the SO_2 emissions rate will improve from 19 lb/MWhr burning oil in the old boilers to 0.074 lb/MWhr burning natural gas in the new combustion turbines. Furthermore, the NO_x emissions rate will improve from 3.3 lb/MWhr to 0.76 lb/MWhr after the re-powering project.

Carbon monoxide concentrations, measured as the second highest nonoverlapping 8-hour average, decreased 26% between 1975 and 1981 and continued to decrease through 1998 (see Figure 1.10). The improvements were readily apparent at heavily trafficked urban sites at which the vehicle density did not change (U.S. Environmental Protection Agency 1983, 2000). Therefore, it is apparent that the improvements are due primarily to the gradual replacement of older model cars with newer ones that emit far less CO per vehicle.

Ambient ozone concentrations decreased erratically between 1975 and 1991, but have moved sideways since then, as shown in Figure 1.11 (U.S. Environmental Protection Agency 2000). As measured by the composite average (at 471 sites) of the second-highest daily maximum 1-hour average, O_3 concentrations decreased from 0.15 ppm in

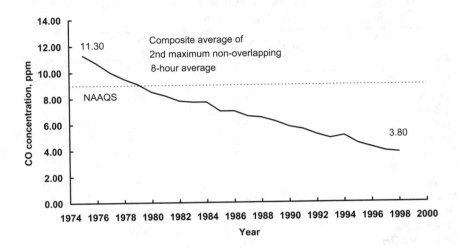

Figure 1.10
Trend in U.S. carbon monoxide levels, 1975–1998.
(Adapted from U.S. Environmental Protection Agency, EPA-454/R-00-003, 2000.)

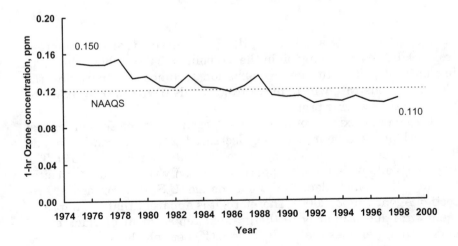

Figure 1.11
Trend in the composite average of the U.S. second-highest daily
maximum 1-hour ozone level, 1975–1998.
(Adapted from U.S. Environmental Protection Agency, EPA-454/R-00-003, 2000.)

1975 to slightly below the NAAQS in 1998. However, a number of cities are still above the 0.12 ppm standard.

Ambient lead (Pb) concentrations decreased over 90% from 1975 to 1990. This decrease correlates well with the decrease in lead contained in gasoline over the same time period. Even though the NAAQS for lead was not promulgated until 1978, regulations restricting the lead content of gasoline were issued in the early 1970s. As measured at 202 urban sites, the quarterly average atmospheric lead concentrations decreased from 0.91 $\mu g/m^3$ in 1975 to 0.07 $\mu g/m^3$ in 1990, and to 0.04 $\mu g/m^3$ in 1998.

Based on the observed trends in the data for the past several years, it appears that national air quality is improving. Emissions are decreasing, as are ambient concentrations. This does not mean that efforts to continue to improve air quality should be relaxed; the underlying forces that tend to increase air pollution—growing population and increasing energy use—are still present. In addition, even at current air quality levels, there may be long-term dangers due to air pollution. As a nation, we are capable of dealing with our current air pollution problems, and, with continued dedication to the task, we should be capable of dealing with them in the future.

◆ PROBLEMS ◆

1.1 In 1989, what fraction of U.S. NO_x emissions was contributed by highway transportation? By electric utilities? What were these fractions in 1998?

1.2 The secondary NAAQS for SO_2 is 0.5 ppm (for a 3-hour averaging time). Calculate the equivalent concentration in $\mu g/m^3$ at standard temperature and pressure (STP: $T = 25$ C, $P = 1$ atm).

1.3 Assuming compliance with federal NSPSs, predict the daily rates of emissions of particulates, NO_x, and SO_x from a coal-fired power plant producing 1000 MW of electrical power at an overall thermal efficiency of 40%. Assume coal contains 1.5% sulfur and has a heating value of 10,000 Btu/lb.

1.4 Assume that gasoline is burned with 99% efficiency in a car engine, with 1% remaining in the exhaust gases as VOCs. If the engine exhausts 16 kg of gases ($MW = 30$) for each kg of gasoline ($MW = 100$), calculate the fraction of VOCs in the exhaust. Give your answer in parts per million.

1.5 Which emits more SO_x per unit amount of electricity produced: a coal-fired plant with a 90% efficient SO_2 scrubber or an oil-fired plant with no scrubber? Assume the coal has a heat content of 13,000 Btu/lb and contains 3.5% sulfur, and that the oil has a heat content of $6.0(10)^6$ Btu/barrel (42 gallons) and contains 0.9% sulfur. Assume the oil has a specific gravity of 0.92.

1.6 Estimate the daily emissions of particulates from a solid-waste incinerator emitting at 0.18 g/dscm. The incinerator burns 50 tonnes per day (1 tonne = 1 metric ton = 1000 kg) and exhausts gases in a ratio of 20 kg of gases per kg of feed. Assume the gases exit at 200 C and 1 atm. Also assume that the gases have an average molecular weight of 30, and contain 12% CO_2 and 8% H_2O as emitted.

1.7 A stack was sampled for 8 minutes. The gas velocity (at stack conditions) through the sampler was 15.0 m/s and the area of the sample probe was 5.07 cm^2. The net mass of particulates collected was 0.430 grams.

a. Calculate the concentration of particulates in the stack in $\mu g/m^3$.

b. The temperature in the stack is 300 F and the pressure is 725 mm Hg. If the total flow rate of gas out of the stack is 190 standard cubic meters per second, calculate the total mass emissions of particulates out of the stack in kg/day.

1.8 A previously calibrated rotameter is being used to measure a gas flow. The old calibration curve is a straight line described by the following equation:

$$Q = 1.10RR - 0.50$$

where

Q = flow, liters/minute

RR = rotameter reading, dimensionless

The rotameter was calibrated using ethane (C_2H_6) at 1.40 atm and 60 F. The gas presently being used is methane (CH_4) at 1.10 atm and 100 F.

a. Calculate the true volumetric flow rate of methane in the rotameter when the rotameter reads 4.0.

b. Calculate the mass flow rate of methane in g/min.

1.9 An orifice meter was calibrated for dry air at $T = 90$ F and $P = 1.3$ atm. The calibration data are as follows:

ΔP, inches of water	Q, ft^3/minute
0	0
0.5	2.0
1.0	3.0
2.0	4.2
3.0	5.0
4.0	5.5
5.0	5.8

a. Plot the calibration curve for this meter.

b. Derive a mathematical relationship describing the calibration curve.

c. Calculate the volumetric flow rate when the meter reads 4.5 inches of water.

d. Calculate the mass flow rate of air in g/min.

1.10 Assume you are using the meter of Problem 1.9 to measure a flow rate of argon at 150 F and 0.80 atm.

a. What is the volumetric flow rate of argon when the meter reads 4.5 inches of water?

b. What is the mass flow rate of argon in g/min?

1.11 The one-hour standard for CO is 35 ppm. Calculate the equivalent concentration in mg/m^3.

1.12 Which is denser: dry air or humid air? Assume each is at the same temperature and pressure. Prove your answer.

1.13 Calculate the density of a mixture of 90% air and 10% SO_2 at a temperature of 150 C and pressure of 1.10 atm.

1.14 Calculate the density of combustion gases with the following composition: 75% N_2, 5.5% O_2, 11% CO_2, 8.4% H_2O, and 0.1% CO. The temperature is 350 F and the pressure is 755 mm Hg.

1.15 Refer to Example 1.11. Assume all the numbers are the same except that over a 20-minute sampling time, 40.0 grams of water are collected. Calculate the NO_2 concentration in the stack.

1.16 Assuming that the average stack gas velocity in Problem 1.15 is 10.0 m/s and the stack is circular with a 3.0 meter diameter, calculate the mass emissions rate of NO_2 in kg/day.

1.17 Assuming Eq. (1.29) holds with $a = -0.03$ km^{-1} and $b = 0.0025$ m^3/km-μg, estimate the visibility limit in air with (a) 100 $\mu g/m^3$ of particles, and (b) 500 $\mu g/m^3$ of particles. Assume $\sigma_a = \sigma_s$.

1.18 Assuming complete equilibrium, calculate the level of HbCO in blood when the air contains 100 ppm CO.

1.19 Exposure to air containing 600 ppm CO for 30 minutes results in an actual HbCO blood concentration of 10%. Estimate the percentage of the equilibrium saturation value of HbCO.

1.20 If on a particular day, the 24-hour average concentrations of particles and SO_2 are equal to the annual average values given in Table 1.11, calculate the PSI for New York and Chicago, and verbally describe the air.

1.21 If on a particular day, the 24-hour average concentration of particles and SO_2 are twice as high as the annual average values given in Table 1.11, calculate the PSI for Los Angeles, Houston, and Honolulu, and verbally describe the air.

1.22 For the following three coals burned in large electric power plants, determine the SO_2 emission rates (lb/million Btu) and scrubber efficiencies (%) required under federal NSPSs.

   a. 4.0% S and 10,000 Btu/lb
   b. 2.25% S and 13,000 Btu/lb
   c. 1.0% S and 15,000 Btu/lb

1.23 If coal (b) of Problem 1.22 is burned in a 40%-efficient 400-MW plant, estimate the allowable rate of SO_2 emissions (kg/day) under federal NSPSs.

1.24 For an SO_2 emission rate of 22,000 kg/day and an exhaust gas flow rate of 5.0 million m^3/hour (after the scrubber) measured at 150 C and 1 atm pressure, calculate the concentration of SO_2 in the exhaust gases (ppm).

1.25 Assume that the NO_2 concentration in a house with a gas stove is 150 µg/m^3. Calculate the equivalent concentration in ppm at STP.

1.26 U.S. energy use was approximately 80 quads (quadrillion Btu) in 1992. About 90% of this energy was supplied by burning fossil fuels (coal, oil, and gas). Assuming the average fuel composition and energy content are CH_2 and 15,000 Btu/lb, respectively, calculate the U.S. contribution to global carbon emissions in 1992. Give your answer in billions of metric tons.

1.27 A stack (T = 550 F and P = 750 mm Hg) was sampled using EPA Method 5. The total gas volume that flowed through the dry gas meter was 2.785 cubic meters (at T = 60 F and P = 800 mm Hg). The mass of particles collected was 1.50 g. Also, 72.0 g of H_2O was collected in impingers.

   a. Calculate the concentration of PM in the stack (µg/m^3) at stack conditions.
   b. The stack gas exits at an average velocity of 20 m/s, and the stack has an inside diameter of 5 meters. What is the emission rate of PM, in kg/day?
   c. What is the PM concentration in µg/dscm?

1.28 A stack gas flowing at 80 m^3/min (at T = 25 C and 1 atm) contains approximately 75% N_2, 5% O_2, 8% H_2O, and 12% CO_2.

   a. Calculate the gas molecular weight.
   b. The gas contains 650 ppm SO_2; calculate the emission rate of SO_2 in kg/day.

1.29 That portion of the atmosphere known as the ozone layer can be found at a height of about 30 km above the earth's surface in the middle latitudes. In this portion of the stratosphere, the temperature is –40 C and the pressure is 10 mbar. Assume that the air in the ozone layer is about 78% nitrogen, 20.8% oxygen, and 0.2% ozone; calculate the density of air in this part of the atmosphere. Give your answer in milligrams/cubic centimeter.

1.30 The concentration of CO_2 in the air in 1800 was about 280 ppm and reached about 300 ppm by 1900. Calculate the percent increases that occurred from 1800 to 1900 and from 1900 to 1999.

REFERENCES

Abelson, P. Editorial comment. *Science, 289*, Aug. 25, 2000.

Adams, R. M., Hamilton, S. A., and McCari, B. A. "An Assessment of the Economic Effects of Ozone on U.S. Agriculture," *Journal of the Air Pollution Control Association, 35*(9), September 1985, p. 938.

Bacastow, R. B., Keeling, C. D., and Whorf, T. P. "Measurements of the Concentration of Carbon Dioxide at Mauna Loa Observatory, Hawaii," in *Carbon Dioxide Review: 1982*, W. C. Clark, Ed. New York: Oxford University Press, 1982.

Barnola, J. M., Raynaud, D., Korotkevich, Y. S., and Lorius, C. "Vostok Ice Core Provides 160,000 Year Record of Atmospheric CO_2," *Nature*, October 1987, p. 329.

Brandt, C. S., and Heck, W. W. "Effects of Air Pollutants on Vegetation," in *Air Pollution*, Vol. I (2nd ed.), A. C. Stern, Ed. New York: Academic Press, 1968.

Brown, J. C., Jr., and Cooper, C. D. *The Effects of Relaxed Emissions Standards and Heavy Duty Vehicles on Projected Highway Emissions Through 1990*, a paper presented at the 71st Annual Meeting of the Air Pollution Control Association, Houston, TX, June 25–30, 1978.

Bubenick, D. V., Record, F. A., and Kindya, R. J. *Acid Rain: An Overview of the Problem*, a paper presented at the 74th Annual Meeting of the American Institute of Chemical Engineers, New Orleans, LA, November 8–12, 1981.

Council on Environmental Quality. "First Annual Report of the Council on Environmental Quality–1970." Washington, D.C.: The White House, 1971.

Crawford, M. *Air Pollution Control Theory*. New York: McGraw Hill, 1976.

Faiz, A., Sinha, K., Walsh, M., and Varma, A. "Automotive Air Pollution: Issues and Options for Developing Countries," in *Policy, Research and External Affairs Working Papers*, WPS 492, Infrastructure and Urban Development Department, World Bank, Washington, D.C., August 1990.

Faiz, A., Weaver, C., Sinha, K., Walsh, M., and Carbajo, J. *Air Pollution from Motor Vehicles: Issues and Options for Developing Countries*, a report for the World Bank, Washington, D.C., April 1992.

Flavin, C. "Slowing Global Warming: A Worldwide Strategy," Worldwatch Paper 91, Worldwatch Institute, Washington, D.C., 1989.

Florida Administrative Code, Chapter 17–2, "Air Pollution," 1982.

40 CFR 50. Code of Federal Regulations, Title 40, Part 50, *National Primary and Secondary Ambient Air Quality Standards*, Office of Federal Register, Washington, D.C., July 1, 1991.

40 CFR 52. Code of Federal Regulations, Title 40, Part 52, *Approval and Promulgation of Implementation Plans*, Office of Federal Register, Washington, D.C., July 1, 1991.

40 CFR 58. Code of Federal Regulations, Title 40, Part 58, *Ambient Air Quality Surveillance*, Office of Federal Register, Washington, D.C., July 1, 1982.

40 CFR 60. Code of Federal Regulations, Title 40, Part 60, *Standards of Performance for New Stationary Sources*, Office of Federal Register, Washington, D.C., July 1, 1991.

40 CFR 61. Code of Federal Regulations, Title 40, Part 61, *Emission Standards for Hazardous Air Pollutants*, Office of Federal Register, Washington, D.C., July 1, 1991.

Friedlander, S. K. *Smoke, Dust and Haze*. New York: Wiley, 1977.

Goldsmith, J. R. "Effects of Air Pollution on Human Health," in *Air Pollution*, Vol. I (2nd ed.), A. C. Stern, Ed. New York: Academic Press, 1968.

Hasselblad, V., Eddy, D. M., and Kotchmar, D. J. "Synthesis of Environmental Evidence: Nitrogen Dioxide Epidemiology Studies," *Journal of the Air Pollution Control Association, 42*(5), May 1992, p. 662.

Jensen, K. F. and Dochinger, L. S. "Response of Eastern Hardwood Species to Ozone, Sulfur Dioxide, and Acid Precipitation," *Journal of the Air Pollution Control Association, 39*(6), June 1989, p. 852.

Keeling, C. D., and Whorf, T. P. http://cdiac.esd.ornl.gov/trends/co2/sio-mlo.html.

Krier, J. E., and Ursin, E. *Pollution and Policy*. Los Angeles: University of California Press, 1977.

Lear, G. U.S. Environmental Protection Agency. Personal communication, August 2000.

Leggett, J., Ed. *Global Warming*. Oxford: Oxford University Press, 1990.

Likens, G. E. "Acid Precipitation," *Chemical and Engineering News, 54*(48), November 22, 1976.

Lippmann, M. "Health Effects of Ozone—A Critical Review," *Journal of the Air Pollution Control Association, 39*(5), May 1989, p. 672.

Masters, G. M. *Introduction to Environmental Engineering and Science*. Englewood Cliffs, NJ: Prentice-Hall, 1991.

Mathai, C. V. "Visibility and Fine Particles," *Journal of the Air and Waste Management Association, 40*(11), November 1990, p. 1486.

McLaughlin, S. B. "Effects of Air Pollution on Forests—A Critical Review," *Journal of the Air Pollution Control Association, 35*(5), May 1985, p. 512.

Middleton, W. E. K. *Vision through the Atmosphere*. Toronto: University of Toronto Press, 1952.

Molburg, J. "A Graphical Representation of the New NSPS for Sulfur Dioxide," *Journal of the Air Pollution Control Association, 30*(2), February 1980.

Molina, M. J., and Rowland, F. S. "Stratospheric Sink for Chlorofluoromethanes: Chlorine Atom Catalysed Destruction of Ozone," *Nature, 249*, 1974, p. 810.

National Research Council. *Airborne Particles*, Subcommittee on Airborne Particles, Committee on Medical and Biologic Effects of Environmental Pollutants. Baltimore: University Park Press, 1979.

Orlando Sentinel. "Inuits Swear World's Climate Getting Warmer," story by Associated Press, p. A7, November 16, 2000.

Robinson, E. "Effect on the Physical Properties of the Atmosphere," in *Air Pollution*, Vol. I (2nd ed.), A. C. Stern, Ed. New York: Academic Press, 1968.

Singer, S. F., Ed. *Global Climate Change*. New York: Paragon House, 1989.

Stokinger, H. E., and Coffin, D. L. "Biologic Effects of Air Pollution," in *Air Pollution*, Vol. I (2nd ed.), A. C. Stern, Ed. New York: Academic Press, 1968.

Stott, P. A., Tett, S. F. B., Jones, G. S., Allen, M. R., Mitchell, J. F. B., and Jenkins, G. J. "External Control of 20th Century Temperature by Natural and Anthropogenic Forcings," *Science, 290*(5499), December 15, 2000, pp. 2133–2136.

Studt, T. "Making Air Pollution Models Fit Reality Closer," *R & D Magazine*, December 1991.

Tebbens, B. D. "Gaseous Pollutants in the Air," in *Air Pollution*, Vol. I (2nd ed.), A. C. Stern, Ed. New York: Academic Press, 1968.

Te Brake, W. H. "Air Pollution and Fuel Crises in Preindustrial London, 1250–1650," *Technology and Culture, 16*(3), 1975.

U.S. Environmental Protection Agency. *Compilation of Air Pollutant Emission Factors* (5th ed.), AP-42 (includes Supplements 1-13), Washington, D.C., 1977.

———. *Research Summary—Acid Rain*, EPA-600/8-79-082, Washington, D.C., October 1979.

———. *Projecting Future Sea Level Rise*, EPA-230/9-83-007, Washington, D.C., October 1983.

———. *National Air Pollutant Emission Estimates, 1940–1990*, EPA-450/4-91-026, Research Triangle Park, NC, November 1991.

———. *National Air Quality and Emissions Trends Report, 1998*, EPA-454/R-00-003, Research Triangle Park, NC, March 2000.

Wark, K., and Warner, C. F. *Air Pollution—Its Origin and Control* (2nd ed.). New York: Harper & Row, 1981.

Weiner, J. *The Next One Hundred Years: Shaping the Fate of Our Living Earth*. New York: Bantam Books, 1990.

Zahm, Alan. Florida Dept. of Environmental Protection. Personal communication, August 2000.

Zwiers, F. W., and Weaver, A. J. "The Causes of 20th Century Warming," *Science, 290*(5499), December 15, 2000, pp. 2081–2083.

◆◆◆ CHAPTER 2

WHAT IS PROCESS DESIGN?

Theoretical understanding, comprehension of practical and economic limitations, common sense, ability to do original and hard work—these are the requirements for a good design engineer, and they must be used in the approach to any design problem.

Max S. Peters, 1958

◆ 2.1 INTRODUCTION ◆

In the broadest sense, *design* is the development of a plan to accomplish a particular goal. Throughout this text, we will use the term *process design* to describe the sequence of steps from the planning stage to the equipment specification stage of an air pollution control project. The major steps in the design sequence are the preliminary problem definition, the final problem definition, and a series of decision points that consist of alternatives and their associated subproblems. At each decision point, an engineer must evaluate the alternatives and choose the one that is the most technically and economically feasible. The thorough evaluation of each alternative at a decision point requires the solution of all of the subproblems associated with that alternative.

We will illustrate the design sequence using the following hypothetical situation.

The Acme Corporation produces laminates for the building industry. One operation involves the drying of a granular material that contains a significant fraction of very fine dust. The exhaust from the dryer is clearly visible and may exceed allowable particulate emission limits.

The preliminary problem in this situation is simply that the exhaust is visible and company management is concerned from both aesthetic and regulatory standpoints. The engineer assigned the pre-

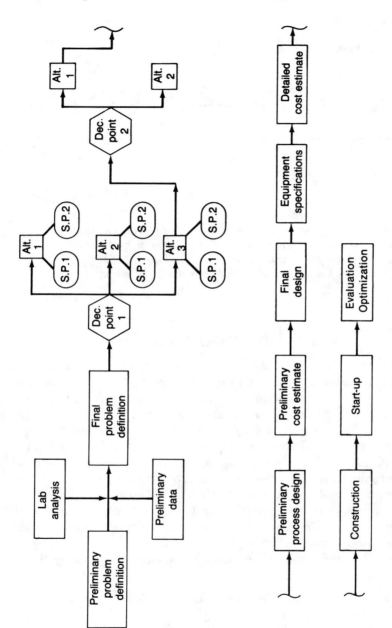

Figure 2.1
Steps in the design process.

liminary problem initiates the design sequence shown schematically in Figure 2.1.

Preliminary material balances and air flow measurements indicate that the emission rates from the dryer do exceed allowable limits. The calculated emission rates are verified by laboratory analysis of dryer exhaust samples. At this point, a decision is made that an emission control system must be installed on the dryer. Final problem definition establishes the design basis for the project. The design basis will include the exhaust flow rate, inlet and exit particulate loading, operating temperature and pressure, and projected variation in all operating parameters.

At decision point 1 (D.P.1), alternative control techniques are evaluated that might include (1) electrostatic precipitation, (2) filtration, and (3) wet scrubbing. (These particulate control methods are discussed in detail in Chapters 5, 6, and 7, respectively.) Each alternative process will provide an acceptable control level, but each has several subproblems (S.P.) associated with it. The dust from the dryer might be flammable, and precipitation would then be ruled out because of safety considerations. Filtration might be impractical because of the adhesion properties of the dust. Wet scrubbing might not be acceptable in some areas because of the problems associated with disposal of the sludge.

If wet scrubbing (alternative 3) is selected at D.P.1, then D.P.2 will involve the selection of a type of scrubber. Subproblems at D.P.2 that lead to the selection of a type of scrubber will include pressure drop requirements, liquid recirculating rates, materials of construction, and power requirements.

An important decision point in every pollution control project is the evaluation of methods for disposing of collected pollutants; alternatives include recycling or reuse, incineration, and disposal in a landfill. The decision point for disposal of pollutants has a major impact on the overall economic feasibility of a project.

The completion of all decision points leads to the final design, from which complete equipment specifications can be prepared and a project cost estimate can be developed. Construction, startup, and evaluation complete the design process.

◆ 2.2 GENERAL DESIGN CONSIDERATIONS ◆

PROCESS FLOW SHEETS

A process flow sheet provides a graphic description of a process and can vary in complexity from a simple block diagram to a detailed schematic showing instrumentation and stream operating conditions. The degree of detail is determined by the stage of development of the process and by the intended use of the flow sheet.

A flow sheet is a valuable tool at every step of the design process. In the initial development state, a simple qualitative flow sheet is used in developing the connectivity of the process. *Connectivity* refers to how individual components of the process are joined and how the overall process connects to the entire manufacturing facility. As an illustration, consider a project for recovering a mixture of toluene and ethyl acetate from a dryer exhaust using a fixed-bed carbon solvent recovery system. Figure 2.2 shows a sketch of a preliminary flow sheet of a fixed-bed system to serve as a starting point in the project. Figure 2.2 is completely qualitative and shows only the basic components of any fixed-bed adsorption system.

As the process is developed, the flow sheet is modified to include additional requirements such as carbon drying with hot air after steam regeneration. The process is complicated by the fact that ethyl acetate is slightly soluble in water. If the recovered solvent mixture is to be burned as fuel, no further separation beyond decantation is necessary. However, as shown in the detailed flow sheet of Figure 2.3, a distillation step will be necessary to return a dry solvent for reuse in the original process.

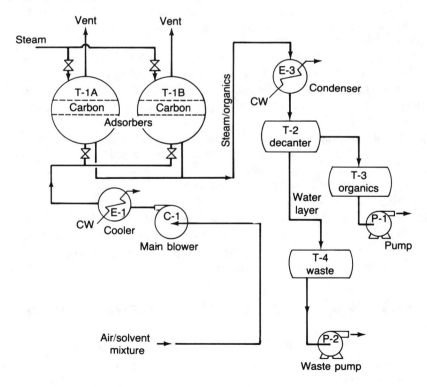

Figure 2.2
Preliminary flow sheet for a fixed-bed carbon adsorption/solvent recovery system.

The detailed flow sheet can be used to prepare an equipment list, which is then used in the preparation of the preliminary cost estimate. It is often desirable to include stream flow conditions on the detailed flow sheet. This information can be included by flagging particular streams or by showing stream conditions in a box below the flow sheet. The latter procedure makes it easier to compare stream conditions.

MATERIAL BALANCES AND ENERGY BALANCES

Equipment selection and sizing require a complete knowledge of all material and energy flow to and from each unit. The law of conservation of mass and energy, which is the basis for material and energy balance calculations, can be expressed as

$$\text{accumulation} = \text{input} - \text{output} + \text{net generation} \qquad (2.1)$$

For steady-state operation, all operating parameters are time independent, and Eq. (2.1) becomes

$$0 = \text{input} - \text{output} + \text{net generation} \qquad (2.2)$$

In the majority of cases, Eq. (2.2) describes material and energy balances around pollution control equipment that is designed for steady-state operation. Some exceptions to this generalization are encountered in the design of incinerators, direct-fired dryers, and adsorbers owing to the heat generation and/or pollutant accumulation within the units.

The following steps are helpful in performing material and energy balance calculations.

1. Draw a sketch of the process.
2. Identify and label all entering and exiting streams.
3. Label all pertinent data on the sketch.
4. Draw a dashed envelope around that portion of the process involved in the balance.
5. Select a suitable basis for the calculation.

▶ ▶ ▶ ▶ ▶ ▶ ▶ ▶ ▶

Example 2.1

Exhausts from two storage bins at a fiberglass plant are combined and passed through a cyclone that provides 95% particulate removal on a weight basis. The following measurements were made.

Exhaust from bin A:
flow rate—3000 acfm (actual ft^3/min) dry air
loading—15 grains of SiO_2 per standard cubic foot (gr/scf)
   [assume standard P and T are: 1 atm and 77 F]
pressure—10 pounds per square inch absolute (psia)
temperature—90 F

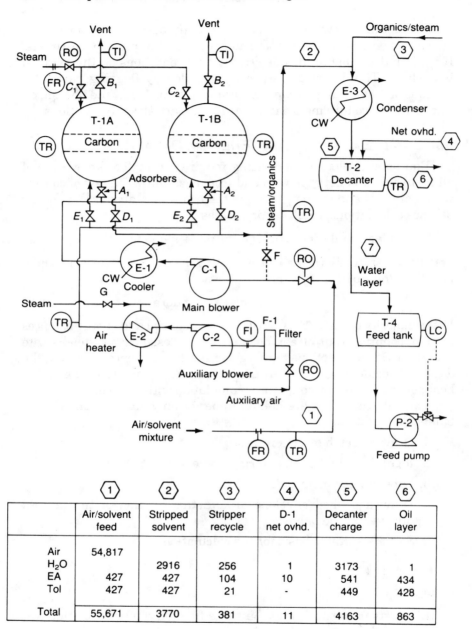

Figure 2.3
Detailed flow sheet for fixed-bed carbon adsorption/solvent recovery
including solvent drying by distillation.
(Adapted from Fair, 1969.)

	① Air/solvent feed	② Stripped solvent	③ Stripper recycle	④ D-1 net ovhd.	⑤ Decanter charge	⑥ Oil layer
Air	54,817					
H₂O		2916	256	1	3173	1
EA	427	427	104	10	541	434
Tol	427	427	21	-	449	428
Total	55,671	3770	381	11	4163	863

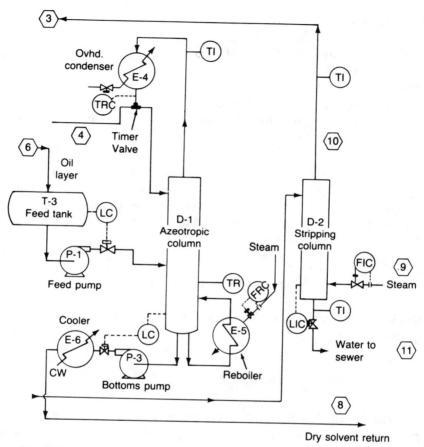

Material balance lb/hr average

	7	8	9	10	11
	Water layer	D-1 Bottoms	Stripping steam	Stripper ovhd.	Stripper bottoms
	3172	-	635	256	3551
	107	424	-	104	3
	21	428	-	21	-
	3300	852	635	381	3554

Exhaust from bin B:
flow rate—2500 acfm dry air
loading—10 grains of Na_2CO_3 per scf
pressure—11 psia
temperature—110 F

Find the rate of solids discharge (collection of solids) from the cyclone, and the concentration of solids remaining in the air being exhausted from the cyclone.

Solution

First, we draw a sketch of the process and label the operating conditions on the sketch. It is necessary to perform a balance around point C (the junction of streams A and B) to find the total solids in the feed to the cyclone. We draw an envelope around point C, and select a basis of one hour of operation.

Air balance:

$$Q_A \, \rho_A + Q_B \, \rho_B = Q_F \, \rho_F$$

where

Q = volumetric flow rate of air, in cubic feet per minute (cfm)

ρ = density of air, lb/ft^3

Solids balance:

$$Q_A C_{SA} + Q_B C_{SB} = Q_F C_{SF} = \dot{M}_{SF}$$

where

Q = volumetric flow rate of air, in cubic feet per minute (cfm)

C = concentration of solids, in stream (grains/ft^3)

A,B,F = subscripts identifying each stream

$\dot{M}_{SF}$ = mass flow rate of solids in stream F

Note that Q, C, and ρ must have consistent units. That is, Q can be expressed in standard cubic feet per minute (scfm), C in grains/scf, and ρ in lb/scf; or Q can be expressed in actual cubic feet per minute (acfm), C in gr/acf, and ρ in lb/acf. Often, particulate matter (PM) measurements are reported as gr/scf, and regulations are written in terms of gr/scf. So, we will first convert the air flow rates to scfm.

$$Q_A = 3000 \frac{acf}{min} \times \frac{10 \text{ psia}}{14.7 \text{ psia}} \times \frac{(77 + 460)R}{(90 + 460)R} = 1993 \text{ scfm}$$

$$Q_B = 2500 \frac{acf}{min} \times \frac{11 \text{ psia}}{14.7 \text{ psia}} \times \frac{(77 + 460)R}{(110 + 460)R} = 1762 \text{ scfm}$$

Note that since the air flow rates have been converted to scfm, $\rho_A = \rho_B = \rho_F$; the air balance simplifies to:

$$Q_A + Q_B = Q_F = 3755 \text{ scfm}$$

Also note that no air exits from the bottom of the cyclone (it all goes out with stream E). Thus, $Q_E = Q_F$.

Now, we return to the solids balance. The solids flow rate into the cyclone is:

$$\dot{M}_{SF} = 1993 \text{ scfm} \times 15 \text{ gr/scf} + 1762 \text{ scfm} \times 10 \text{ gr/scf} = 47{,}515 \text{ gr/min}$$

$$= 47{,}515 \frac{\text{gr}}{\text{min}} \times \frac{1 \text{ lb}}{7000 \text{ gr}} \times \frac{60 \text{ min}}{1 \text{ hr}} = 407 \frac{\text{lb}}{\text{hr}}$$

Since the cyclone is 95% efficient, the solids collection rate is:

$$\dot{M}_{SD} = 0.95 \times 407 \text{ lb/hr} = 387 \text{ lb/hr}$$

The concentration of solids in the air exhausted from the cyclone is:

$$C_{SE} = \frac{0.05 \times 407 \text{ lb/hr}}{3755 \text{ scf/min}} \times \frac{1 \text{ hr}}{60 \text{ min}} \times \frac{7000 \text{ gr}}{1 \text{ lb}} = \frac{0.63 \text{ gr}}{\text{scf}}$$

Sketch for Example 2.1

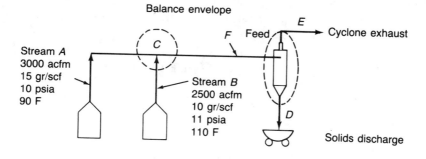

At this point, we can list some generalizations governing the solution of material and energy balance equations.

1. For a direct solution of simultaneous equations with n unknowns, n independent equations are required.

2. The n required equations for a material balance can consist of one overall balance plus $C - 1$ component balances, where C is the number of components.

3. If an energy balance and material balance are considered simultaneously, an additional independent equation can be written for the overall enthalpy balance around the system.

With the cost of energy increasing significantly in the past few years, pollution control engineers must work even harder to lower energy consumption in control equipment. Minimization of energy use can be accomplished through the following:

1. Better equipment design leading to increased efficiencies
2. Better equipment selection for specific control applications
3. Optimization of equipment operation through frequent evaluation of equipment performance

A basic understanding of energy fundamentals is a prerequisite for good design. The following section presents a brief review of the general energy balance equation as applied to flow systems. For a detailed derivation of the basic energy relationships, refer to any introductory thermodynamics text.

THE TOTAL ENERGY EQUATION FOR STEADY-STATE FLOW SYSTEMS

For a steady-state flow system, referring to Eq. (2.2), the total energy entering a defined system boundary must be equal to the energy leaving the system minus the net generation within the system. Again, it is necessary to define the system and its boundaries in such a way as to make maximum use of the information available on the system.

Refer to the system in Figure 2.4 and assume steady state; we can write a total energy balance as

$$\dot{M}_a \left(U_a + Z_a + K_a \right) = W - Q + \dot{M}_b \left(U_b + Z_b + K_b \right) \qquad (2.3)$$
$$+ \dot{M}_b \left(P_b V_b \right) - \dot{M}_a \left(P_a V_a \right)$$

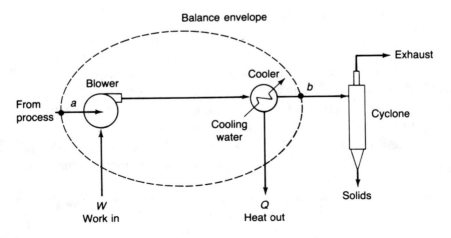

Figure 2.4
Energy balance around a portion of an air pollution control system.

where

U_a, U_b = internal energy of fluid per unit mass at points a and b, ft-lb$_f$/lb$_m$

Z_a, Z_b = potential energy of fluid per unit mass at points a and b, ft-lb$_f$/lb$_m$

K_a, K_b = kinetic energy per unit mass at points a and b, ft-lb$_f$/lb$_m$

W = mechanical work done by the system on the surroundings, ft-lb$_f$/hr (this work is positive by definition)

Q = heat absorbed by the system from the surroundings, ft-lb$_f$/hr (this heat is positive by definition)

$\dot{M}_a, \dot{M}_b$ = entering and exiting mass flow rates, which are equal under steady-state conditions, lb$_m$/hr

$P_a V_a, P_b V_b$ = work done *on* the fluid and *by* the fluid upon entering and exiting the system, ft-lb$_f$/lb$_m$

P_a, P_b = pressure, lb$_f$/ft^2

V_a, V_b = volume per unit mass, ft^3/lb$_m$

We can rewrite Eq. (2.3) in a more usable form by incorporating the concept of **enthalpy** H, where

$$H = U + PV \qquad (2.4)$$

Enthalpy is a physical property of the fluid and is a *point function*; that is, the enthalpy of a substance is a function of the conditions at a point and not a function of the path to that particular point. We do not use absolute enthalpies; rather, we use a difference in enthalpy between a desired point and a standard reference point. The enthalpies of water and steam can be obtained from standard steam tables (Green 1984), and the enthalpy of air can be determined from standard psychrometric tables (Danielson 1973; Perry 1978), as presented in Appendix B. At or near atmospheric pressure, air behaves nearly ideally and its enthalpy is virtually independent of pressure. Thus, air enthalpy can be calculated by

$$\Delta H = H_b - H_a = \int_{T_a}^{T_b} C_P dT \qquad (2.5)$$

where

C_P = specific heat at constant pressure, Btu/lb$_m$-°F

T = absolute temperature, °R

For temperatures below 150 C, it is sufficiently accurate to use

$$\Delta H = C_{P(\text{avg})} (T_b - T_a) \qquad (2.6)$$

where $C_{P(\text{avg})}$ = average value at $(T_b + T_a)/2$

Equation (2.3) can now be simplified to

$$\dot{M}\left(\Delta H + \frac{\Delta\left(v^2\right)}{2g_c\alpha} + \Delta Z\right) = Q - W \qquad (2.7)$$

where

g_c = Newton's law conversion factor, 32.174 ft-lb_m/lb_f-sec^2

α = kinetic energy correction factor, which permits the average velocity v to be substituted for point velocity u (note that α is usually assumed to be 1.0 for turbulent flow)

Example 2.2

Dry air passes through a 10-hp blower and heat exchanger system similar to that shown in Figure 2.4, at a rate of 20 lbmol/min. The air enters the blower at 1 atm and 450 F, and exits the heat exchanger at 1.2 atm and 125 F. The average air velocity at the blower intake is 150 ft/sec, and the velocity at the exchanger exit is 250 ft/sec. The overall efficiency of the blower and driver is 50%. Assuming the system is well insulated, calculate the rate of heat removal in the heat exchanger in Btu/hr.

Solution

We select a basis of 1.0 lb/sec dry air flow, and draw an envelope indicating the boundaries of the system. Since the difference in height between point a and point b is small, we can assume $Z_a = Z_b$, and the potential energy term drops out.

$$\dot{M} = 20\,\frac{\text{lbmol}}{\text{min}} \times \frac{29\;\text{lb}_m}{1\;\text{lbmol}} \times \frac{1\;\text{min}}{60\;\text{sec}} = 9.67\;\text{lb}_m\,/\sec$$

We can calculate the external work added to the air stream as

$$W = 0.50 \times -10\;\text{hp} \times 550\,\frac{\text{ft-lb}_f}{\text{sec-hp}} = -2750\,\frac{\text{ft-lb}_f}{\text{sec}}$$

By definition, work done *on* a system is negative.

We use $C_{P(\text{avg})}$ to calculate ΔH.

$$T_{\text{avg}} = \frac{450\,\text{F} + 125\,\text{F}}{2} = 287.5\;\text{F}$$

From Appendix B, $C_{P(\text{avg})} = 0.24$ Btu/lb_m-°F

$$\Delta H = C_{P(\text{avg})}\,(T_b - T_a)\,(778\;\text{ft-lb}_f/\text{Btu})$$
$$\Delta H = 0.24(125 - 450)778 = -60{,}700\;\text{ft-lb}_f/\text{lb}_m$$

Assuming $\alpha = 1.0$,

$$\frac{\Delta\left(v^2\right)}{2g_c\alpha} = \frac{(250)^2 - (150)^2}{2 \times 32.17 \times 1.0} = 621.7 \text{ ft-lb}_f/\text{lb}_m$$

Substituting various values into Eq. (2.7), we obtain

$$9.67 \frac{\text{lb}_m}{\text{sec}}\left[-60{,}700\frac{\text{ft-lb}_f}{\text{lb}_m} + 621.7\frac{\text{ft-lb}_f}{\text{lb}_m} + 0\right]$$

$$= Q - \left(-2750\frac{\text{ft-lb}_f}{\text{sec}}\right)$$

$$Q = -584{,}000\frac{\text{ft-lb}_f}{\text{sec}}$$

To convert to Btu/hr,

$$Q = -584{,}000\frac{\text{ft-lb}_f}{\text{sec}} \times 3600\frac{\text{sec}}{\text{hr}} \times \frac{1 \text{ Btu}}{778 \text{ ft-lb}_f}$$

$$Q = -2{,}700{,}000 \text{ Btu/hr}$$

The value of Q is negative because heat is removed from the system.

◀ ◀ ◀ ◀ ◀ ◀ ◀ ◀ ◀

◆ 2.3 ENGINEERING ECONOMICS ◆

Pollution control problems are usually open-ended; there are usually several viable solutions. As indicated earlier in this chapter, a major decision point in the design process is the selection of the most technically and economically feasible solution alternative. Although a detailed discussion of economic principles as applied to decision making is beyond the scope of this text, we will present several basic procedures for comparing alternative investments that are valuable tools for process engineers.

OPTIMIZING FIXED CAPITAL AND OPERATING COSTS

To develop a comparison procedure for alternatives, we must define a *base case*. The **base case** is the alternative that will provide the desired control results at the least total cost. The base case should represent a system operating under nearly optimum conditions. A trade-off usually exists between *fixed capital* and *operating cost*, as illustrated by Figure 2.5 for a typical fabric filter. In this case, the trade-off is between fabric area (capital cost) and pressure drop (operating cost). Optimum conditions from an economic standpoint would be in the region of minimum total cost. The process design that pro-

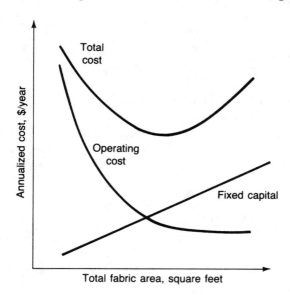

Figure 2.5
Typical cost relationships for fabric filters.

vides the lowest total cost may differ from the one that provides the best operations—that is, the design that optimizes ease of operation and stability of removal levels and minimizes maintenance. These operational considerations might dictate a design somewhat removed from the economic optimum. Thus, it is obvious that process engineers must be careful to define which "optimum" they are referring to.

DEPRECIATION

All physical assets such as pollution control equipment decrease in value with time owing to deterioration and/or obsolescence. Deterioration is the result of physical wear and, in many cases, of corrosion. Obsolescence can be due to technological advances, or changes in control requirements. In either case, the end result is eventual equipment retirement and the necessity for replacement capital. Depreciation is recognized as a legitimate operating cost and can be claimed as such in computing income taxes. Keep in mind that depreciation is a non-cash cost.

Depreciation can be calculated by one of several methods; company policy normally determines which method is used. All methods require that a *useful service life* and *salvage value* be estimated for each item of equipment. In the preparation of preliminary cost estimates, the usual practice is to assume straight-line depreciation calculated by the following equation:

$$d = \frac{V_R - V_S}{n} \qquad \textbf{(2.8)}$$

where

d = annual depreciation, dollars/year

V_R = initial cost of equipment, dollars

V_S = salvage or scrap value, dollars

n = service life, years

The asset value after a years in service is found by

$$V_a = V_R - da \qquad \textbf{(2.9)}$$

where

V_a = asset value, dollars

a = time in service, years

A further simplification often used in preliminary cost estimates is to assume a zero salvage value and an average service life of 10 years. Under these conditions, the annual depreciation per year is 10% of the initial cost.

A new company might wish to depreciate an asset rapidly during the first few years to minimize the company's taxes during its early development period. In this case, the declining-balance method is often used. In this method, the annual depreciation is a fixed percentage of the asset value at the beginning of the depreciation year. The asset value at the end of a years of service is

$$V_a = V_R \left(1 - f\right)^a \qquad \textbf{(2.10)}$$

where f = fixed percentage factor, dimensionless

The asset value at the end of the estimated service life n is then

$$V_S = V_R \left(1 - f\right)^n \qquad \textbf{(2.11)}$$

For depreciable assets placed in service before 1981, the Internal Revenue Service permits the use of a declining-balance factor equal to twice the first-year straight-line rate.* The Economic Recovery Tax Act of 1981 stipulated that all depreciable property placed in service in 1981 or later must be depreciated at rates published in ACRS (Accelerated Capital Recovery System) tables. All property is classified in 3-, 5-, 10-, or 15-year service-life groups, with no salvage value. The majority of process equipment, including pollution control equipment, falls in the 5-year group. Depreciation rates for equipment placed in service after 1986 for the 5-year group were:

---

*Referred to as the double-declining-balance method.

Year	Rate
1	20%
2	32%
3	24%
4	16%
5	8%

Straight-line depreciation was the only acceptable alternative method permitted by the 1981 act. In 1986, however, modifications were made to the Tax Reform Act of 1981 and the use of the double-declining-balance method was reinstated by the IRS for assets having service lives of 3 to 10 years. These changes are now referred to as the Modified Accelerated Cost Recovery System (MACRS). For a detailed discussion of depreciation and tax effects, consult the text by Steiner (1992). IRS Publication 534 gives a list of asset classes and permissible depreciation rates. Humphreys and English (1992) provide a good review of ACRS and MACRS depreciation stipulations.

Example 2.3

Companies A and B bought identical venturi scrubbers in 1992 that cost $75,000 each. In both applications, the service life was estimated to be 5 years with zero salvage value. The corporate income tax rate for both companies was 50%. Company A used straight-line depreciation and Company B used the MACRS method. How much more money did Company B save over the first 3 years of service based on its depreciation procedure?

Solution

Depreciation claimed by Company A:

$$d_A = \frac{\$75,000}{5} \times 3 = \$45,000$$

Depreciation claimed by Company B: The MACRS depreciation in the first three years totals 76% of the initial cost. The depreciation charged during those three years is

$$d_B = \$75,000 \times 0.76 = \$57,000$$

Since the corporate tax rate for each company is 50%, Company B saved

$$0.50 \times (\$57,000 - \$45,000) = \$6,000$$

As a practical matter, all companies use the most advantageous method of depreciation allowed by law for tax purposes, whereas engineers typically use straight-line depreciation for evaluation of alternative cases.

INCREMENTAL RATE OF RETURN ON INVESTMENTS

A frequently used measure of the profitability of an investment is the *rate of return on the investment* ROI, defined by the following:

$$\text{ROI} = \frac{P}{I} \times 100 \qquad\qquad (2.12)$$

where

ROI = return on investment, percent

P = annual profit from investment (income − expenses), dollars

I = total investment, dollars

Incremental ROI refers to the incremental return based on the incremental profit and incremental investment between two equipment alternatives. In this procedure, alternatives are compared with the acceptable alternative requiring the least investment as the base case.

Often, when a company must invest in pollution control equipment, all the alternative solutions cost money and none may generate any true profit. However, faced with having to make an investment to comply with the law, the company must choose the best alternative among several, each of which solves the pollution problem. In this case, the concept of incremental ROI still can be applied, with the cost savings of one alternative compared with another being treated as a "profit." It is interesting to note that, nowadays, certain pollution control credits are traded on the open market. Thus, a company may wish to invest in an alternative that exceeds the requirements, and then sell the credits on the open market!

▶ ▶ ▶ ▶ ▶ ▶ ▶ ▶ ▶

Example 2.4

A company must purchase a cyclone to control dust from a foundry operation. The lowest bid on a cyclone that would meet all control requirements is for a carbon-steel cyclone with an installed cost of $40,000. The cyclone has a service life of 5 years. A bid was also received for a stainless-steel cyclone that is guaranteed for 10 years and would lower maintenance costs by $1100 per year. The installed cost of the stainless-steel cyclone is $60,000. Both cyclones are estimated to have zero salvage value. If the company currently receives a 12% return before taxes on all investments, which cyclone should be purchased?

Solution

$$\text{depreciation on carbon-steel cyclone} = \frac{\$40,000}{5} = \$8000/\text{year}$$

$$\text{depreciation on stainless-steel cyclone} = \frac{\$60,000}{10} = \$6000/\text{year}$$

total yearly savings with stainless-steel cyclone
$$= (\$8000 - 6000) + \$1100 = \$3100/\text{year}$$
incremental investment $= \$60{,}000 - 40{,}000 = \$20{,}000$

$$\text{Incremental ROI} = \frac{\$3100}{20{,}000} \times 100 = 15.5\%$$

Since the incremental return exceeds company requirements, the higher bid should be accepted. Note that, in this case, even if the incremental return did not appear to be acceptably high, further consideration must be given to the stainless-steel cyclone because, presumably, a second carbon-steel cyclone would need to be purchased after five years.

COMPARISON OF SEVERAL ALTERNATIVES

Where there are three or more alternatives to be compared, the incremental analysis should be made as follows:

1. Select the acceptable unit with the lowest installed cost as the base case and designate it case 1 or the base case.

2. Designate higher-cost units case 2, case 3, etc., in order of increasing cost.

3. Calculate the incremental ROI between cases 1 and 2. If the ROI is acceptable, then case 2 now becomes the base case. If the ROI is not acceptable, case 1 remains the base case and case 2 is discarded.

4. Calculate the ROI between case 3 and the base case, which now may be either case 2 or case 1. If the ROI is acceptable, case 3 is the new base case. If the ROI is unacceptable, case 3 is discarded.

5. Continue this process until all cases have been evaluated. This procedure ensures that each increment of investment will provide an acceptable return.

PAYOUT PERIOD

A simple measure of profitability is the length of time required to recover the depreciable fixed capital investment for a project. This period of years is called the *payout period* and is defined by the following equation.

$$\text{payout period} = \frac{\text{fixed capital investment}}{(\text{annual profit} + \text{annual depreciation})_{\text{avg}}} \quad \textbf{(2.13)}$$

The payout period can be adjusted to include interest costs by adding such costs over the life of the project to the fixed capital investment.

♦ ## 2.4 CONTROL EQUIPMENT ♦
COST ESTIMATION

Company management must budget capital for pollution control equipment and its operation. It is essential that accurate cost estimates are available to establish expenditure priorities that assist in the overall corporate decision process. Certainly, a major restraint on all equipment design is the overriding necessity to maintain a satisfactory corporate profit structure.

Factors affecting equipment costs include company policies, local and federal government regulations, and national economic conditions. Individual companies can establish strict design standards that limit equipment selection. Union contracts and company agreements with fabricators can also indirectly affect costs. Government regulations can have a significant effect on costs through depreciation and tax credit policies.

Cost estimates can be classified by the degree of accuracy required. Three levels of accuracy normally recognized in estimating equipment costs are:

1. Order of magnitude estimate for conceptual planning: Accuracy ± 50%

2. Preliminary estimate for the initial go or no go decision on a project: Accuracy ± 25%

3. Final detailed engineering estimate for preparation of budget requests: Accuracy ± 5%

When available, cost data are usually several months to several years out of date, and must be adjusted for the effect of inflation. Cost data for many types of air pollution control equipment are presented in later chapters. Extrapolation of equipment costs to a later date is accomplished by using one of several published cost indexes. A **cost index** is a ratio of the cost of an item or equipment group at a specific time to the cost of the item at some base time in the past. Indexes are published by the government for labor and materials (both for retail and wholesale). The most familiar index is the Consumer Price Index (CPI), which provides a measure of inflation each year. In equipment cost estimation, an equipment or process cost index is used to update equipment cost as follows:

$$\text{cost}_{\text{present}} = \text{cost}_{\text{past}} \left(\frac{\text{present index}}{\text{past index}} \right) \qquad (2.14)$$

Three indices frequently used in estimating the cost of process and air pollution control equipment are shown in Table 2.1. Brief descriptions of the Marshall-Swift and Chemical Engineering cost indices are given in the April 1985 issue of the journal *Chemical Engineering*. A discussion of the Vatavuk Air Pollution Control Cost Index (VAPCCI) is given in the December 1995 issue of that same journal. It should be noted here that the VAPCCI gives quarterly values for eleven different types of air pollution control equipment; however, the values shown for this index in Table 2.1 are combined annual averages of the first and second quarters of each year for five selected types of control equipment. Values of the VAPCCI for individual types of control equipment (such as wet scrubbers, carbon adsorbers, etc.) may be obtained from the EPA Web site: www.epa.gov/ttn/catc/products.html#cccinfo.

Table 2.1 Equipment Cost Indexes

Year	Marshall-Swift Equipment Cost[1] Index (1926=100)	Chemical Engineering Plant Cost Index[1] (1958 = 100)	Vatavuk Air Pollution Control Cost Index[2] (1989 [1st qtr] = 100)[3]
1980	659.6	261.2	
1981	721.3	297.0	
1982	745.6	314.0	
1983	760.8	316.9	
1984	780.4	322.7	
1985	789.6	325.3	
1986	797.6	318.4	
1987	813.6	323.8	
1988	852.0	342.5	
1989	895.1	355.4	100.8
1990	915.1	357.6	104.3
1991	930.6	361.3	108.0
1992	943.1	358.2	108.4
1993	964.2	359.2	110.8
1994	993.4	328.1	101.4[3]
1995	1027.5	381.1	104.7
1996	1039.1	381.7	105.9
1997	1056.8	386.5	108.3
1998	1061.9	389.5	109.1
1999	1068.3	390.6	109.5
2000	1080.6	392.6	110.0

[1] Adapted from indices published monthly in *Chemical Engineering*.
[2] Average of 1st and 2nd quarters for five major types of air pollution control equipment.
[3] Index was re-normalized to 100 in the 1st quarter of 1994.

EQUIPMENT COSTS

Process equipment costs can be correlated with size or capacity by the following relationship.

$$\text{cost}_B = \text{cost}_A \left(\frac{\text{capacity}_B}{\text{capacity}_A} \right)^b = \text{cost}_A \left(\frac{\text{size}_B}{\text{size}_A} \right)^{b'} \qquad \textbf{(2.15)}$$

where

capacity_B = equipment throughput, in standard units of cfm, gpm, etc.

size_B = equipment size, in standard units of ft^2, ft^3, etc.

b, b' = constants

The exponents b and b' vary from 0.5 to 0.8, and will average between 0.6 and 0.7 for many types of equipment. Typical exponent values for various types of air pollution control equipment are shown in Table 2.2.

Table 2.2 Cost Exponents for Air Pollution Control Equipment

Equipment Type	Size Range	Cost Exponent
Dry Cyclones	5000–100,000 cfm	0.65
Multiclones	10,000–200,000 cfm	0.65
Scrubbers		
Impingement	5000–90,000 cfm	0.80
Gravity Spray	5000–200,000 cfm	0.62
Centrifugal	5000–100,000 cfm	0.76
Venturi, Low Energy	5000–100,000 cfm	0.76
Venturi, High Energy	5000–100,000 cfm	0.72
Electrostatic		
Precipitators	10,000–1,000,000 cfm	0.62
Reverse Pulse		
Filters	1000–100,000 cfm	0.60
Adsorbers, Fixed-Bed	2000–20,000 cfm	0.70

Adapted from Peters and Timmerhaus, 1991.

Example 2.5

A carbon-steel cyclone designed for a throughput of 5000 scfm was purchased in 1997 for $15,000. Estimate the cost of a 10,000-scfm carbon-steel cyclone purchased in 2000.

Solution

We use the Marshall-Swift (M-S) equipment cost index from Table 2.1 to extrapolate the cost of a 5000 scfm unit to 2000. We find the cost exponent for cyclones in Table 2.2

M-S index in 1997 = 1056.8

M-S index in 2000 = 1080.6

Cost exponent = 0.65

$$\text{Cost}_{10,000\text{ scfm, }2000} = \$15,000\left(\frac{10,000}{5000}\right)^{0.65}\left(\frac{1080.6}{1056.8}\right) = \$24,068$$

Due to estimation accuracy limits, all such answers should be rounded to no more than three significant figures; this answer should be reported as $24,100 (or possibly $24,000). Using either the Chemical Engineering Index or the VAPCCI results in an estimate of $23,900 (or $24,000).

◀ ◀ ◀ ◀ ◀ ◀ ◀ ◀ ◀

2.5 Preliminary Fixed Capital
◆   Cost Estimates for   ◆
Pollution Control Projects

Cost analyses for complete industrial plants have shown that a preliminary Total Installed Cost (TIC) estimate can be developed from an estimate of the Delivered Equipment Cost (DEC) for the major items of equipment. The direct and indirect installation costs are estimated as percentages of the DEC, and all costs are added together to produce the TIC. The DEC is a function of the f.o.b. purchased equipment cost (PEC) and some add-ons. The term f.o.b. stands for free-on-board, and refers to the cost of the equipment that the manufacturer charges the customer, including any charges for loading the equipment onto a train or truck, but not including the shipping charges or anything else. A similar approach can be used for estimating the installed cost of air pollution control equipment. Average installed-cost factors for pollution control equipment are given in Table 2.3. These factors are meant to serve as a general guide and must be adjusted to fit specific installations.

▶ ▶ ▶ ▶ ▶ ▶ ▶ ▶ ▶

Example 2.6

The delivered cost for a 1000 scfm fabric filter is $22,000. The unit is to be installed outdoors on an existing concrete pad. Estimate the total fixed capital investment for the system.

Solution

We use Table 2.3, omitting those items not required. Based on the minimum information available in the problem, we can omit the costs for foundation and supports and painting. The total installed cost (also called the fixed capital investment) is then

total = $22,000(2.17 − 0.04 − 0.02) = $46,420 or $46,400 (rounded)

◀ ◀ ◀ ◀ ◀ ◀ ◀ ◀ ◀

Table 2.3 Average Installed-Cost Factors for Estimating Capital Costs

Direct Costs	Precipi-tators	Scrubbers	Fabric Filters	Inciner-ators	Adsorbers
1. Equipment Costs					
a. Control Device and Auxiliary Equipment (f.o.b.)	As Req'd. (P)				
b. Instruments and Controls	0.10				
c. Taxes	0.03				
d. Freight	0.05				
Subtotal (DEC)	1.00 = 1.18 × P				
2. Installation Direct Costs					
a. Foundations and Supports	0.04	0.12	0.04	0.08	0.08
b. Erection and Handling	0.50	0.40	0.50	0.14	0.20
c. Electrical	0.08	0.01	0.08	0.04	0.08
d. Piping	0.01	0.05	0.01	0.02	0.05
e. Insulation	0.02	0.03	0.07	0.01	0.02
f. Painting	0.02	0.01	0.02	0.01	0.01
g. Site Preparation	As Req'd.				
h. Facilities and Buildings	As Req'd.				
Subtotal (× DEC)	1.67	1.62	1.72	1.30	1.44
Indirect Costs					
3. Installation Indirect Costs					
a. Engineering and Supervision	0.20	0.10	0.10	0.10	0.10
b. Construction and Field Expenses	0.20	0.10	0.20	0.05	0.05
c. Construction Fee	0.10	0.10	0.10	0.10	0.10
d. Start-up	0.01	0.01	0.01	0.02	0.02
e. Performance Test	0.01	0.01	0.01	0.01	0.01
f. Model Study	0.02				
g. Contingencies	0.03	0.03	0.03	0.03	0.03
Total (× DEC)	2.24	1.97	2.17	1.61	1.75

Adapted from Neveril et al., 1978; Heinsohn and Kabel, 1999.

♦ 2.6 ANNUAL OPERATING COST ESTIMATES ♦

We can prepare preliminary annual operating cost estimates by using Table 2.4. The major operating cost items for many installations are labor, utilities, and depreciation. For many types of equipment, utility costs will far exceed all other costs. On the other hand, in the case of wet scrubbers and spray dryers used to remove acid gases, chemical reagent costs may equal or exceed other operating costs. For equipment requiring component replacement at regular intervals, such as carbon replacement in adsorbers and catalyst replacement in catalytic incinerators, replacement cost on an annual basis is included under *direct operating costs*. On major replacement items, such as carbon and catalysts that might be replaced at 3- to 5-year intervals, the replacement cost can be annualized by considering the total replacement cost as the present worth of an ordinary annuity. The annual cost is then the annual payment of this annuity, which can be calculated as follows:

$$\text{AOC} = C_R \left[\frac{i(1+i)^n}{(1+i)^n - 1} \right] \tag{2.16}$$

where

AOC = annualized operating cost, dollars

C_R = replacement cost of component, dollars

i = annual discrete interest rate

n = replacement interval, years

The expression enclosed in brackets is referred to as the *capital recovery factor*.

Table 2.4 Annual Equipment Operating Costs

Direct Operating Costs
Labor, hours/year × cost per hour
Supervision, 15% of operating labor
Maintenance, 5% of fixed capital investment (FCI)
Utilities
Electricity, kWh/year × cost per kWh
Steam, lb/year × cost per pound
Cooling water, gal/year × cost per gal

Indirect Operating Costs
Labor overhead, 60% of total labor
Taxes, 1% of FCI
Insurance, 1% of FCI
Depreciation, as required and allowed

♦ Problems ♦

2.1 A textile finishing process involves drying a fabric that has been treated with a volatile solvent. The wet fabric entering the dryer contains 45% solvent, and the dried fabric contains 3% of the entering solvent. Solvent-free air enters the dryer at a rate of 8 pounds per pound of solvent-free fabric.

a. Calculate the percentage of solvent in the dried fabric by weight.

b. Calculate the concentration of solvent in the dryer exhaust in mg/m^3 if the exhaust is at 160 F and 1.0 atmosphere.

2.2 A company uses a 50% solution of NaOH to maintain the pH of water circulated in one of its furnace exhaust scrubbers. The cost of the solution is $200 per ton f.o.b. (free-on-board). The supplier says it would prefer to sell the company a 65% NaOH solution and the company says the water content is immaterial as long as the equivalent costs are the same including freight. If the freight is $15 per ton, what is the maximum price per ton the company should pay for the 65% solution?

2.3 An exhaust stream from a reactor consists of 90% air and 10% NH_3 (by volume). The exhaust is treated in a packed scrubber as shown in Figure P2.3. Calculate the rate of NH_3 leaving the vent stack, in pounds per day.

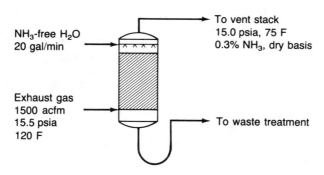

Figure P2.3

2.4 Exhaust gas from a chemical process is fed to a direct-flame incinerator as shown in Figure P2.4. The fan motor draws 5.0 horsepower and the blower/driver is 60% efficient. The fuel is methane (with a lower heating value = 21,575 Btu/lb_m), and it is flowing at 90 lb_m/hr.

$$P_A = 1.0 \text{ atm} \qquad V_A = 25 \text{ ft/sec}$$
$$P_B = 1.0 \text{ atm} \qquad V_B = 75 \text{ ft/sec}$$

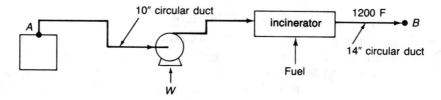

Figure P2.4

Calculate the following:

a. The enthalpy difference of the exhaust between points A and B

b. The cost of operating the system for 24 hours if power costs 9¢/kWh and fuel costs $7.00/10^6 Btu$

2.5   A control process for a glass furnace will remove boron from the 1400 F furnace by cooling the exhaust to 275 F in an evaporative cooler to condense B_2O_3. The cooled exhaust will then pass through a reverse-jet filter to remove suspended dust, including the B_2O_3 particulates. Draw a preliminary flow sheet for the process, indicating any auxiliary equipment needed. List any operating difficulties you think might arise with this system.

2.6   As an alternative to the process described in Problem 2.5, the exhaust will be treated in a UOP-type wet scrubber. The UOP scrubber is a tray type unit in which there is a layer of plastic spheres on each tray to improve gas–liquid contact. The liquid from the scrubber will be clarified in a settling basin and recirculated to the scrubber. Sludge will be pumped from one end of the settling basin. Draw a flow sheet for this process including pumps and other auxiliaries. List any problem areas you think may exist.

2.7   The delivered equipment cost for a fixed-bed carbon adsorption system with a capacity of 5000 acfm was $100,000 in 1994. Estimate the installed cost of a similar system with a capcity of 7500 acfm in 2000.

2.8   The incinerator system shown in Figure P2.4 can be equipped with a catalyst bed and the same degree of hydrocarbon removal accomplished at a temperature of 750 F. The catalyst and installation will cost $25,000, and will have a service life of 3 years. If the salvage value of the catalyst is $6000, fuel costs $7.00 per million Btu, and the company requires a 15% ROI before taxes, should the catalyst be installed?

2.9   The exhaust gas from a rotary kiln hazardous waste incinerator must be cooled from 1200 C to 300 C before entering a spray dryer for acid gas removal. The engineer in charge of this project investigated several alternatives to cool the exhaust gas and has concluded that the two most practical solutions would be either the installation of a waste heat boiler or the installation of a spray injection system in which water is sprayed directly into the exhaust duct upstream of the spray dryer. Based on the following data, which option should be selected?

Waste Heat Boiler

Cost installed	$2 million
Service life	10 years
Operating cost excluding depreciation	$0.4 million/yr
Generated steam value	$0.4 million/yr

Water Injection System

Cost installed	$0.5 million
Service life	5 years
Operating cost excluding depreciation	$0.15 million/yr

Note: The additional water vapor added to the exhaust gas will increase the size of the spray dryer by 50%. The original spray dryer cost was estimated to be $1.0 million with a service life of 10 years.

2.10 A dryer similar to the one described in Problem 2.1 is processing a nonwoven sheeting material that has just passed through a benzene wash to remove surface contaminants. The dryer operating conditions are as follows:

- sheeting dry weight is 4.0 ounces (oz) per square yard (yd^2)
- sheeting enters dryer at a rate of 150 yd^2/min
- sheeting enters dryer with 50% benzene by weight and exits with 1.8% benzene
- "pure" air enters dryer at 220 F and 1 atm
- exhaust gas leaves dryer at 180 F and 1 atm, containing 5.0% benzene (volume percent)

Since benzene is a hazardous air pollutant (HAP), it has been decided to install a direct-flame thermal oxidizer to destroy the benzene. Insurance regulations require that the incoming stream to the oxidizer (the dryer exhaust) contain no more than 25% of the lower explosive limit (LEL) of any combustible gases. The LEL of benzene is 1.4% by volume in air.

a. Calculate the additional air needed to dilute the dryer exhaust prior to feeding it into the oxidizer (lb/min and scfm).

b. In your opinion, where should the air be added to the system— in front of the dryer or after? Give reasons for your answer.

c. If the oxidizer operates at 1600 F, and the cost of buying natural gas to heat air is $8.00/million Btu of heat added to the air, calculate the additional operating cost incurred due to adding this extra dilution air ($/hour).

2.11 A gas containing equal parts of methane, ethane, and ammonia flows at a constant rate through a laboratory water-based absorption unit, which absorbs 96% of the ammonia and retains it in the liquid. No methane or ethane is absorbed into the water, and no water evaporates into the gas. Initially, there was exactly 5.00 kg of water in the absorber, and at the end of 4 hours of operation, the liquid mass is 5.25 kg. Calculate the molar flow rate (mol/hr) of the gas stream coming into the absorber, and the mole fraction of ammonia in the exit gas stream.

REFERENCES

Danielson, J. A., Ed. *Air Pollution Engineering Manual* (2nd ed.). EPA Publication AP-40, Research Triangle Park, NC, 1973.

Fair, J. R. *Mixed Solvent Recovery and Purification*, Case Study 7, Washington University Design Series, St. Louis, 1969.

Green, D. W., Ed. *Engineering Manual* (6th ed.). New York: McGraw-Hill, 1984.

Heinsohn, R. J., and Kabel, R. L. *Sources and Control of Air Pollution.* Upper Saddle River, NJ: Prentice-Hall, 1999.

Humphreys, K. K., and English, L. M., Eds. *Project and Cost Engineers' Handbook* (3rd ed.). New York: Marcel Dekker, 1993.

Neveril, R. B., Price, J. U., and Engdahl, K. L. "Capital and Operating Costs of Selected Air Pollution Control Systems—V," *Journal of the Air Pollution Control Association, 28*(12), December 1978.

Perry, R. H. *Industrial Ventilation* (15th ed.). Lansing, MI: American Conference of Governmental Industrial Hygienists, 1978.

Peters, M. S., and Timmerhaus, K. D. *Plant Design and Economics for Chemical Engineers* (4th ed.). New York: McGraw-Hill, 1991.

Steiner, H. M. *Engineering Economic Principles.* New York: McGraw-Hill, 1992.

 CHAPTER 3

PARTICULATE MATTER

The dust in smaller particles arose,
Than those which fluid bodies do compose.

John Arbuthnot, "On A Dusty Day," c. 1725

♦ 3.1 INTRODUCTION ♦

Particulate matter (PM) constitutes a major class of air pollution. PM (often called particulates) comes in a variety of shapes and sizes, and can be either liquid droplets or dry dusts, with a wide range of physical and chemical properties. Particulates are emitted from many different sources, including both combustion and noncombustion processes in industry, mining and/or construction activities, motor vehicles, and refuse incineration. Natural sources of particulates include volcanoes, forest fires, windstorms, pollen, ocean spray, and so forth. Some information on particulate causes, sources, and effects was presented in Chapter 1. In this chapter, we present information about certain characteristics of particles and particulate behavior in fluids, with emphasis on characteristics that pertain to the important engineering task of separating and removing particles from a stream of gas.

♦ 3.2 CHARACTERISTICS OF PARTICLES ♦

Before attempting the design of any collection device, we must obtain information about the particles, the gas stream, and the process conditions. Important particulate characteristics include size, size distribution, shape, density, stickiness, corrosivity, reactivity, and toxicity. Gas stream characteristics of importance are pressure, temperature, viscosity, humidity, chemical composition, and flammability. Process conditions include gas flow rate, particulate loading (mass concentration of particles in the gas stream), removal efficiency

99

requirements, and allowable pressure drop. Many of the devices in use today for separating particulates from gas exploit the vast difference in the physical size of particles and gas molecules. In the next few sections, we will further explore this characteristic and its implications.

One of the most important characteristics of a suspension of particles is the size distribution of the particles. Figure 3.1 illustrates the wide range of sizes (five orders of magnitude) of common particle dispersoids. [The common unit of measurement for small particles is the micrometer (µm), often referred to as the *micron*.] As can be expected with such a wide range of sizes, one type of collection device might be better suited than others for a specific particulate dispersoid. Furthermore, a single collection device is always more efficient in collecting larger particles and less efficient in collecting smaller ones. Thus, to calculate the overall collection efficiency of a device, it is imperative to have good information on the size distribution of the particles.

The overall efficiency η can be calculated on a basis of total number of particles collected or total mass collected. Generally, regulations are written based on mass collection, and efficiencies are calculated on a mass basis, as shown in Eq. (3.1).

$$\eta = \frac{\dot{M}_i - \dot{M}_e}{\dot{M}_i} = \frac{L_i - L_e}{L_i} \tag{3.1}$$

where

η = overall collection efficiency (fraction)

$\dot{M}_i$ = total mass input rate, g/s or equivalent

$\dot{M}_e$ = total mass emission rate, g/s or equivalent

L_i = particulate loading in the inlet gas to the device, g/m^3

L_e = particulate loading in the exit gas stream, g/m^3

When the particulate size distribution is known, and the efficiency of the device is known as a function of particle size, the overall collection efficiency can be predicted as follows:

$$\eta = \sum \eta_j m_j \tag{3.2}$$

where

η_j = efficiency of collection for the jth size range

m_j = mass fraction of particles in the jth size range

The mass collection efficiency can be vastly different from the number efficiency because large particles are much more massive than small particles (mass varies with the cube of the diameter). This is illustrated in Example 3.1.

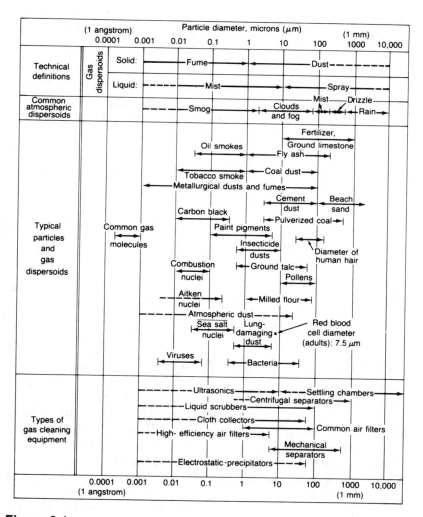

Figure 3.1

Characteristics of particles and particle dispersoids.

(Adapted from Lapple, 1961.)

Example 3.1

A particle dispersoid consists of 300 spherical particles: 100 1-μm particles, 100 10-μm particles, and 100 100-μm particles. A device is 10% efficient on the 1-μm particles, 50% efficient on the 10-μm particles, and 99% efficient on the 100-μm particles. All particles have the same density, and the 1-μm particles have unit mass. Calculate the number efficiency and the mass efficiency of collection.

Solution

We construct the following table. Note that the mass of a sphere increases with the cube of its diameter, so that the total mass of all particles is virtually equal to that of the 100-μm particles, or $100(10)^6$ mass units.

d_j,μm	η_j	Number of Particles Collected	Mass of Particles Collected
1	0.10	10	10(1)
10	0.50	50	50(1000)
100	0.99	99	$99(10^6)$
	Totals:	159	$99(10)^6$

The overall number efficiency is 159/300 = 0.53 or 53%, but the overall mass efficiency is $99(10)^6/100(10)^6 = 0.99$ or 99%.

For simplicity, all previous discussion has been based on an implicit assumption that all particles are spherical. In fact, most particles are not spherical, and they can be quite irregularly shaped. When choosing the most characteristic "diameter" of an irregularly shaped object with one or two dimensions significantly larger than the other, we must remember that our objective is to remove particles from a flowing stream of gas. Therefore, our choice of a size dimension should be based on the particles' behavior in the gas, and not on some microscopic visual examination. For example, a sheet of notebook paper will settle irregularly to the ground, but a ball of crumpled notebook paper will fall rapidly. Particle density is also important. For example, a golf ball and a ping pong ball have approximately the same diameter, but behave quite differently when tossed into the air.

*The **aerodynamic diameter** of a specific particle is the diameter of a sphere with unit density (density of water) that will settle in still air at the same rate as the particle in question* (Friedlander 1977). The aerodynamic diameter will be defined mathematically later in this chapter in Eq. (3.23), but for now it is sufficient to realize that the aerodynamic diameter is the proper diameter to apply to particles, whether discussing collection efficiencies or settling in the lungs. The aerodynamic diameter automatically results from aerodynamic classifiers such as cascade impactors.

As mentioned earlier, we must have adequate information on the size distribution of particulates in the exhaust stream to be controlled. A good means of obtaining this information is by use of a cascade impactor. A **cascade impactor** is a device that separates and sizes suspended particulates in a manner similar to the way that sieves separate and size samples of sand. However, a cascade impactor separates particles by their aerodynamic diameters rather than their physical diameters. Air with particles is drawn through a series of stages that consist of slots and impaction plates (see Figure 3.2). Each successive stage has narrower slots and closer plates so that each successive stage captures increasingly smaller particles. The masses of particles collected on all stages are then used to determine the size distribution of the particulate stream.

Particulate distributions are often skewed toward much higher numbers in the smaller size ranges—even when considered on a mass basis. A typical (normalized) plot of mass fraction of particles versus size is presented in Figure 3.3. A skewed distribution such as Figure 3.3 is considerably less convenient to work with than a normal (Gaussian) distribution. Fortunately, in many cases, if one plots mass frac-

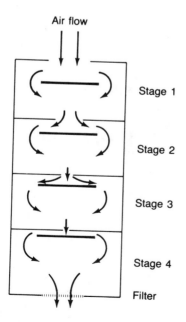

Figure 3.2
Schematic diagram of a cascade impactor.

tion versus the logarithm of the particle diameter, the result is a log-normal distribution, as shown in Figure 3.4.

A log-normal distribution such as shown in Figure 3.4 can be characterized by two parameters: the (geometric) mean and the (geometric) standard deviation. In the case of a log-normal distribution, the relationships between the mean and the standard deviation are

$$\log\left(d_{84.1}\right) = \log\left(d_{50}\right) + \log\sigma_g \qquad \textbf{(3.3)}$$

and

$$\log\left(d_{15.9}\right) = \log\left(d_{50}\right) - \log\sigma_g \qquad \textbf{(3.4)}$$

where

$d_{84.1}$ = diameter such that particles constituting 84.1% of the total mass of particles are smaller than this size

d_{50} = geometric mean diameter

$d_{15.9}$ = diameter such that particles constituting 15.9% of the total mass of particles are smaller than this size

σ_g = geometric standard deviation

Equations (3.3) and (3.4) can be written together as

$$\sigma_g = \frac{d_{84.1}}{d_{50}} = \frac{d_{50}}{d_{15.9}} \qquad \textbf{(3.5)}$$

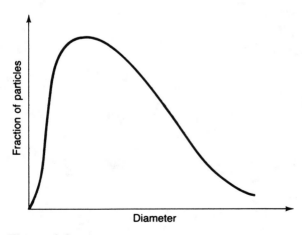

Figure 3.3
A skewed particle distribution.

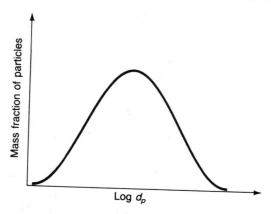

Figure 3.4
A log-normal particle distribution.

A representation even more convenient than Figure 3.4 is a plot of the particulate distribution data on log-probability paper. By definition, if the distribution plots linearly, it is *log-normal*, and d_{50} and σ_g can be used to characterize it. The linear plot is more convenient for interpolation when only a few actual data points are available. Example 3.2 illustrates these techniques.

▶ ▶ ▶ ▶ ▶ ▶ ▶ ▶ ▶

Example 3.2

The following data were obtained from a cascade impactor (a particulate sampling device that classifies the particles by size). Show that the distribution is log-normal, and find d_{50} and σ_g.

Size Range, μm	0–2	2–5	5–9	9–15	15–25	>25
Mass, mg	4.5	179.5	368	276	73.5	18.5

Solution

We prepare a table of size range versus cumulative percent less than the stated size, as follows:

Size Range, μm	Mass Fraction in Size Range, m_j	Cumulative Percent Less Than Top Size
0–2	0.0049	0.5
2–5	0.195	20.0
5–9	0.400	60.0
9–15	0.300	90.0
15–25	0.080	98.0
>25	0.020	100

A plot of these data is presented in Figure 3.5, from which values of $d_{84.1}$, d_{50} and $d_{15.9}$ can be read as 13.5, 7.8, and 4.6 microns. Thus, $\sigma_g = 13.5/7.8 = 1.73$, or $7.8/4.6 = 1.70$. (Owing to inaccuracies of reading such plots, σ_g values rarely agree perfectly.) Our final answers are

$d_{50} = 7.8\ \mu m$

$\sigma_g = 1.72$ (an average of our two calculated values)

Example 3.3

Assuming that Figure 3.5 is characteristic of the particulates, predict the overall efficiency of a device that has the following efficiency versus size relationship:

Size Range, µm	$d_{j(avg)}$, µm	η_j,%
0–2	1	10
2–4	3	30
4–6	5	60
6–10	8	80
10–14	12	90
14–20	17	95
20–30	25	98
30–50	40	99
>50	>50	100

Solution

From Figure 3.5, we determine the fraction of particles in each size range. We multiply the fraction in each size range by the efficiency for that size, and sum the results to give an overall efficiency for the device on the particulate dispersoid.

Size Range, µm	$d_{j(avg)}$, µm	η_j	Cumulative Percent Less Than Top Size (from Figure 3.5)	m_j	$\eta_j m_j$, %
0–2	1	10	0.5	0.005	0.05
2–4	3	30	10	0.095	2.85
4–6	5	60	30	0.20	12.0
6–10	8	80	67	0.37	29.6
10–14	12	90	86	0.19	17.1
14–20	17	95	96	0.10	9.5
20–30	25	98	99.4	0.034	3.33
30–50	40	99	99.8	0.004	0.40
>50	>50	100	100	0.002	0.2
					$\eta = 75\%$

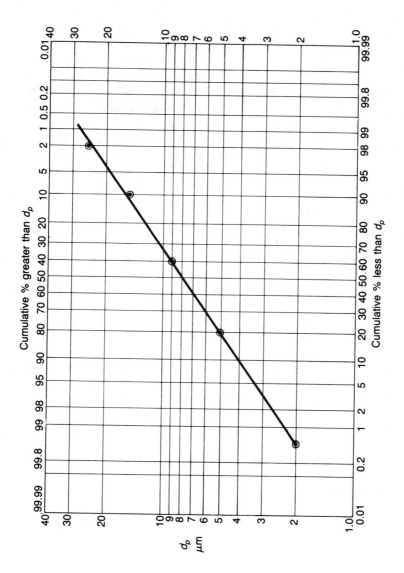

Figure 3.5
Plot of data for Example 3.2 showing a log-normal distribution.

◆ 3.3 PARTICULATE BEHAVIOR IN FLUIDS ◆

Particulates often must be separated from a fluid as part of a pollution control system. In water and wastewater treatment, particles of sludge are removed primarily by sedimentation tanks and/or filters. In air pollution control, particles can be removed by gravity settlers, centrifugal settlers, fabric filters, electrostatic precipitators, or wet scrubbers. In all of these devices, particles are separated from the surrounding fluid by the application of one or more forces. These forces, which include gravitational, inertial, centrifugal, and electrostatic forces, cause the particles to accelerate away from the direction of the mean fluid flow, toward the direction of the net force. The particles must then be collected and removed from the system to prevent ultimate re-entrainment into the fluid. Thus, design and operation of particulate pollution control equipment require a basic understanding of the characteristics of particles and of the dynamics of particles in fluids.

THE DRAG FORCE

Consider a particle in motion relative to a fluid. Either the particle or the fluid or both can be moving relative to an absolute frame of reference. The fluid exerts a drag force on the particle, which acts to oppose the relative velocity of the particle. The drag force can be represented as

$$F_D = C_D A_p \rho_F v_r^2 / 2 \tag{3.6}$$

where

F_D = drag force, N

C_D = drag coefficient

A_p = projected area of particle, m^2

ρ_F = density of fluid, kg/m^3

v_r = relative velocity, m/s

Usually, the drag coefficient must be determined experimentally because it is a strong function of particle shape and of the flow regime as characterized by the Reynolds number. For a particle, the **Reynolds number** is defined as

$$\text{Re} = \frac{d_p v_r \rho_F}{\mu} \tag{3.7}$$

where

d_p = particle diameter, m

μ = fluid viscosity, kg/m-s

Figure 3.6 is a plot of the drag coefficient as a function of the Reynolds number for three simple shapes.

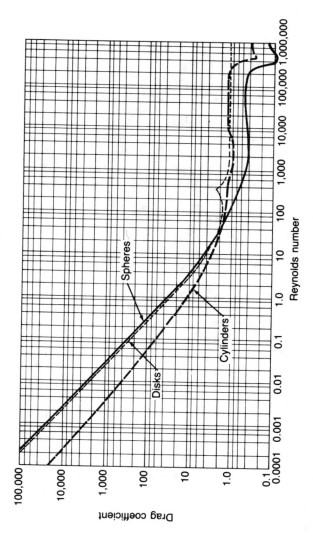

Figure 3.6

Drag coefficients for spheres, disks, and cylinders.

(Adapted from Lapple and Shepherd, 1940.)

The following discussion and equations apply to spheres at Reynolds numbers less than 1—the flow regime known as the *Stokes regime*. Theodore and Buonicore (1976) present more detailed equations that are valid for Reynolds numbers up to 10^5 (that is, where other models for the drag force are applicable). For rigid spheres and for Reynolds numbers less than 1, analytical integration of the equations of motion results in **Stokes law**:

$$F_D = 3\pi\mu d_p v_r \tag{3.8}$$

Substituting Eqs. (3.8) and (3.7) into Eq. (3.6) results in

$$C_D = 24/\text{Re} \tag{3.9}$$

Equation (3.9) approximates the solid line in Figure 3.6 for Reynolds numbers less than 1 (that is, in the Stokes regime). For Reynolds numbers greater than 1, Figure 3.6 or its statistical curve fit (Theodore and Buonicore 1976) should be used to obtain the drag coefficient C_D for spheres. The statistical fit of the curve in Figure 3.6 is

$$\log C_D = 1.35237 - 0.60810\,(\text{LRe}) - 0.22961\,(\text{LRe})^2$$
$$+ 0.098938\,(\text{LRe})^3 + 0.041528\,(\text{LRe})^4 - 0.032717\,(\text{LRe})^5 \tag{3.10}$$
$$+ 0.007329\,(\text{LRe})^6 - 0.0005568\,(\text{LRe})^7$$

where LRe = log(Re)

Equation (3.10) provides a good fit of the "spheres" curve in Figure 3.6 for all Reynolds numbers greater than 1.0, but its complexity makes it useful only in computer applications.

Another empirical approach recommended by Theodore and Buonicore (1976) is to divide Figure 3.6 into three regions based on Reynolds number. The general equation for drag coefficient is then given by

$$C_D = \alpha \text{Re}^{-\beta} \tag{3.11}$$

where α and β are constants (given in Table 3.1)

There is a lower limit for particle size in Stokes law. When the particle is large relative to the *mean free path* λ of the gas, the fluid can be modeled as a continuum. However, when the particle diameter is of the same magnitude as λ, the particle no longer "senses" the fluid as a continuum, but as discrete molecules. The particle is able to "slip" between the gas molecules, and this slippage reduces the effective drag on the particles predicted by Stokes law. The *Cunningham correction factor* corrects the Stokes drag coefficient for this effect. The correction factor is reported by Wark and Warner (1981) as

$$C = 1 + 2.0\frac{\lambda}{d_p}\left[1.257 + 0.40\,\exp\left(-0.55 d_p/\lambda\right)\right] \tag{3.12}$$

Reynolds Number	α	β
<2.0	24.0	1.0
2–500	18.5	0.6
500–200,000	0.44	0.0

Table 3.1 Constants for Eq. (3.11)

The **mean free path** can be obtained from the kinetic theory of gases as

$$\lambda = \frac{\mu}{0.499 P \sqrt{8MW / \pi RT}} \qquad (3.13)$$

where

λ = mean free path, m

P = absolute pressure, Pa

R = universal gas constant, 8314 J/kgmol-°K

MW = molecular weight, kg/kgmol

T = absolute temperature, °K

μ = absolute viscosity, kg/m-s

The variation of the Cunningham correction factor with particle size is shown in Table 3.2. For particles smaller than 1 micron, the slip correction factor is always significant, but rapidly approaches 1.0 as particle size increases above 5 microns. The corrected drag coefficient C'_D is the Stokes drag coefficient divided by the Cunningham correction factor. Because the Reynolds number is almost always small when C differs significantly from 1.0, we can correct Eq. (3.9) as follows:

$$C'_D = \frac{C_D}{C} = \frac{24}{C(\text{Re})} \qquad (3.14)$$

d_p, μm	C
0.01	22.5
0.05	5.02
0.10	2.89
0.50	1.334
1.0	1.166
2.0	1.083
5.0	1.033
10.0	1.017

Table 3.2 Cunningham Correction Factor at 1 atm and 25 C

EXTERNAL FORCES

For a relative motion to exist between a fluid and a freely suspended particle, at least one external force must exist. Considering only one net external force, which is opposed by the drag force F_D, Newton's second law of motion can be written as

$$F_e - F_D = M_p \frac{dv_r}{dt} \qquad (3.15)$$

where

F_e = net external force

M_p = mass of the particle

For a spherical particle in the Stokes regime, Eq. (3.8) and the product of volume times density for a sphere can be substituted into Eq. (3.15) to yield

$$\frac{dv_r}{dt} + \frac{18\mu}{\rho_p d_p^2} v_r = \frac{F_e}{M_p} \qquad (3.16)$$

where ρ_p = particle density

The grouping $\rho_p d_p^2 / 18\mu$ has dimension of time, and is a *characteristic time* τ for Eq. (3.16). The slip-corrected characteristic time τ' is merely τC. This characteristic time is the distinguishing parameter of particle dispersoids. Two seemingly different systems (with different particle sizes and densities, and different fluids) will behave in the same manner if τ' is the same for both systems. We can rewrite Eq. (3.16) as

$$\frac{dv_r}{dt} + \frac{v_r}{\tau'} = \frac{F_e}{M_p} \qquad (3.17)$$

Equation (3.17) is the basic differential equation governing the motion of a particle in a fluid (in the Stokes regime). The term F_e / M_p is the net external force per unit mass applied to the particle. The net external force can be due to an electrostatic field, a centrifugal field, a gravitational field, and so forth. In the next section, we will use Eq. (3.17) to derive a relationship for the Stokes settling velocity of a particle in quiescent air (that is, in a pure gravitational field).

GRAVITATIONAL SETTLING

If a particle is released into quiescent air with an initial downward velocity of zero, and with only gravitational force acting on the particle, the equation of motion is

$$\frac{dv_r}{dt} + \frac{v_r}{\tau'} = \left(\frac{\rho_p - \rho_F}{\rho_p} \right) g \qquad (3.18)$$

where g is the gravitational constant

The term $(\rho_p - \rho_F)/\rho_p$ corrects for the buoyancy of the particle in the fluid. For a solid particle in a gas, this term is nearly unity and can often be ignored. The solution to Eq. (3.18) is

$$v_r = v_t\left[1 - \exp(-t/\tau')\right] \tag{3.19}$$

where v_t is the terminal settling velocity, given by

$$v_t = \tau'g = \frac{C\rho_p d_p^2}{18\mu}g \tag{3.20}$$

Note that τ' has units of time, so that any consistent set of units can be used for density, diameter, and viscosity.

From a plot of Eq. (3.19), we can easily estimate the time when terminal velocity is attained (see Figure 3.7). From Figure 3.7, we can see that after four characteristic times, the particle's velocity is virtually equal to its terminal velocity.

We often ignore the transient (acceleration) portion of a particle's settling time. In Table 3.3, we have calculated the terminal velocity and the characteristic time for several particles. Because τ' is small, the terminal velocity is attained in a few milliseconds or less, so we are justified in ignoring the transient portion of the settling time.

For a particle larger than 10–20 microns settling at its terminal velocity, the Reynolds number is too high for the Stokes regime analysis to be valid. For these larger particles, empirical means are required to obtain the settling velocity; that is, we must use either direct experimental results or empirical models in place of Eq. (3.9) in the development of equations analogous to Eq. (3.20). For example, for

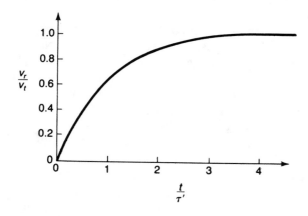

Figure 3.7
Dimensionless particle velocity versus dimensionless time.

Table 3.3 Settling Velocities and Characteristic Times for Unit Density
Spheres in Air at 25 C and 1 atm

d_p, μm	C	v_t, m/s	τ', s
0.1	2.89	$8.6(10)^{-7}$	$8.8(10)^{-8}$
0.5	1.334	$1.0(10)^{-5}$	$1.0(10)^{-6}$
1.0	1.166	$3.5(10)^{-5}$	$3.6(10)^{-6}$
5.0	1.033	$7.8(10)^{-4}$	$7.9(10)^{-5}$
10.0	1.017	$3.1(10)^{-3}$	$3.1(10)^{-4}$

$2 < \text{Re} < 500$, the terminal velocity [using Eq. (3.11) for C_D] is given by Theodore and Buonicore (1976) as

$$v_t = \frac{0.153 d_p^{1.14} \rho_p^{0.71} g^{0.71}}{\mu^{0.43} \rho_F^{0.29}} \tag{3.21}$$

Also, for $500 < \text{Re} < 200{,}000$, the terminal velocity is given by those authors as

$$v_t = 1.74 \left(\rho_p d_p / \rho_F \right)^{0.5} g^{0.5} \tag{3.22}$$

The terms in Eqs. (3.21) and (3.22) must have a consistent set of units; one consistent is ft/sec for v_t, ft for d_p, lb_m/ft^3 for ρ, ft/sec^2 for g, and lb_m/ft-sec for μ.

For quick reference, it is usually more convenient (and often more accurate) to use a chart (such as Figure 3.8) portraying actual experimental settling velocities. Note that the top and left-side axes must be used for larger particles, and the bottom and right-side axes are used for smaller particles.

In Section 3.2, we introduced the term *aerodynamic diameter*. We can now define this term mathematically. The motion of a Stokes particle depends only on its value of τ'. From the definition of τ', we can see that τ' is the ratio of the particle properties ($C \rho_p d_p^2$) to a fluid property (viscosity). Thus, all systems with the same value of τ' should exhibit the same aerodynamic behavior regardless of their separate values of μ, ρ_p, and d_p. The **aerodynamic diameter** *is defined as the diameter of a unit density sphere* ($\rho_p = \rho_w = 1000$ kg/m³) *that has the same settling velocity as the particle in question.* (In Eq. (3.23), any consistent set of units can be used; one such set is indicated.)

$$d_a = \sqrt{\frac{18 \mu v_t}{C \rho_w g}} \tag{3.23}$$

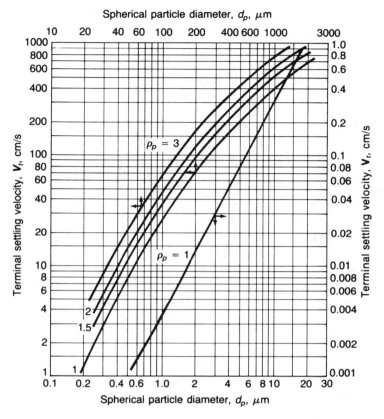

Figure 3.8
Terminal settling velocity of spherical particles in air at STP (particle
density given in g/cc).
(Adapted from Wark and Warner, 1981.)

where

d_a = aerodynamic diameter, m

μ = gas viscosity, kg/m-s

v_t = settling velocity, m/s

ρ_w = density of water, kg/m^3

g = gravitational acceleration, m/s^2

COLLECTION OF PARTICLES BY IMPACTION, INTERCEPTION, AND DIFFUSION

When a flowing fluid approaches a stationary object such as a fabric filter thread, a large water droplet, or a metal plate, the fluid flow

streamlines will diverge around that object. Because of their inertia, particles in the fluid will not follow streamlines exactly, but will tend to continue in their original directions. If the particles have enough inertia and are located close enough to the stationary object, they will collide with the object, and can be collected by it. This phenomenon is depicted in Figure 3.9.

Impaction of particles occurs when the center of mass of a particle that is diverging from the fluid streamlines strikes a stationary object. **Interception** occurs when the particle's center of mass closely misses the object, but, because of its finite size, the particle strikes the object. Collection of particles by **diffusion** occurs when small particulates (which are subject to random motion about the mean path and would usually miss the object even considering their finite size) happen to "diffuse" toward the object while passing near it. Once striking the object by any of these means, particles are collected only if there are short-range forces (van der Waals, electrostatic, chemical, and so forth) strong enough to hold them to the surface.

A simple means of explaining impaction is with the concept of *stopping distance*. If a sphere in the Stokes regime is projected with an initial velocity v_0 into a motionless fluid, its velocity as a function of time (ignoring all but drag forces) is

$$v = v_0 e^{(-t/\tau')} \tag{3.24}$$

The total distance traveled by the particle before it comes to rest (the stopping distance) is

$$x_{\text{stop}} = \int_0^\infty v \, dt = v_0 \tau' \tag{3.25}$$

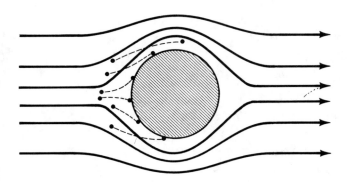

Figure 3.9
Collection of particles on a stationary object.

If the particle stops before striking the object, it can then be swept around the object by the altered fluid flow. Since τ' is small, x_{stop} is also small. For instance, if a 1.0-μm particle with unit density is projected at 10 m/s into air, it will stop after traveling 36 microns (less than half the diameter of a human hair).

An impaction parameter N_I can be defined as the ratio of the stopping distance of a particle (based on upstream fluid velocity) to the diameter of the stationary object, or

$$N_I = \frac{x_{stop}}{d_o} \qquad (3.26)$$

If N_I is large, most of the particles will impact the object. If N_I is very small, most of the particles will follow the fluid flow around the object.

◆ 3.4 OVERVIEW OF PARTICULATE CONTROL EQUIPMENT ◆

There are several different classes of particulate control equipment, including mechanical separators (such as gravity settlers, or cyclones), fabric filters, electrostatic precipitators, and wet scrubbers. In this section, we will briefly introduce each of the major types of particulate control equipment (each will be discussed in detail in subsequent chapters).

A **gravity settler** is merely a large chamber in which the gas velocity is slowed, allowing particles to settle out by gravity. A **cyclone** removes particles by causing the entire gas stream to flow in a spiral pattern inside of a tube. Owing to centrifugal force, the larger particles move outward and collide with the wall of the tube. The particles slide down the wall and fall to the bottom of the cyclone, where they are removed. The cleaned gas flows out of the top of the cyclone.

A **fabric filter** (baghouse) operates on the same principle as a vacuum cleaner. Air carrying dust particles is forced through a cloth bag. As the air passes through the fabric, the dust accumulates on the cloth, providing a cleaned air stream. The dust is periodically removed from the cloth by shaking or by reversing the air flow.

An **electrostatic precipitator** applies electrical force to separate particles from the gas stream. A high voltage drop is established between electrodes, and particles passing through the resulting electrical field acquire a charge. The charged particles are attracted to and collected on an oppositely charged plate, and the cleaned gas flows through the device. Periodically, the plates are cleaned by rapping to shake off the layer of dust that accumulates. The dust is collected in hoppers at the bottom of the device.

A **wet scrubber** employs the principles of impaction and interception of dust particles by droplets of water. The larger, heavier water droplets are easily separated from the gas by gravity. The solid particles can then be independently separated from the water, or the water can be otherwise treated before re-use or discharge.

Although broad generalities are often best avoided, some general statements might help to put the various types of particulate control equipment into perspective before embarking on a detailed discussion of each type in the next few chapters. Mechanical collectors are typically much less expensive than the others, but are typically only moderately efficient. They are much better for large particles than for fine dusts, and are often used as precleaners for the more-efficient final control devices, especially when the dust loading is high. Fabric filters tend to have very high efficiencies but are costly. Fabric filters usually are limited to dry, low-temperature conditions, but can handle many different types of dusts.

Electrostatic precipitators (ESPs) can handle very large volumetric flow rates at low pressure drops, and can achieve very high efficiencies. However, ESPs are costly and are relatively inflexible to changes in process operating conditions. Wet scrubbers can also achieve high efficiencies and have the major advantage that some gaseous pollutants can be removed simultaneously with the particulates. However, wet scrubbers can be very costly to operate (owing to a high pressure drop), and they produce a wet sludge that can present additional disposal problems.

All things considered, there is no way to decide, a priori, what the best system is. Each air pollution control problem is unique and demands an engineered solution. Thus, there is ample opportunity for design engineers to exercise their creativity, even when dealing with small problems and/or "off the shelf" equipment.

The overall collection efficiency of a system composed of two or more devices in series is not simply the sum nor the product of the efficiencies of each device. Each device's efficiency is based on the mass loading of particles entering that device, but the overall system efficiency is based on the total mass collected as a fraction of the total mass entering the first device.

The simplest way to approach this problem is to define the penetration of a device as the mass fraction that is not collected (that is, the fraction that penetrates through the device). Thus,

$$Pt = 1 - \eta \qquad (3.27)$$

where Pt = penetration (fraction)

The overall penetration of a system is simply the product of the penetrations of all of the individual devices; that is,

$$Pt_o = \prod_{i=1}^{n} Pt_i \qquad (3.28)$$

where

Pt_o = overall penetration

Pt_i = penetration of device i

The overall collection efficiency of the system is simply $1 - Pt_o$, as depicted in Figure 3.10. In applying this approach, care must be used in determining the efficiency of any downstream device because the larger, easier-to-collect particles will be removed by the upstream device.

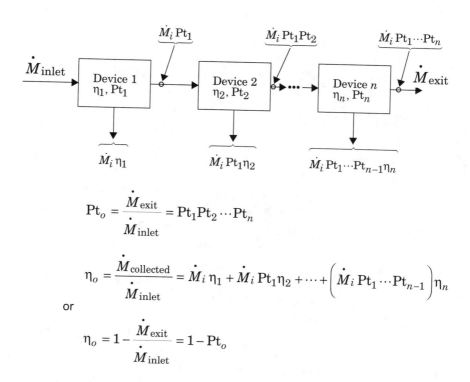

Figure 3.10

Efficiencies and penetrations of a particulate control system with several devices in series.

◆ Problems ◆

3.1 A stream of gas flowing at 3000 acfm has a particulate loading of 50 gr/ft^3 (grains/cubic foot). What is the allowable exit loading if 95% collection efficiency is required? What is the mass of particles collected, in kg/day?

3.2 A coal with 6% ash and a heating value of 13,000 Btu/lb is burned in a new power plant. Assume that 30% of the ash falls out as bottom ash in the furnace. Calculate the efficiency of an ESP required to meet federal new source performance standards.

3.3 The following data were obtained from a cascade impactor:

Size Range, μm	0–4	4–8	8–16	16–30	30–50	>50
Mass, mg	25	125	100	75	30	5

Is the distribution log-normal? If so, estimate σ_g and d_{50}.

3.4 Consider a cyclone that has the following theoretical efficiency versus particle size relationship:

Size Range, μm	0–2	2–4	4–7	7–10	10–15	15–25	25–40	40–60	>60
Efficiency, %	10	25	45	70	85	95	97	98	100

Calculate the overall efficiency of this cyclone for the particle distribution of Problem 3.3.

3.5 Consider an ESP that has the following theoretical efficiency versus particle size relationship:

Size Range, μm	0–2	2–4	4–7	7–10	10–15	15–25	>25
Efficiency, %	50	80	95	97	98	99	100

Calculate the overall efficiency of this ESP for the particle distribution of Problem 3.3.

3.6 For the particle distribution given in Figure 3.5, calculate the overall efficiency of the cyclone of Problem 3.4.

3.7 For the particle distribution given in Figure 3.5, calculate the overall efficiency of the ESP of Problem 3.5.

3.8 A particle distribution is log-normal with a d_{50} of 8.0 microns, and a σ_g of 3.0. Plot this distribution and determine the mass percent of particles below 2.0 microns.

3.9 A particle distribution is log-normal with a d_{50} of 6.0 microns and a σ_g of 2.0. Plot this distribution and determine the mass percent of particles below 2.0 microns.

3.10 Calculate the Cunningham correction factor for a particle with a 0.03-micron diameter in air at 1 atm pressure and 150 C.

3.11 Calculate the Cunningham correction factor for a particle with a 0.03-micron diameter in air at 1 atm pressure and 0 C.

3.12 Assuming Stokes behavior, calculate the terminal settling velocity in standard air for the following particles: (a) diameter = 20 µm, specific gravity = 2.0; (b) diameter = 5 µm, specific gravity = 0.8; and (c) diameter = 40 µm, specific gravity = 2.5.

3.13 Assuming Stokes behavior, calculate the terminal settling velocity in standard air for the following particles: (a) diameter = 10 µm, specific gravity = 1.2; and (b) diameter = 4 µm, specific gravity = 2.0.

3.14 Check your result for the 40-µm particle of Problem 3.12, using Eq. (3.21). Explain any differences in the results.

3.15 A gas stream with a particulate loading of 20.0 g/m³ is passed through a 70%-efficient cyclone followed by a 95%-efficient ESP. Calculate the overall efficiency of the system.

3.16 Particulate removal on a certain gas stream must be 98.5% efficient to meet standards. If a 60%-efficient cyclone precleaner is used with a wet scrubber, what is the required efficiency of the scrubber?

3.17 Calculate the overall efficiency of a system composed of the cyclone of Problem 3.4 followed by the ESP of Problem 3.5, operating on the following particle distribution:

Size, µm	0–2	2–4	4–7	7–10	10–15	15–25	>25
Weight Percent, %	6	23	30	18	13	8	2

Remember to consider that the size distribution of the remaining particles will change after having passed through the cyclone, and will be different when entering the ESP.

3.18 Assume that in the ambient air near an industrial plant on a certain day, a sample of the suspended particulate matter has a size distribution as follows:

Size Range, µm	0–2	2–6	6–10	10–16	16–30	>30
Mass, mg	15	20	45	40	20	10

A simple high-volume sampler yields a value of 80 µg/m³ for total suspended particulates for this air. Predict what result a PM-10 sampler would yield.

3.19 A stream of air at 25 C and 1 atm is laden with particulate matter (unit density spheres) of only two sizes: 1 µm and 5 µm. The stream flows through a pipe at an average velocity of 20 m/s. The pipe exhausts into a water mist scrubber where the water droplets average 100 µm in diameter. Based only on a comparison of their impaction numbers, what will be the ratio of particles collected (5 µm to 1 µm)?

3.20 A gas stream flowing at 1000 cfm with a particulate loading of 400 gr/ft³ discharges from a certain industrial plant through an 80% efficient cyclone. A recent law requires that the emissions from this stack be limited to 10.0 lb/hr, and the company is considering adding a wet scrubber after the cyclone. What is the required efficiency of the wet scrubber?

3.21 Dusty air at a fertilizer plant flows through a 70% efficient cyclone and then through an ESP. The inlet air to the cyclone has a dust loading of 50 grains/cubic foot.

a. In order to meet a control standard of 98.5% collection efficiency for the fertilizer plant as a whole, what is the allowable concentration of dust (in grains/cubic foot) in the air that exits from the ESP?

b. The outlet air from the ESP actually contains dust at 0.50 grain/cubic foot. Calculate the efficiency of the ESP.

3.22 A 5-micron droplet of water is being carried in standard air toward a 500-micron drop of water at a relative velocity of 40 m/s. Will the two drops collide? Support your answer with calculations.

3.23 Based on Figure 3.5, what percentage of the mass of particles is in the size range 4 to 9 micrometers?

3.24 Estimate the terminal settling velocity of a unit density sphere with a diameter of 2 μm falling through standard air. First use Figure 3.8, then repeat using either Table 3.3 or one of the equations from the text. Are there any significant differences? Now repeat the process for a 200 μm unit density sphere. Explain any significant differences.

3.25 A sample of dust was taken aerodynamically. The dust is characterized as being log-normal with a geometric mean of 6.0 μm. Also, 5% of the particles were larger than 25 μm. Estimate the PM-2.5 fraction (that is, the mass percent of this dust that is less than 2.5 μm).

REFERENCES

Friedlander, S. K. *Smoke, Dust and Haze.* New York: Wiley, 1977.

Lapple, C. E. *Stanford Research Institute Journal, 5*(95), 1961.

Lapple, C. E., and Shepherd, C. B. "Calculation of Particle Trajectories," *Industrial and Engineering Chemistry, 32*(5), 1940.

Theodore, L., and Buonicore, A. J. *Industrial Air Pollution Control Equipment for Particulates.* Cleveland, OH: CRC Press, 1976.

Wark, K., and Warner, C. F. *Air Pollution—Its Origin and Control* (2nd ed.). New York: Harper & Row, 1981.

 **CHAPTER 4**

CYCLONES

Centrifugal or cyclone collectors are widely used for the separation and recovery of industrial dusts from air or process gases. The usual type of cyclone is simple to construct and is very low in first costs compared with other types of dust collecting equipment.

C. B. Shepherd and C. E. Lapple, 1939

♦ 4.1 INTRODUCTION ♦

The quote opening this chapter is as true today as it was over 60 years ago. Cyclone separators have been used in the United States for about 100 years, and are still one of the most widely used of all industrial gas-cleaning devices. The main reasons for the widespread use of cyclones are that they are inexpensive to purchase, they have no moving parts, and they can be constructed to withstand harsh operating conditions. Typically, a particulate-laden gas enters tangentially near the top of the cyclone, as shown schematically in Figure 4.1. The gas flow is forced into a downward spiral simply because of the cyclone's shape and the tangential entry. Another type of cyclone (a vane-axial cyclone) employs an axial inlet with fixed turning vanes to achieve a spiraling flow. Centrifugal force and inertia cause the particles to move outward, collide with the outer wall, and then slide downward to the bottom of the device. Near the bottom of the cyclone, the gas reverses its downward spiral and moves upward in a smaller, inner spiral. The cleaned gas exits from the top through a "vortex-finder" tube, and the particles exit from the bottom of the cyclone through a pipe sealed by a spring-loaded flapper valve or a rotary valve.

Much of the research for this chapter was done by Mr. Jeff Nangle as an independent study while he was a graduate student in the Environmental Engineering program at the University of Central Florida. His work is hereby acknowledged.

123

Cyclones by themselves are generally not adequate to meet stringent air pollution regulations, but they serve an important purpose. Their low capital cost and their nearly maintenance-free operation make them ideal for use as precleaners for more expensive final control devices such as baghouses or electrostatic precipitators (see Figure 4.2). In addition to use for pollution control work, cyclones are used extensively in process industries; for example, they are used for recovering and recycling certain catalysts in petroleum refineries and for recovering freeze-dried coffee in food-processing plants. An excellent animation of cyclone operation and other information about cyclones (including a web calculator for cyclone design) are available at http://www.aerosols.wustl.edu/aaqrl/Courses/CYCOPCRESP/index.html.

In the past, cyclones have often been regarded as low-efficiency collectors. However, efficiency varies greatly with particle size and with cyclone design. During the past decade, advanced design work has greatly improved cyclone performance. Current literature from some of the cyclone manufacturers advertises cyclones that routinely achieve 90% or greater efficiency for particles larger than 10 microns, and even some that achieve 99% efficiency for particles greater than

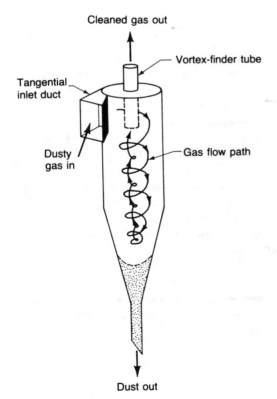

Figure 4.1
Schematic flow diagram of a standard cyclone.

Figure 4.2
A cyclone (center) in use as a precleaner for a pulse-jet baghouse (bottom-left) at an asphaltic concrete plant.

5 microns (typically, these are cyclones designed for collecting particles of high density).

In general, as efficiencies increase, operating costs increase (primarily because of the higher pressure drops required). Therefore, three broad categories of cyclones are available: high efficiency, conventional, and high throughput. Generalized efficiency curves for these three types of cyclones are presented in Figure 4.3.

Advantages of cyclones:

1. Low capital cost

2. Ability to operate at high temperatures

3. Low maintenance requirements because there are no moving parts

Disadvantages of cyclones:

1. Low efficiencies (especially for very small particles)

2. High operating costs (due to pressure drop)

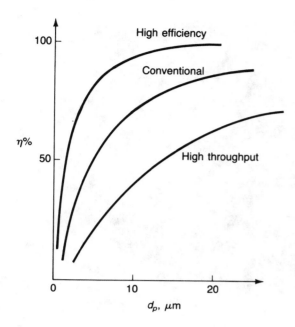

Figure 4.3
General relationship of collection efficiency versus particle size for cyclones.
Note: Efficiency versus size curves represent broad generalizations, not exact relationships.

STANDARD CYCLONE DIMENSIONS

Extensive work has been done to determine in what manner dimensions of cyclones affect performance. In two classic studies, which are still used today, Shepherd and Lapple (1939, 1940) determined "optimum" dimensions for cyclones. All dimensions were related to the body diameter of the cyclone so that their results could be applied generally. Subsequent investigators reported similar work, and the so-called standard cyclones were born. Table 4.1 summarizes the dimensions of standard cyclones of the three types mentioned previously. Figure 4.4 illustrates the various dimensions used in Table 4.1.

◆ 4.2 THEORY ◆

COLLECTION EFFICIENCY

A very simple model can be used to determine the effects of both cyclone design and operation on collection efficiency. In this model, the gas spins through a number of revolutions N_e in the outer vortex. The value of N_e can be approximated by

$$N_e = \frac{1}{H}\left[L_b + \frac{L_c}{2}\right] \qquad (4.1)$$

Table 4.1   Standard Cyclone Dimensions

	High Efficiency		Conventional		High Throughput	
Cyclone Type	(1)	(2)	(3)	(4)	(5)	(6)
Body Diameter, D/D	1.0	1.0	1.0	1.0	1.0	1.0
Height of Inlet, H/D	0.5	0.44	0.5	0.5	0.75	0.8
Width of Inlet, W/D	0.2	0.21	0.25	0.25	0.375	0.35
Diameter of Gas Exit, D_e/D	0.5	0.4	0.5	0.5	0.75	0.75
Length of Vortex Finder, S/D	0.5	0.5	0.625	0.6	0.875	0.85
Length of Body, L_b/D	1.5	1.4	2.0	1.75	1.5	1.7
Length of Cone, L_c/D	2.5	2.5	2.0	2.0	2.5	2.0
Diameter of Dust Outlet, D_d/D	0.375	0.4	0.25	0.4	0.375	0.4

Columns (1) and (5) adapted from Stairmand, 1951; columns (2), (4), and (6) adapted from Swift, 1969; column (3) adapted from Lapple, 1951.

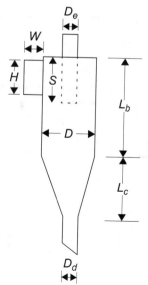

Figure 4.4
Dimensions of a standard cyclone.
(Adapted from Lapple, 1951.)

where

N_e = number of effective turns

H = height of inlet duct, m or ft

L_b = length of cyclone body, m or ft

L_c = length (vertical) of cyclone cone, m or ft

To be collected, particles must strike the wall within the amount of time that the gas travels in the outer vortex. The *gas residence time* in the outer vortex is

$$\Delta t = 2\pi R N_e / V_i \qquad (4.2)$$

where

Δt = gas residence time

R = cyclone body radius

V_i = gas inlet velocity

The maximum radial distance traveled by any particle is the width of the inlet duct W. Assume that centrifugal force quickly accelerates the particle to its terminal velocity in the outward (radial) direction. Terminal velocity is achieved when the opposing drag force equals the centrifugal force. The terminal velocity that will just allow a particle to be collected in time Δt is

$$V_t = W / \Delta t \qquad (4.3)$$

where V_t = particle terminal velocity in the radial direction

The particle terminal velocity is a function of particle size. As shown in Chapter 3, the terminal velocity of a particle is equal to its characteristic time multiplied by the external force. Assuming Stokes regime flow and spherical particles under a centrifugal force, we obtain

$$V_t = \frac{d_p^2 \left(\rho_p - \rho_g \right) V_i^2}{18 \mu R} \qquad (4.4)$$

where

V_t = terminal velocity, m/s

d_p = diameter of the particle, m

ρ_p = density of the particle, kg/m^3

μ = gas viscosity, kg/m-s

ρ_g = gas density, kg/m^3

Substitution of Eq. (4.2) into Eq. (4.3) eliminates Δt. Then, setting Eq. (4.3) equal to Eq. (4.4) to eliminate V_t and rearranging to solve for particle diameter, we obtain

$$d_p = \left[\frac{9\mu W}{\pi N_e V_i \left(\rho_p - \rho_g \right)} \right]^{1/2} \qquad \textbf{(4.5)}$$

Theoretically, d_p is the size of the smallest particle that will be collected if it starts at the inside edge of the inlet duct. Thus, in theory, all particles of size d_p or larger should be collected with 100% efficiency. In practice, this relationship is modified slightly [see Eq. (4.6), next section].

Note that in Eqs. (4.2) to (4.5), the units must be consistent. One consistent set is m for d_p, R, and W; m/s for V_i and V_t; kg/m-s for μ; and kg/m^3 for ρ_p and ρ_g. An equivalent set in U.S. customary units is ft for d_p, R, and W; ft/sec for V_i and V_t; lb$_m$/ft-sec for μ; and lb$_m$/ft^3 for ρ_p and ρ_g.

◆ 4.3 DESIGN CONSIDERATIONS ◆

COLLECTION EFFICIENCY

From Eq. (4.5), we can see that, in theory, the smallest diameter of particles collected with 100% efficiency is directly related to gas viscosity and inlet duct width, and inversely related to number of effective turns, inlet gas velocity, and density difference between the particles and the gas. In practice, collection efficiency does, in fact, depend on these parameters. However, the model leading to Eq. (4.5) has a major flaw: it predicts that *all* particles larger than d_p will be collected with 100% efficiency, which is not correct.

The semi-empirical relationship developed by Lapple (1951) to calculate a "50% cut diameter," d_{pc}, which is the diameter of particles collected with 50% efficiency, is

$$d_{pc} = \left[\frac{9\mu W}{2\pi N_e V_i \left(\rho_p - \rho_g \right)} \right]^{1/2} \qquad \textbf{(4.6)}$$

where d_{pc} = diameter of particle collected with 50% efficiency

Note the similarity between Eqs. (4.5) and (4.6). Lapple then developed a general curve for standard conventional cyclones, from which we can predict the collection efficiency for any particle size (see Figure 4.5). If the size distribution of particles is known, the overall collection efficiency of a cyclone can be predicted by using Figure 4.5. Theodore and DePaola (1980) have fitted an algebraic equation to Figure 4.5, which makes Lapple's approach more precise and more convenient for application to computers. The efficiency of collection of any size of particle is given by

$$\eta_j = \frac{1}{1 + \left(d_{pc} / \bar{d}_{pj} \right)^2} \qquad \textbf{(4.7)}$$

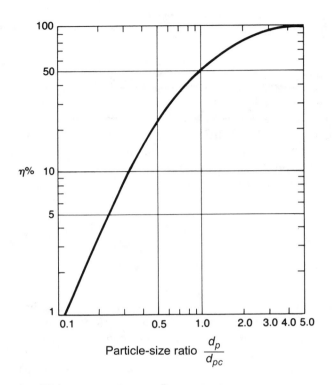

Figure 4.5
Particle collection efficiency versus particle size ratio for standard conventional cyclones.

where

η_j = collection efficiency for the jth particle size range

$\overline{d_{p_j}}$ = characteristic diameter of the jth particle size range

The overall efficiency of the cyclone is a weighted average of the collection efficiencies for the various size ranges; that is,

$$\eta_o = \sum \eta_j m_j \qquad (4.8)$$

where

η_o = overall collection efficiency

m_j = mass fraction of particles in the jth size range

The use of Eqs. (4.6), (4.7), and (4.8) is illustrated in Example 4.1.

Example 4.1

Consider a conventional cyclone of standard proportions as described by Lapple (1951), with a body diameter of 1.0 meter. For air with a flow rate of 150 m³/min at $T = 350$ K and 1 atm, containing particles with a density of 1600 kg/m³ and a size distribution as given below, calculate the overall collection efficiency.

Particle Size Range, μm	Mass Percent in Size Range
0–2	1.0
2–4	9.0
4–6	10.0
6–10	30.0
10–18	30.0
18–30	14.0
30–50	5.0
50–100	1.0

Solution

First, we calculate d_{pc} from Eq. (4.6). From Appendix B, Table B.2 (after converting units), the viscosity is 0.075 kg/m-hr, and the density is 1.01 kg/m³. The inlet velocity is the volumetric flow divided by the inlet area (H × W):

$$V_i = \frac{150 \text{ m}^3}{\text{min}} \times \frac{1}{(0.5 \text{ m})(0.25 \text{ m})} = \frac{1200 \text{ m}}{\text{min}}$$

For a Lapple standard cyclone, $N_e = 6$ from Eq. (4.1).

$$d_{pc} = \left[\frac{9\left(0.075\dfrac{\text{kg}}{\text{m-hr}}\right)(0.25 \text{ m})}{2\pi(6)\left(1200\dfrac{\text{m}}{\text{min}}\right)\left(60\dfrac{\text{min}}{\text{hr}}\right)\left[(1600-1)\dfrac{\text{kg}}{\text{m}^3}\right]} \right]^{1/2}$$

$$d_{pc} = 6.26 \times 10^{-6} \text{ m} = 6.3 \text{ μm}$$

Next, we determine the collection efficiency for each size range from Figure 4.5 or Eq. (4.7). The arithmetic midpoint of the range is often used as the characteristic particle size. It is convenient to construct the following table (see next page).

j	Size Range, μm	$\bar{d}_{pj}$,μm	$\bar{d}_{pj}/d_{pc}$	η_j	m_j, %	Percent Collected $\eta_j m_j$, %
1	0–2	1	0.159	0.02	1.0	0.02
2	2–4	3	0.476	0.18	9.0	1.62
3	4–6	5	0.794	0.39	10.0	3.9
4	6–10	8	1.27	0.62	30.0	18.6
5	10–18	14	2.22	0.83	30.0	24.9
6	18–30	24	3.81	0.94	14.0	13.2
7	30–50	40	6.35	0.98	5.0	4.9
8	50–100	75	11.9	0.99	1.0	1.0
						68.1

Finally, as shown in the table,

$$\eta_o = \sum_{j=1}^{8} \eta_j m_j = 68.1\%$$

The overall collection efficiency of this cyclone for this particular air/particulate mixture is approximately 68%, or 0.68.

◀ ◀ ◀ ◀ ◀ ◀ ◀ ◀

The effects of a change in operating conditions can be quantified using Eq. (4.6). Recall that the fractional penetration is

$$Pt = 1 - \eta_o \qquad (4.9)$$

where Pt = fractional penetration

If conditions change such that one variable changes and all others remain constant, the relationships shown in Table 4.2 can be used to estimate the new penetration.

▶ ▶ ▶ ▶ ▶ ▶ ▶ ▶

Example 4.2

Estimate the new efficiency of the cyclone of Example 4.1 if (a) the air flow rate is increased to 200 m³/min or (b) the air temperature is increased to 400 K.

Solution

(a)

$$Pt_2 = Pt_1 \left(\frac{Q_1}{Q_2}\right)^{0.5} = 0.32\left(\frac{150}{200}\right)^{0.5} = 0.28$$

Thus,

$$\eta_2 = 0.72 = 72\%$$

(b) An increase in air temperature has two main effects: it increases the volumetric flow rate to 171 m^3/min and it increases the air viscosity to 0.083 kg/m-hr. The overall effect on penetration is the product of the individual effects.

$$Pt_2 = Pt_1 \left(\frac{Q_1}{Q_2}\right)^{0.5} \left(\frac{\mu_2}{\mu_1}\right)^{0.5} = 0.32 \left(\frac{150}{171}\right)^{0.5} \left(\frac{0.083}{0.075}\right)^{0.5}$$

$$Pt_2 = 0.315 \approx 0.32$$

Thus,

$$\eta_2 = 0.68 = 68\% \quad \text{(no change)}$$

◀ ◀ ◀ ◀ ◀ ◀ ◀ ◀ ◀

More complicated (but more accurate) models of cyclones have been developed. Kalen and Zenz (1974) proposed the existence of a "saltation velocity" in the cyclone to explain why collection efficiency was sometimes observed to decrease with an increase in inlet velocity. Their semi-empirical correlation for the saltation velocity as applied and extended by Koch and Licht (1977) is

$$V_s = 2.055\psi \left[\frac{(W/D)^{0.4}}{\left[1 - (W/D)\right]^{0.333}}\right] D^{0.067} V_i^{0.667} \tag{4.10}$$

where

V_s = saltation velocity, ft/sec

D = cyclone body diameter, ft

ψ = velocity function, ft/sec

V_i = inlet velocity, ft/sec

Table 4.2   Effects of Operating Condition Changes on Cyclone Performance

Variable	Change	Efficiency	Relationship
Gas Flow Rate	Increase	Increase	$Pt_2/Pt_1 = (Q_1/Q_2)^{0.5}$
Gas Viscosity	Increase	Decrease	$Pt_2/Pt_1 = (\mu_2/\mu_1)^{0.5}$
Density Difference	Increase	Increase	$Pt_2/Pt_1 = \left[\dfrac{(\rho_p - \rho_g)_1}{(\rho_p - \rho_g)_2}\right]^{0.5}$
Dust Loading	Increase	Increase	$Pt_2/Pt_1 = (L_1/L_2)^{0.18}$

The function ψ in Eq. (4.10) is given by

$$\psi = \left[4g\mu \frac{\left(\rho_p - \rho_g\right)}{3\rho_g^2} \right]^{0.333} \tag{4.11}$$

where

g = gravitational constant, 32.2 ft/sec^2

μ = gas viscosity, lb$_m$/ft-sec

ρ_p, ρ_g = particle and gas density, lb$_m$/ft^3

According to the saltation model, the maximum collection efficiency occurs at $V_i = 1.25V_s$, with re-entrainment of particles occurring at higher inlet velocities. Generally, cyclone inlet velocity should be 50 to 100 ft/sec (15 to 30 m/s).

Leith and Licht (1972) and Dietz (1979) have published methods for other cyclone models that predict collection efficiency more accurately than the Lapple model, but require considerably more effort to use. For many situations, the Lapple model is quite acceptable.

PRESSURE DROP

The other major consideration for cyclones (besides efficiency) is pressure drop. Generally, higher efficiencies are obtained by forcing the gas through the cyclone at higher velocities. However, this results in an increased pressure drop. Since increased pressure loss through the cyclone requires increased fan work, there is an economic trade-off that must be considered in the design process.

Although several pressure drop models exist, the approach of Shepherd and Lapple is the simplest to use and its accuracy is comparable to that of the other methods (Koch and Licht 1977). The Shepherd and Lapple equation is

$$H_v = K \frac{HW}{D_e^2} \tag{4.12}$$

where

H_v = pressure drop, expressed in number of inlet velocity heads

K = a constant that depends on cyclone configuration and operating conditions

Theoretically, K can vary considerably, but for air pollution work with standard tangential-entry cyclones, values of K are in the range of 12 to 18 (Caplan 1962). Licht (1984) recommends that K simply be set equal to 16, and states that, although more complicated methods are available, none are superior to Eq. (4.12).

The number of inlet velocity heads calculated from Eq. (4.12) can be converted to a static pressure drop as follows:

$$\Delta P = \frac{1}{2}\rho_g V_i^2 H_v \qquad \textbf{(4.13)}$$

where

ΔP = pressure drop, N/m^2 or Pa

ρ_g = gas density, kg/m^3

V_i = inlet gas velocity, m/s

Cyclone pressure drops range from about 0.5 to 10 velocity heads (250 to 4000 Pa or 1 to 16 inches of water). Once the pressure drop has been calculated, the fluid power requirement can be obtained as

$$\dot{w}_f = Q\Delta P \qquad \textbf{(4.14)}$$

where

$\dot{w}_f$ = work input rate into the fluid (fluid power), W

Q = volumetric flow rate, m^3/s

Depending on the units of ΔP, ρ, V_i, and Q in Eqs. (4.13) and (4.14), unit conversion constants might be required in each equation. The actual power required is simply the fluid power divided by a combined fan/motor efficiency. Example 4.3 illustrates (in SI units) the calculations of cyclone pressure drop and the corresponding power consumed by the moving fluid. Example 4.4 illustrates the calculations in U.S. customary units, and extends the considerations to include a typical combined fan/motor efficiency to determine the right motor size.

▶ ▶ ▶ ▶ ▶ ▶ ▶ ▶ ▶

Example 4.3

For the cyclone of Example 4.1, assume a K value of 15 and calculate (a) the cyclone pressure drop in kPa, and (b) the fluid power consumed in the cyclone in kW.

Solution

(a)

$$H_v = 15\frac{(0.5)(0.25)}{(0.5)^2} = 7.5$$

$$V_i = \frac{1200 \text{ m}}{\text{min}} \times \frac{1 \text{ min}}{60 \text{ s}} = 20 \text{ m/s}$$

$$\Delta P = \frac{1}{2} \times 1.01\frac{\text{kg}}{\text{m}^3} \times 400\frac{\text{m}^2}{\text{s}^2} \times \frac{1 \text{ N}}{1 \text{ kg-m/s}^2} \times 7.5$$

$$\Delta P = 1515 \text{ N/m}^2 = 1.515 \text{ kPa}$$

(b)

$$\dot{w}_f = \frac{150 \text{ m}^3}{\text{min}} \times 1515 \text{ N/m}^2 \times \frac{1 \text{ min}}{60 \text{ s}}$$

$$\dot{w}_f = 3788 \frac{\text{N-m}}{\text{s}} = 3788 \text{ J/s} = 3.79 \text{ kW}$$

◀ ◀ ◀ ◀ ◀ ◀ ◀ ◀

▶ ▶ ▶ ▶ ▶ ▶ ▶ ▶ ▶

Example 4.4

For a 90%-efficient cyclone processing 9000 cfm of air with a pressure drop of 9.0 inches of water, calculate the fluid power in units of horsepower. Furthermore, estimate the size of the electric motor needed to run a fan to move the air through this cyclone.

Solution

$$\dot{w}_f = kQ\Delta P$$

where k must have units of hp/(ft^3/min) (in. H$_2$O). Thus,

$$k = \frac{1 \text{ hp}}{550 \text{ ft-lb}_f/\text{sec}} \times \frac{0.0361 \text{ psi}}{1 \text{ in. H}_2\text{O}} \times \frac{144 \text{ in.}^2}{1 \text{ ft}^2} \times \frac{1 \text{ min}}{60 \text{ sec}}$$

$$k = 0.0001575 \text{ hp-min}/\left(\text{ft}^3\text{-in. H}_2\text{O}\right) \text{ [hp/cfm-in. H}_2\text{O]}$$

and

$$\dot{w}_f = 0.0001575 \times 9000 \times 9.0 = 12.8 \text{ hp}$$

Assuming a typical efficiency of 70% for the combination of the fan and motor,

$$\dot{w}_F = \frac{12.8 \text{ hp}}{0.70} = 18.2 \text{ hp}$$

A 20-hp motor should be chosen for this system.

◀ ◀ ◀ ◀ ◀ ◀ ◀ ◀

OTHER CONSIDERATIONS

When a large volume of gas must be treated, a single cyclone might be impractical. Several cyclones can be used in parallel; one form of parallel cyclones is the multiple-tube cyclone shown in Figure 4.6. Reasonably high efficiencies (90% for 5- to 10-micron particles) are obtained by the small diameter (15 to 60 cm) tubes.

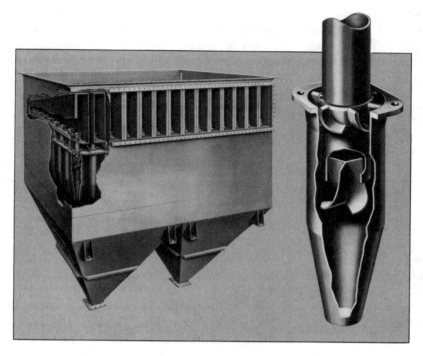

Figure 4.6
Axial inlet multi-tube or "multi" cyclone.
(Courtesy of Research-Cottrell, Inc., Somerville, NJ.)

Using cyclones in series increases overall efficiency, but at the cost of a significant increase in pressure drop. Furthermore, since most of the larger particles are removed in the first cyclone, the second cyclone usually has a lower overall efficiency than the first. Scroll-shaped entrances and modified inlet areas have been used to achieve higher efficiencies.

When designing a cyclone, we use a trial-and-error procedure in which we first choose a body diameter, and then calculate d_{pc} and the efficiency. If the efficiency is too low, we choose a smaller diameter and iterate. If the efficiency is acceptable, we check the pressure drop constraint. If the ΔP is too high, we must either choose a different type of cyclone, or split the flow of gas between two cyclones in parallel. The design procedure is illustrated in Example 4.5.

▶ ▶ ▶ ▶ ▶ ▶ ▶ ▶ ▶

Example 4.5

Design a conventional Lapple cyclone to function as a precleaner on a gas stream that flows at 120 m³/min. The cyclone must achieve a minimum overall efficiency of 70% for the following particulate distribution, with a maximum allowable ΔP of 3000 Pa. The particulate density is 1500 kg/m^3, the gas density is 1.0 kg/m^3, and the gas viscosity is 0.07 kg/m-hr. Specify your final choice of body diameter, overall cyclone efficiency, inlet gas velocity, and pressure drop. (Assume $K = 14$)

Size Range, μm	Mass Percent in Size Range
0–2	2.0
2–4	18.0
4–10	30.0
10–20	30.0
20–40	15.0
40–100	4.0
>100	1.0

Solution

This trial-and-error solution is easily achieved using a spreadsheet. Note that many of the variables in Eqs. (4.1), (4.6), (4.7), and (4.8) are functions of one key parameter—the cyclone body diameter, D. Set up the spreadsheet with all calculations based on the cell where we input body diameter. Then, all that needs to be done is to input different values of D until a feasible solution is obtained.

By a feasible solution, we mean one that meets all the listed constraints—in this case, an efficiency greater than or equal to 70%, and a pressure drop less than or equal to 3000 Pa. Often, the solution can be optimized somewhat depending on whether a smaller pressure drop or larger efficiency is more important, but keep in mind that the final cyclone diameter that is specified should result in a cyclone that can be readily purchased or built without incurring extraordinary costs for special sizing.

Since the problem specifies a conventional Lapple cyclone, choose a type 3 cyclone from Table 4.1; the dimensional relationships are:

$$H = 0.5\,D \quad W = 0.25\,D \quad D_e = 0.5\,D \quad L_b = 2\,D \quad L_c = 2\,D$$

Other formulas to be coded into the spreadsheet are:

$$V_i = Q/(HW)$$

and Eqs. (4.1), (4.6), (4.7), (4.8), (4.12), and (4.13). We must also remember to include conversion factors (such as 60 min/hr) to make the units work, especially in Eq. (4.6).

Once the spreadsheet is built, start simply by entering a value of D, say 1.0 m. If the efficiency is too low, then choose a smaller D. If the pressure drop is too high, then choose a larger D. Keep trying values of D until a feasible solution is obtained. If it turns out that a feasible solution does not exist, one or more constraints must be relaxed, or a different type of cyclone must be chosen.

A spreadsheet to solve this problem is displayed in Figure 4.7. Using this spreadsheet, four values of D were entered and the following results were obtained:

D,m	η, %	ΔP, Pa	ΔP, in. H_2O
1.0	61.8	896	3.6
0.9	66.5	1366	5.5
0.8	71.4	2188	8.8
0.7	76.5	3732	15.0

Based on the results shown above, a diameter of 0.8 meters is selected.

Input Data								
Q =	120	m^3/min		**Prelim Calculations**				
Effic tgt =	70	%		H =	0.4	m		
Max dP =	3000	Pa		W =	0.2	m		
part dens =	1500	kg/m^3		V =	1500	m/min		
gas dens =	1	kg/m^3		De =	0.4	m		
gas visc =	0.07	kg/m-hr		Lb =	1.6	m		
K =	14			Lc =	1.6	m		
Assume D								
D (meters) =	0.8	← Here is where you input values of D						
eq 4.1, Ne =	6							
eq 4.6, dpc =	4.98	microns						
					By eq. 4.7			
Efficiency Calculations:		size range	dp avg	dp/dpc	"Eta" j	mass %	% collected	
		0–2	1	0.20	0.04	2	0.08	
		2–4	3	0.60	0.27	18	4.80	
		4–10	7	1.41	0.66	30	19.93	
		10–20	15	3.01	0.90	30	27.02	
		20–40	30	6.03	0.97	15	14.60	
		40–100	70	14.06	0.99	4	3.98	
		>100	100	20.09	1.00	1	1.00	
Pressure drop calcs:					Overall	Effic =	71.40	%
eq 4.12	Hv =	7.00						
eq 4.13	dP =	2187.50	Pa					
or	dP =	8.79	in H2O					

Figure 4.7
Spreadsheet solution for Example 4.5.

◆ 4.4 Costs ◆

Single cyclones are very inexpensive, having capital costs at least an order of magnitude less than final control devices such as baghouses and electrostatic precipitators. Purchase costs depend on collection efficiency, gas throughput, design pressure drop, and the materials of construction. Some representative 1982 cost data for cyclones are presented in Table 4.3. Installation costs and ductwork connections are often more expensive than the cyclone itself. Costs for multi-tube cyclones are approximately $1.00/cfm for flow rates up to 1000 cfm, $0.60/cfm for 5000 cfm, and $0.50/cfm for 10,000 cfm or greater (Neveril 1978).

The key parameter used for estimating the cost of a cyclone system is the area of the inlet to the cyclone. Since the area is a function of the cyclone diameter, one might expect the cost to be directly correlated to diameter, but for some reason, authors in the past have correlated the cost to inlet area (Neveril 1978; Vatavuk 1990). The following equations are based on data gathered in 1988.

For a package system consisting of a carbon-steel cyclone with supports, a hopper (or drum for small cyclones), and a fan and motor to pull the gas through the cyclone, the purchased cost can be estimated from:

$$P_c = 6520\, A^{0.903} \qquad (4.15)$$

where

P_c = cost of the cyclone system, 1988 dollars, f.o.b. manufacturer

A = cyclone gas inlet area, ft^2 (correlation valid for $0.20 \le A \le 2.64\ \text{ft}^2$)

Cyclones often must have a (fairly heavy duty) rotary air lock discharge valve to safely discharge the dust from the bottom leg into a collection hopper. The purpose of the air lock valve is to prevent air at atmospheric pressure from leaking into the cyclone from the bottom and blowing dust back into the cleaned gas exhaust stream. The cost of the rotary air lock valve can be significant and is given by Eq. (4.16):

$$P_v = 273\, A^{0.0965} \qquad (4.16)$$

where

P_v = cost of the rotary air lock valve, 1988 dollars, f.o.b. manufacturer

A = cyclone inlet area, ft^2 (valid for $0.35 < A < 2.64\ \text{ft}^2$)

Equation (4.16) is valid only down to 0.35 ft^2 because smaller cyclones, which often empty into a 55-gallon drum through a flapper valve, operate with a low enough pressure drop that they do not require a rotary air lock valve.

The total purchased cost (in 1988 dollars) of a cyclone system is the sum of P_c and P_v. The cost can be escalated to current dollars using the indexes introduced in Chapter 2. Here, and in other chapters, we have

Table 4.3 Representative 1982 Costs for Single Cyclones (without accessories, not installed)

Air Flow Rate cfm	Collection Efficiency* %	Pressure Drop in. H_2O	Unit Cost $/cfm
3000	90	2	1.20
	90	6	0.70
	90	10	0.60
5000	90	6	0.66
	70	6	0.58
10,000	90	2	1.00
	90	6	0.50
	90	10	0.38
20,000	90	2	1.00
	90	6	0.48
	90	10	0.33
	70	6	0.32
50,000	90	2	1.15
	90	6	0.45
	90	10	0.32
50,000– 100,000	90	6	0.45
	70	6	0.22

*Collection efficiencies are for 5-micron particles with a density of 2700 kg/m^3.
Adapted from Heumann, 1983.

chosen to present the cost equations in the original year from which the data were obtained in order to retain the clarity of the original work. More recent cost information is available for many of the technologies, and where the more recent costs have changed significantly (other than due to inflation), they are presented. In general, it is always recommended that an engineer obtain current price quotes during the detailed design before committing finally to a course of action.

In order to estimate the total installed cost (TIC) of a cyclone system, a good rule of thumb is to multiply the total purchased cost by a factor of two. Even though the installation work is not complex, since P is typically a low number to begin with, just a small amount of engineering, labor, and ductwork can add up to a big percentage of the purchased cost.

▶ ▶ ▶ ▶ ▶ ▶ ▶ ▶ ▶

Example 4.6

Estimate the 1988 total installed cost for the cyclone of Example 4.4. Do this in two different ways and compare your answers. Assume the inlet area is 1.75 ft^2.

Solution

(a) From Table 4.3, a cyclone processing 10,000 cfm of air with 90% particulate removal and a pressure drop of between 6 and 10 inches of water will have a unit cost of between $0.50 and $0.38 per cfm. We can interpolate pressure drop to obtain the estimated cost for a cyclone with a ΔP of 9.0 inches of water and processing 10,000 cfm.

$$\text{unit cost} = \frac{\$0.50}{\text{cfm}} + \frac{(0.50 - 0.38)}{(6 - 10)} \times (9.0 - 6.0) = \frac{\$0.41}{\text{cfm}}$$

Similarly, for a 90% efficient cyclone with a ΔP of 6 inches of water, we can interpolate air flow rate to find the unit cost for a cyclone with a flow rate of 9000 cfm.

$$\text{unit cost} = \frac{\$0.66}{\text{cfm}} + \frac{(0.50 - 0.66)}{(10,000 - 5000)} \times (9000 - 5000) = \frac{\$0.53}{\text{cfm}}$$

Based on the above two estimates, we might estimate the unit cost of this particular cyclone at $0.44/cfm, and the extended cost as:

$$\text{extended cost} = \frac{\$0.44}{\text{cfm}} \times 9000 \text{ cfm} = \$3960$$

However, this is only the purchased cost of the basic cyclone. It does not include accessories and is in 1982 dollars. Doubling this estimate to account for supports, fan, and motor, and doubling it again to account for installation costs, and then escalating it to 1988 dollars using the Marshall and Swift Index, our final estimate using this method is:

$$\text{TIC} = \$3960 \times 2 \times 2 \times (852/745.6) = \underline{\$18,100}$$

(b) Using equation (4.15) to estimate the cyclone system cost

$$P_c = 6520 \times 1.75^{0.903}$$

$$P_c = \$10,303$$

Assuming that we need a rotary air lock valve,

$$P_v = 273 \times 1.75^{0.0965}$$

$$= \$288$$

Our estimate of the total installed cost in 1988 dollars using this second method is:

$$\text{TIC} = 2 \times (10,303 + 288) = \underline{\$21,200}$$

◆ PROBLEMS ◆

4.1 For the particulate-laden gas stream given in Example 4.1, calculate the overall efficiency of a high-efficiency Swift cyclone (see Table 4.1) with a body diameter of 1.00 meter. Assume that a scroll modification on the inlet increases N_e to 9 turns.

4.2 For the following particle size distribution, calculate the efficiency of a Lapple standard conventional cyclone with a body diameter of 0.50 meters. The particulate density ρ_p = 1200 kg/m^3, the gas density ρ_g = 0.90 kg/m^3, the gas viscosity μ = 0.06 kg/m-hr, and the inlet gas velocity V_i = 25 m/s.

Size Range, μm	Mass Percent in Size Range
0–4	3.0
4–10	10.0
10–20	30.0
20–40	40.0
40–80	15.0
>80	2.0

4.3 Assume the efficiency of the cyclone of Problem 4.2 is 90%, and calculate the new efficiency if (a) the inlet gas velocity decreases to 15 m/s, or if (b) the particulate density changes to 1000 kg/m^3.

4.4 Recalculate the efficiency of the cyclone of Problem 4.2 for the following particle size distribution.

Size Range, μm	Mass Percent in Size Range
0–2	4.0
2–6	16.0
6–12	20.0
12–20	20.0
20–40	20.0
40–80	18.0
>80	2.0

4.5 Calculate the pressure drop for the cyclone of Problem 4.1, assuming K = 16. Give your answer in kPa and in in. H$_2$O.

4.6 Design a Lapple standard cyclone to function as a precleaner. The particles have a density of 1250 kg/m^3, and have the same size distribution as in Problem 4.4, but the air is at 150 F and 1 atm, and is flowing at 2000 acfm. The cyclone need be only 65–75% efficient. Specify your final choice of body diameter and the corresponding efficiency.

4.7 Calculate the pressure drop for the cyclone of Problem 4.2, assuming K = 14.

4.8 Assume that a certain type of cyclone has a particle cut diameter of 6.0 microns; (a) calculate the overall efficiency of this cyclone on the particles of Problem 4.4 and (b) calculate the overall efficiency of two of these cyclones placed in series.

4.9 Calculate the saltation velocity (m/s) for the cyclone of Problem 4.2. Should re-entrainment be a problem for this cyclone?

4.10 Design a single cyclone (Swift conventional) to handle 20,000 cfm (at 250 F and 1 atm) of air contaminated with the particles of Problem 4.2. The required efficiency must be between 75% and 80%, and the maximum allowable pressure drop is 8.0 in. H_2O.

4.11 A Swift high-efficiency cyclone is being compared with a Swift high-throughput cyclone of the same diameter. To evaluate the "cost" of the increased efficiency, estimate and compare the pressure drops for each. Assume $K = 16$ for the high-efficiency model and $K = 12$ for the high-throughput model. The volumetric gas flow rate is the same for both cyclones.

4.12 Estimate the annual electricity cost to run a fan to push 25,000 cfm of air through a device that has a pressure drop of 2500 N/m^2. Assume a fan/motor efficiency of 0.60. Electricity costs $0.08/kWh, and the fan runs 7800 hours per year.

4.13 Estimate the f.o.b. cost (in 2002 dollars) for a cyclone with an inlet area of 2.0 ft^2.

4.14 Estimate the total installed cost (in 2002 dollars) for a cyclone that processes 20,000 cfm of air at 90% efficiency with a pressure drop of 6.0 in. H_2O.

4.15 Estimate the annualized operating cost for the cyclone of Problem 4.14 assuming the cyclone operates (a) 4000 hours per year, and (b) 8000 hours per year. Include as operating costs only electricity (fan efficiency = 0.6, electricity costs $0.08/kWh) and depreciation (assume straight-line depreciation, a 10-year life, and no salvage value). Assume the cyclone was bought and installed for $9800 five years ago.

4.16 A flour mill has been required to install particulate control equipment to remove 90% of its emissions. A testing firm has determined a size distribution of the flour as follows:

Size Range, μm	Mass Percent in Size Range
0–2	10
2–10	20
10–20	20
20–40	25
40–70	20
>70	5

A consulting firm, B. S. Schuters, Inc., has recommended a two-stage cyclone (two cyclones in series) for this problem, and has come up with the following design:

$$W_i = 20 \text{ inches}$$
$$H_i = 30 \text{ inches}$$
$$N_e = 8 \text{ turns}$$
$$V_i = 70 \text{ ft/sec}$$
$$\rho_p = 62.4 \text{ lb}_m/\text{ft}^3$$

The viscosity of the air is $1.2(10)^{-5}$ lb$_m$/ft-sec. Verify or refute this recommendation.

4.17 For a Swift conventional cyclone with a body diameter of 1.0 m and an airflow rate of 250 m^3/min (at 394 K and 1 atm), calculate the overall collection efficiency for the particle size distribution given below. The particle density is 1500 kg/m^3.

Particle Size Range, μm	Mass Percent in Size Range
0–5	2
5–10	10
10–20	20
20–50	30
50–70	20
70–90	10
>90	8

4.18 Design a Lapple standard cyclone to clean a dusty airstream flowing at 10,000 acfm (at 200 F and 1 atm). The required efficiency must be between 75% and 85% with a maximum allowable pressure drop of 10 in. H$_2$O. The particle density is 1200 kg/m^3 and the particle size distribution is given below.

Particle Size Range, μm	Mass Percent in Size Range
0–5	10
5–15	30
15–30	40
30–50	15
>50	5

4.19 For a Swift high-throughput cyclone with a body diameter of 0.5 m, calculate the pressure drop expected for a gas flowing at 100 m^3/min (at 250 F and 1 atm). Assume $K = 12$. Also calculate the fluid power used to move this gas.

4.20 Develop a spreadsheet to aid you in the design of a Swift high-efficiency, a Swift conventional, and a Swift high-throughput cyclone. Using the data in Problem 4.10, and for a fixed cyclone diameter of 1.8 m, compare the efficiency and pressure drop for these three cyclones.

4.21 Compare a Lapple conventional cyclone with a Swift conventional cyclone, each with a diameter of 1.2 meters. Standard air flows into the cyclone at 250 m^3/min. The PM in the air stream has a density of 1500 kg/m^3. The size distribution is the same as in Example 4.1. Calculate the efficiency and pressure drop for each cyclone.

4.22 Criticize the design for PM control that calls for two identically sized cyclones in series.

REFERENCES

Caplan, K. J. "Source Control by Centrifugal Force and Gravity," in *Air Pollution,* vol. 11, A. C. Stern, Ed. New York: Academic Press, 1962.

Dietz, P. W. *Collection Efficiency of Cyclone Separators,* General Electric Corporate Research and Development, Report No. 79CRD244, December 1979.

Heumann, M. "Understanding Cyclone Dust Collectors," *Plant Engineering,* May 26, 1983.

Kalen, B., and Zenz, F. "Theoretical Empirical Approach to Saltation Velocity in Cyclone Design," *American Institute of Chemical Engineers Symposium Series, 70*(137), 1974.

Koch, W. H., and Licht, W. "New Design Approach Boosts Cyclone Efficiency," *Chemical Engineering, 84*(24), November 7, 1977.

Lapple, C. E. "Processes Use Many Collector Types," *Chemical Engineering, 58*(5), May 1951.

Leith, D., and Licht, W. "The Collection Efficiency of Cyclone Type Particle Collectors—A New Theoretical Approach," *American Institute of Chemical Engineers Symposium Series, 126*(68), 1972.

Licht, W. "Control of Particles by Mechanical Collectors," Chapter 13 in *Handbook of Air Pollution Technology,* S. Calvert and H. M. Englund, Eds. New York: Wiley, 1984.

Neveril, R. B. *Capital and Operating Costs of Selected Air Pollution Control Systems,* EPA–450/5–80–002. Washington, D.C.: U.S. Environmental Protection Agency, December 1978.

Shepherd, C. B., and Lapple, C. E. "Flow Pattern and Pressure Drop in Cyclone Dust Collectors," *Industrial and Engineering Chemistry, 31*(8), 1939.

———. "Flow Pattern and Pressure Drop in Cyclone Dust Collectors," *Industrial and Engineering Chemistry, 32*(9), 1940.

Stairmand, C. J. "The Design and Performance of Cyclone Separators," *Transactions of Industrial Chemical Engineers, 29,* 1951.

Swift, P. "Dust Control in Industry," *Steam Heating Engineering, 38,* 1969.

Theodore, L., and DePaola, V. "Predicting Cyclone Efficiency," *Journal of the Air Pollution Control Association, 30*(10), October 1980.

Vatavuk, William M. "Pricing Equipment for Air Pollution Control," *Chemical Engineering, 97* (5), May 1990.

Vatavuk, W. M., and Neveril, R. B. "Part V: Estimating the Size and Cost of Gas Conditioners," *Chemical Engineering,* January 26, 1981.

◆◆◆ CHAPTER 5

ELECTROSTATIC
PRECIPITATORS

Precipitator design has also acquired new importance during the past decade . . . strong enforcement of clean air standards can require curtailment or even complete shutdown of entire production units. Hence, in this sense, precipitator design and engineering practice is now of equal importance to that of the production equipment itself.

Harry J. White, 1984

◆ 5.1 INTRODUCTION ◆

The process of electrostatic precipitation involves (1) the ionization of contaminated air flowing between electrodes, (2) the charging, migration, and collection of the contaminants (particles) on oppositely charged plates, and (3) the removal of the particles from the plates. The particles can be either dry dusts or liquid droplets. The air flows through the electrostatic precipitator (ESP), but the particles are left behind on the plates. The material is knocked off or washed off the plates, and is collected in the bottom of the ESP. The ESP is unique among air pollution control devices in that the forces of collection act only on the particles and not on the entire air stream. This phenomenon typically results in a high collection efficiency with a very low air pressure drop.

In the early 1900s, Dr. F. G. Cottrell, then an instructor at the University of California at Berkeley, was approached to help find a solution to an operating problem at a nearby H_2SO_4 plant at Pinole, California (Danielson 1973). First, he tried a centrifuge to collect acid mist, but it was not successful. In 1906, all of his notes and models were destroyed in the San Francisco earthquake. Discouraged but not willing to quit, Cottrell rejected an offer to head the Chemistry Department at Texas A&M so he could follow up on his idea of collecting the acid mist by electrostatic precipitation. In 1907, Cottrell developed a small ESP (100–200 acfm) that worked satisfactorily at the Pinole plant, and a new industry was born. During the following ten years, several other ESPs were installed for applications such as removal of cement kiln dust, lead smelter fumes, tar, and pulp and paper alkali salts. These ESPs treated airflow rates of up to 300,000 acfm.

In about 1920, pulverized-coal furnaces were introduced in response to the needs of the electric power industry (White 1977). Pulverized-coal furnaces were superior to the then-current stoker-fired power boilers, but they presented a new problem: much of the coal ash from pulverized-coal furnaces was in the form of tiny particles (fly ash) suspended in the combustion gases. Fortunately, ESPs were available and were very well suited for the task of removing these particles; the first ESP on a coal-fired power plant was placed in service in 1923. After several years of troubleshooting various problems, a collection efficiency of 90% was achieved. Since that time, more than 2000 ESPs, treating more than 1.3 billion acfm, have been installed for fly-ash collection in the United States. Although the electric power industry accounts for about 80% of all the ESP capacity installed in the United States (U.S. Environmental Protection Agency 1985), many other industries use this technology. Other major users include pulp and paper (7% of all ESPs), iron and steel (3%), cement and other rock products (3%), and nonferrous metals (1%).

In the 1930s and 1940s, fly-ash ESPs were built that achieved efficiencies near 95%. By the 1950s, guarantees were being made for efficiencies of 97–98%. By the mid-1970s, ESP specifications were often above 99.5% efficiency. Modern ESPs have been designed for efficiencies greater than 99.9%. Keep in mind that the 0.4% improvement in collection efficiency (from 99.5% to 99.9%) represents an 80% decrease in emissions. A photo of an ESP installed on a 364 MW coal-fired power plant is shown in Figure 5.1, and a cutaway view of an ESP is shown in Figure 5.2. In addition to dry ESPs (the focus of this chapter), wet ESPs can be used when there is a potential for explosion, when the particulates are sticky or are liquid droplets, or when the dry dust has an extremely high resistivity (see Section 5.3).

Electrostatic precipitators have several advantages and disadvantages in comparison with other particulate control devices.

Advantages

1. Very high efficiencies, even for very small particles
2. Can handle very large gas volumes with low pressure drop
3. Dry collection of valuable materials, or wet collection of fumes and mists
4. Can be designed for a wide range of gas temperatures
5. Low operating costs, except at very high efficiencies

Disadvantages

1. High capital costs
2. Will not control gaseous emissions
3. Not very flexible, once installed, to changes in operating conditions
4. Take up a lot of space
5. Might not work on particulates with very high electrical resistivity

◆ 5.2 THEORY ◆

Consider a dusty airflow in a rectangular channel defined by two parallel plates as shown in Figure 5.3. Consider only the half-channel between the charging wires and the plate, with width $D/2$ and height H.

With a few assumptions, we can derive the basic equation used in ESP design—the *Deutsch equation* (first derived in 1922). The assumptions are

1. Gases (and particles) move in the x direction at constant velocity u, with no longitudinal mixing.
2. The particles are uniformly distributed in the y and z directions at every x location.

Figure 5.1

An ESP installed on a 364 MW coal-fired power plant (the coal has about 15% ash, and the ESP treats roughly 1 million acfm at 99.6% efficiency; the plate height is 14.5 m with 30 cm spacing, 50 channels, 5 sections in the direction of flow, and 10 electrical sections total).

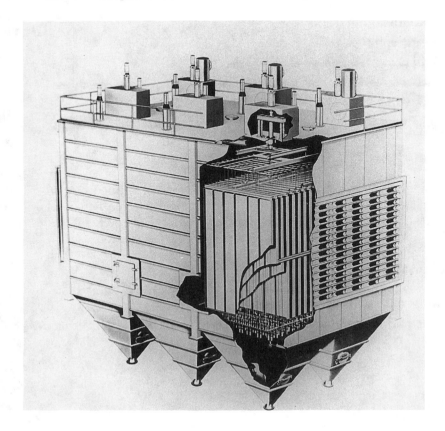

Figure 5.2
Cutaway view of an electrostatic precipitator.
(Courtesy of the Western Precipitation Division, Joy Manufacturing Co., Los Angeles, CA.)

    3. The charging and collecting fields are constant and uniform; the particles quickly attain terminal velocity w in the y direction.

    4. Re-entrainment of collected particles is negligible.

The concentration of particles will decrease with x because of the net migration of particles to the plate. A material balance on particles flowing into and out of a very short cross-section of the channel (located at an arbitrary x distance) shows that the difference between the mass of particles flowing into and out of the slice must equal the mass of particles removed at the plate. Thus,

$$uH\frac{D}{2}C_x - uH\frac{D}{2}C_{x+\Delta x} = \text{mass removed} \qquad (5.1)$$

where

    u = gas velocity, m/min

    H = plate height, m

D = channel width, m

C = particle concentration or loading, g/m^3

However, the mass removed is simply equal to the flux of particles in the y direction times the area normal to the flux. Thus, Eq. (5.1) becomes

$$uH\frac{D}{2}\left(C_x - C_{x+\Delta x}\right) = wC_{x+\Delta x/2}H\Delta x \tag{5.2}$$

where w = drift velocity (terminal velocity in the y direction), m/min.

Dividing through Eq. (5.2) by Δx, and taking the limit as Δx approaches zero, we obtain

$$\frac{-uHD}{2}\frac{dC}{dx} = wHC \tag{5.3}$$

Equation (5.3) can be separated and integrated from 0 to L (channel length, m) to give

$$\ln\left(\frac{C_L}{C_0}\right) = \frac{-2wHL}{uHD} \tag{5.4}$$

or

$$\ln\left(\frac{C_L}{C_0}\right) = \frac{-wA_p}{Q_c} \tag{5.5}$$

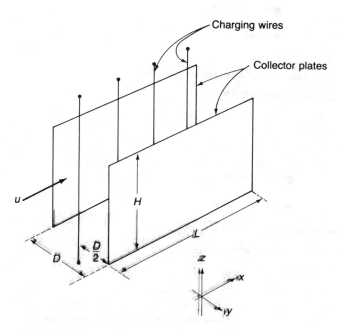

Figure 5.3
Schematic diagram of airflow between two ESP plates.

where

A_p = area of one plate (two sided), m^2

Q_c = volumetric gas flow in one channel, m^3/min

For the whole ESP, we can use the total collection area and the total gas flow rate, and Eq. (5.5) can be written as

$$\frac{C_L}{C_0} = e^{(-wA/Q)} \qquad (5.6)$$

Using the usual definition of collection efficiency, Eq. (5.6) becomes **the Deutsch equation:**

$$\boxed{\eta = 1 - e^{(-wA/Q)}} \qquad (5.7)$$

where η = fractional collection efficiency

Note that any consistent set of units can be used for w, A, and Q (for example, ft/min, ft^2, and ft^3/min).

The drift velocity w in an electrical force field can be calculated in a manner similar to that used in Chapter 3 for calculating the terminal settling velocity in a gravitational field. In short, the terminal drift velocity equals the product of the characteristic time of the particle in the gas multiplied by the electrostatic force per unit mass, or

$$\boxed{w = \tau' F_E} \qquad (5.8)$$

where

F_E = electrostatic force per unit mass on the particle, N/kg

τ' = slip-corrected characteristic time, s

According to White (1977), F_E can be as high as 3000 times the force of gravity for a 1-μm particle.

The total electrostatic force on a particle is the product of the charge on the particle multiplied by the collecting field strength, or

$$\boxed{M_p F_E = q E_{co}} \qquad (5.9)$$

where

M_p = mass of the particle, kg

q = charge on the particle, coulombs (C)

E_{co} = collecting field strength, V/m

Particles in the ESP are charged by two mechanisms. The most common ESP application uses a negative corona charging field. The high negative voltage (up to 100,000 V) ionizes gas molecules, generating free electrons. The electrons flow toward the grounded plates. They strike and become attached to electronegative gases, creating negative ions. In field charging, the negative ions are driven by the electrical field onto dust particles that intercept the field lines. This is more effective for particles larger than one micron. Particles smaller than about 0.2 microns are charged more effectively by diffusional charging; that is, the charging of small particles is a result of collisions of gas ions and small particles due to the random motions of each.

The theoretical saturation charge on a spherical particle is given by

$$q = \pi d_p^2 \varepsilon_0 K E_{ch} \tag{5.10}$$

where

d_p = particle diameter, m

ε_0 = permittivity of free space, $8.85(10)^{-12}$ C/V-m

K = a constant given by Eq. (5.11)

E_{ch} = charging field strength, V/m

The constant K is defined by

$$K = 3\varepsilon / (\varepsilon + 2) \tag{5.11}$$

where ε = dielectric constant for the particle relative to free space

Note that for many particles, the constant K ranges from 1.5 to 2.4.

Combining Eqs. (5.8), (5.9), and (5.10), and using the definition of τ' from Chapter 3, we can solve for the theoretical drift velocity of a spherical particle in an ESP as follows:

$$w = \frac{C d_p}{3\mu} \varepsilon_0 K E_{ch} E_{co} \tag{5.12}$$

where

C = the Cunningham correction factor

μ = gas viscosity, kg/m-s

The units for all of the parameters in Eq. (5.12) have been defined previously. From Eq. (5.12), we can see that the theoretical drift velocity is proportional to the particle diameter and (approximately) to the square of the field strength, because the charging and collecting fields are approximately equal.

Equation (5.12) is reasonably accurate for smooth, spherical particles subjected to constant gas flows and electrostatic fields. However, the effects of randomly shaped particles of various sizes, variations in the field strength, nonuniformity of the gas flow distribution, and re-entrainment of particles from the walls and/or re-entrainment during rapping combine to make the theoretical drift velocity unreliable for use in Eq. (5.7) for the design of an ESP. In practice (especially when very high efficiency is required), an effective drift velocity w_e is obtained from pilot studies or from previous experience with similar ESP applications. The effective drift velocity can account for particle penetration due to gas sneakage (by-passing), re-entrainment, and rapping losses, none of which are accounted for by the theoretical drift velocity model. With w_e in place of w in Eq. (5.7), we can calculate the required total plate area.

Although Eq. (5.7) serves well enough for initial process design, ESP manufacturers have more detailed (proprietary) models that they use for final design. In this text, we will use Eq. (5.7) with the effective drift velocity, w_e. Typical values of w_e are 2 to 20 cm/s for coal fly ash, 6 to 8 cm/s for sulfuric acid mists, 6 to 7 cm/s for dry cement dusts, and 6 to 14 cm/s for blast furnace dusts (Wark and Warner 1981).

♦ 5.3 DESIGN CONSIDERATIONS ♦

The complete design of an ESP includes sizing and determining the configuration of the plates; calculating the needed electrical energization; determining the structural needs; and specifying the rapping, dust removal, and performance-monitoring systems. We will consider only the first two items in detail.

PLATE SIZING

The plates in an ESP are typically taller than they are long, and are placed in parallel in several sections (see Figure 5.2, which shows three sections in the direction of flow and two sections across). The area for use in Eq. (5.7) is the collection plate area (or active plate area) rather than the total plate area. Consider one section of n plates in parallel across the entire width of the ESP. The gas flows through the "ducts" (spaces between the plates), so the $n - 2$ interior plates all have both sides collecting dust ("active"), while the two exterior plates each can only utilize one side. Thus, there are $n - 1$ active plates in this section. For an ESP with N_s sections in the direction of flow, the total collection area is the total number of active plates times the double-sided area per plate:

$$A = A_p (n - 1) N_s = A_p (N - N_s) \qquad (5.13)$$

where

A_p = two-sided plate area (= $2HL_p$)

n = number of plates in parallel across the width of the ESP

N = total number of plates in the ESP

N_s = number of sections in the direction of flow

Equation (5.13) can be used to estimate the number of plates required given the dimensions of a plate as shown in the next example.

Example 5.1

(a) Calculate the total collection area for a 98% efficient ESP that is treating 10,000 m³/min of air. The effective drift velocity is 6.0 m/min. (b) Assuming the plates are 6 m high and 3 m long and that there are two sections in the direction of flow, calculate the number of plates required.

Solution

Rearranging Eq. (5.7),

$$\ln(1 - \eta) = \frac{-Aw_e}{Q}$$

$$A = \frac{-Q}{w_e}\ln(1 - \eta)$$

$$A = \frac{-10{,}000 \text{ m}^3/\text{min}}{6.0 \text{ m/min}}\ln(0.02) = 6520 \text{ m}^2$$

From Eq. (5.13) with $N_s = 2$,

$$A = A_p(N - 2)$$

Rearranging,

$$N = \frac{A}{A_p} + 2 = \frac{6520}{3 \times 6 \times 2} + 2 = 183.1 = 184 \text{ plates}$$

Because each section must have an integral number of plates, round up to 184 plates (92 in each section in the direction of flow).

CORONA

A field (or corona) must be established to charge particles. **Corona** *is the ionization of gas molecules by high-energy electrons in the region of a strong electric field.* The excess electrons generated by the corona are readily attached onto electronegative gases such as oxygen or SO_2. In turn, the negative ions that are produced are adsorbed onto particles, which then migrate to the grounded plates. Typically, the discharge electrodes (wires) are energized while the collecting plates are grounded, but the wires can establish either a positive or negative corona. Negative corona (in which the wires have a negative charge) has inherently better voltage/current characteristics, and is used more frequently. However, negative corona produces more ozone than does positive corona. For this reason, positive corona, even though less efficient, is used for all indoor air-cleaning applications.

PARTICULATE RESISTIVITY

In addition to size and size distribution, another important property of the particles is *resistivity*. Once particles have migrated to a plate, they are considered to be collected. However, collected particles can be re-entrained into the gas, thus lowering the net ESP efficiency.

The **resistivity** *of fly ash is a measure of its resistance to electrical conduction*. Resistivity is extremely important because it can vary widely, and because it strongly influences particle collection efficiency. Once collected, particles begin to lose their charge to the plate. This transfer of charge completes the electrical circuit, produces current flow, and allows maintenance of the voltage drop between the wires and the plates. If the resistivity is too low (that is, the dust is a good conductor), the electrostatic charge is drained off too quickly and the dust is re-entrained into the gas. If the resistivity is too high (that is, the dust is a good insulator), the charge does not drain off at the collecting plates. In this situation, (1) a "back corona" develops, reducing the ionization and migration of particles in the gas, and (2) the particles remain strongly attracted to the plate and are difficult to "rap" off.

The resistivity of a material is determined experimentally by establishing a current flow through a slab (of known geometry) of the material (see Figure 5.4). It is important to make resistivity measurements of freshly collected dust in the actual flue gas stream produced from burning the particular coal to be used. Thus, such measurements should be made in the field rather than in the laboratory. Resistivities measured in the lab on the "same" dust can be from 100 to 1000 times greater than field resistivities (White 1984). The resistivity P is simply the resistance times the area normal to the current flow divided by the path length, as shown in Eq. (5.14).

$$P = \frac{RA}{l} = \frac{V}{i}\frac{A}{l}$$

(5.14)

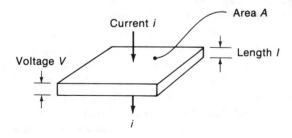

Figure 5.4
Determination of resistivity.

where

P = resistivity, ohm-cm

R = resistance, ohm

A = area normal to the current flow, cm^2

l = path length in the direction of current flow, cm

V = voltage, V

i = current, A

The resistivity P of materials ranges from 10^{-3} to 10^{14} ohm-cm; for coal fly ashes, P usually ranges from 10^8 to 10^{13} ohm-cm (White 1977); the resistivity of dry cement dust can exceed 10^{13} ohm-cm (U.S. Environmental Protection Agency 1985). ESP design and operation are difficult for resistivities above 10^{11} ohm-cm.

The major factors influencing fly-ash resistivity are temperature and chemical composition (of the fly ash and of the combustion gases). The conductivity of the dust layer is derived from two effects: volume conduction through the material itself, and surface conduction via adsorbed gases or liquids. Volume conduction decreases with increased temperature, whereas surface conduction increases with T, as shown in Figure 5.5. Therefore, resistivity (which is the inverse of conductivity) has a distinct maximum value. Unfortunately for power boiler operators, this maximum occurs at about 250–350 F.

The temperature of the maximum resistivity is unfortunate because operators often cannot reduce ESP temperatures below 250 F without risking the condensation of sulfuric acid on some of the cold surfaces. On the other hand, increasing the temperature above 350 F results in unnecessary loss of heat out the stack, which represents a monetary loss.

Resistivity decreases with increased coal sulfur content (as shown in Figure 5.6) because of increased adsorption of conductive gases by the fly ash. For example, resistivity changes were responsible for

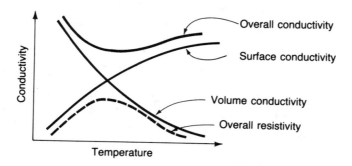

Figure 5.5
Temperature/conductivity relationship.

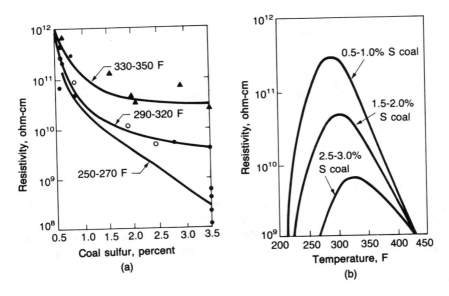

Figure 5.6
Variation of fly-ash resistivity with coal sulfur content and flue gas temperature.
([a] Adapted from White, 1977; [b] adapted from Bump, 1977.)

Res ↓ Sulf ↑
Sulf ↓   Emis ↑ => Res ↑ = Emis ↑
too high

increased fly-ash emissions when power plants switched from high-sulfur coal to low-sulfur coal to reduce SO_2 emissions. Increases in resistivity can be partially offset by adding chemicals to the flue gas. Later in this section we will further discuss this technique, known as *flue gas conditioning*.

A highly resistive dust increases the occurrence of sparking in the precipitator and forces a lower operating voltage. A serious back corona can develop, which reduces both particle charging and collection. The effects of resistivity are more significant above 10^{11} ohm-cm, but can be accounted for in design by the effective drift velocity. Figure 5.7 presents the effect of resistivity on effective drift velocity. The figure is based on performance data for full-scale installations of fly-ash ESPs.

▶ ▶ ▶ ▶ ▶ ▶ ▶ ▶ ▶

Example 5.2

Estimate the total collection areas required for two 99% efficient fly-ash ESPs that treat 8000 m^3/min. The ash resistivities are (a) $1.6(10)^{10}$ ohm-cm, and (b) $2.5(10)^{11}$ ohm-cm.

Solution
  (a)

reading graph

$$\log\left[1.6(10)^{10}\right] = 10.2$$

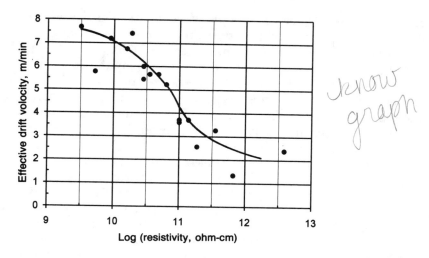

Figure 5.7
Effect of fly-ash resistivity on effective drift velocity in an ESP.
(Adapted from White, 1984.)

From Figure 5.7, $w_e = 6.8$ m/min. Rearranging Eq. (5.7) (the Deutsch equation), we obtain

$$A = \frac{Q}{w_e}\ln\left(\frac{1}{1-0.99}\right) = \frac{8000}{6.8}\ln(100) = 5418 \text{ m}^2$$

(b)

$$\log\left[2.5(10)^{11}\right] = 11.4$$

From Figure 5.7, $w_e = 3.0$ m/min. Thus,

$$A = \frac{8000}{3}\ln(100) = 12,280 \text{ m}^2$$

INTERNAL CONFIGURATION

The design of the internal configuration of an ESP is often ignored in textbooks, possibly because this part of the design involves more art than science. The even distribution of gas flow through the ducts is very important to the proper operation of an ESP, as are uniform plate spacing, proper electrode arrangement, "trueness" of plates (plates must be flat and parallel such that all points between two adjacent plates are equidistant), slopes of hoppers, adequate numbers of electrical sections, and many other features. Some of the basic structural considerations are discussed by Schneider et al. (1975).

Although there have been improvements in computer models for ESP design, reliance is still placed on experience and pilot-scale studies. However, White (1977) notes that pilot-scale ESPs tend to operate much better per unit size than full-scale ESPs (thus, we recommend caution when scaling up results to full-scale units). Some practical design parameters are listed in Table 5.1.

Using the information given in Table 5.1 and a basic understanding of the configuration of an ESP, we can specify the basic geometry of an ESP. For instance, the overall width of the precipitator is virtually equal to the number of ducts for gas flow times the duct (channel) width, increased by a little extra for width of the plates themselves and for the gaps between the outside plates and the walls. The number of ducts (which is equal to one less than the number of plates in

Table 5.1   Selected Design Parameters for Fly-Ash ESPs and Typical Values

Parameter	Range of Values
Drift Velocity w_e	1.0–10 m/min
Channel (Duct) Width D	15–40 cm
Specific Collection Area SCA (Plate Area/Gas Flow)	0.25–2.1 $m^2/(m^3/min)$
Gas Velocity u	1.2–2.5 m/s (70–150 m/min)
Aspect Ratio R (Duct Length/Plate Height)	0.5-1.5 (not less than 1.0 for $\eta > 99\%$)
Corona Power Ratio P_c/Q (Corona Power/Gas Flow)	1.75–17.5 $W/(m^3/min)$
Corona Current Ratio I_c/A (Corona Current/Plate Area)	50–750 $\mu A/m^2$
Power Density versus Resistivity	

Ash Resistivity, ohm-cm	Power Density, W/m^2
$10^4 - 10^7$	43
$10^7 - 10^8$	32
$10^9 - 10^{10}$	27
10^{11}	22
10^{12}	16
10^{13}	10.8

Plate Area per Electrical Set A_s	460–7400 m^2
Number of Electrical Sections	
a. In the Direction of Gas Flow, N_s	2–8
b. Total, N_t	1–10 sections/(1000 m^3/min)

Adapted from White, 1977; Szabo, 1982.

parallel across the width, i.e., $N_d = n - 1$) is related to the gas flow rate, gas linear velocity, and duct geometry by

$$N_d = \frac{Q}{uDH} \tag{5.15}$$

where

N_d = number of ducts

Q = total volumetric gas flow rate into the ESP, m^3/min

u = linear gas velocity in the ESP, m/min

D = channel width (plate separation), m

H = plate height, m

At the start of the design, Eq. (5.15) can be used to estimate N_d by assuming a value for H and choosing representative values of u and D. Another use for this equation is to calculate the gas velocity in the ducts after all the other parameters have been specified.

The overall length of the precipitator is given by

$$L_o = N_s L_p + (N_s - 1) L_s + L_{en} + L_{ex} \tag{5.16}$$

where

L_o = overall length, m

N_s = number of electrical sections in the direction of flow

L_p = plate length, m

L_s = spacing between electrical sections, m

L_{en} = entrance section length, m

L_{ex} = exit section length, m

The spacing between sections can be 0.5–1.5 meters, and the entrance and exit length each can be several meters long. Plates for large fly-ash ESPs are often 6–12 meters high and 1–4 meters long (in direction of gas flow). The ESP height can be 1.5 to 3 times the plate height due to hoppers, superstructure, controls, and so forth.

The number of electrical sections (in the direction of flow) ranges between 2 and 8, and depends on the aspect ratio and the plate dimensions. However, the number of sections must be sufficient to provide the minimum total collection area required but not a great excess of area. The number of sections can be estimated by

$$N_s = RH/L_p \tag{5.17}$$

where

N_s = number of sections in the direction of flow (an integer)

R = aspect ratio (total plate length/plate height)

When the numbers of ducts and sections have been specified, the actual collection area can be calculated as

$$A_a = 2HL_pN_sN_d \qquad (5.18)$$

where A_a = actual collection area, m^2

During the design process, several plate sizes and number of ducts are tried until one combination is found such that A_a is equal to (or slightly greater than) the required collection area, and the final geometry of the ESP is "reasonable" (e.g., not a "cigar box" shape).

ESP performance improves with increasing sectionalization. Here we refer to grouping all the parallel plates across the width of the ESP into two or more electrical sections, as well as increasing the number of sections in the direction of flow. The reader should again refer to Figure 5.2, which depicts an ESP with six total sections, three in the direction of flow and two across the width. As explained by White (1977), there are several fundamental reasons for this improvement. Electrode alignment and spacing are more accurate for smaller sections. Smaller rectifier sets are more stable and can operate at higher voltages. Large numbers of electrical sections allow for meeting the overall efficiency targets even if one or more sections are inoperable. However, adding extra sections increases the capital cost.

Theodore and Reynolds (1983) have developed a method to calculate the probability of whether an ESP will be out of compliance owing to the failure of a specific number of sections, given information about the configuration of the ESP and the collection efficiency of each section. This type of calculation is helpful in the design stage in weighing the costs of increased sectionalization versus the risk of noncompliance owing to an assumed number of electrical section failures.

▶ ▶ ▶ ▶ ▶ ▶ ▶ ▶ ▶

Example 5.3

For a 99% efficient precipitator treating 20,000 m^3/min of gas needing a total collection area of 14,000 m^2, estimate the overall width, length, and height of the ESP. Use typical values for plate height, channel width, gas velocity, and aspect ratio. Assume plates are available in heights from 6–12 meters and are 3 meters long.

Solution

Choosing typical values from Table 5.1, we assume H = 12 m, D = 25 cm, u = 100 m/min, and the aspect ratio = 1.0. From Eqs. (5.15), (5.17), and (5.18), we obtain

$$N_d = \frac{20{,}000 \text{ m}^3/\text{min}}{(100 \text{ m/min})(0.25 \text{ m})(12 \text{ m})} = 67 \text{ ducts}$$

$$N_s = \frac{(1.0)(12)}{3} = 4 \text{ sections}$$

$$A_a = 2(12)(3)(4)(67) = 19,296 \text{ m}^2$$

However, 19,296 m^2 is excessively higher than the required 14,000 m^2. In this case, it seems best to change the plate size. Since the aspect ratio must remain above 1.0 for this high efficiency, we try a plate height of 10 m. Thus,

$$N_d = \frac{20,000}{100(0.25)10} = 80 \text{ ducts}$$

$$N_s = \frac{(1.0)(10)}{3} = 3.3 \text{ sections (round up to 4)}$$

and

$$A_a = 2(10)(3)(4)(80) = 19,200 \text{ m}^2$$

which is not significantly better. Continuing this process, we might end up with a design calling for plates 8 m × 3 m, 100 ducts, and 3 sections. The final design plate area with three sections would be 14,400 m^2, and the final aspect ratio would be about 1.1.

We can now check some of the other design parameters for our design against the typical ranges given in Table 5.1.

$$\text{specific collection area} = \frac{14,400 \text{ m}^2}{20,000 \text{ m}^3/\text{min}} = 0.72 \frac{\text{m}^2}{\text{m}^3/\text{min}}$$

$$\text{plate area per electrical set} = \frac{14,400}{3} = 4800 \text{ m}^2$$

The plate area per set is not beyond reason. However, rather than have one set extend across the entire width of the ESP, we could divide the width into 2, 3, or 4 parallel chambers, thus giving 6, 9, or 12 independent electrical sets. This additional sectionalization is prudent. In the case of electrical failure of one or two sections, the latter approach would allow continued operation of the ESP at a reasonably high efficiency. If the first approach were followed, electrical outage of just one set would reduce operating collection area by 33%.

In summary, our estimate of the overall width of the ESP is 100 × 0.25 m = 25 m. The overall length is about 18–20 meters, assuming entrance and exit lengths of 3–5 meters each. The overall height is about 16 m, which allows space below the plates for the hoppers and dust transport system, and space above the plates for the rappers and transformer-rectifier sets.

◀ ◀ ◀ ◀ ◀ ◀ ◀ ◀ ◀

PLATES AND WIRES

The type and positioning of the collecting plates and the charging wires can be major factors in the operation and maintenance of an ESP. The plates are usually steel sheets with stiffeners. Baffles are added to reduce turbulence (and thus reduce dust re-entrainment) in the vicinity of the plates. The plates should be true (perfectly flat) and should be hung straight and parallel so that the spacing between plates at any point is uniform to within 0.5 cm (Lewandowski 1978).

The discharge electrodes in older ESPs in the United States are wires (of about 2.5 mm diameter) kept taut by weights and positioned through guides to prevent excess swaying. The wires tend to be high-maintenance items. Corrosion can occur near the top of the wires because of air leakage and acid condensation. Also, long weighted wires tend to oscillate. The middle of the wire can approach the plate quite closely, causing increased sparking and wear. In the past, European designs favored rigid, mast-type supports for the wires, and many used barbs on the wires, or serrated strips instead of round wires. Figure 5.8 presents some drawings of plates and wires. Companies on both continents have begun using rigid electrodes because they have advantages over either wires or wire-frame (mast-type electrodes).

REMOVAL OF PARTICLE DUST

After collection, the particles must be removed periodically so that the ESP can continue to function properly. Particle removal is accomplished by rapping the plates, causing a vibration that knocks off the layers of dust. The dust falls into hoppers and is then discharged through pneumatic tubes or screw conveyors to a loading facility. The wires also collect some dust; they also are rapped or vibrated periodically. The plates remain energized during rapping.

The two basic approaches to rapping are the American approach and the European approach (Frenkel 1978). In most American designs, the plates are rapped by a falling weight. The intensity of the rap is easily adjusted by varying the height from which the weight is dropped, or by adjusting the acceleration-field strength. In a typical European design, rapping is accomplished by a fixed-size rotating hammer. Thus, to adjust the rapping intensity, the hammers physically must be changed. Generally, one rapping unit is provided for every 1200 to 1600 square feet of collection area (U.S. Environmental Protection Agency 1985). Both designs allow for convenient adjustment of the rapping interval, which can vary from 1 to 10 minutes.

ESP hoppers are discussed in detail by Schneider et al. (1975). Hoppers catch the falling dust and provide temporary storage. Most hoppers have a pyramidal shape that converges to either a round or square discharge. Hopper walls must be steeply sloped (usually greater than 60%) to prevent dust caking and bridging. Also, hoppers

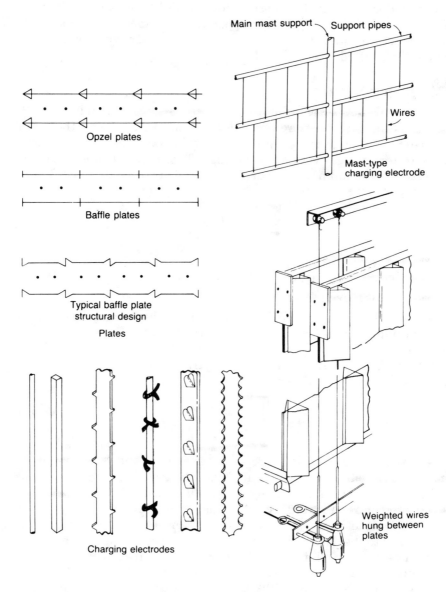

Figure 5.8
Schematic diagrams of various charging and collecting electrodes.

are often heat traced because warm ash flows much better than cold ash. Usually, about 60–70% of the dust is removed through the first (inlet) set of hoppers. However, in case of failure of the first electrical set, the dust load is transferred to the next downstream hopper. Therefore, liberal sizing of the hoppers is recommended. Proper support structure must be provided so that a hopper will not collapse when filled with dust.

POWER CONSUMPTION

There are two sources for operating power consumption in an ESP: corona power and pressure drop, with corona power being the main source. Even though the gas pressure drop is low (typically less than 2 cm of water), the gas volume flow is high. Therefore, we must also consider the cost of fan power needed to pull the air through an ESP.

Corona power can be approximated by the equation

$$P_c = I_c V_{avg} \qquad \text{(5.19)}$$

where

P_c = corona power, W

I_c = corona current, A

V_{avg} = average voltage, V

Even though voltages in ESPs are very high, the current flow due to gas ion migration is low, so the power consumption is not unreasonably high.

According to White (1977), the effective drift velocity is related to the corona power as follows:

$$w_e = \frac{kP_c}{A} \qquad \text{(5.20)}$$

where the terms have been defined previously, and k is simply an adjustable constant. For well-built fly-ash ESPs, k is in the range from 0.5 to 0.7 for units of w_e in ft/sec and P_c/A in W/ft^2. The ratio P_c/A is called the **power density**. Although the power density often increases (sometimes by as much as a factor of ten) from the inlet of the airflow to the outlet, the overall power density (total corona power/ total plate area) is a fairly stable and representative parameter. Typical values of the overall power density are 1–2 W/ft^2 (U.S. Environmental Protection Agency 1985). By substituting Eq. (5.20) into Eq. (5.7) (the Deutsch equation), the corona power can be related to the collection efficiency as

$$\eta = 1 - e^{(-kP_c/Q)} \qquad \text{(5.21)}$$

White (1977) has shown Eq. (5.21) with $k = 0.55$ for P_c/Q in units of W/cfs of gas flow to be reasonably accurate for efficiencies up to about 98.5%. For efficiencies above 98.5%, the required corona power increases rapidly for an increase in efficiency, as shown in Figure 5.9.

▶ ▶ ▶ ▶ ▶ ▶ ▶ ▶ ▶

Example 5.4

An ESP is to be designed to treat 9000 m³/min of gas to remove particles at (a) 98% efficiency, or (b) 99.8% efficiency. Using Figure 5.9, estimate the required corona power in kW.

Solution

(a)

$$P_c = \frac{100 \text{ W}}{1000 \text{ acfm}} \times 9000 \frac{\text{m}^3}{\text{min}} \times \frac{35.3 \text{ ft}^3}{1 \text{ m}^3} \times \frac{1 \text{ kW}}{1000 \text{ W}} = 31.8 \text{ kW}$$

(b)

$$P_c = \frac{330 \text{ W}}{1000 \text{ acfm}} \times 9000 \times 35.3 \times \frac{1}{1000} = 105 \text{ kW}$$

Note that the 1.8% increase in collection efficiency (a tenfold reduction in emissions) requires a tripling of corona power. Other important parameters (such as plate area) also would change.

◀ ◀ ◀ ◀ ◀ ◀ ◀ ◀ ◀

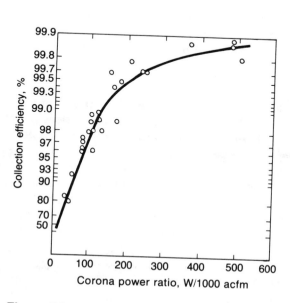

Figure 5.9

Collection efficiency as a function of corona power ratio, based on actual data. (Adapted from White, 1984.)

FLUE GAS CONDITIONING

Existing ESPs.   When there is a sudden decline in performance of an ESP on a coal-fired power plant due to switching to a low-sulfur coal, fly-ash emissions increase and the plant can be subject to fines, derating, or shutdown. Short of abandoning the ESP and installing a new baghouse or a new, larger ESP, one possible solution is to change the fly-ash resistivity.

Earlier in this section, we discussed the strong influence of temperature and chemical composition on fly-ash resistivity, and thus on collection efficiency. However, the inlet gas temperature to an existing ESP often cannot be decreased because of the possibility of acid condensation and corrosion. Ducting can sometimes be rearranged such that the gas temperature entering the precipitator can be raised, but this procedure incurs both capital and operating costs. Therefore, by elimination, chemical conditioning of the flue gas (the addition of small quantities of chemicals such as SO_3 gas or sodium or ammonium salts) is the most practical approach. There are several vendors and systems commercially available and, happily, this approach often works very well with reasonably small expense.

Flue gas conditioning is extremely important for ESPs in the cement industry. In this case, the conditioning chemical is simply water! Moisture conditioning can be accomplished by steam injection or by liquid water spray into the dusty gas stream. Proper spray nozzle design and spacing, and careful temperature control are crucial. If too much water is injected, the dust will cake on the interior of the ESP (U.S. Environmental Protection Agency 1985). Figure 5.10 demonstrates the effectiveness of adding moisture to cement dust exhaust streams.

New Systems.   For new designs, we must consider not only the traditional ESPs on the cold (downstream) side of the air preheater but also hot-side ESPs, cold-side ESPs with chemical conditioning, and baghouses. Hot-side ESPs (those located upstream of the air preheater) have the advantage of higher temperatures that reduce ash resistivity, but they require larger installations to handle the increased volumetric gas flow. In the 1970s, over 200 hot-side ESPs were installed or designed, representing over 53,000 MW of capacity (Andes et al. 1983). However, a follow-up survey showed several significant operating and maintenance problems to be associated with hot-side ESPs. In fact, no new hot-side ESPs are currently being specified for fly-ash applications.

Baghouses have the advantage of good efficiency for any type of dust, but they incur increased operating costs owing to pressure drop and especially bag replacement. Cold-side ESPs with chemical flue gas conditioning retain all the operating and maintenance advantages of the traditional cold-side ESPs, and they can achieve high collection efficiencies.

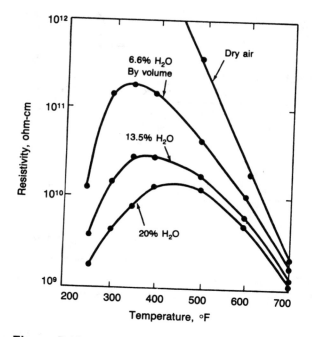

Figure 5.10
Effect of moisture on resistivity of cement kiln dust.
(Adapted from U.S. Environmental Protection Agency, 1985.)

Breisch (1979) compared the preceding four options for an 800 MW unit burning low-sulfur coal and meeting a 99.5% collection efficiency requirement. He found that cold-side precipitation with chemical conditioning was most economical. His results are summarized in Table 5.2. However, economic comparisons can be very site specific, and each case should be judged individually.

◆ 5.4 COSTS ◆

The free-on-board (f.o.b.) cost of an ESP is a fairly predictable function of collection plate area. However, the installed cost can vary considerably owing to construction location, time of year, available space, and so forth (Schneider et al. 1975). Vatavuk (1990) gathered cost quotes from ESP vendors for a number of units and regressed them against collection area. He developed Eq. (5.22) for estimating the purchase price of an ESP. Equation (5.22) was originally developed for predicting a price for an ESP in 1988 dollars. However, recent information from the EPA indicates that prices from ESP manufacturers hardly changed at all in the decade that followed (U.S. Environmental Protection Agency 1996). That may have been a result of improvements in the technology, reductions in steel prices, market

Table 5.2   Comparison of Four Systems for Particulate Collection for Low-Sulfur Coal Fly Ash

	Cold-Side ESP	Hot-Side ESP	Baghouse	Cold-Side ESP with Chemical Condition
Gas Volume (10^3 acfm)	2800	4458	2800	2800
Temperature, F	300	750	300	300
Collecting Surface Area (1000 ft^2)	1884	1500	1400	887
Total Investment				
($ million)	27.9	23.5	16.3	14.7
($/ft^2 collecting surface)	14.8	15.6	14.6	16.6
Total Annual Cost, including amortization				
($ million)	5.75	5.30	3.89	3.16

Adapted from Breisch, 1979.

competition, or some combination of factors. Nevertheless, Eq. (5.22) can be used to predict prices in 1998 dollars.

$$P = aA^b \qquad\qquad (5.22)$$

where

P = purchase price, f.o.b. manufacturer, **in 1998 dollars**

A = net plate area, ft^2

a and b = constants, as follows:

for 10,000 ft^2 < A < 50,000 ft^2, a = 962 and b = 0.628

for 50,000 ft^2 < A < 1,000,000 ft^2, a = 90.6 and b = 0.843

Although installation costs vary, some average factors for ESPs are presented in Table 5.3. The factors are applied to the delivered equipment cost (DEC) to estimate the total installed cost (TIC).

The major direct operating costs are due to electrical power consumption (primarily corona power and fan power) and maintenance (parts and labor). Operating labor has been estimated at 0.5–2 person-hours per shift (Neveril et al. 1978b). Typical maintenance items include rappers, wire electrodes, electrical controls, and ash-handling systems.

Table 5.3   Average Cost Factors for Estimating the Total Installed Cost
(TIC) of ESPs

1. Delivered Equipment Cost	
a. Control Device and Auxiliary Equipment (f.o.b.)	as required (P)
b. Instruments and Controls	0.10
c. Taxes	0.03
d. Freight	0.05
Subtotal (DEC)	$\overline{1.18}$ (P)
2. Installation Costs	
Direct	
a. Foundations and Supports	0.04
b. Erection and Handling	0.50
c. Electrical	0.08
d. Piping, Insulation, Painting	0.05
e. Site Preparation; Facilities and Buildings	as required
Indirect	
a. Engineering and Supervision	0.20
b. Construction and Field Expense	0.20
c. Construction Fee	0.10
d. Start-up and Performance Testing	0.02
e. Contingencies	0.03
Total (TIC)	$\overline{2.22}$ × DEC
(excluding site preparation and any buildings required)	

Adapted from Neveril et al., 1978(a).

◆ PROBLEMS ◆

5.1 A 98% efficient ESP is to treat a gas stream flowing at 5000 m³/ min. If the effective drift velocity is 6.0 cm/s, calculate (a) the plate area (required and actual) in m², and (b) the number of plates if each plate is 6 m by 10 m, and there are 3 sections in the direction of flow.

5.2 Assume that the ESP of Problem 5.1 is built with an actual collection area of 5400 m². Now suppose that the actual drift velocity turns out to be 5.0 cm/s and the actual gas flow turns out to be 5500 m³/min. Calculate the actual efficiency.

5.3 An ESP must treat 500,000 acfm with 99% efficiency. Assuming an effective drift velocity of 0.4 ft/sec, calculate the required plate area in ft², and the number of plates if each is 20 ft tall by 10 ft long, and there are 4 sections in the direction of flow.

5.4 After the ESP of Problem 5.3 has been built, the gas flow increases to 700,000 acfm. Calculate the new efficiency.

5.5   A new coal-fired power plant burns 5000 tons/day of coal with 7.0% ash and a heating value of 12,000 Btu/lb$_m$. Forty percent of the ash drops out of the furnace as slag. Calculate the efficiency required of an ESP to remove fly ash if the plant is to meet the 1980 Federal NSPS for particulates.

5.6   A 98% efficient ESP follows the Deutsch equation. If the gas flow rate changes and the efficiency drops to 94%, calculate the ratio of the new gas flow rate to the old gas flow rate.

5.7   For a fixed gas flow rate and drift velocity, calculate the ratio of plate areas for two precipitators—one at 90% efficiency and the other at 99% efficiency.

5.8   Provide a reasonable design for a 99% efficient ESP treating 7500 m^3/min of gas. The particles have an effective drift velocity of 10.0 cm/s. Specify the total plate area, channel width, number and size of plates, number of electrical sections (total and in the direction of flow), and total corona power to be supplied, and estimate the overall dimensions.

5.9   Estimate the DEC and TIC of an ESP with 50,000 m^2 of net plate area. Calculate your answers in 1998 dollars, then estimate the 2002 TIC.

5.10 Provide a reasonable design for a 99.4% efficient ESP treating 30,000 m^3/min of gas. The dust has a resistivity of $7.1(10)^{10}$ ohm-cm. Specify the same items as in Problem 5.8.

5.11 Estimate the DEC and TIC (in 1998 dollars) of the ESP of Problem 5.1, assuming the net plate area is 5760 m^2.

5.12 A coal-burning power plant has two options for fly-ash removal from its stack gases. In either case, a total efficiency of 99.5% must be achieved for a total flow rate of 1,000,000 acfm. The options are as follows:

Option A: Use only an ESP. Assume an effective precipitation velocity of 0.3 ft/sec, and a ΔP of 0.2 in. H$_2$O.

Option B: Use a multi-cyclone precleaner with an 85% efficiency and a pressure drop of 3.0 in. H$_2$O, followed by an ESP with a pressure drop of 0.2 in. H$_2$O. Because the cyclone will remove most of the larger particles, assume an effective drift velocity of 0.25 ft/sec for this ESP.

For each option specify

a. Collection area of the ESP (assume the Deutsch equation is applicable)

b. Capital cost (TIC) in 1998 dollars

c. Power consumption (fan + corona power)

d. Annual power costs (7800 hours of operation per year)

e. Annual operating costs (include only power costs and depreciation; exclude labor, taxes, and all other costs)

Assume the following cost data:
Installed cyclone costs = $1.50/acfm (1998 dollars)
Electricity costs = $0.06/kWh
Fan efficiency = 60%
Depreciation of capital investment = 15 year straight-line method

5.13 An ESP with a total plate area of 5000 m^2 treats 8000 m^3/min of air containing particulate matter of essentially one diameter. If the electrostatic force per unit mass on a particle is 2000 times the force of gravity and the slip-corrected characteristic time, τ', for this particle-in-air system is $5.1(10)^{-6}$ sec, calculate the efficiency of the ESP.

5.14 Design an ESP that is 98.5% efficient in treating 8050 m^3/min of gas. Assume the plates are 6.1 m high and 4.1 m long. The effective drift velocity of the particles is 5.2 m/min. The design should include the plate design (total area, layout, number), channel width, gas velocity, number of ducts, aspect ratio, number of sections, and actual area.

5.15 Consider a 98% efficient ESP which removes fly ash from combustion gases flowing at 100 m^3/sec. The resistivity of the ash is $1.0 (10)^{10}$ ohm-cm.

a. Calculate the total collection area required.

b. Assuming that each plate is 8 m by 4 m, calculate the number of plates required.

5.16 An ESP must treat 300,000 acfm with 99% efficiency.

a. For an effective drift velocity of 0.25 ft/sec, calculate the required collection area.

b. Assuming the plates are 15 feet tall and 6 feet long, and are arranged with three sections in the direction of flow, calculate the aspect ratio.

c. Estimate the total number of plates in this ESP.

d. Estimate the cost of this ESP (f.o.b. manufacturer and TIC), both in 1998 dollars.

e. Calculate the fan power if the average pressure drop is 0.6 in. H_2O, and the fan/motor efficiency is 70%.

5.17 An ESP is treating the flue gas from a coal combustion unit and achieving a 97% PM removal efficiency. Suddenly the efficiency drops to 92%, but the gas flow rate has remained exactly the same.

a. Estimate the ratio of the new effective drift velocity (w_e) to the old w_e.

   b. Under the *old* conditions (while achieving 97% efficiency), the fly-ash resistivity was measured at 3.16 $(10)^{10}$ ohm-cm. Estimate the value of the *new* drift velocity, m/min.

5.18 An ESP on a power plant has been designed for 98.5% removal efficiency. At this level the PM emissions were estimated to be 2000 kg/day. Before the plant begins construction, the state air pollution control agency changes the standard and says that no more than 1000 kg/day of emissions will be permitted. What is the new design efficiency?

REFERENCES

Andes, G. M., Cummings, W. E., Link, S. A., and Steinbach, P. H. "Survey Reveals Hot-Side Precipitator Performance," *Pollution Engineering, 15*(3), March 1983.

Breisch, E. W. "Method and Cost Analysis of Alternative Collectors for Low Sulfur Coal Fly Ash," Symposium on the Transfer and Utilization of Particulate Control Technology, Vol. 1, *Electrostatic Precipitators*, EPA-600/7-79-44a, compiled by F. P. Venditti et al., Denver Research Institute, Denver, CO, February 1979, pp. 121–129.

Bump, R. L. "Electrostatic Precipitators in Industry," *Chemical Engineering, 84*, January 17, 1977.

Danielson, J. A., Ed. *Air Pollution Engineering Manual* (2nd ed.), AP-40. Washington, DC: U.S. Environmental Protection Agency, May 1973.

Frenkel, D. I. "Tuning Electrostatic Precipitators," *Chemical Engineering, 85*, June 19, 1978.

Lewandowski, G. A. "Specifying Mechanical Design of Electrostatic Precipitators, *Chemical Engineering, 85*, June 19, 1978.

Neveril, R. B., Price, J. U., and Engdahl, K. L. "Capital and Operating Costs of Selected Air Pollution Control Systems–I," *Journal of the Air Pollution Control Association, 28*(8), August 1978(a).

_____. "Capital and Operating Costs of Selected Air Pollution Control Systems–V," *Journal of the Air Pollution Control Association, 28*(12), December 1978(b).

Schneider, G. G., Horzella, T. I., Cooper, J., and Striegl, P. J. "Selecting and Specifying Electrostatic Precipitators," *Chemical Engineering, 82*, May 26, 1975.

Szabo, M. F. "Electrostatic Precipitators," in *Specifying Air Pollution Control Equipment,* R. A. Young and F. L. Cross, Eds. New York: Dekker, 1982.

Theodore, L., and Reynolds, J. "ESP Bus Section Failures: Design Considerations," *Journal of the Air Pollution Control Association, 33*(12), December, 1983.

U.S. Environmental Protection Agency. *Operation and Maintenance Manual for Electrostatic Precipitators*, EPA-625/1-85-017, Research Triangle Park, NC, September 1985.

———. *OAQPS Control Cost Manual* (5th ed.). EPA 453/B-96-001, Research Triangle Park, NC, February 1996.

Vatavuk, W. M. *Estimating Costs of Air Pollution Control*. Chelsea, MI: Lewis Publishers, 1990.

Wark, K., and Warner, C. F. *Air Pollution—Its Origin and Control* (2nd ed.). New York: Harper & Row, 1981.

White, H. J. "Electrostatic Precipitation of Fly Ash—Parts I, II, III, and IV," *Journal of the Air Pollution Control Association, 27*, Nos. 1, 2, 3, 4, January-April 1977.

———. "Control of Particulates by Electrostatic Precipitation," Chapter 12 in *Handbook of Air Pollution Technology*, S. Calvert and H. M. Englund, Eds. New York: Wiley, 1984.

◆◆◆ **CHAPTER 6**

FABRIC FILTERS

Given this extremely low sulfur content [0.3% in coal] and the mild Australian climate, air heater gas outlet temperatures on efficient boilers can run as low as 220 F. At these conditions fly ash is highly resistive and precipitator performance has been unsatisfactory . . . [and so for fly-ash collection] the Australians are turning to fabric filtration.

A. C. Leutbecher, 1978

◆ 6.1 INTRODUCTION ◆

Fabric filtration is a well-known and accepted method for separating dry particles from a gas stream (usually air or combustion gases). In fabric filtration, the dusty gas flows into and through a number of filter bags placed in parallel, leaving the dust retained by the fabric. The fabric itself does some filtering of the particles; however, the fabric is more important in its role as a support medium for the layer of dust that quickly accumulates on it. The dust layer is responsible for the highly efficient filtering of small particles for which baghouses are known. (This dust-layer effect is more important for woven fabrics than for felted fabrics.)

There are many different types of fabrics, different ways of weaving them into various sizes of bags, different ways of configuring bags in a baghouse, and different ways of flowing the air through the bags. Extended operation of a baghouse requires that the dust be periodically cleaned off the cloth surface and removed from the baghouse. The three common types of baghouses, classified by the method used for cleaning the dust from the bags, are reverse-air, shaker, and pulse-jet baghouses, each of which will be described in more detail later in this chapter. Although the detailed mechanical design of a baghouse is usually left to the vendor, air pollution control engineers must be familiar with the theory and practice of baghouses to be able to make the process design.

177

Advantages of baghouses:

1. They have very high collection efficiencies even for very small particles.

2. They can operate on a wide variety of dust types.

3. They are modular in design, and modules can be preassembled at the factory. They can operate over an extremely wide range of volumetric flow rates.

4. They require reasonably low pressure drops.

Disadvantages of baghouses:

1. They require large floor areas.

2. Fabrics can be harmed by high temperatures or corrosive chemicals.

3. They cannot operate in moist environments; fabric can become "blinded."

4. They have potential for fire or explosion.

The advantages of baghouses outweigh the disadvantages in many cases. From 1970–1982, orders for fabric filter systems rose from 25–30% to about 45–50% of the industrial gas cleaning market (at the expense of electrostatic precipitators) (McIlvaine 1983). The use of baghouses has become widely accepted and often preferred for particulate matter collection in the U.S. electric utility industry. This industry's use of baghouses grew from almost nothing in 1975 to 57 installations (on 16,060 MW of generating capacity) in 1989 (Cushing et al. 1990).

♦ 6.2 THEORY ♦

Consider a brand-new woven filter cloth, as depicted in Figure 6.1. The fibers are about 100–150 microns in diameter, and the open spaces between the fibers can be as large as 50–75 microns. The spaces are occupied by tiny, randomly oriented fibrils. Initially, when this clean cloth is put in service, the collection efficiency will be low because a large portion of the dust will pass directly through (penetrate) the cloth. However, owing to impaction, interception, and diffusion, dust particles will quickly build up on the fibrils and "bridge across" the gaps. Once these interstitial holes are filled and a particulate layer has formed on the cloth, the filtration efficiency will increase substantially.

The particulate layer is a very efficient filter, but, as might be expected, it increases the resistance to gas flow. The pressure drop through a baghouse at a given gas flow rate is given by

$$\Delta P = \Delta P_f + \Delta P_p + \Delta P_s \tag{6.1}$$

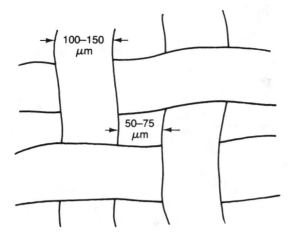

Figure 6.1
A new, clean woven filter cloth.

where

ΔP = total pressure drop

ΔP_f = pressure drop due to the fabric

ΔP_p = pressure drop due to the particulate layer

ΔP_s = pressure drop due to the baghouse structure

The pressure drop due to the structure usually is low, and it will be ignored in the following discussion.

From Darcy's equation for fluid flow through porous media, equations can be written individually for the fabric and the particulate layer; that is,

$$\Delta P_f = \frac{D_f \mu V}{60 K_f} \qquad (6.2)$$

and

$$\Delta P_p = \frac{D_p \mu V}{60 K_p} \qquad (6.3)$$

where

ΔP_f, ΔP_p = pressure drop, N/m^2

D_f, D_p = depth (in the direction of flow) of the filter and the particulate layer, respectively, m

μ = gas viscosity, kg/m-s

V = superficial filtering velocity, m/min

K_f, K_p = permeability of the filter and the particulate layer, m^2

60 = conversion factor, s/min

The superficial filtering velocity V, also known as the *air/cloth ratio*, is equal to the volumetric gas flow rate divided by the cloth area; that is,

$$V = \frac{Q}{A} \qquad (6.4)$$

where

Q = volumetric gas flow rate, m^3/min

A = cloth area, m^2

As the filter operates, the depth of the dust layer, D_p, increases. In fact, for a constant filtering velocity and a constant mass concentration of dust (often referred to as dust loading), D_p should increase linearly with time; that is,

$$D_p = \frac{LVt}{\rho_L} \qquad (6.5)$$

where

L = dust loading, kg/m^3

t = time of operation, min

ρ_L = bulk density of the particulate layer, kg/m^3

Substituting Eq. (6.5) into Eq. (6.3), and then adding Eqs. (6.2) and (6.3), we obtain

$$\Delta P = \left(\frac{D_f \mu}{60 K_f} \right) V + \left(\frac{\mu}{60 K_p \rho_L} \right) (LVt) V \qquad (6.6)$$

Next, we divide through Eq. (6.6) by V, and define the filter drag S and the areal dust density W as follows:

$$S = \frac{\Delta P}{V} \qquad (6.7)$$

$$W = LVt \qquad (6.8)$$

where

S = filter drag, N-min/m^3 or Pa-min/m

W = areal dust density, kg/m^2 of fabric

The areal dust density is the mass of dust per unit area of fabric. Equation (6.7) can be rewritten as

$$S = K_1 + K_2 W \qquad (6.9)$$

where

$$K_1 = \frac{D_f \mu}{60 K_f}$$

$$K_2 = \frac{\mu}{60 K_P \rho_L}$$

The units on K_1 are N-min/m^3 or Pa-min/m, and the units on K_2 are N-min/kg-m or Pa-min-m/kg. The linear model represented by Eq. (6.9) is often called the **filter drag model.** While the *form* of this two-parameter model is very useful, the evaluation of K_1 for clean fabric is inappropriate. Also, the evaluation of K_2 from its defining parameters is inconvenient. Therefore, we rewrite Eq. (6.9) as

$$S = K_e + K_s W \qquad (6.10)$$

where

K_e = extrapolated clean cloth filter drag, N-min/m^3

K_s = "slope" constant for the particular dust, gas, and fabric involved, N-min/kg-m

Both K_e and K_s are determined empirically from pilot tests on a dusty gas that is similar to the one for which the design is being made. Note that in U.S. customary units, if S and K_e are expressed in units of in. H$_2$O-min/ft, then V would be in units of ft/min, W would be in units of lb$_m$/ft^2, and K_s would be in units of in. H$_2$O-min-ft/lb$_m$. Note that K_s also can be expressed in units of inverse time.

Typical plots of filter drag versus areal dust density are shown in Figure 6.2. Note that each curve in the figure has a nonlinear portion that starts at a different value of W. The residual mass of dust on the fabric at the start of a new filter cycle is strictly a function of the intensity of the previous cleaning cycle. The curves are initially nonlinear because initial flow through the fabric is not uniform; that is, the previous cleaning cycle generally dislodges the dust cake in irregular chunks, leaving some parts of the bag very clean and others still quite dusty.

On the average, after a fabric has been through several cleaning cycles, it retains a residual amount of dust W_r. This residual dust yields a residual filter drag S_r that can be quite different from the clean cloth drag K_1. Thus, Eq. (6.9) is not pertinent, and we use Eq. (6.10) with empirically determined constants.

▶ ▶ ▶ ▶ ▶ ▶ ▶ ▶ ▶

Example 6.1

Based on the following test data for a clean fabric, predict the design pressure drop in a baghouse after 70 minutes of operation with $L = 5.0$ g/m^3 and $V = 0.9$ m/min.

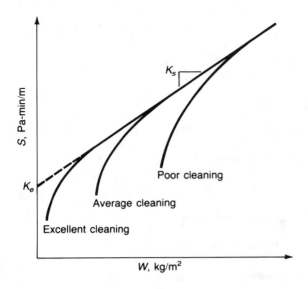

Figure 6.2
Typical filter drag versus dust density curves for different degrees of cleaning.

Test Data		
Time, min	ΔP, Pa	Constants
0	150	$V = 0.9$ m/min
5	380	$L = 5.0$ g/m^3
10	505	
20	610	
30	690	
60	990	

Solution

First, we use the test data to generate a plot of filter drag versus areal dust density. The data to be plotted are

$S = \Delta P/V$, Pa-min/m	$W = LVt$, g/m^2
167	0
422	22.5
561	45
678	90
767	135
1100	270

A plot of these data (see the figure) shows an initial characteristic curvature, which should be ignored in obtaining the slope and intercept.

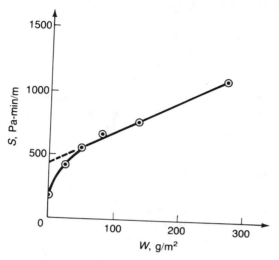

Filter drag versus areal dust density for the data of Example 6.1.

From a linear least-squares fit of the last four data points, the values of the constants K_e and K_s are 455 Pa-min/m and 2.381 Pa-min-m/g, respectively. Note that the units of K_s can be reduced to reciprocal time as follows:

$$2.381 \frac{\text{Pa-min-m}}{\text{g}} \times \frac{1 \text{ N/m}^2}{1 \text{ Pa}} \times \frac{1 \text{ kg-m/s}^2}{1 \text{ N}}$$
$$\times \frac{1000 \text{ g}}{1 \text{ kg}} \times \frac{60 \text{ s}}{1 \text{ min}} = 1.43(10)^5 \text{ s}^{-1}$$

However, it is often more convenient to use K_s in units of pressure-time-distance/mass as shown in the following calculations.

Knowing the coefficients of the filter drag model, we can predict the design pressure drop.

$$W = 5.0 \frac{\text{g}}{\text{m}^3} \times 0.9 \frac{\text{m}}{\text{min}} \times 70 \text{ min} = 315 \frac{\text{g}}{\text{m}^2}$$

$$S = 455 \frac{\text{Pa-min}}{\text{m}} + \left(2.381 \frac{\text{Pa-min-m}}{\text{g}} \right)\left(315 \frac{\text{g}}{\text{m}^2} \right) = 1205 \frac{\text{Pa-min}}{\text{m}}$$

$$\Delta P = 0.9 \frac{\text{m}}{\text{min}} \times 1205 \frac{\text{Pa-min}}{\text{m}} = 1085 \text{ Pa}$$

◀ ◀ ◀ ◀ ◀ ◀ ◀ ◀ ◀

Actually, K_s is not independent of the filtering velocity. According to Dennis and Klemm (1978), K_s varies with the square root of V. Thus, K_s for a given fabric filtration system can be adjusted for a change in the filtering velocity as

$$K_{s_2} = K_{s_1} \left(\frac{V_2}{V_1} \right)^{1/2} \tag{6.11}$$

This relationship can be used in design work when pilot data were taken at a V that is different from the design V, or it can be used in the analysis of a proposed flow increase in an existing baghouse.

♦ 6.3 DESIGN CONSIDERATIONS ♦

Designing a baghouse is somewhat unusual in that the collection efficiency is generally not a concern of the designer! The reason is that a well-designed, well-maintained fabric filter that is operated properly generally collects particles ranging from submicron sizes to those several hundred microns in diameter at efficiencies of greater than 99% (Turner et al. 1987a). With a high collection efficiency as a "given," baghouse design involves optimizing the filtering velocity V to balance capital costs (baghouse size) versus operating costs (pressure drop). Major factors that affect the selection of the design V include prior experience with similar dusts, fabric characteristics, particle characteristics, and gas stream characteristics.

REVERSE-AIR AND SHAKER BAGHOUSES

Both reverse-air and shaker baghouses have been widely used for many years, and a vast amount of design and operating information is available for a variety of dusts. Table 6.1 provides some data on recommended maximum filtering velocities for various dusts. This recommended maximum design filtering velocity is the V that we will select when starting out on a baghouse design problem. In a baghouse with N compartments, this is V_{N-1}, the filtering velocity through the $N-1$ compartments left on-line while one compartment is off-line for cleaning. As can be seen from the table, V is usually from about 2 to 4 ft/min. However, these values of V should be adjusted in specific cases depending on dust loading, fineness of the dust, and other factors. For example, one should decrease the tabular values of V by 10–15% for dust loadings greater than 40 gr/ft^3, and increase the design V by as much as 20% for dust loadings less than 5 gr/ft^3. Similarly, for particulate sizes less than 3 μm (or greater than 50 μm), the tabulated value of V should be decreased (or increased) by as much as 20% (Turner et al. 1987a). Values of V that are too high can lead to excessive particle penetration, blinding of the fabric, and reduced bag life (U.S. Environmental Protection Agency 1986).

Fabric selection (including type of weave) is important and is based in part on its particle release properties. The fabric must be matched properly with the gas stream characteristics, as well as with the type of particulate. The commonly used fabrics have very different abilities regarding operating temperatures and chemical content of the gas stream as shown in Table 6.2.

Table 6.1   Maximum Filtering Velocities for Various Dusts in Shaker or Reverse-Air Baghouses

Dusts	Maximum Filtering Velocity, cfm/ft^2 or ft/min
Activated Charcoal, Carbon Black, Detergents, Metal Fumes	1.50
Aluminum Oxide, Carbon, Fertilizer, Graphite, Iron Ore, Lime, Paint Pigments, Fly Ash, Dyes	2.0
Aluminum, Clay, Coke, Charcoal, Cocoa, Lead Oxide, Mica, Soap, Sugar, Talc	2.25
Bauxite, Ceramics, Chrome Ore, Feldspar, Flour, Flint, Glass, Gypsum, Plastics, Cement	2.50
Asbestos, Limestone, Quartz, Silica	2.75
Cork, Feeds and Grain, Marble, Oyster Shell, Salt	3.0–3.25
Leather, Paper, Tobacco, Wood	3.50

Adapted from Danielson, 1973; Turner et al., 1987(a).

Table 6.2   Temperature and Chemical Resistance of Some Common Industrial Fabrics

Fabric	Recommended Maximum Temperature, °F	Chemical Resistance	
		Acid	Base
Dynel	160	Good	Good
Cotton	180	Poor	Good
Wool	200	Good	Poor
Nylon	200	Poor	Good
Polypropylene	200	Excellent	Excellent
Orlon	260	Good	Fair
Dacron	275	Good	Fair
Nomex®	400	Fair	Good
Teflon®	400	Excellent	Excellent
Glass	550	Good	Good

Adapted from Kraus, 1979; Buonicore and Davis, 1992.

Both reverse-air and shaker baghouses are constructed with several compartments. When it is time to clean the bags, one compartment is isolated from the dusty gas flow. In the reverse-air system, clean air is blown through the bags in the isolated compartment in the direction opposite the normal flow to dislodge the particulate layer. In a shaker baghouse, the bags are shaken to dislodge the previously collected dust. In both cases, chunks of agglomerated dust fall into a hopper below the compartment. The dust is periodically removed from the hopper and disposed of or reused if applicable. Figure 6.3 is a cutaway view of a simple shaker baghouse, and Figure 6.4 is a photograph of a large industrial baghouse system.

Generally, reverse airflow is a gentle but somewhat less effective method of cleaning filter bags. However, for certain fabrics (especially glass fiber fabrics), the repeated flexing that occurs with shaking results in quick wear and early bag failure. A recent innovation in cleaning—namely, sonic horns—has allowed electric utilities to more effectively clean woven glass bags in reverse-air baghouses. In this technique, a sonic blast from horns mounted inside the baghouse helps dislodge more dust than is normally removed via reverse airflow. It appears that reverse air cleaning with sonic assist is now the method of choice for baghouses at large-scale utilities burning pulverized coal (Cushing et al. 1990).

The number of compartments chosen during the design depends on the total flow to be filtered, the available (or desired) maximum pressure drop ΔP_m, the filtration time t_f desired between two cleanings of the same compartment, and the time required to clean one compartment t_c. The time interval between cleanings of any two compartments is the run time t_r, (portrayed schematically with t_c in Figure 6.5). The filtration time t_f is the elapsed time from the moment one compartment is returned to service until that same compartment is removed for cleaning again (after all the other compartments have been cleaned in rotation). The time t_f is related to t_r and t_c as

$$t_f = N\left(t_r + t_c\right) - t_c \qquad (6.12)$$

where

t_f = filtration time, min

t_r = run time, min

t_c = cleaning time, min

N = total number of compartments

Figure 6.5 portrays the time variation of pressure drop in an operating baghouse during a time that includes the consecutive cleaning of two compartments. Note that when one compartment is taken off-line for cleaning, all of the gas must then flow through the remaining compartments. Consequently, the total pressure drop increases suddenly.

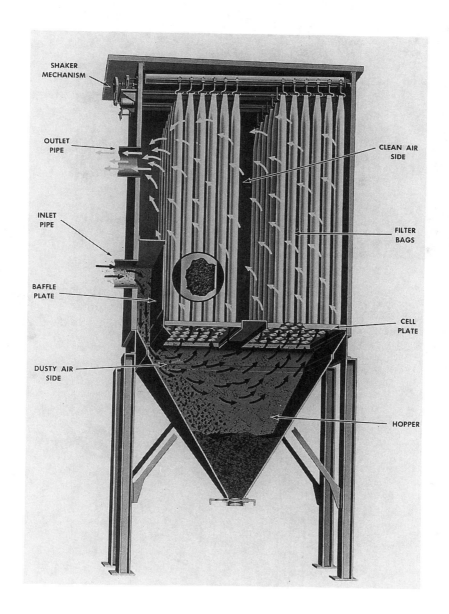

Figure 6.3
Cutaway view of a shaker baghouse.
(Courtesy of Wheelabrator-Frye, Inc., Pittsburgh, PA.)

Figure 6.4
A large industrial shaker baghouse system. (This system was designed for foundry shakeout and sand handling plus an electric melt furnace. There are fifteen units in three systems of five units each. Design airflow is 150,000 cfm total.)
(Photo courtesy of American Filter Co., Louisville, KY)

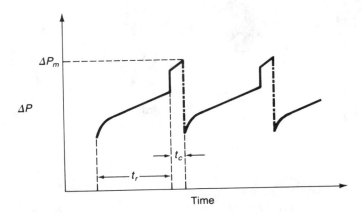

Figure 6.5
Variation of pressure drop with time in a compartmented baghouse.

Just as the pressure drop reaches its maximum allowable value, the cleaned compartment is returned to service, and the pressure drop decreases suddenly. Furthermore, the gas flow rates are different in each compartment, with the cleanest one having the highest flow rate.

In practice, ΔP_m can range from less than 6 up to about 20 inches of water, t_f can range from 30 minutes to 2 hours, and t_c can range from 1 to 5 minutes. As usual in design work, when starting on a completely fresh problem with no constraints (for example, no existing equipment), designers have considerable freedom of choice. How one decides, a priori, on the "best" values of ΔP_m, t_f, t_c, and the number of compartments is a matter of experience, common sense, and insight. If too few compartments are used, an excessive number of bags must be used to provide sufficient filter area when one compartment is being cleaned. Designs with many compartments have higher capital costs and can be more difficult to maintain. As a general guide, Table 6.3 presents information on the number of compartments typically used as a function of the net cloth area (the fabric area that is left on-line while one compartment is being cleaned). However, it is important to always strive to provide some flexibility in the design. Once constructed, a baghouse will be operated as best suits the needs of operations personnel. Such operation might not always be at the calculated design point.

Net Cloth Area, ft^2	Number of Compartments
1–4000	2
4000–12,000	3
12,000–25,000	4–5
25,000–40,000	6–7
40,000–60,000	8–10
60,000–80,000	11–13
80,000–110,000	14–16
110,000–150,000	17–20
>150,000	>20

Table 6.3   Number of Compartments as a Function of Net Cloth Area

Note: Net cloth area = Q_{design}/V_{design} = fabric left on-line, even when one compartment is down for cleaning.

▶ ▶ ▶ ▶ ▶ ▶ ▶ ▶ ▶

Example 6.2

Estimate the net cloth area for a shaker baghouse that must filter 40,000 cfm of air with 10 grains of flour dust per cubic foot of air. Also specify the number of compartments to be used and calculate the total number of bags required if each bag is 8 feet long and 6 inches in diameter.

Solution

From Table 6.1, the recommended maximum V for flour is 2.5 ft/min. Thus, the net cloth area is

$$A = \frac{Q}{V} = \frac{40,000}{2.5} = 16,000 \text{ ft}^2$$

From Table 6.3, we might specify five compartments. To meet the design filtering velocity when filtering with one compartment off-line, there must be 4000 ft^2 of fabric in each compartment, for 20,000 ft^2 total. The fabric area of one bag is approximately

$$\pi(0.5)8 = 12.6 \text{ ft}^2/\text{bag}$$

Thus, the total number of bags is

$$\frac{20,000}{12.6} = 1587 \text{ bags}$$

Physical considerations of arranging an equal number of bags in each compartment might dictate a slightly larger total number of bags.

As mentioned previously, we must rely on experience to a large extent in specifying some of the key design parameters for compartmented baghouses. However, the total airflow rate and the maximum allowable pressure drop are interdependent, and are related to the number of compartments, the filtration time, and the cleaning time. Crawford (1976) has developed a detailed mathematical model to predict the filtration time and a cleaning cycle when given a maximum pressure drop constraint. (Proprietary computer models are used by the major baghouse vendors.) However, we will now discuss a simpler approach that is easier to use and that gives comparable results to some of the more complex methods.

Consider a baghouse with N compartments in parallel filtering air with a total flow rate Q. The compartments are cleaned in sequence (1, 2, . . . , N). At no time can the pressure drop across the baghouse exceed a given maximum value ΔP_m. Because the compartments are in parallel, all of the compartments will experience ΔP_m simultaneously. Ideally, each compartment handles the same total volume of air over one entire filtration time. Thus, the total dust accumulated in a compartment (just before its cleaning) can be calculated by assuming that it has experienced an average flow rate over the entire cycle.

In reality, at any given time, the flow rate through each compartment will differ from the others because each compartment will have a different amount of dust accumulated in it at that time in the cycle. The flow rate through the cleanest compartment will be the greatest, and that through the dirtiest compartment will be the smallest. Furthermore, the relative flow distribution through the compartments changes during the cycle as newly cleaned compartments come on-line. However, to calculate dust load, we can assume that the flow rates through all compartments are the same—equal to the total flow rate divided by the number of compartments on-line. When all compartments are filtering, the flow rate through one compartment is

$$Q_N = \frac{Q}{N} \qquad (6.13)$$

where

Q_N = flow rate through one compartment, cfm

Q = total flow rate through the baghouse, cfm

When one compartment is off-line for cleaning, the flow rate through each of the other compartments is

$$Q_{N-1} = \frac{Q}{N-1} \qquad (6.14)$$

Similarly, the filtering velocities corresponding to Eqs. (6.13) and (6.14), respectively, are

$$V_N = \frac{Q_N}{A_c} = \frac{Q}{NA_c} \qquad (6.15)$$

and

$$V_{N-1} = \frac{Q_{N-1}}{A_c} = \frac{Q}{(N-1)A_c} \qquad (6.16)$$

where

V_N, V_{N-1} = average filtering velocities, ft/min

A_c = cloth area in one compartment, ft^2

Note that V_{N-1} is, in fact, the design filtering velocity.

From Figure 6.5, we can see that ΔP_m occurs at the end of the cleaning time of an arbitrary compartment, say $j - 1$, just before placing compartment $j - 1$ back into service. At that time, compartment j (which is next to be cleaned) has been on-line for a time t_j, equal to the filtration time minus one run time, or

$$t_j = t_f - t_r \qquad (6.17)$$

where t_j = time that compartment j has been on-line (just before compartment $j - 1$ is returned to service), min

Substituting Eq. (6.12) for t_f into Eq. (6.17), we obtain

$$t_j = t_f - t_r = (N - 1)(t_r + t_c) \tag{6.18}$$

During the time t_j, the cloth in compartment j has accumulated an areal dust density W_j, given by

$$W_j = (N - 1)(V_N L t_r + V_{N-1} L t_c) \tag{6.19}$$

where

W_j = areal dust density, lb_m/ft^2

L = particulate loading, lb_m/ft^3

Thus, the filter drag in compartment j is

$$S_j = K_e + K_s W_j \tag{6.20}$$

where S_j = filter drag, in. H_2O-min/ft

Up to this point, we have used the average filtering velocity satisfactorily. However, in calculating the pressure drop, we must know the actual filtering velocity V_j in compartment j at time t_j. As mentioned previously, V_j is lower than the average velocity V_{N-1} because compartment j has the most dust in it at time t_j.

At first glance, it might seem that V_j would depend on several parameters. However, when we programmed Crawford's (1976) equations for a detailed mathematical description of a multicompartment baghouse and tested numerous cases, we found that the ratio of V_j/V_{N-1} depends almost entirely on only the number of compartments. This relationship is given in Table 6.4. Thus, we can estimate the actual filtering velocity as

$$V_j = f_N V_{N-1} \tag{6.21}$$

where

V_j = actual filtering velocity in compartment j at time t_j, ft/min

f_N = correction factor (from Table 6.4)

Thus, finally, the pressure drop ΔP_j, which is equal to the maximum allowable pressure drop ΔP_m, is

$$\Delta P_j = \Delta P_m = S_j V_j \tag{6.22}$$

where ΔP_j = pressure drop for compartment j, in. H_2O

If we know the number of compartments, the maximum allowable pressure drop, and the time required for cleaning, our objective is to select a run time and filtration time such that ΔP_m is approached but

Table 6.4   Ratio of Actual Filtering Velocity V_j to Average Filtering Velocity V_{N-1} in a Multicompartment Baghouse

Total Number of Compartments N	$f_N = V_j/V_{N-1}$
3	0.87
4	0.80
5	0.76
7	0.71
10	0.67
12	0.65
15	0.64
20	0.62

Notes: V_{N-1} is the average filtering velocity calculated from Eq. (6.16).
  V_j is the actual filtering velocity in the dirtiest compartment (next to be cleaned) while the former dirtiest compartment is off-line for cleaning.

not exceeded. With these variables known, we can solve Eqs. (6.16), (6.21), (6.22), (6.20), (6.19), and (6.18), in that order. Conversely, if we are starting with a desired filtration time t_f, we can calculate the maximum pressure drop that must be supplied by solving Eq. (6.18) for t_r, and then solving Eqs. (6.19), (6.20), (6.21), and (6.22) in sequence.

▶ ▶ ▶ ▶ ▶ ▶ ▶ ▶ ▶

Example 6.3

For the baghouse of Example 6.2, assume that the filter drag model holds with $K_e = 1.00$ in. H_2O-min/ft and $K_s = 0.003$ in. H_2O-min-ft/grain. Also, assume that a compartment can be cleaned and returned to service in 4 minutes. For a filtration time of 60 minutes, calculate the maximum pressure drop that must be supplied.

Solution

With all five compartments in service, $V_N = 40,000/20,000 = 2.0$ ft/min. With only four compartments on-line, $V_{N-1} = 40,000/16,000 = 2.5$ ft/min. From Eq. (6.12) for a cleaning time of 4 minutes, we obtain t_r as

$$t_r = \left(t_f + t_c\right)/N - t_c$$

$$= \frac{60 + 4}{5} - 4 = 8.8 \text{ min}$$

Next, we calculate the accumulated areal dust density from Eq. (6.19). Thus,

$$W_j = 4 \times \left(\frac{10 \text{ gr}}{\text{ft}^3}\right)\left[\left(\frac{2.0 \text{ ft}}{\text{min}}\right)(8.8 \text{ min}) + \left(\frac{25 \text{ ft}}{\text{min}}\right)(4 \text{ min})\right]$$

$$= 1104 \text{ gr/ft}^2$$

The maximum allowable filter drag from Eq. (6.20) is

$$S_j = 1.00 + 0.003(1104)$$
$$= 4.31 \text{ in. } H_2O\text{-min/ft}$$

Using $f_N = 0.76$ from Table 6.4, we obtain V_j from Eq. (6.21) as

$$V_j = (0.76)(2.5) = 1.9 \text{ ft/min}$$

Finally

$$\Delta P_m = (4.31)(1.9) = 8.2 \text{ in. } H_2O$$

◀ ◀ ◀ ◀ ◀ ◀ ◀ ◀

PULSE-JET BAGHOUSES

Pulse-jet baghouses, introduced only about 35 years ago, have now captured about one-half of the industrial air filtration market. The pulse-jet method is shown schematically in Figure 6.6. In this method, air is filtered through the bags from the outside to the inside. A cage inside each bag prevents the bag from collapsing. The bags are closed at the bottom and clamped into a clean air plenum at the top. The bags are cleaned by short (30–100 millisecond) blasts of high-pressure (90–100 psi) air. The pulse of air is directed through a venturi and sets up a shock wave that flexes the bags and snaps off the collected particulate layer. Each bag is pulsed every few minutes.

A major advantage of the pulse-jet method is that it allows the cleaning of some of the bags while dusty air continues to flow through the baghouse. There are no compartments and thus no extra bags, which reduces the size and cost. Because bags are replaced from the top, there is no need to provide walkways between rows of bags, further reducing the size of the baghouse. Also, there are no moving parts exposed to the dusty air. Another major advantage of pulse-jet cleaning is that felted fabrics can be used at much higher air-to-cloth ratios, typically two to three times higher than with conventional cleaning methods. Table 6.5 gives some data on filtering velocities for pulse-jet applications. The higher filtering velocities reduce the net cloth area required, and further reduce size and capital costs. In recent years, there has been a trend toward significantly lower air-to-cloth ratios than those shown in Table 6.5, especially for applications in the conservative electric power industry. Despite this fact, pulse-jet baghouses still typically require only about half the footprint of reverse-air baghouses, an important consideration in space-limited sites. In addition, use of a lower V (3–4 ft/min for fly-ash collection) results in lower average pressure loss, less-frequent cleanings, longer bag life, and lower outlet emissions (Belba et al. 1992).

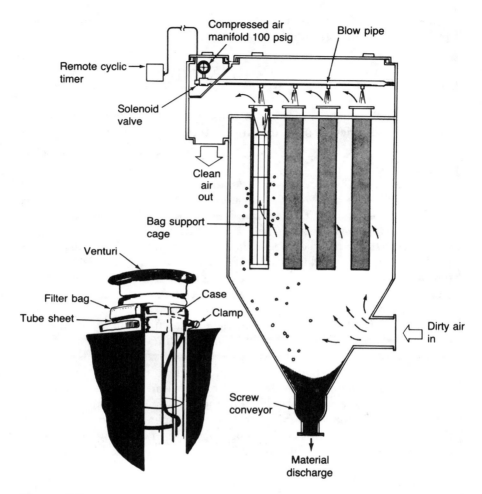

Figure 6.6
Schematic diagram of a pulse-jet baghouse.

▶ ▶ ▶ ▶ ▶ ▶ ▶ ▶ ▶ ▶

Example 6.4

Estimate the number of bags required in a pulse-jet baghouse to filter the same airstream as in Example 6.2.

Solution

From Table 6.5, we choose a V of 13 ft/min. Therefore, the net cloth area is

$$A = \frac{40,000}{13} = 3077 \text{ ft}^2$$

Since there are no compartments, the net area is equal to the total area, and the number of bags is

$$\frac{3077}{12.6} = 245 \text{ bags}$$

Again, the physical positioning of equal numbers of bags in rows and columns might dictate a slightly larger total number of bags.

Pulse-jet methods have some disadvantages. Because there is still gas flow into the bag when a pulse of cleaning air ripples the fabric, much of the same dust that is knocked off the bag is re-suspended and goes right back onto the same bag or one of its neighbors. In fact, the net dust removed per pulse can be as low as 1% of the total amount of dust on the bag (Leith and Ellenbecker 1980). When using pulse-jet cleaning, the length of the bags is limited because the pulse of air quickly dissipates its energy. Initially, even with shorter bags (8 feet long), the bottom third of the bags was not properly cleaned. This problem was alleviated in the late 1970s with the development of a patented diffuser tube. (A diffuser tube is a perforated metal cylinder that fits inside the bag and forces the pulse of air to travel farther before being dissipated.) Turner (1981), has reported good, uniform cleaning using the diffuser tube, resulting in lower overall pressure drops and increased bag life.

Particles penetrate through pulse-jet systems in a different way than in conventional systems; most of the dust penetration occurs when the fabric snaps back onto its support cage after a cleaning pulse (Leith and First 1977). Such penetration can be greatly reduced by gradually

Table 6.5   Maximum Filtering Velocities* for Various Dusts or Fumes in Pulse-Jet Baghouses

Dusts or Fumes	Maximum Filtering Velocity, cfm/ft^2 or ft/min
Carbon, Graphite, Metallurgical Fumes, Soap, Detergents, Zinc Oxide	5–6
Cement (Raw), Clay (Green), Plastics, Paint Pigments, Starch, Sugar, Wood Flour, Zinc (Metallic)	7–8
Aluminum Oxide, Cement (Finished), Clay (Vitrified), Lime, Limestone, Gypsum, Mica, Quartz, Soybean, Talc	9–11
Cocoa, Chocolate, Flour, Grains, Leather Dust, Sawdust, Tobacco	12–14

*Decrease velocities by 1 ft/min if the loading is great or if the particles are very small.
Adapted from Danielson, 1973; Theodore and Buonicore, 1976.

reducing the air pressure at the end of a cleaning pulse instead of using the usual sharp cutoff (Leith, Gibson, and First 1978). However, such pulse modification requires an increased usage of compressed air.

Compressed air usage is a major operating expense for pulse-jet baghouses. Typically, the volumetric flow rate of the compressed air is about 0.2–0.8% of the flow rate of the filtered air when both are corrected to the same temperature and pressure. The cost for compression power can equal that of the main fan power. The equation for calculating compressor power is

$$\dot{w} = \frac{1}{\eta}\frac{\gamma}{\gamma-1}P_1 Q_1 \left[\left(\frac{P_2}{P_1}\right)^{(\gamma-1)/\gamma} - 1\right] \tag{6.23}$$

where

$\dot{w}$ = actual compressor power, kW

η = compressor efficiency

γ = ratio of heat capacities (C_p/C_v) of the gas being compressed (for air, $\gamma = 1.4$)

P_1, P_2 = initial and final pressures (absolute), kPa

Q_1 = volumetric flow into the compressor (at inlet conditions), m^3/s

▶ ▶ ▶ ▶ ▶ ▶ ▶ ▶ ▶

Example 6.5

A pulse-jet baghouse filters 20.0 m^3/s of air at 150 C and 1 atm (101.3 kPa). Assuming a ratio (after correcting to STP) of 0.7% for compressed air to filtered air, calculate the compressor power required (in kW) if the air is to be supplied at 100 psig (792 kPa). Assume the compressor efficiency is 50%.

Solution

$$Q_1 = 20.0\frac{m^3}{s} \times \frac{298\ K}{423\ K} \times 0.007 = 0.0986\frac{m^3}{s}$$

$$\dot{w} = \frac{1}{0.50}\left(\frac{1.4}{0.4}\right)101.3\ kPa \times \frac{1000\ N/m^2}{1\ kPa}$$

$$\times \frac{0.0986\ m^3}{s}\left[\left(\frac{792}{101.3}\right)^{0.4/1.4} - 1\right]$$

$$= 6.99(10)^4\,(0.7988)\frac{N\text{-}m}{s}$$

$$= 5.589(10)^4\,\frac{N\text{-}m}{s} \times \frac{1\ kJ}{1000\ N\text{-}m} \times \frac{1\ kW}{1\ kJ/s}$$

$$= 55.9\ kW$$

◀ ◀ ◀ ◀ ◀ ◀ ◀ ◀ ◀

Pressure drop through a pulse-jet baghouse has been empirically related to filtering velocity, pulse pressure, and areal dust density deposited during one filtration cycle (Leith and First 1977). One model, developed for fly ash collected on polyester bags, is

$$\Delta P = 2.72 \Delta W^{0.45} P^{-1.38} V^{2.34} \qquad \textbf{(6.24)}$$

where

ΔP = pressure drop across the bag and dust deposit, cm H_2O

V = filtering velocity, cm/s

ΔW = areal dust density added between two consecutive cleanings ($\Delta W = LVt_f$), mg/cm^2

P = pulse pressure, atm

Leith and Ellenbecker (1980) developed a theoretical model to predict the equilibrium pressure drop in a pulse-jet baghouse. Although the model might need further refinement, it appears to work well for at least one set of data. At the present time, however, prediction of operating pressure drop in a pulse-jet baghouse is largely empirical. Based on data from 35 pulse-jet baghouses installed on pulverized coal-fired boilers, Belba, Grubb, and Chang (1992) developed an empirical relationship to predict baghouse pressure drop based on the air-to-cloth ratio

$$\Delta P = 1.7 \; V \pm 40\% \qquad \textbf{(6.25)}$$

where

ΔP = flange-to-flange pressure drop, in. H_2O

V = air-to-cloth ratio (filtering velocity), ft/min

Equation (6.25) was developed from actual operating data covering the ranges $1.6 < V < 6.7$ ft/min, and $3.0 < \Delta P < 10.4$ in. H_2O.

Engineers are continually striving to find ways to achieve operating cost savings, such as reducing the power required by the pulse air compressor. The power for the pulse air compressor can be reduced by reducing pulse pressure or by increasing the time between pulses. However, from Eq. (6.24), we can see that either of these actions will result in an increased pressure drop, which increases the main fan power and can negate the savings in compressor power. Furthermore, at high values of V or ΔW, there is a risk of an uncontrollable increase in pressure drop (Bakke 1974). Caution must be exercised when adjusting these variables on an operating pulse-jet baghouse.

OTHER CONSIDERATIONS

The major design considerations for baghouses are type of cleaning method, type of fabric, fabric area, and total pressure drop. Other important considerations are listed in Table 6.6. High humidity in a warm exhaust gas is a problem, especially in the winter. Condensation of moisture on the fabric and dust layer rapidly fills the void spaces and greatly increases pressure drop.

As mentioned previously, the use of fabric filtration to control emissions from coal-fired power plants has grown dramatically during the last three decades. All types of baghouses (reverse-air, shaker, and pulse-jet) are used. Emissions are typically well below the NSPS standard of 0.03 lb/million Btu, and in some cases are less than 0.01 lb/million Btu. The designs tend to be conservative (low V) to ensure that power generation is not limited by baghouse problems. Typical values of V being used in power plant baghouses are 1.5–2.0 ft/min for reverse-air, 2.5–3.0 ft/min for shaker, and 3–5 ft/min for pulse-jet applications. Typical values of average pressure drop are 4–8 in. H_2O.

Table 6.6   Some Design Considerations for Baghouse Systems

Consideration	Comments
Temperature and Humidity	Fabrics have different maximum allowable temperatures. Operation above these temperatures can rapidly degrade bags. Low temperatures can cause condensation of acid and/or blinding of the fabric with wet dust. Wet particulate matter can bridge over in hoppers. As temperatures increase, both gas viscosity and gas volumetric flow rate increase. Both tend to increase pressure drop requirements.
Chemical Nature of Gas	Different fabrics have different resistances to acids or alkalies.
Fire/Explosion	Some fabrics are flammable; some dusts are explosive.
Bag Arrangement	It is important to consider maintenance; arranging bags in straight rows is better than dense packing. Providing walkways every few rows might be a very good investment. Spacing of a few inches between bags is a sufficient operating clearance.
Dust Handling	The dust removal rate (mass and volume), conveyor system (pneumatic tube or screw conveyor), and hopper slope (dust must flow out by gravity) should all be considered.
Fan Location	A clean-air-side fan (a "pull-through" baghouse) saves on maintenance of the fan and allows the use of a more efficient fan with backward-curved blades. However, this method requires an airtight structure. Furthermore, a stronger structure is generally required.

According to Belba et al. (1992), bag life decreases as coal sulfur content increases and as the filtering velocity increases. Glass fiber bags have shorter useful lifetimes than some of the synthetic fiber bags (1–2 versus 3–5 years). Individual bags fail at an average annual rate of about 1% of installed bags.

Innovations in baghouse technology include electrostatically enhanced dust collection and sonic-assisted cleaning. Electrostatic enhancement involves running a charging wire down the central axis of each bag. Sonic cleaning uses high-decibel, low-frequency blasts from air horns installed inside the baghouse. Both techniques enable better dust collection at lower pressure drops.

In addition to collecting fly ash, there have been a number of applications where baghouses have been used in conjunction with dry scrubbing—injection of a slurried or powdered sulfur dioxide absorbent (such as lime or sodium bicarbonate)—to control sulfur dioxide emissions simultaneously with fly-ash emissions. Fabric filtration has also found favor in dry-scrubbing applications for HCl on medical and/or hazardous waste incinerators.

♦ 6.4 Costs ♦

The annualized cost of owning and operating a fabric filter baghouse can be very high. The (annualized) capital cost due to depreciation and lost interest is about 35% of the annual operating cost (AOC); fan power costs are about 15%, replacement bag purchases are about 15%, and operating and maintenance labor are about 35% of the AOC (Smith 1974.) Sometimes the collected dust can be recycled into the product or, perhaps, sold to another industry. In those cases, a credit should be taken in the calculation of operating costs. If the dust cannot be reused, then disposal costs must be considered. These can be significant especially if the dust contains any hazardous components, such as leachable heavy metals.

The capital cost for a baghouse system can be estimated based on gross cloth area. The cost also depends on whether the baghouse is a shaker, reverse-air, or pulse-jet baghouse, whether it is made of mild or stainless steel, how big it is, and whether or not it is externally insulated. Furthermore, standardized baghouses, which are constructed from pre-built modules, are much less expensive than custom-designed systems. Turner et al. (1998) have reported the following correlations for baghouse prices (in December 1998 dollars). The basic baghouse is considered to be constructed of mild steel and is uninsulated. The equations are

$$BBP = a_1 + b_1 GCA \tag{6.26}$$

where

BBP = basic baghouse price (excluding bags), 1998 dollars

GCA = gross cloth area, ft^2

a_1, b_1 = constants (from Table 6.7)

$$SSA = a_2 + b_2 GCA \qquad \textbf{(6.27)}$$

where

SSA = stainless-steel add-on, 1998 dollars

a_2, b_2 = constants (from Table 6.7)

$$INS = a_3 + b_3 GCA \qquad \textbf{(6.28)}$$

where

INS = insulation add-on, 1998 dollars

a_3, b_3 = constants (from Table 6.7)

The total purchase price (P) is the baghouse price plus the cost of the bags. Tables 6.8 and 6.9 list some data for the costs of various bags and cages.

The delivered equipment cost (DEC) is higher than the purchase price because the purchaser must pay for instruments and controls, sales tax, and freight. These charges can amount to 18% of the DEC (Neveril, Price, and Engdahl 1978a,b); that is,

$$DEC = 1.18 \, (P) \qquad \textbf{(6.29)}$$

Furthermore, the installation costs (both direct and indirect) are approximately equal to the DEC. Turner et al. (1998) report that for fabric filtration systems, the total installed cost (TIC) equals 2.19 times the delivered equipment cost, or

Table 6.7   Values of Cost Coefficients for Baghouses for Use in Equations (6.26)–(6.28)

Coefficient[1]	Shaker, small[2]	Shaker, large[3]	Reverse-Air[4]	Pulse-Jet[5]
		Type of Baghouse		
(BBP) a_1	$23,040	$96,230	$37,730	$13,540
b_1	$5.79/ft^2	$3.33/ft^2	$4.62/ft^2	$8.88/ft^2
(SSA) a_2	$13,970	$51,280	$26,220	$1,811
b_2	$2.68/ft^2	$1.43/ft^2	$2.00/ft^2	$4.25/ft^2
(INS) a_3	$6,295	$26,330	$13,010	-$195
b_3	$1.23/ft^2	$0.57/ft^2	$0.89/ft^2	$2.74/ft^2

[1]All prices are for December 1998; all b_1 coefficients are based on gross cloth area.
[2]Applicable range of cloth area is 5,000 to 30,000 ft^2—do not use outside of range.
[3]Applicable range of cloth area is 30,000 to 70,000 ft^2—do not use outside of range.
[4]Applicable range of cloth area is 10,000 to 100,000 ft^2—do not use outside of range.
[5]Applicable range of cloth area is 4,000 to 20,000 ft^2—do not use outside of range.

Adapted from Turner et al., 1998.

Table 6.8 Fabric Filter Bag Prices, $/ft^2 (in 1998 $)

Type of Cleaning	Bag Diameter (inches)	Type of Material[a]									
		PE	PP	NO	HA	FG	CO	TF	P8	RT	NX
Pulse jet, TR[b]	4-1/2 to 5-1/8	0.75	0.81	2.17	1.24	1.92	NA	12.21	4.06	2.87	20.66
	6 to 8	0.67	0.72	1.95	1.15	1.60	NA	9.70	3.85	2.62	NA
Pulse jet, BR[b]	4-1/2 to 5-1/8	0.53	0.53	1.84	0.95	1.69	NA	12.92	3.60	2.42	16.67
	6 to 8	0.50	0.60	1.77	0.98	1.55	NA	9.00	3.51	2.30	NA
Shaker, Strap top	5	0.63	0.88	1.61	1.03	NA	0.70	NA	NA	NA	NA
Shaker, Loop top	5	0.61	1.01	1.53	1.04	NA	0.59	NA	NA	NA	NA
Reverse air with rings	8	0.63	1.52	1.35	NA	1.14	NA	NA	NA	NA	NA
	11-1/2	0.62	NA	1.43	NA	1.01	NA	NA	NA	NA	NA
Reverse air w/o rings	8	0.44	NA	1.39	NA	0.95	NA	NA	NA	NA	NA
	11-1/2	0.44	NA	1.17	NA	0.75	NA	NA	NA	NA	NA

NA = Not applicable
[a]Materials: PE = 16-oz. polyester; PP = 16-oz. polypropylene; NO = 14-oz. Nomex; HA = 16-oz. homopolymer acrylic; FG = 16-01 fiberglass with 10% Teflon; CO = 9-oz. cotton; TF = 22-oz. Teflon felt; P8 = 16-oz. P84; RT = 16-oz. Ryton; NX = 16-oz. Nextel
[b]Bag removal methods: TR = Top bag removal (snap in); BR = Bottom bag removal
Note: For shakers and reverse-air baghouses, all bags are woven. For pulse-jet baghouses, all bags are felts except for fiberglass, which is woven. All prices are for finished bags, and prices can vary from one supplier to another.

Adapted from Turner et al., 1998.

Table 6.9 Prices for Pulse-Jet Bag Cages, $/Cage (in 1998 $)

	4½ in. × 8 ft cages	5⅝ in. × 10 ft. cages
mild steel	$ = 4.26 $e^{0.0522A}$ in 100 cage lots	$ = 3.08 $A^{0.525}$ in 100 cage lots
	$ = 3.42 $e^{0.0593A}$ in 500 cage lots	$ = 2.52 $A^{0.569}$ in 500 cage lots
stainless steel	$ = 4.85 + 1.57 A in 100 cage lots	$ = 8.85 + 1.23 A in 100 cage lots
	$ = 3.85 + 1.57 A in 500 cage lots	$ = 8.85 + 1.23 A in 500 cage lots

Notes: A = single-bag fabric area, ft^2
For snap-band collar with built-in venturi, add $6.00 per cage for mild steel and $13.00 per cage for stainless steel.

Adapted from Turner et al., 1998.

$$\text{TIC} = 2.19 \text{ DEC} \tag{6.30}$$

Example 6.6 illustrates the use of the preceding equations.

Example 6.6

Estimate the total installed cost (in 1998 dollars) of the baghouse of Example 6.2. Assume the baghouse is uninsulated and is made from mild steel. Assume polyester bags with strap tops are used.

Solution

BBP =	23,040 + 5.79(20,000) =	$138,840
SSA =		0
INS =		0
Total:		$138,840
Bags =	$0.63/ft² × 20,000 ft² =	$ 12,600
P =		$151,440
DEC =	1.18 × $151,440	$178,699
TIC =	2.19 × $178,699	$391,350

The 1998 total installed cost is approximately $391,000.

◆ PROBLEMS ◆

6.1 From the following test data, estimate the values of K_s and K_e for the filter drag model.

Limestone Dust Loading   1.00 g/m³
Fabric Area   1.00 m²
Air Flow Rate   0.80 m³/min

Time, min	5	10	15	20	25	30
Filter ΔP, Pa	330	490	550	600	640	700

Give your answers in units of (Pa-min-m)/g and (Pa-min)/m, respectively.

6.2 Engineer Smith was conducting a test on a certain filter cloth in conjunction with the design of a baghouse. Air with a flow rate of 1.0 ft³/min and a temperature of 27 C was blown through a 0.5 ft² piece of filter cloth. The dust loading was 3.0 grains/ft³. Engineer Smith set up instruments to automatically record the ΔP across the cloth as a function of time, and then turned on the experiment and left the room. On returning, Smith found that the recorder had malfunctioned slightly and had recorded only the following three measurements. Smith's boss is in a hurry for an answer, so the experiment cannot be repeated. Make your best estimate of a filter drag model (numerical values and units for K_e and K_s) for this case.

Data	
Time, hours	ΔP, kPa
0	1.0
1.0	3.0
2.0	4.0

6.3 How does an increase in gas temperature affect pressure drop in a baghouse, assuming the mass flow rate of particles and the molar flow rate of gas are constant?

6.4 Using the data of Problem 6.1, design a shaker baghouse to filter 4000 m^3/min of air that contains 1.50 g/m^3 of limestone dust. Assume that the cleaning time for one compartment is 3.0 minutes and that the available ΔP_m is 2000 Pa. In your design, specify the filtering velocity, the number of compartments, the cloth area per compartment, and the filtration time t_f.

6.5 For $K_e = 0.577$ in. H_2O-min/ft and $K_s = 0.00986$ in. H_2O-min-ft/gr, design a reverse-air baghouse to filter 20,000 cfm of air with 2.5 gr/ft³ of flour. Assume a cleaning time of 3.0 minutes and a filtration time of 60 minutes. In your design, specify the number of compartments, the filtering velocity, the cloth area per compartment, and the total number of bags required if each bag is 10 ft long and 1 ft in diameter. Also, specify the maximum pressure drop that will be experienced during the run.

6.6 What fabric or fabrics might be best suited for filtering particles from (a) a 180 F gas stream that contains ammonia, and (b) a 250 F gas stream that contains SO_2?

6.7 A pulse-jet baghouse is to be designed to filter the air of Problem 6.5. Calculate the number of bags required if each bag is 8 ft long and 6 in. in diameter.

6.8 Calculate the compressor power for the baghouse of Problem 6.7. Assume the ratio of pulse air to filtered air is 0.6%. Calculate the fan power for this baghouse assuming an average ΔP of 4.0 in. H_2O and a fan efficiency of 60%.

6.9 Estimate the permeability for the dust of Problem 6.5 if the bulk density of the dust layer is 12.0 kg/m^3 and the viscosity of the air is 0.050 lb_m/ft-hr. Give your answer in ft^2.

6.10 Calculate the numerical values of K_e and K_s in Problem 6.5 in SI units.

6.11 A pulse-jet baghouse is desired for a finished cement plant. Calculate the number of bags required to filter 8000 m^3/min of air with a dust loading of 3.0 g/m^3. Each bag is 3.0 m long with a 0.3 m diameter. If the average pressure drop is 1.0 kPa and the main fan is 60% efficient, calculate the fan power in kW. If the pulse air volumetric flow rate is 0.5% of the filter airflow rate and the pulse air pressure is 6.0 atm, calculate the power drawn by a 50% efficient compressor (in kW).

6.12 Estimate the total installed cost (in 1998 dollars) for an insulated, mild-steel reverse-air baghouse. The baghouse has 11 compartments with 8000 ft^2 in each compartment, and has 8-inch glass fiber bags, with rings.

6.13 Estimate the total installed cost (in 1998 dollars) of the baghouse of Problem 6.11, assuming it is mild steel, is uninsulated, and uses 6-inch Nomex bags on cages that snap in from the top, with built-in venturis.

6.14 Estimate the 1998 total installed cost for the baghouse of Problem 6.4, assuming it is mild steel and is insulated. Assume the baghouse uses polypropylene bags with loop tops.

6.15 A certain pilot-scale baghouse follows the filter drag model

$$S = K_e + K_s W$$

with

$$K_e = 0.50 \frac{\text{in. H}_2\text{O}}{\text{ft/min}}$$

and

$$K_s = 0.001 \frac{\text{in. H}_2\text{O-ft-min}}{\text{gr}}$$

a. Predict the pressure drop for a full-scale baghouse after 30 minutes of operation on an airstream with a dust loading of 30 gr/ft^3 and an air-to-cloth ratio of 3.0 ft/min.

b. Calculate the number of bags required in the baghouse to clean the airstream, which is flowing at 100,000 acfm. Each bag is 1 ft in diameter and 15 ft long.

6.16 A manufacturing plant has been required to install a particulate control system to remove 93% of its emissions. Your tests of the dust show a size distribution as follows:

Size Range, μm	Weight, %	Size Range, μm	Weight, %
0–2	20	20–40	25
2–10	20	40–70	15
10–20	10	>70	10

The total airflow is 10,000 cfm at 90 F and 1 atm. The mass loading is 50 gr/ft^3. The particle density is the same as that of water.

a. Based on a visual inspection of the particulate distribution, recommend a system that has a good chance of solving this problem economically, and state your reasons.

b. Make preliminary design calculations to size the major pieces of equipment for the system you have recommended. (Note: This is not a detailed design; make reasonable assumptions where necessary.)

Notes

1. If the system includes a cyclone, specify only the cyclone diameter, and estimate the efficiency and the pressure drop.

2. If the system includes an ESP, specify only the efficiency and the total plate area.

3. If the system includes a baghouse, use the values of K_e and K_s from Problem 6.15 and specify the type of baghouse, the filtering velocity, and the fabric area, and estimate the pressure drop.

4. Estimate the size of the fan (hp) required for your system, assuming a ΔP in the ducts of 2.0 in. H$_2$O.

5. Do not estimate costs in this problem.

6.17 Specify the design filtering velocity for a shaker-type baghouse that is to be used to clean soap dust from a stream of air flowing at 50,000 acfm. Specify the number of compartments, and calculate the average filtering velocity when one compartment is off-line for cleaning and when all compartments are on-line.

6.18 Calculate the number of bags needed in a pulse-jet baghouse to filter 1200 m^3/min of air contaminated with limestone dust. The bags are 12 ft long and 8 in. in diameter. Specify an arrangement for the bags (number of rows and columns).

6.19 Based on the following test data for the fabric with $L = 5$ g/m^3 and $V = 1$ m/min, a reverse-air baghouse is being designed. The baghouse, which treats 1000 m^3/min of dusty air (with an actual loading of 4.6 g/m^3), contains three compartments with a cloth area of 500 m^2 per compartment. Determine the maximum allowable pressure drop for a filtration time of 80 minutes. Assume that a compartment can be cleaned and returned to service in 4 minutes.

Test Data			
Time, min	ΔP, Pa	Time, min	ΔP, Pa
0	100	30	680
5	350	40	825
10	460	60	1000
20	600		

6.20 A reverse-air baghouse is to be designed to remove fly ash from a power plant exhaust gas stream flowing at 300,000 acfm. The airstream is at 300 F and has a significant content of SO_2 gas. Specify (a) the type of fabric, (b) the superficial filtering velocity, ft/min, (c) number of compartments, (d) cloth area per compartment, ft^2.

6.21 A pulse-jet baghouse is being designed to filter PM from 20,000 acfm (at 77 F and 1 atm) of air coming from a commercial tortilla-making factory (the plant processes corn and wheat).

a. Specify the number of bags required if each bag is 10 ft long and 8 in. in diameter.

b. Calculate the compressor power (in kW) if the pulse air is supplied at 110 psig, and the volume used is 0.6% of the filtered air. The compressor is 50% efficient.

c. Calculate the fan power (in kW) if the average pressure drop is 6 in. H_2O. Assume the fan is 70% efficient.

6.22 A shaker-type baghouse is to be designed to remove corn dust from the exhaust airstream of a cattle-feed processing plant feed dryer. The air is flowing at 145,000 acfm, at 1 atm and 220 F. Specify (a) type of fabric, (b) superficial filtering velocity, ft/min, (c) number of compartments, (d) cloth area per compartment, ft^2.

6.23 A cyclone followed by a pulse-jet baghouse has been proposed to filter PM from 24,000 acfm (at 77 F and 1 atm) of air coming from a sugar refinery.

a. Assuming a Lapple standard conventional cyclone, specify the body diameter required to achieve an inlet velocity of approximately 6000 ft/min.

b. Calculate the system fan power (in hp) if the average pressure drop is 4 in. H_2O in the cyclone, 2 in. H_2O in the duct work, and 6 in. H_2O in the baghouse. Assume a typical fan efficiency.

c. Calculate the daily operating cost of the fan just based on electricity usage (electricity costs $0.09/kWh).

d. Calculate the cost of running the compressor for the pulse-jet baghouse, assuming that the ratio of pulse air to filtered air is 0.008, and the compressor supplies air at 105 psig.

REFERENCES

Bakke, E. "Optimizing Filtration Parameters," *Journal of the Air Pollution Control Association, 24*(12), December 1974.

Belba, V. H., Grubb, W. T., and Chang, R. "The Potential of Pulse-Jet Baghouses for Utility Boilers—Part I: A Worldwide Survey of Users," *Journal of the Air and Waste Management Association, 42*(2), February 1992, p. 209.

Buonicore, A. J., and Davis, W. T., Eds. *Air Pollution Engineering Manual.* New York: Van Nostrand Reinhold, 1992.

Crawford, M. *Air Pollution Control Theory.* New York: McGraw-Hill, 1976.

Cushing, K. M., Merritt, R. L., and Chang, R. L. "Operating History and Current Status of Fabric Filters in the Utility industry," *Journal of the Air and Waste Management Association, 40*(7), July 1990, p. 1051.

Danielson, J. A., Ed. *Air Pollution Engineering Manual* (2nd ed.), AP–40. Washington, D.C.: U.S. Environmental Protection Agency, May 1973.

Dennis, R., and Klemm, H. "Modeling Coal Fly Ash Filtration With Glass Fabrics," *Third Symposium on Fabric Filters for Particulate Collection,* EPA–600/7–78–087. Washington, D.C.: U.S. Environmental Protection Agency, June 1978.

Kraus, M. N. "Baghouses: Separating and Collecting Industrial Dusts," *Chemical Engineering, 86*(8), April 9, 1979.

Leith, D., and Ellenbecker, M. J. "Theory for Pressure Drop in a Pulse-Jet Cleaned Fabric Filter," *Atmospheric Environment, 14*(7), 1980.

Leith, D., and First, M. W. "Pressure Drop in a Pulse-Jet Fabric Filter," *Filtration,* September 14, 1977.

Leith, D., Gibson, D. D., and First, M. W. "Performance of Top and Bottom Inlet Pulse Jet Fabric Filters," *Journal of the Air Pollution Control Association, 28*(7), July 1978.

Leutbecher, A. C. "Australian Experience, Filtration of Flyash from Very Low Sulfur Coals," in *Third Symposium on Fabric Filters for Particulate Collection,* EPA–600/7–78–087. Washington, D.C.: Environmental Protection Agency, 1978.

McIlvaine, R. W. "Market Trends for Air Pollution Control Equipment," *Journal of the Air Pollution Control Association, 33*(3), March 1983.

Neveril, R. B., Price, J. U., and Engdahl, K. L. "Capital and Operating Costs of Selected Air Pollution Control Systems–I," *Journal of the Air Pollution Control Association, 28*(8), August 1978a.

———. "Capital and Operating Costs of Selected Air Pollution Control Systems—V," *Journal of the Air Pollution Control Association, 28*(12), December 1978b.

Smith, G. L. "Engineering and Economic Considerations in Fabric Filtration," *Journal of the Air Pollution Control Association, 24*(12), December 1974.

Theodore, L., and Buonicore, A. J. *Industrial Air Pollution Control Equipment for Particulates.* Cleveland: CBS Press, 1976.

Turner, J. H. *New Pathways for Fabric Filtration,* a paper presented at the Florida Section of the Air Pollution Control Association Annual Meeting, Palm Coast, FL, September 1981.

Turner, J. H., McKenna, J. D., Mycock, J. C., Nunn, A. B., and Vatavuk, W. M. "Chapter 5—Fabric Filters" in *OAQPS Control Cost Manual* (5th ed.), EPA 453/B-96-001, U.S. EPA, Research Triangle Park, NC, December 1998.

Turner, J. H., Viner, A. S., McKenna, J. D., Jenkins, R. E., and Vatavuk, W. M. "Sizing and Costing of Fabric Filters—Part I: Sizing Considerations," *Journal of Air Pollution Control Association, 37*(6), June 1987(a), p. 749.

———. "Sizing and Costing of Fabric Filters—Part II: Costing Considerations," *Journal of the Air Pollution Control Association, 37*(9), September 1987(b), p. 1105.

U.S. Environmental Protection Agency. *Operation and Maintenance Manual for Fabric Filters,* EPA–625/1–86–020, Research Triangle Park, NC, June 1986.

Vatavuk, W. M., and Neveril, R. B. "Part XI: Estimate the Size and Cost of Baghouses," *Chemical Engineering, 89*(6), March 22, 1982.

CHAPTER 7

PARTICULATE SCRUBBERS

A agoa tudo lava (water washes everything).

Old Portuguese proverb

◆ 7.1 INTRODUCTION ◆

Wet collection devices for fumes, mists, and suspended dusts are called *scrubbers*. This class of pollution control equipment collects particles by direct contact with a liquid (usually water). There are a multitude of scrubber designs on the market; most of them can be grouped according to the liquid contacting mechanism used. In addition, scrubbers can be broadly classified as low-, moderate-, or high-energy units. Energy requirements can be expressed as the pressure drop across the scrubber (in inches or centimeters of water) or by the level of contacting power (in horsepower per cfm or kilowatt-hours per volume of gas treated). The most common units for contacting power are hp/1000 cfm or kWh/1000 m^3. Liquid circulation rates are given in gallons per 1000 cubic feet (1 gal/1000 ft^3 = 0.134 L/m^3). In this chapter, we will briefly describe the most frequently used scrubber designs, and will also discuss some typical operating conditions.

SPRAY-CHAMBER SCRUBBERS

In spray-chamber scrubbers, particulate-laden air is passed through a circular or rectangular chamber and contacted with a liquid spray produced by spray nozzles. (Scrubbers that use spray nozzles are also called *preformed spray scrubbers*.) Droplet size is controlled to optimize particle contact and to provide easy droplet separation from the airstream. Figure 7.1 shows spray chamber arrangements employing countercurrent and cross-flow contact between gas and liquid. Baffles are sometimes used to improve gas-spray contact.

Liquid requirements usually range from 10 to 20 gal/1000 ft^3 and satisfactory droplet size is obtained with a liquid pressure of 35 to 50

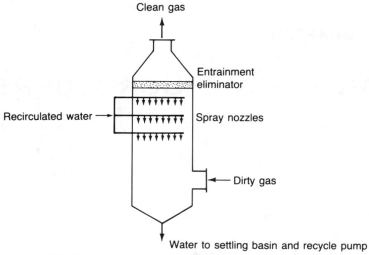

(a)   Vertical spray chamber (countercurrent flow)

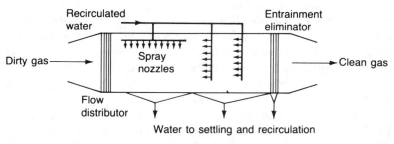

(b)   Horizontal spray chamber (cross–flow)

Figure 7.1
Typical spray chamber arrangements.

psi in the spray nozzles. Nozzles providing a cone spray geometry are the most effective. Recirculated water must be sufficiently settled or filtered to prevent excessive nozzle fouling. Nozzle cleaning and replacement represent a major part of the required maintenance for spray chambers.

Spray chambers are capable of 90% efficiency for particles larger than 8 microns. They have been used effectively in a wide variety of applications including paper dust control in paper towel manufacture and latex aerosol control in fiberglass production. For use in fiberglass production, it is necessary to maintain a suspended particulate level in the recirculated water less than 1500 mg/L to prevent significant re-entrainment.

Power consumption for spray chambers is low. They typically operate with a relatively low ΔP, usually in the range from 1 to 4 inches of water or 0.5 to 2 hp/1000 cfm at a gas velocity of 3 to 6 feet per second.

CYCLONE SPRAY CHAMBERS

Collection efficiency can be improved by modifying a typical spray chamber to introduce the influent tangentially as shown in Figure 7.2, then higher superficial gas velocities are possible. The added centrifugal force permits good droplet separation and allows the use of a smaller droplet size, which improves collection efficiency. Inlet velocities of 150 to 250 ft/sec and superficial velocities to 10 ft/sec are typical, providing collection efficiencies of 95% for particles larger than 5 microns. Liquid recirculation rates for cyclone spray chambers usually range from 3 to 6 gal/1000 ft^3. Pressure drops range from 4 to 8 inches of water, with power inputs from 1 to 3.5 hp/1000 cfm.

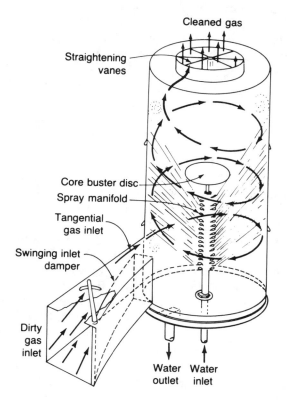

Figure 7.2
Cyclone spray chamber.
(Adapted from *Air Pollution Manual—Part II*, 1968.)

ORIFICE AND WET-IMPINGEMENT SCRUBBERS

Figure 7.3 shows illustrations of an orifice scrubber and two impingement scrubbers. (Orifice scrubbers are also called *self-induced* or *gas atomized spray scrubbers*.) In both of these designs, the combined air-droplet mixture impinges on a solid surface. In orifice scrubbers, the air impinges on the liquid surface and then on a series of baffles. In impingement scrubbers, the air-droplet mixture flows through a perforated tray containing a layer of liquid and froth and then impinges on a plate mounted directly above the perforations.

Orifice scrubbers have the advantage of low water recirculation rates (primarily for makeup of evaporation losses), which are typically about 0.5 gal/1000 ft^3. At least one manufacturer supplies a unit with an adjustable orifice for higher impingement velocities. Adjustment is accomplished with a movable cone mounted in the inlet duct of the scrubber. Pressure drops in the range of 8 to 12 inches of water are common at a contacting power input of 2 to 4 hp/1000 cfm. Efficiencies of 90% have been reported for particles larger than 2 microns (*Air Pollution Manual—Part II* 1968).

Impingement or perforated plate scrubbers operate in the same pressure drop range as orifice scrubbers. Liquid requirements usually range from 2 to 3 gal/1000 ft^3 at a contacting power of 2 to 5 hp/1000 cfm. Efficiencies of 97% are possible for 5-μm particles. One possible modification of the impingement design is to use a layer of plastic spheres on the perforated tray. These spheres are partially fluidized and the abrasive action helps to clean adhering particulate matter. This design has been used successfully in moderate fouling applications such as emission control on lime kilns and chemical recovery boilers.

VENTURI AND VENTURI JET SCRUBBERS

Wet scrubber efficiency is a strong function of the relative velocity between the liquid droplet and suspended particles and the particle size, as will be discussed in Section 7.2. High relative velocities are obtained in venturi scrubbers by injecting water at low pressure into the throat of a venturi through which air is passing at 150 to 500 feet per second. The venturi scrubber shown in Figure 7.4 can achieve collection efficiencies above 98% for particles larger than 0.5 microns at high throat velocity (high pressure drop). Power input can range from 3 to 12 hp/1000 cfm with a liquid recirculation rate of 3 to 10 gal/1000 ft^3. The venturi design has the advantage of simplicity and is easy to install and maintain, but typically requires large pressure drops.

Venturi jet scrubbers are a modification of the venturi design in which water at high velocity is jetted into the throat of a venturi, inducing a draft of 2 to 4 inches of water. Water rates in the range from 30 to 80 gal/1000 ft^3 at nozzle pressures of 50 to 150 psig are used. Efficiencies of 92% are possible for 1.0-μm particles, significantly lower than for the high-pressure-drop venturi scrubber.

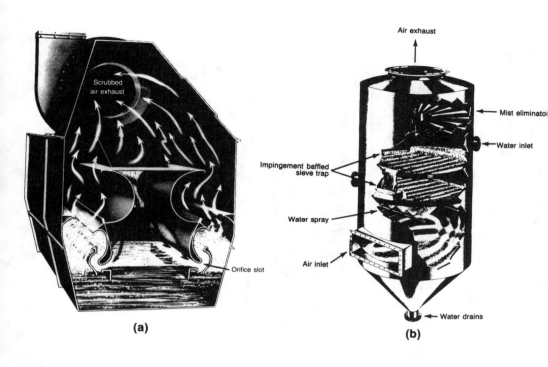

(a)

(b)

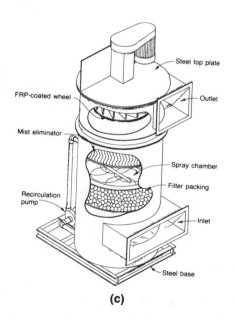

(c)

Figure 7.3

Orifice and impingement scrubbers: (a) orifice scrubber; (b) impingement scrubber; (c) packed tray impingement.

(a) Courtesy of American Air Filter Company; (b) courtesy of Peabody Process Systems, Inc., Norwalk, CT; (c) courtesy of Midwest Air Products Co., Inc., Owosso, MI.

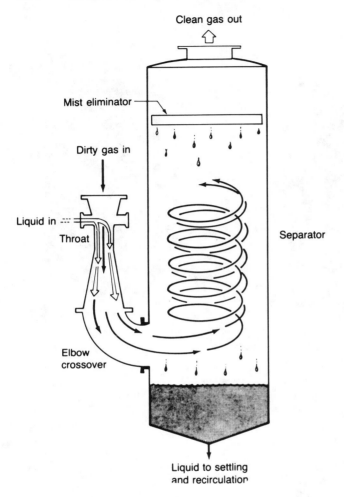

Figure 7.4
Typical venturi scrubber with a cyclone separator configuration.

OTHER DESIGNS

Several scrubbers are available that use mechanically enhanced gas–liquid contact. A typical design uses a motor driven impeller located at the gas–liquid contacting zone, such as shown in Figure 7.5. Disintegration scrubbers use rotating cages to produce small diameter, high-velocity droplets. A major disadvantage of mechanical scrubbers is high maintenance cost.

Packed towers (such as described in Chapter 13) and wetted fibrous filters are occasionally used for particulate control. Again, a major disadvantage of these units is high maintenance cost.

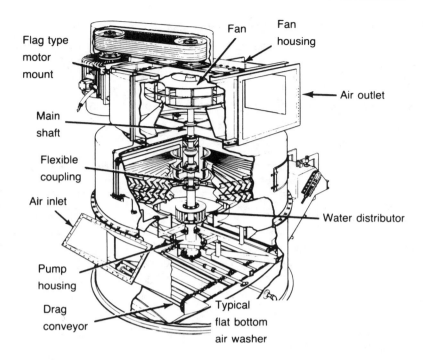

Flag type motor mount

Fan

Fan housing

Air outlet

Main shaft

Flexible coupling

Air inlet

Water distributor

Pump housing

Drag conveyor

Typical flat bottom air washer

Figure 7.5
Mechanical wet scrubber design.
(Courtesy of Centri-Spray Corporation.)

TYPICAL WET SCRUBBER APPLICATIONS

Now that we have looked at several scrubber designs, it is useful to indicate a few of the many applications for scrubbers. Table 7.1 lists typical scrubber applications in several process industries along with a general range of operating conditions.

ADVANTAGES AND DISADVANTAGES OF WET SCRUBBERS

When selecting air pollution control equipment for a specific emission control problem, we must consider the advantages and disadvantages of each class of equipment. Some of the major advantages and disadvantages of wet scrubbers are as follows:

Advantages

1. Can handle flammable and explosive dusts with little risk.

2. Provide gas absorption and dust collection in a single unit.

3. Can handle mists.

4. Provide cooling of hot gases.

5. Collection efficiency can be varied.

6. Corrosive gases and dusts can be neutralized.

Disadvantages

1. High potential for corrosion problems.

2. Effluent liquid can create water pollution problems.

Table 7.1 Typical Applications of Wet Scrubbers

Source	Type Scrubber (See Notes)	Flow Rate acfm	Pressure Drop in. H_2O
Incinerators			
Hazardous Waste	HEV/PT	1,000–100,000	45–50
	F/C MV	1,000–100,000	35–45
Hospital Waste	HEV	500–50,000	45–50
	F/C MV	500–50,000	35–40
Liquid Waste	HEV/PT	500–40,000	40–45
Gaseous Waste	PT	100–100,000	5–15
Pulp/Paper Applications			
Lime Sludge Kiln	MEV or	2,000–100,000	18–30
	PAV	2,000–60,000	10–15
Lime Slaker	LEV or	500–20,000	6–12
	FBS or	500–20,000	6–8
	ST	500–20,000	6–10
Dissolving Tank	LEV or	500–30,000	8–16
	MP or	500–30,000	4–8
	FBS or	500–30,000	6–8
	ST	500–30,000	4–8
	LEV/PT	500–30,000	10–20
Bleach Plant	PT or	1,000–60,000	6–10
	FBS	1,000–60,000	6–8
BRN. Stock Washer	PT or	1,000–60,000	6–8
	CS or	1,000–60,000	8–10
	FBS	1,000–60,000	6–8
Chemical Industry			
Soluble Gas	PT or	50–60,000	3–20
	FBS or	50–60,000	6–12
	WF	50–60,000	3–10
Dust	LEV or PAV	50–60,000	6–20
	FBS	50–60,000	6–12

Notes: CS = cyclonic spray scrubber; F/C MV = flue-gas condensation, multiple-throat venturi; FBS = fluidized-bed scrubber (tray or catenary grid); HEV = high-energy venturi; LEV = low-energy venturi; MEV = medium-energy venturi; MP = mesh pad; PAV = pump-aided venturi; PT = packed tower; ST = spray tower; WF = wetted filter

3. Protection against freezing required; off gas might require reheating to avoid visible plume.

4. Collected particulate may be contaminated and may not be recyclable.

5. Disposal of waste sludge may be very expensive.

♦ 7.2 THEORY AND DESIGN CONSIDERATIONS ♦

The mechanism of particle collection by inertial impaction was discussed in Chapter 3. Inertial impaction of particles on the surface of liquid droplets is the dominant particulate control mechanism in most industrial wet scrubber applications. Diffusional processes are most effective for particles smaller than 0.1 micron, so they will not be considered in the following discussion. Refer to the work by Crawford (1976) and Calvert (1984) for a discussion of diffusional mechanisms.

The design of wet scrubbers is based on models, developed for various gas–liquid contacting processes, that predict the *penetration* for a given particle diameter. **Penetration** Pt_d *is defined as the fraction of particles of a specified diameter d_p that are not captured.* Penetration is related to collection efficiency by

$$Pt_d = 1 - \eta_d \qquad (7.1)$$

where η_d = fractional efficiency for particles having a diameter d_p

The overall penetration Pt_o is found from

$$Pt_o = \sum (Pt_d \times M_d) \qquad (7.2)$$

where M_d = mass fraction of particles at specified diameter

In effect Pt_o is found by summing the products of the penetration for each particle size times the mass fraction of that particle size. The procedure is simplified by assuming that most industrial aerosols have a log-normal particle size distribution, as was shown in Figure 3.4.

In the following discussion, we will use penetration models for spray chambers and venturi scrubbers to illustrate the design calculations for preformed spray and gas atomization units.

SPRAY CHAMBERS

As shown in Figure 7.1, spray chambers are simple in design and operation. The gas-liquid contact can be countercurrent, co-current, or cross-current. The primary particle collection mechanism in spray chambers is inertial impaction on droplets in preformed sprays.

Calvert (1977) developed the following equation for particle penetration in a countercurrent vertical spray chamber.

$$\text{Pt}_d = \exp\left(-\frac{3Q_L V_{t_d} z \eta_d}{4Q_G r_d \left(V_{t_d} - V_G\right)}\right) = \exp\left(-\frac{A_d V_{t_d} \eta_d}{Q_G}\right) \qquad (7.3)$$

where

Pt_d = penetration of a given particle size (ranging from 0 to 1.0)

Q_L = volumetric liquid flow rate, m^3/s

Q_G = volumetric gas flow rate, m^3/s

V_G = superficial gas velocity, cm/s

V_{t_d} = terminal settling velocity of droplets, cm/s

η_d = fractional collection efficiency of a single droplet, 0 to 1.0

r_d = droplet radius, cm

z = length of scrubber contact zone, cm

A_d = cross-sectional area of all the droplets in the scrubber, cm^2

In Eq. (7.3), A_d is given by

$$A_d = \frac{3Q_L z}{4 r_d \left(V_{t_d} - V_G\right)} \qquad (7.4)$$

The single droplet target efficiency is estimated from

$$\eta_d = \left(\frac{K_p}{K_p + 0.7}\right)^2 \qquad (7.5)$$

Calvert defined the impaction parameter, K_p, as

$$K_p = \frac{C \rho_p d_p^2 V_{p,d}}{9 \mu_G d_d} = \frac{\rho_w d_a^2 V_{p,d}}{9 \mu_G d_d} \qquad (7.6)$$

where

C = Cunningham correction factor, dimensionless

ρ_p = particle density, g/cm^3

d_p = physical particle diameter, cm

$V_{p,d}$ = particle velocity (relative to droplet), cm/s

d_d = droplet diameter, cm

μ_G = gas viscosity, poise

d_a = aerodynamic particle diameter, cm

ρ_w = water density, g/cm^3

Note the similar grouping of parameters in K_p and in the characteristic time (see Chapter 3). Also note that in a vertical countercurrent spray tower, the velocity of the particle relative to the droplet is simply equal to the terminal settling velocity of the droplet. That is, $V_{p,d}$ = $V_{t,d}$. This is most easily understood by visualizing the two limiting cases: (1) droplets falling through still air, and (2) stationary droplets suspended by the dusty air flowing upwards at a gas velocity equal to the droplet terminal settling velocity.*

Equation (7.3) is based on the assumption that the spray droplets are of uniform size and immediately reach the terminal settling velocity. The cross-sectional area of all droplets in the scrubber A_d assumes no droplets reach the scrubber walls. Calvert (1984) suggests that Q_L/Q_G be multiplied by 0.2 to correct for wall effects and other losses of suspended droplets.

To apply Eq. (7.3) to spray chamber design, we must know the particle size distribution in the inlet gas stream. We first calculate the penetration for each particle size range, and then calculate the overall penetration Pt_o, as was illustrated in Example 3.3. For a given inlet particle size distribution, there is an optimum set of operating parameters to provide the highest overall efficiency.

Langmuir and Blodgett (1946) showed that the droplet target efficiency η_d is given by the following dimensionless group.

$$\eta_d = \frac{k V_{p,d} V_{t_p}}{g d_d} \tag{7.7}$$

where

k = an empirical constant

V_{t_p} = terminal velocity of particles, cm/s

g = gravitational acceleration, 980 cm/s^2

d_d = collection or droplet diameter, cm

Equation (7.7) indicates that decreasing the droplet size will increase the target efficiency. Stairmand (1964) has shown that there is in fact an optimum droplet size, and efficiency is not increased by decreasing droplet size smaller than the optimum. Figure 7.6 shows a plot of the target efficiency as a function of droplet size for several particle sizes of a material having a density of 2 g/cm^3. These data indicate that spray chambers should operate with droplet sizes in the range from 500 to 1000 microns. An optimum droplet size exists because very small droplets are rapidly accelerated to the gas velocity, thus decreasing the value of $V_{p,d}$, whereas very large droplets have much less surface area (for the same total mass of water).

For cross-flow chambers, Eq. (7.3) can be modified to

$$Pt_d = \exp\left(\frac{-3Q_L z \eta_d}{4 Q_G r_d}\right) = \exp\left(-\frac{A_d V_{p,d} \eta_d}{Q_G}\right) \tag{7.8}$$

*Our thanks for this visualization to Dr. Tom Overcamp of Clemson University.

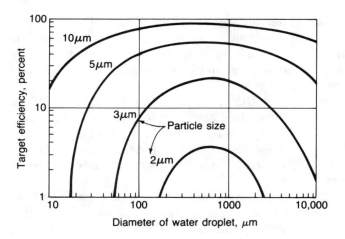

Figure 7.6
Target efficiency as a function of droplet diameter for various particle sizes (ρ_p = 2 g/cm^3).
(Adapted from *Air Pollution Manual—Part II*, 1968.)

Calvert (1974) published plots of solutions to Eqs. (7.3) and (7.8), which are shown in Figure 7.7. In these figures, d_{pc}, the aerodynamic particle diameter for which the collection efficiency is 50%, is plotted against the chamber height, with both droplet diameter and liquid-to-gas ratio as parameters. Note that below a certain chamber height, d_{pc} increases very rapidly, meaning that Pt_d approaches 100%. Figure 7.7 is valid only for air and water near standard conditions.

▶ ▶ ▶ ▶ ▶ ▶ ▶ ▶ ▶

Example 7.1

Calculate the penetration of an 8-µm unit-density particle through a vertical countercurrent spray chamber if the operating conditions are Q_L/Q_G = 1 L/m^3, V_G = 20 cm/s, d_d = 300 µm and z = 3 m. Assume atmospheric pressure and 25 C.

Solution

For this problem, we use Eqs. (7.3), (7.5), and (7.6). From Figure 3.8, V_{t_d} = 120 cm/s. Ignoring the settling velocity of an 8-µm particle, $V_{p,d} = V_{t_d}$. Thus,

$$V_{p,d} = 120 \text{ cm/s}$$

$$Q_L/Q_G = 0.001 \text{ m}^3/\text{m}^3$$

$$z = 300 \text{ cm}$$

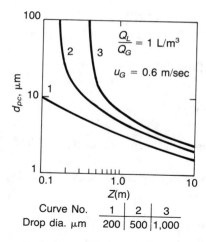

Curve No.   1 | 2 | 3
Drop dia. μm   200 | 500 | 1,000

Performance cut diameter predictions
for typical vertical countercurrent spray

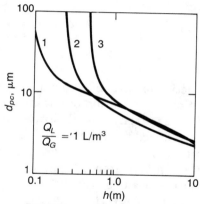

Performance cut diameter predictions
for typical cross-current spray

Figure 7.7
Predicted cut diameters for countercurrent and cross-flow spray chambers.
(Adapted from Calvert, 1974.)

First, we find K_p from Eq. (7.6).

$$K_p = \frac{1.0 d_a^2 V_{p,d}}{9 \mu_G d_d}$$

From Table B.2 in Appendix B, $\mu_G = 0.044$ lb$_m$/hr-ft or 0.00018 poise.
Assuming $d_a = d_p$, we obtain

$$K_p = \frac{(0.0008)^2 (120)}{9 (0.00018)(0.030)} = 1.58$$

From Eq. (7.5) $\eta_d = \left(\dfrac{1.58}{1.58 + 0.7}\right)^2 = 0.480$

From Eq. (7.3) $Pt_d = \exp\left[\dfrac{-3(0.001)(120)(300)(0.48)}{4(0.015)(120-20)}\right]$

$$Pt_d = \exp(-8.64) = 0.00018$$

Thus, assuming 100% spray utilization, the scrubber is 99.98% efficient for 8-µm particles.

If we assume that only 20% of the spray forms effective droplets, then $Q_L/Q_G = 0.2 \times 0.001 = 0.0002$. Thus,

$$Pt_d = \exp\left[(-8.64)(0.2)\right] = \exp[-1.728] = 0.178$$

For this case, the collection efficiency is 82%. This result is probably more realistic than the former result. This example demonstrates the importance of nozzle selection and location.

◀ ◀ ◀ ◀ ◀ ◀ ◀ ◀ ◀

VENTURI SCRUBBERS

Venturi scrubbers use inertial impaction of suspended particles on water droplets formed by gas atomization. Atomization occurs as liquid is introduced into the high-velocity gas stream, as illustrated in Figure 7.4. Sufficient liquid must be introduced to give thorough gas–droplet contact, and the gas velocity at the point of initial gas–liquid contact must be high enough to produce complete atomization.

Particle penetration through a venturi scrubber can be approximated by the following equation, developed by Calvert et al. (1972).

$$Pt_d = \exp\left\{\dfrac{Q_L V_G \rho_L d_d}{55 Q_G \mu_G}\left[-0.7 - K_p f + 1.4 \ln\left(\dfrac{K_p f + 0.7}{0.7}\right) + \dfrac{0.49}{0.7 + K_p f}\right]\dfrac{1}{K_p}\right\}$$

$$(7.9)$$

where

K_p = inertial impaction parameter calculated from Eq. (7.6) for the gas velocity at the throat entrance, dimensionless

f = empirical factor ($f = 0.25$ for hydrophobic particles; $f = 0.50$ for hydrophilic particles)

Gas atomization produces a wide distribution of droplet size; however, Eq. (7.9) can be solved with satisfactory results using the Sauter mean droplet diameter d_d, which is found with the Nukiyama–Tanasawa (1938) relationship:

$$d_d = \frac{58,600}{V_G}\left(\frac{\sigma}{\rho_L}\right)^{0.5} + 597\left(\frac{\mu_L}{(\sigma\rho_L)^{0.5}}\right)^{0.45}\left(1000\frac{Q_L}{Q_G}\right)^{1.5} \quad \text{(7.10)}$$

where

d_d = Sauter mean droplet diameter, μm

ρ_L = density of the liquid, g/cm^3

σ = liquid surface tension, dyne/cm

μ_L = liquid viscosity, poise

▶ ▶ ▶ ▶ ▶ ▶ ▶ ▶ ▶

Example 7.2

Calculate the collection efficiency of a venturi scrubber for 1-μm particles if the operating conditions are V_G = 50 m/s at the throat entrance, and Q_L/Q_G = 1 L/m^3. Assume atmospheric pressure and 20 C (at these conditions, σ = 72 dyne/cm and μ_L = 1 cp). Also assume f = 0.5 for this case.

Solution

For this problem, we use Eqs. (7.6), (7.9), and (7.10). First, we find the droplet diameter. At 20 C, σ = 72 dyne/cm and μ_L = 0.01 poise. Therefore,

$$d_d = \frac{58,600}{50\times100}\left(\frac{72}{1.0}\right)^{0.5} + 597\left(\frac{0.01}{\sqrt{72}}\right)^{0.45}$$

$$\times\left(1000\times0.001\frac{\text{m}^3}{\text{m}^3}\right)^{1.5}$$

$$= 99.4 + 28.7 = 128.1 \text{ μm}$$

$$K_p = \frac{d_a^2 V_p}{9\mu_G d_d} = \frac{(0.0001)^2\times50\times100}{9\times0.00018\times0.0128} = 2.41$$

$$\text{Pt}_d = \exp\left\{\frac{0.001\text{ m}^3}{55\text{ m}^3}\times\frac{50\times100\times0.0128}{0.00018}\right.$$

$$\times\left[-0.7-(2.4)(0.5)+1.4\ln\left(\frac{(0.5)(2.4)+0.7}{0.7}\right)\right.$$

$$+\left.\left.\frac{0.49}{0.7+2.4(0.5)}\right]\frac{1}{2.4}\right\}$$

$$= \exp\left\{6.46\left[-0.7-1.2+1.4(0.998)+0.26\right]\frac{1}{2.4}\right\}$$

$$= \exp(-0.646) = 0.52$$

This scrubber will capture $(1.00 - 0.52)100 = 48\%$ of 1-μm particles. This low efficiency is due to the low inlet velocity, which gives a low pressure drop (see Example 7.3).

Pressure Loss in Venturi Scrubbers.   Pressure loss in large industrial venturi scrubbers is due primarily to droplet acceleration and can be estimated by the equation of Yung et al. (1977):

$$\Delta P = 2\rho_L V_G^2 \left(\frac{Q_L}{Q_G}\right)\left(1 - X^2 + \sqrt{X^4 - X^2}\right) \tag{7.11}$$

where

ΔP = pressure loss, dyne/cm^2

V_G = gas velocity, cm/s

X = dimensionless throat length

In Eq. (7.11), X is given by

$$X = \frac{3l_t C_D \rho_G}{16 d_d \rho_L} + 1 \tag{7.12}$$

where

l_t = venturi throat length, cm

C_D = drag coefficient for droplets with Sauter mean diameter, dimensionless

Hesketh (1979) suggests that for droplets with Reynolds numbers in the range from 10 to 500, the drag coefficient is given by

$$C_D = \frac{24}{\text{Re}} + \frac{4}{(\text{Re})^{1/3}} \tag{7.13}$$

where Re = droplet Reynolds number from Eq. (3.7), dimensionless

Example 7.3

Estimate the pressure drop across the venturi scrubber described in Example 7.2 if the throat length is 30 cm.

Solution

From Eq. (3.7), Re for the liquid droplet is

$$\text{Re} = \frac{d_d V_G \rho_G}{\mu_G}$$

Assuming the ideal gas law holds, $\rho_G = 0.0012$ g/cm^3

$$\text{Re} = \frac{0.0128 \text{ cm} \times 50 \times 100 \text{ cm}}{0.00018 \text{ g/cm-s}} \times 0.0012 \frac{\text{g}}{\text{cm}^3} = 427$$

$$C_D = \frac{24}{427} + \frac{4}{(427)^{1/3}} = 0.588$$

$$X = \frac{3l_t C_D \rho_G}{16 d_d \rho_L} + 1$$

$$= \frac{3 \times 30 \times 0.588 \times 0.0012}{16 \times 0.0128 \times 1} + 1$$

$$= 1.31$$

By Eq. (7.11),

$$\Delta P = 2 \times 1.0 \times (50 \times 100)^2 \times 0.001 \left(1.0 - (1.31)^2 + \sqrt{(1.31)^4 - (1.31)^2} \right)$$

$$= 19{,}620 \text{ dyne/cm}^2$$

$$= 1962 \text{ N/m}^2$$

$$= 20.0 \text{ cm H}_2\text{O}$$

$$= 7.9 \text{ in. H}_2\text{O}$$

This is quite low for a venturi scrubber.

CONTACTING POWER APPROACH
IN WET SCRUBBER DESIGN

Lapple and Kamack (1955) showed that the efficiency of a wet scrubber can be related to the energy expended in producing the actual gas–liquid contact. The more energy expended, the more turbulent the contacting process and the more efficient the overall collection. This work indicated that scrubbers of different geometry would provide similar collection efficiencies if operated at the same gas–liquid contact energy level.

Semrau (1963,1980) extended the work of Lapple and Kamack by defining **contacting power** as the *energy dissipated per unit volume of gas treated,* which he related to scrubber efficiency by

$$\eta = 1 - \exp(-N_t) \qquad \textbf{(7.14)}$$

where N_t is the *number of transfer units,* calculated from

$$N_t = \alpha P_T^\gamma \qquad \textbf{(7.15)}$$

where

N_t = number of transfer units, dimensionless

P_T = contacting power, hp/1000 cfm or kWh/1000 m^3

α = coefficient of P_T, (hp/1000 cfm)$^{-\gamma}$ or (kWh/1000 m^3)$^{-\gamma}$

γ = exponent of P_T, dimensionless

Contacting power should be determined from the friction loss across the wetted portion of the scrubber. Semrau calls this loss the *effective friction loss*. Pressure loss due to changes in the gas stream kinetic energy should not be included. The total contacting power P_T includes energy supplied by the gas, the liquid, and mechanical devices such as a rotor to increase contact. In the majority of scrubbers, gas-phase contacting power dominates.

Figure 7.8 shows a plot of P_T versus N_t for a venturi scrubber collecting a metallurgical fume, and Figure 7.9 illustrates the performance of a venturi scrubber collecting fly ash as a function of effective friction loss. Friction loss ΔP in inches of water can be converted to P_T in hp/1000 cfm by multiplying friction loss by 0.1575.

The contacting power concept is a very simplistic approach, but it does correlate scrubber performance satisfactorily; it is useful in extrapolating performance of a particular scrubber design over a range of operating conditions. The major difficulty lies in the accurate determination of the effective friction loss or energy that is actually used in the gas–liquid contacting mechanism.

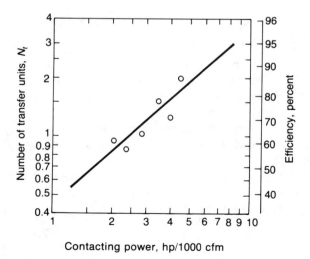

Figure 7.8
Performance curve for a venturi scrubber collecting a metallurgical fume.
(Adapted from Semrau, 1963.)

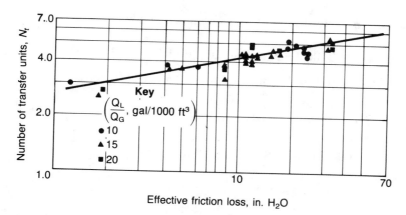

Figure 7.9
Performance curve for a venturi scrubber collecting fly ash.
(Adapted from Semrau, 1980.)

▶ ▶ ▶ ▶ ▶ ▶ ▶ ▶ ▶

Example 7.4

A venturi scrubber used to control a metallurgical process has been tested at two effective friction loss levels with the following results:

Friction Loss, in. H_2O	Overall Collection Efficiency
12.7	56.0%
38.1	89.0%

Estimate the contacting power required, in hp/1000 cfm, to attain 97% overall efficiency if the scrubber is operating at 80 F.

Solution

For this problem, we use Eqs. (7.14) and (7.15). First, we convert the friction loss to contacting power in hp/1000 cfm.

$$(P_T)_{56\%} = (12.7 \text{ in. } H_2O)(0.1575) = 2.00 \text{ hp/1000 cfm}$$

$$(P_T)_{89\%} = (38.1)(0.1575) = 6.00 \text{ hp/1000 cfm}$$

Next, we use Eq. (7.14) to determine the number of transfer units N_t at each efficiency.

$$(N_t)_{56\%} = \ln\left(\frac{1}{1-\eta}\right) = \ln\frac{1}{1-0.56} = 0.821$$

$$(N_t)_{89\%} = \ln\left(\frac{1}{1-0.89}\right) = 2.207$$

$$(N_t)_{97\%} = \ln\left(\frac{1}{1-0.97}\right) = 3.506$$

Next, we use Eq. (7.15) to find α and γ simultaneously.

$$0.821 = \alpha(2.00)^{\gamma} \quad \text{or} \quad \ln \alpha + \gamma \ln 2.00 = \ln(0.821) \quad \text{(A)}$$

$$2.207 = \alpha(6.00)^{\gamma} \quad \text{or} \quad \ln \alpha + \gamma \ln 6.00 = \ln(2.207) \quad \text{(B)}$$

Subtracting Eq. (A) from Eq. (B),

$$0.989 = \gamma(\ln\ 6.00 - \ln\ 2.00) = \gamma \ln \frac{6.00}{2.00}$$

$$\gamma = 0.900$$

$$\alpha = \frac{2.207}{(6.00)^{0.900}} = 0.440$$

Finally, we use Eq. (7.15) to calculate the required contacting power.

$$3.506 = 0.440 P_T^{0.900}$$

$$P_T = \left(\frac{3.506}{0.440}\right)^{1.111} = 10.0 \text{ hp/1000 cfm}$$

◀ ◀ ◀ ◀ ◀ ◀ ◀ ◀ ◀

◆ 7.3 OTHER CONSIDERATIONS ◆

ELIMINATION OF LIQUID ENTRAINMENT

In any process involving turbulent gas–liquid contact, there will be some entrainment of liquid in the exiting gas stream. Provision must be made to remove at least 95% of the liquid carryover to avoid recontamination of the exit gas.

In preformed spray scrubbers, louver-type mist eliminators are frequently used. Since the majority of entrained liquid droplets are in the size range from 150 to 300 microns, relatively low velocities (5 to 15 m/s) will provide satisfactory mist elimination at pressure drops in the range from 0.5 to 1.5 cm H_2O. Figure 7.10 shows several impaction-type mist eliminators.

Venturi scrubbers are usually followed by a cyclone chamber with a mesh mist eliminator to remove entrainment, as was shown in Figure 7.4. Droplet sizes in the range from 75 to 150 microns are effectively removed at cyclone inlet velocities of 25 to 40 m/s. Water droplets can be assumed to behave as solid particles in cyclones and the design equations presented in Chapter 4 can be used to size cyclone droplet separators. For a detailed discussion of mist and entrainment elimination, refer to the work of Strauss (1977).

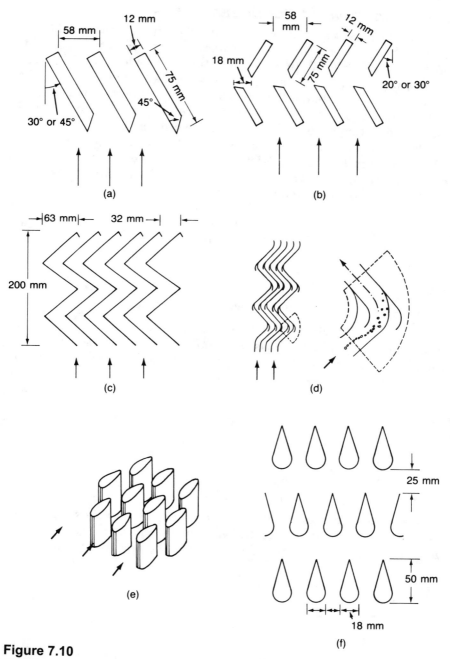

Figure 7.10
Simple inertial separators: (a) single row of baffles; (b) two rows of baffles; (c) W-pattern separator; (d) wave separator (Euroform pattern); (e) vertical rod teardrop shape in horizontal flow system; (f) horizontal rod with teardrop shape in vertical flow system.
(Adapted from Strauss, 1977.)

HUMIDIFICATION OF SCRUBBED GASES

When air passes through a wet scrubber, it becomes humidified; that is, liquid water evaporates and increases the water vapor content of the air. At the same time, the air stream becomes cooled because the latent heat of vaporization of the water is provided by sensible heat from the air stream. Because the contact between air and water in a scrubber is intimate (lots of surface area exposure at phase interfaces), the exiting air is usually saturated with water vapor.

The term *humidity* refers to the amount of water vapor in the air. **Absolute humidity** $\mathcal{H}$ *is defined as the mass of water vapor carried by a unit mass of dry air.* As such, the humidity at one atmosphere total pressure is calculated as follows:

$$\mathcal{H} = \frac{18\overline{P}_{H_2O}}{29\left(1 - \overline{P}_{H_2O}\right)} \tag{7.16}$$

where

$\mathcal{H}$ = absolute humidity, lb_m water vapor/lb_m dry air

$\overline{P}_{H_2O}$ = partial pressure of water vapor, atm

The **relative humidity** $\mathcal{H}_r$ *is the ratio (expressed as a percentage) of the partial pressure of the vapor actually present in gas to the vapor pressure of liquid water at the gas temperature.* Thus,

$$\mathcal{H}_r = \frac{\overline{P}_{H_2O}}{(P_v)_{H_2O}} \times 100\% \tag{7.17}$$

where

$\mathcal{H}_r$ = relative humidity, %

$(P_v)_{H_2O}$ = vapor pressure of liquid water, atm

The **percentage humidity** $\mathcal{H}_p$ *is the ratio (expressed as a percentage) of the actual humidity to the saturation humidity at the same temperature,* and thus is slightly different from the relative humidity.

$$\begin{aligned}
\mathcal{H}_p &= \frac{\mathcal{H}}{\mathcal{H}_s} \times 100 \\
&= \frac{\mathcal{H}_r\left[1 - (P_v)_{H_2O}\right]}{1 - \overline{P}_{H_2O}}
\end{aligned} \tag{7.18}$$

where

$\mathcal{H}_p$ = percentage humidity, %

$\mathcal{H}_s$ = saturation humidity, lb_m moisture/lb_m dry air

Note that below about 100 F, $\overline{P}_{H_2O}$ is small enough that percentage humidity and relative humidity are approximately equal. Another useful relation is that between gas phase mole fraction and humidity, or

$$y_{H_2O} = \frac{\overline{P}_{H_2O}}{P} = \frac{\mathscr{H}/18}{(1/29) + (\mathscr{H}/18)} \tag{7.19}$$

where P = total pressure, atm

Since $\mathscr{H}/18$ is often much less than $1/29$, y_{H_2O} can be approximated as $\mathscr{H}(29/18)$.

Consider an adiabatic scrubber in which unsaturated air enters at temperature T, contacts water at temperature T_s, and exits as a saturated airstream at temperature T_s. Choosing T_s as the datum temperature, an enthalpy balance yields

$$\dot{M}_A\left(C_{p_A} + \mathscr{H}C_{p_v}\right)(T - T_s) = \dot{M}_A\left(\mathscr{H}_s - \mathscr{H}\right)\lambda_s \tag{7.20}$$

where

$\dot{M}_A$ = mass flow rate of dry air, lb_m/min

C_{p_A} = specific heat of dry air, Btu/lb_m

C_{p_v} = specific heat of water vapor, Btu/lb_m

λ_s = latent heat of vaporization of water (at T_s), Btu/lb_m

Given any particular values of T and $\mathscr{H}$, Eq. (7.20) can be solved for the unique T_s ($\mathscr{H}_s$ and λ_s are functions of T_s) that results for that particular case. However, an easier procedure has been developed. Equation (7.20) can be rearranged as follows:

$$\frac{\mathscr{H} - \mathscr{H}_s}{T - T_s} = -\frac{C_{p_A} + \mathscr{H}C_{p_v}}{\lambda_s} \tag{7.21}$$

For various values of T_s, since both $\mathscr{H}_s$ and λ_s are known, Eq. (7.21) can be solved for unique pairs of points (T given $\mathscr{H}$ or $\mathscr{H}$ given T), and the resulting lines can be plotted on T versus $\mathscr{H}$ coordinates; these lines are called *adiabatic cooling lines* (ABC lines). On rectangular coordinates, each line has a slope of $-(C_{p_A} + \mathscr{H}C_{p_v})/\lambda_s$ and is slightly curved because of the $\mathscr{H}C_{p_v}$ term. On slightly distorted coordinates, the ABC lines become straight and parallel, and a very convenient chart results—the *psychrometric chart*.

Figure 7.11 is a simplified version of a psychrometric chart (complete psychrometric charts for two temperature ranges are given in Appendix B). The ABC lines are the straight upper-left-to-lower-right diagonals, and each is labeled with its adiabatic saturation temperature T_s (also called the *wet-bulb temperature*). The gas temperature (also called the *dry-bulb temperature*) is plotted on the x axis, and the

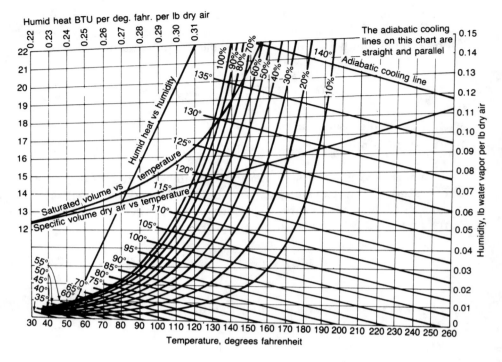

Figure 7.11
Psychrometric chart.
(Adapted from McCabe et al., 1985)

absolute humidity is plotted on the right-side y axis. In Figure 7.11, the left-most curved line is the 100% percentage humidity line; lesser percentage humidity lines are located to its right and also curve upward. Other information can be obtained from a psychrometric chart, but we will restrict our attention to humidity and saturation temperature. The following example illustrates the use of the chart. For teaching purposes, we will use Figure 7.11 in the next example; however, the charts in Appendix B are more accurate and complete.

▶ ▶ ▶ ▶ ▶ ▶ ▶ ▶ ▶ ▶

Example 7.5

A stream of air flowing at 100,000 acfm (150 F, 1 atm, and 20% percentage humidity) is to be scrubbed. Assuming that the air becomes saturated, estimate the exiting gas temperature and volumetric flow rate. Also, estimate the flow rate of water required to replace the water lost to evaporation.

Solution

Read vertically up from 150 F and find the point of intersection with the 20% humidity line; then read horizontally over to the right axis. We find

$$\mathscr{H} = 0.042 \text{ lb}_m \text{H}_2\text{O/lb}_m \text{ dry air}$$

Next, from the point of intersection of 150 F and 20% humidity, read diagonally up and to the left to the 100% saturation line. Finally, read vertically down to the x axis. We find

$$T_s = 107 \text{ F}$$

T_s also could be determined by linear interpolation between the two nearest ABC lines at the 100% saturation line.

From the intersection of the $T_s = 107$ F line and the 100% humidity line, read over to the right axis to obtain

$$\mathscr{H}_s = 0.053 \text{ lb}_m / \text{lb}_m$$

At $\mathscr{H} = 0.042$, the mole fraction of water vapor is

$$y_{\text{H}_2\text{O}} = \left(\frac{29}{18}\right) 0.042 = 0.068$$

From the ideal gas law, the mass flow rate of *dry* air is

$$\dot{M}_A = \frac{29 \text{ lb}_m}{\text{lbmol}} \times \frac{(1 \text{ atm})(100{,}000 \text{ ft}^3/\text{min})}{(0.73 \text{ atm-ft}^3/\text{lbmol-°R})(610 \text{ R})} \times (1 - 0.068)$$

$$\dot{M}_A = 6070 \text{ lb}_m/\text{min}$$

The increase in humidity equals the water evaporated. Thus, the flow rate of water required to replace the water lost by evaporation is

$$\dot{M}_w = (0.053 - 0.042) \frac{\text{lb H}_2\text{O}}{\text{lb dry air}} \times 6070 \frac{\text{lb dry air}}{\text{min}}$$

$$\dot{M}_w = 67 \text{ lb}_m/\text{min} = 8.0 \text{ gpm}$$

The exiting gas flow rate can be calculated as

$$\dot{M}_A = \tfrac{1}{29}(6070 \text{ lb}_m/\text{min}) = 209.3 \text{ lbmol/min}$$

$$\dot{M}_w = \tfrac{1}{18}(0.053)(6070 \text{ lb}_m/\text{min}) = 17.9 \text{ lbmol/min}$$

$$Q = \frac{[(209.3+17.9)\text{lbmol/min}](0.73 \text{ atm-ft}^3/\text{lbmol-°R})(567 \text{ R})}{1.0 \text{ atm}}$$

$$= 94{,}000 \text{ acfm}$$

Alternatively, the exiting volumetric flow rate can be estimated directly from the psychrometric chart. First, we read up from 107 F to the curved line labeled "saturated volume vs. temperature." Next, we read over to the left axis, and find the volume (in ft^3/lb dry air) is 15.5 ft^3/lb. Thus, the total volumetric flow rate is

$$Q = 15.5 \frac{ft^3}{lb_m} \times 6070 \frac{lb_m}{min} = 94,100 \text{ acfm}$$

◄ ◄ ◄ ◄ ◄ ◄ ◄ ◄ ◄

♦ 7.4 COSTS ♦

A preliminary estimate of the installed cost of a wet scrubber can be calculated using Figure 7.12 and Table 2.3, along with a current cost index to update the total cost.

Wet scrubbers have a high corrosion potential and it is frequently necessary to use stainless steel or other corrosion-resistant materials for construction. Costs developed from Figure 7.12 should be multiplied by 1.9 for 304 stainless-steel construction, by 2.7 for 316 stainless-steel, and by 1.7 for fiber-reinforced plastic construction. If the venturi, the crossover and elbow, and the entrainment separator are constructed of different materials, Cheremisinoff and Young (1976) suggest using 12%, 10%, and 78%, respectively, for the fractions of total structural metal in these sections when estimating the cost of the complete system. (Refer to Cheremisinoff and Young for a discussion of recommended materials of construction if acid gases are present in the airstream.)

The major operating costs for wet scrubbers are blower and recirculating pump power costs, maintenance, sludge disposal, operating labor, and process water. Blower operating cost can be estimated from either the contacting power or pressure drop across the unit, adjusted for 65% blower efficiency. Recirculating pump energy cost is estimated assuming a 50 psi drop for spray nozzles and a 25 psi drop for gas atomized units at a pump efficiency of 75%. Operating labor requirements range from 1 to 3 hours per unit per shift. With landfill disposal of wet sludges containing compounds classified as hazardous materials now banned, sludge disposal costs have increased dramatically. These costs must be carefully considered in the economic analysis of control alternatives. Refer to Wentz (1989) for a review of sludge disposal methods and attendant costs.

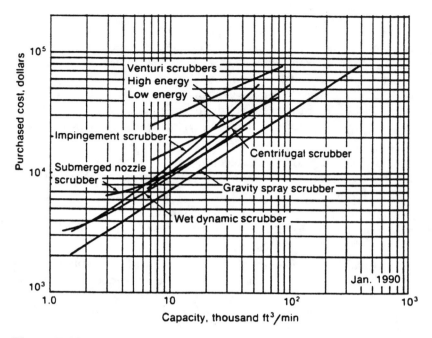

Figure 7.12

Costs (f.o.b.) of wet scrubbers in January 1990.

Note: Wet scrubber costs increased by an average of 15% over the period 1990–2000. Adjust Figure 7.12 accordingly.

(Peters, M. S., and Timmerhaus, K. D., *Plant Design and Economics for Chemical Engineers,* © 1991 McGraw-Hill, Inc. Used by permission.)

♦ Problems ♦

7.1 A vertical countercurrent spray chamber has a 2.0-m contact zone and operates with a liquid/gas ratio of 1.0 L/m^3 and an average droplet diameter of 200 microns. During a test run, the following data were taken on the unit:

average scrubber temperature = 80 F
gas velocity 0.4 m/s
inlet loading 1.5 gr/ft^3

The following particle size distribution was obtained from a cascade impactor.

Size Range, μm	<4	4–8	8–16	16–30	30–50	>50
Mass, mg	25	125	100	80	20	10

Estimate the overall efficiency of the unit, assuming 20% spray utilization.

7.2 The spray chamber of Problem 7.1 treats 20,000 acfm of air at atmospheric pressure, 80 F, and a relative humidity of 50%. Assuming that air leaving the chamber is saturated, and that 2.5% of the recirculated water is withdrawn for sludge removal and for corrosion control, how much makeup water is required in gallons per minute? For an overall scrubbing efficiency of 80%, what is the weight percent solids in the sludge that is removed?

7.3 The venturi scrubber described in Example 7.2 is modified (that is, equipped with a smaller throat and larger fan) to provide a 30% increase in the throat velocity. Assuming that all other operating parameters except liquid droplet size remain constant, calculate the new Pt_d for 1.0-µm particles.

7.4 A venturi scrubber operates such that $N_t = (P_T)^{1/2}$. At a power input of 4 hp/1000 cfm and an inlet dust loading of 1.5 gr/ft^3, what is the outlet loading?

7.5 A venturi scrubber treats 20,000 acfm of air at atmospheric pressure and 77 F. The velocity at the throat entrance is 400 ft/sec, and water is introduced at a rate of 7.5 gal/1000 ft^3 of gas. The venturi throat length is 15 inches. Estimate the pressure drop across the venturi (in in. H$_2$O), and convert the pressure drop to contacting power (in kWh/1000 m^3).

7.6 Estimate the 2001 installed cost of the high-energy venturi scrubber described in Problem 7.5.

7.7 Calculate the collection efficiency for a 15-µm particle in the spray chamber described in Example 7.1. Assume that (a) 50% and (b) 20% of the spray stays suspended.

7.8 A consultant has proposed the following equation to replace the Nukiyama-Tanasawa relationship for an air-water system at "standard" conditions:

$$d_d = \frac{16,400}{V_G} + 1.45 \left(\frac{Q_L}{Q_G} \right)^{1.5}$$

where d_d is in µm, V_G is in ft/sec, Q_L is in gallons per minute (gpm), and Q_G is in 1000 acfm. Compare this equation with your own equations derived from the Nukiyama-Tanasawa relationship for an air-water system at temperatures of 15 C, 25 C, and 40 C. Derive your equations using the same units as the consultant.

7.9 Using your equations developed in Problem 7.8, determine the effect of the liquid-to-gas ratio on the removal efficiency of a venturi scrubber for 1-µm and 6-µm particles. The operating parameters are $V_G = 300$ fps, $T = 15$ C, and $f = 0.5$. Evaluate two liquid-to-gas ratios (Q_L/Q_G): 5 and 10 gal/1000 ft^3.

7.10 Determine the overall removal efficiency of a venturi scrubber for the particle distribution given below. Operating parameters are V_G = 300 fps, Q_L/Q_G = 10 gal/1000 ft^3, and f = 0.5. The operating temperature is 15 C.

Size Range, μm	0–2	2–4	4–8	8–10	10–20	>20
Mass %	15	30	37	6	10	2

7.11 Calculate the collection efficiency of the scrubber of Example 7.1 for 5-μm particles assuming 25% spray utilization.

7.12 Calculate the new pressure drop for the venturi scrubber of Problem 7.3 after the increase in the gas velocity in the throat.

7.13 Using the charts in Appendix B, estimate how much water will evaporate when 15,000 cfm of air at 95 F and 20% relative humidity becomes saturated adiabatically with water. Give your answer in gallons per minute of liquid water.

REFERENCES

Air Pollution Manual—Part II, American Industrial Hygiene Association, Detroit, MI, 1968.

Buonicore, A. J., and Davis, W. T., Eds. *Air Pollution Engineering Manual.* New York: Van Nostrand Reinhold, 1992.

Calvert, S. "Engineering Design of Fine Particle Scrubbers," *Journal of the Air Pollution Control Association, 24*(10), October 1974.

———. "Scrubbing," in *Air Pollution—Vol. IV,* A. C. Stern, Ed. New York: Academic Press, 1977.

———. "Particle Control by Scrubbing," in *Handbook of Air Pollution Technology,* S. Calvert and H. M. Englund, Eds. New York: Wiley, 1984.

Calvert, S., Goldschmid, J., Leith, D. and Mehat, D. "Wet Scrubber System Study," in *Scrubber Handbook—Vol I,* U.S. Department of Commerce, NTIS. PB–213016, August 1972.

Cheremisinoff, P. N., and Young, R. A. *Pollution Engineering Practice Handbook.* Ann Arbor, MI: Ann Arbor Science, 1976.

Crawford, M. *Air Pollution Control Theory.* New York: McGraw-Hill, 1976.

Hesketh, H. E. *Air Pollution Control.* Ann Arbor, MI: Ann Arbor Science, 1979.

Langmuir, I., and Blodgett, K. B. *U.S. Army Air Force Technical Report No. 5418,* U.S. Department of Commerce, OTS PB 27565, February 19, 1946.

Lapple, C. E., and Kamack, H. J. "Performance of Wet Dust Scrubbers," *Chemical Engineering Progress, 51,* March 1955, pp. 110–121.

McCabe, W. L., Smith, J. C., and Harriott, P. *Unit Operations of Chemical Engineering* (4th ed.). New York: McGraw-Hill, 1985.

Nukiyama, S., and Tanasawa, Y. "An Experiment on the Atomization of Liquid by Means of an Air Stream," *Transactions of the Society of Mechanical Engineers, 4*(14), Japan, 1938.

Peters, M. S., and Timmerhaus, K. D. *Plant Design and Economics for Chemical Engineers* (4th ed.). New York: McGraw-Hill, 1991.

Semrau, K. T. "Dust Scrubber Design—A Critique on the State of the Art," *Journal of the Air Pollution Control Association, 13*, December 1963, pp. 587–593.

———. "Practical Process Design of Particulate Scrubbers," in *Industrial Air Pollution Engineering*, V. Casaseno, Ed. New York: McGraw-Hill, 1980.

Stairmand, C. J. "Removal of Dust from Gases," in *Gas Purification Processes*, G. Nonhebel, Ed. London: G. Newnes Ltd., 1964.

Strauss, W. "Mist Elimination," in *Air Pollution—Vol. IV*, A. C. Stern, Ed. New York: Academic Press, 1977.

Wentz, C. A. *Hazardous Waste Management*. New York: McGraw-Hill, 1989.

Yung, C. S., Barbarika, H. F., and Calvert, S. "Pressure Loss in Venturi Scrubbers," *Journal of the Air Pollution Control Association, 27*, 1977, pp. 348–351.

◆◆◆ CHAPTER 8

AUXILIARY EQUIPMENT
HOODS, DUCTS, FANS, AND COOLERS

Design of a ventilation system for an industrial space consists essentially of three problems: (1) determination of the airflow rate and arranging its flow pattern in the space to be served; (2) design of the duct system or its counterpart; and (3) selection of the fan or other air moving equipment.

W. C. L. Hemeon, 1963

◆ 8.1 INTRODUCTION ◆

Auxiliary equipment is often given only cursory attention during the design process; nevertheless, it is very important to the proper functioning of the final system. An indoor process area must be properly ventilated; an outdoor process must be exhausted through a closed duct system. The contaminated air must be moved with proper velocity from the source point in the plant, through the ducts, into a final control device (FCD), and out the exhaust stack. The airstream often must be cooled before routing through an FCD. Proper design or selection of auxiliaries permits economical operation of the pollution control system while meeting performance standards; improper design results in higher-than-necessary costs and possibly unacceptable performance of the system.

◆ 8.2 HOODS ◆

Hoods are used to gather contaminants (gases or particles) from the workplace air. As a hood gathers contaminants, it also collects significant volumes of ambient air. As the distance between a source and a hood increases, so does the resulting total volumetric flow rate of air into the hood. Since the cost of most pollution control systems is proportional to the total volumetric flow rate, the type, size, and location of the hood are important. Proper hood designs protect the workers' breathing zones while allowing them access to the work, and yet minimize airflow requirements. The lower the airflow, the lower the total cost of the ventilation and control system.

The three main types of hoods are enclosures, canopy hoods, and capturing hoods, as shown in Figure 8.1. **Canopy hoods** are common for exhausting heated open-top tanks. They are used mainly for exhausting hot air or for removing excess humidity. Under most conditions, however, they are of limited value. Typically, canopy hoods have much lower airflow rates than capturing hoods, and they will not work to exhaust unheated tanks. They should not be used for venting hazardous components. For "low" canopy hoods (those that are within the lesser of 3 feet or one tank diameter of the tank), the design equations given by Danielson (1973) are straightforward.

For low, circular canopy hoods,

$$Q_h = 4.7(D_h)^{2.33}(\Delta T)^{0.417} \qquad (8.1)$$

where

Q_h = hood exhaust flow rate, cfm

D_h = hood diameter (usually 1–2 ft larger than the tank), ft

ΔT = difference in temperature between the hot source and the ambient air, °F

For low, rectangular canopy hoods,

$$Q_h = 6.2L(W)^{1.33}(\Delta T)^{0.417} \qquad (8.2)$$

where

L = length (longer dimension) of the hood, ft

W = width of the hood, ft

Both L and W should be 1 to 2 feet larger than the dimensions of the source.

For high canopy hoods, a significant volume of air will be entrained along with the fumes, and the design is more complicated. Details are given by Danielson (1973).

The objective of a **capturing hood** is to create "directional air currents of sufficiently high velocity to capture contaminants in the workroom air near the hood" (McDermott 1976). This objective includes not only contaminants released in the direction of the hood, but also those released in the opposite direction. A minimum design capture velocity of 50 to 100 ft/min (for slow release of contaminants into still air) must exist at the most distant point from the hood at which a contaminant from the process could be expected to be found. However, the minimum design capture velocity could be from 500 to 1000 ft/min if contaminants are released with high velocity into turbulent air. The federal Occupational Safety and Health Act (OSHA) has resulted in the establishment of numerous performance and design standards for ventilation systems. There are OSHA standards for specific substances, for specific operations, and for specific hoods.

Enclosure hoods completely enclose the process release point and are designed to provide face velocities (average air velocities into the

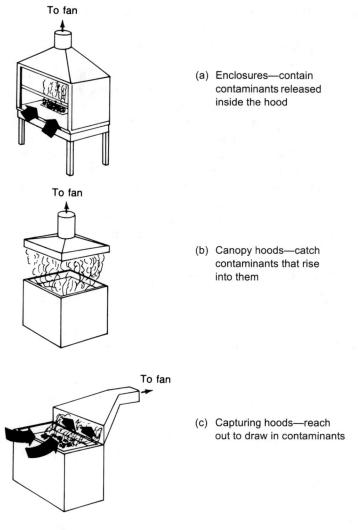

To fan

(a) Enclosures—contain contaminants released inside the hood

To fan

(b) Canopy hoods—catch contaminants that rise into them

To fan

(c) Capturing hoods—reach out to draw in contaminants

Figure 8.1
The three main types of hoods.

hood across the open face area) ranging from 100 to 200 ft/min. By multiplying the face velocity times the face area, the total required volumetric flow rate can be calculated. For capturing hoods, the capture velocity at a certain distance from the hood can be translated directly into a hood face velocity (thus yielding the required volumetric airflow rate) by relationships that depend on the hood geometry and the distance from the hood face (Danielson 1973; McDermott 1976). For controlling particulates, hood velocities are based on standard air (density = 0.075 lb/ft^3). For less dense air, the volumetric flow rate should be increased to maintain a constant mass flow rate of air into the hood (*Industrial Ventilation* 1972).

Slot hoods (hoods with long, narrow rectangular openings) have been used effectively to capture hazardous emissions from open-top tanks such as those used for electroplating processes. On wide tanks, "push/pull" air ventilation systems are often required because hoods by themselves are rarely effective at distances greater than about 2 feet. In a push/pull system, small jets of compressed air are directed across the tank toward the hood. The jet effect entrains air and creates air currents in the direction of the slot.

The pressure loss due to hoods is a strong function of the hood size and shape and the air velocity in the duct leaving the hood. As will be explained in Section 8.3, air flowing in a duct has a **velocity pressure** (VP) that increases with its velocity. The **static pressure** (SP) loss due to turbulence created during air entry into the hood is related to the duct air velocity pressure by a hood entry loss factor F_h, which is multiplied by the duct VP. Some hood entry loss factors are given in Table 8.1, along with corresponding SP losses in units of inches of water (McDermott 1976).

Table 8.1   Hood Types, Entry Loss Factors, and Typical Static Pressure Losses

Hood Type		Entry Loss Factor, F_h (As a Fraction of Duct VP)	Typical Static Pressure Loss, in. H$_2$O	
Name	Shape		duct V = 2000 fpm	V = 4000 fpm
Unflanged		0.90	0.2	0.9
Flanged		0.50	0.1	0.5
Rounded		0.03	0.0	0.03
45° Taper		0.10	0.02	0.1
Slot		1.78 (of slot VP)	2.8 (at a slot velocity of 5000 fpm)	

Adapted from McDermott, 1976.

◆ 8.3 DUCTS ◆

Ducts carry contaminated air from hoods to control equipment and from control equipment to fans. The four basic types of ducts are water cooled, refractory lined, stainless-steel, and carbon-steel. Water cooled and refractory lined ducts are reserved for use at gas temperatures above 1500 F. Generally, stainless-steel ducts are economical for gas temperatures between 1150 and 1500 F, and carbon-steel ducts are adequate for noncorrosive gases below 1150 F (Vatavuk and Neveril 1980b). If the gas is corrosive, stainless steel is required at the lower temperatures. In addition to conveying the gas stream, ducting can act as a heat exchanger to cool hot gases. Since air flowing in ducts experiences pressure loss, it is helpful to review some basic principles of fluid flow as applied to air flowing inside ducts.

When fluid flows through a closed conduit, friction between the fluid and the conduit walls creates a pressure loss. For air, elevation differences are insignificant. Thus, Bernoulli's equation, the mechanical energy balance for incompressible flow (which applies to air at low pressure drops), can be written as

$$\frac{P_1}{\rho} + \frac{v_1^2}{2g_c} + \eta \dot{w} = \frac{P_2}{\rho} + \frac{v_2^2}{2g_c} + h_f \tag{8.3}$$

where

P = static pressure, lb_f/ft^2

ρ = fluid density, lb_m/ft^3

v = fluid average linear velocity, ft/sec

g_c = gravitational constant, $32.2\ lb_m\text{-}ft/lb_f\text{-}sec^2$

η = fan efficiency

$\dot{w}$ = fan power, $ft\text{-}lb_f/lb_m$

h_f = head loss due to friction, $ft\text{-}lb_f/lb_m$

Consider Eq. (8.3) applied between two points in a duct (not including a fan). The term h_f can be converted to units of feet of fluid by multiplying all terms in Eq. (8.3) by g_c/g and by noting that $\rho g/g_c = \gamma$, the specific weight. Thus,

$$\frac{P_1}{\gamma} + \frac{v_1^2}{2g} = \frac{P_2}{\gamma} + \frac{v_2^2}{2g} + H_f \tag{8.4}$$

where

γ = specific weight of the fluid, lb_f/ft^3

g = gravitational acceleration, $32.2\ ft/sec^2$

H_f = head loss due to friction, ft of fluid

Each term in Eq. (8.4) has units of feet of fluid. Therefore, it follows that air traveling at a certain velocity has a *velocity head* according to the relationship

$$H_v = \frac{v^2}{2g} \tag{8.5}$$

where H_v = velocity head of air, ft of air

The term *velocity pressure* is sometimes used interchangeably with *velocity head*. For standard air (in ventilation work, standard air is defined as 70 F, 1.00 atm, and 50% humidity; its density is 0.075 lb_m/ft^3), Eq. (8.5) can be rearranged and written with specific units as

$$V = 4005\sqrt{VP} \tag{8.6}$$

where

VP = velocity pressure, in. H_2O

V = air velocity, ft/min

4005 = a constant to convert head loss into air velocity, $(ft/min)/(in.~H_2O)^{1/2}$

Note that in this chapter we are using two symbols for velocity to avoid confusion in the following discussion: v has units of ft/sec and V has units of ft/min. The constant 4005 is derived as follows: The velocity head H_v is equal to $v^2/2g$, in units of feet of fluid. We can convert the velocity head or velocity pressure to inches of water by using the densities of standard air and water, and the conversion factors 12 in./ft and 60 sec/min (Hemeon 1963). Thus

$$VP = \frac{(V/60)^2}{2(32.2)} \times \frac{0.075}{62.4} \times 12 = \left(\frac{V}{4005}\right)^2$$

For air densities other than standard,

$$VP_{act} = VP_{std} \times \frac{\rho_{act}}{\rho_{std}} \tag{8.7}$$

The **total pressure** (TP) of air at a point in a duct is the static pressure (SP) plus the velocity pressure (VP). Note that SP can be either positive or negative (relative to atmospheric pressure), depending on whether the point in the duct is on the suction side or the discharge side of the fan. In a fluid system, changes in duct geometry can convert SP to VP and vice versa, but TP *always* decreases in the direction of flow in sections of ductwork where the fluid does not pass through any fans (for gases) or pumps (for liquids). (Fans and pumps add mechanical energy to the fluid and increase the TP.) Thus, the mechanical energy balance, Eq. (8.4), can also be written as

$$SP_1 + VP_1 = SP_2 + VP_2 + \text{pressure loss} \tag{8.8}$$

The head loss due to friction H_f is proportional to the square of the flow velocity. For circular ducts or pipes, H_f is given by

$$H_f = \frac{fLv^2}{D_c 2g} \tag{8.9}$$

where

H_f = head loss, ft of fluid

L = length of duct, ft

f = friction factor, dimensionless

D_c = duct diameter, ft

Figure 8.2 (in English customary units) and Figure 8.3 (in SI units) are plots of the frictional pressure loss per unit length of duct for several duct sizes and airflow rates.

In addition to losses caused by friction in straight lengths of duct, energy losses also occur owing to turbulence and friction in bends, fittings, sudden expansions or contractions, and obstructions. These head losses can be calculated by using Eq. (8.9) with an appropriate fitting loss factor in place of fL/D or

$$H_f = K_f \frac{v^2}{2g} \tag{8.10}$$

where K_f = fitting loss factor, dimensionless

Another approach is to represent each fitting as an equivalent length of duct (the length of a straight duct that would result in the same pressure drop as the fitting), and then add the sum of the equivalent lengths to the total actual length. Some fitting loss factors and equivalent lengths are given in Table 8.2.

Still another approach for calculating pressure loss in ducts that has been widely adopted in ventilation work is the velocity pressure method. This method accounts for both wall frictional losses and fluid turbulence losses. Loss factors for hoods (see Table 8.1), straight ducts, and duct fittings are established in terms of velocity pressure. Thus, use of the method is quick and convenient. Table 8.3 presents some data on fitting loss factors in terms of VP, and Figure 8.4 presents friction loss for straight ducts in terms of VP.

In addition to hood entrance losses, duct friction losses, and fitting losses, there is an acceleration loss that comes from accelerating stationary ambient air to the duct velocity. The acceleration loss is separate from the other head losses, and is equal to one velocity pressure:

$$VP_a = \left(\frac{V}{4005}\right)^2 \tag{8.11}$$

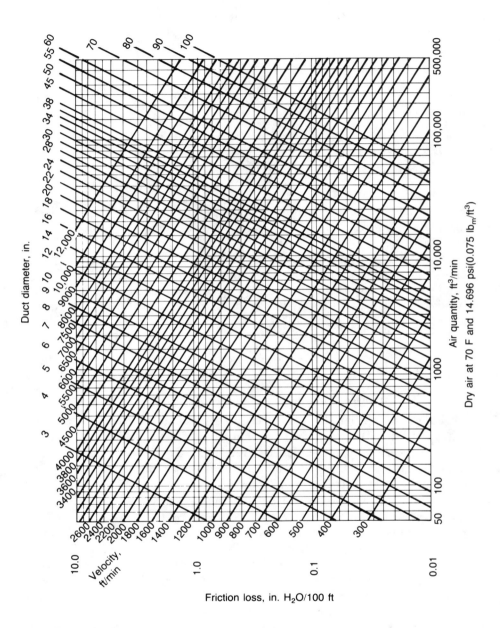

Figure 8.2
Friction losses for air in circular ducts—U.S. customary units.
Note: In each decade on both axes, there are lines at $1.5(10)^n$ and $2.5(10)^n$, where n is the appropriate integral power for that decade—e.g., the first vertical line to the left of the 0.01 axis for friction loss is 0.015 in. H_2O/100 ft, and the third horizontal line above the line for 10,000 ft^3/min (air quantity) is 25,000 ft^3/min.
(Adapted from *ASHRAE Handbook*, 1985.)

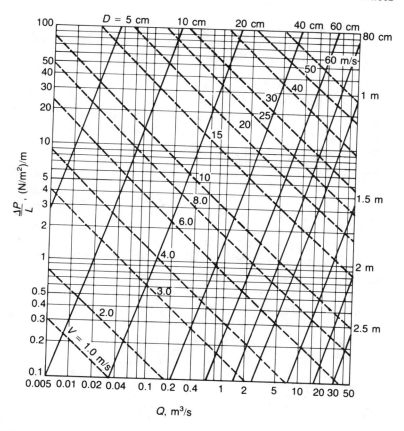

Figure 8.3
Friction losses for air in circular ducts—SI units.
(Adapted from Crawford, 1976.)

Table 8.2   Fitting Pressure Loss Factors and Equivalent Duct Lengths

Fitting	K_f	Equivalent Duct Length (as a multiplier of the duct diameter)
Tee	2.0	45
90° Elbow	0.9	20
60° Elbow	0.6	14
45° Elbow	0.45	10
Branch into Duct		
30° Angle	0.2	10
45° Angle	0.3	18
Sudden Enlargement	0.9	20

Adapted from *Industrial Ventilation*, 1972; Crawford, 1976.

where

VP_a = velocity pressure (pressure loss due to the acceleration of still air to velocity V), in. H_2O

V = air velocity in the duct, ft/min

Typical acceleration losses range from 0.25 to 1.5 inches of water, corresponding to duct velocities of about 2000 to 5000 fpm. The total hood loss is sometimes written as the sum of the hood entry loss plus the air acceleration loss; that is,

$$\Delta P_{hood} = (1.0 + F_h)\,VP_{duct} \qquad (8.12)$$

The pressure loss in a duct increases as the square of the air velocity, so low velocities are preferred. On the other hand, the transportation of certain dusts requires that minimum air velocities be maintained to prevent the dust from settling inside the duct. Table 8.4 can be used as a guideline for choosing an appropriate velocity (Danielson 1973).

It is important to choose the proper air velocity in the duct. For a given flow rate, as the duct diameter is increased, the air velocity will decrease according to the continuity equation. This decrease in velocity decreases the TP losses and fan operating costs. However, we can-

Table 8.3   Fitting Loss Factors in Terms of Velocity Pressure (VP)

Fitting Description	Fraction of VP			
90° Elbow, Circular Duct Radius of Curvature = R Duct Diameters, D_c)				
R = 2.50	0.22			
R = 2.00	0.27			
R = 1.50	0.39			
R = 1.25	0.55			
90° Elbow, Rectangular Duct (Radius of Curvature = R Duct Widths, W)	Aspect Ratio (Duct Height/Width), H/W			
	0.25	0.50	1.00	4.00
R = 0.0 (miter)	1.50	1.32	1.15	0.86
R = 0.5	1.36	1.21	1.05	0.79
R = 1.0	0.45	0.28	0.21	0.19
R = 2.0	0.24	0.15	0.11	0.10
Branch into Duct				
30°	0.18			
45°	0.28			
60°	0.44			
90° (tee)	1.00			

Adapted from *Industrial Ventilation*, 1972.

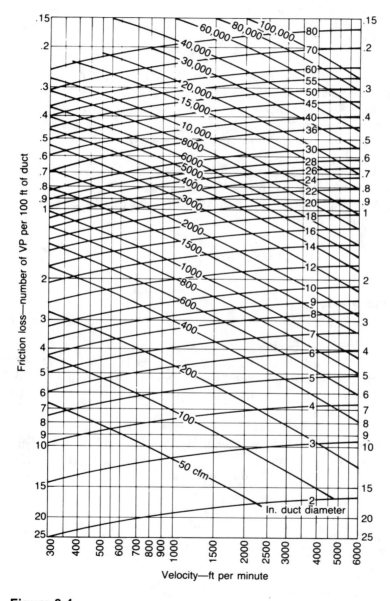

Figure 8.4
Friction losses for air in circular ducts in terms of velocity pressures (VP).
(Adapted from *Industrial Ventilation*, 1972.)

Table 8.4   Minimum Air Velocities in Ducts to Prevent Dust Settling

Type of Dust	Velocity, ft/min
Low Density (Gases, Vapors, Smoke, Flour, Lint)	2000
Medium-Low Density (Grain, Sawdust, Plastic, Rubber)	3000
Medium-High Density (Cement, Sandblast, Grinding)	4000
High Density (Metal Turnings, Lead Dust)	5000

Adapted from Danielson, 1973.

not use overly large ducts because of the requirement of minimum dust transport velocities and the fact that duct installed costs increase with duct diameter. For systems with only gases and vapors, duct velocities in the range of 2000 to 3000 ft/min often result in a good balance between duct costs and fan operating costs (*Industrial Ventilation* 1972).

Recall that the fan operating costs are directly related to the product of the fluid pressure loss times the volumetric flow rate. In Chapter 4, Eq. (4.14) and Examples 4.3 and 4.4 demonstrated that fact. In English customary units, Eq. (4.14) is

$$\dot{w}_f = 0.0001575 Q \Delta P \qquad \text{(8.13)}$$

where

$\dot{w}_f$ = fluid power, hp

Q = airflow rate, acfm

ΔP = pressure loss, in. H_2O

Equation (8.13) is valid regardless of air density as long as both Q and ΔP are determined at the same conditions. This is true because at any air density other than standard, Q is proportional to $1/\rho$, and ΔP is proportional to ρ.

Ducting is often rectangular or square rather than circular because of space limitations, for ease of installation, and so forth. For noncircular ducts, the hydraulic diameter is sometimes used to characterize the duct. The hydraulic diameter can be calculated as

$$D_h = \frac{4A}{P} \qquad \text{(8.14)}$$

where

D_h = hydraulic diameter, ft

A = cross-sectional area, ft^2

P = "wetted" perimeter (inside perimeter of duct cross-section in contact with the fluid), ft

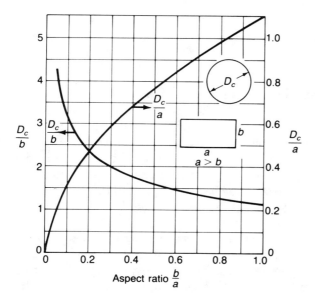

Figure 8.5
Flow-equivalent rectangular and circular ducts.
(Adapted from Crawford, 1976.)

The hydraulic diameter is a good characteristic dimension for square ducts, but the hydraulic diameter is different for rectangles with equal areas but different aspect ratios (ratios of height to width). Therefore, for the calculation of pressure loss in rectangular ducts, several charts similar to Figures 8.2–8.4 would be needed—one for each different aspect ratio. Rather than using this tedious approach, we can use Figure 8.5, developed by Crawford (1976), to equate circular and rectangular ducts based on each carrying an equal flow of air at an equal pressure drop. Figure 8.5 provides for ducts that are truly flow-equivalent. Therefore, the design can be based on circular ducts, for which much data are available. Once the design is finished, rectangular ducts of any required aspect ratio can be substituted to meet spacing needs. This procedure is illustrated in the following example.

▶ ▶ ▶ ▶ ▶ ▶ ▶ ▶ ▶

Example 8.1

Specify the dimensions of (a) a square duct and (b) a 4:1 rectangular duct that are flow-equivalent to a circular duct with a diameter D_c = 36 inches.

Solution

(a) A square duct has an aspect ratio of 1.0. From Figure 8.5, at an aspect ratio of 1.0, we look up to the D_c/b line and read a value of 1.13 from the left ordinate, from which we calculate b = 36/1.13 = 31.8 = 32 in. (a is also 32 inches).

(b) From Figure 8.5, at an aspect ratio of 0.25, $D_c/b = 2.2$ and $b = 16.4$ inches. The dimensions of the duct would then be 64×16 inches.

◀ ◀ ◀ ◀ ◀ ◀ ◀ ◀ ◀

The design of a single-hood, single-exhaust duct system is relatively straightforward. After choosing the airflow rate and the hood (if one is needed), the designer sketches the duct layout as it will fit into the plant. Next, a duct diameter is chosen that will maintain the design duct velocity. Then the designer simply works from the entrance of the hood (or first opening) along the duct system toward the fan, adding up all the pressure losses. The losses due to hood entry and the initial acceleration velocity pressure loss (VP_a) must be included. The same procedure is followed from the outlet of the fan to the exit from the stack. The two sums are then added to give the total pressure loss of the system.

Designing systems for multiple hoods and multiple branches in the ducting is more complex because the total design airflow must be properly distributed among all the hoods and branch ducts. The approach is basically similar to that for a single-duct system. After sketching the layout, we choose a diameter for each duct to maintain design velocities, and then calculate the pressure losses through the hood, ducts, and fittings up to the point where two ducts are joined. At any junction point, the SP in the two ducts must be equal under the design conditions. The branch duct with the greatest calculated SP loss is considered to be the critical path duct.

In practice, only one branch duct will be critical; that is, one branch will have the greatest pressure loss at the design velocity. In operation, the system will "balance" itself; that is, the flow of air will be redistributed such that static pressures are equal at junctions. Actual velocities will be higher than design velocities in the other branches. However, if the designer provides more resistance in some of the ducts, they might have less airflow than desired. Designers have three ways to obtain the proper SP and velocities in ducts (McDermott 1976):

1. Install restrictions, such as adjustable dampers, in branch ducts (this method gives greater flexibility to operators and can accommodate future operating changes most easily).

2. Reduce or increase the diameter of some branch ducts, then recalculate pressure losses. Repeat this procedure until the proper balance is achieved. Check the air velocities to ensure that minimum velocities are maintained.

3. Increase the airflow rate above the minimum required in some hoods to achieve at least the minimum in all.

However, Menkel (1985) points out that option 1 is often not successful in achieving the proper SP and velocities, and options 2 and 3 are preferred.

In the design of ducts, all figures and charts are for standard air. Since all fan rating tables are also developed for standard air, it is convenient to design ducts for standard air. It is helpful to keep in mind that the numerical value of the *volumetric* flow rate of the air (not the mass flow rate) is maintained as constant in this procedure. The TP loss in ducts is calculated based on standard air, as is the rise in TP for a fan. If the pressure drop of a control device must be added to the total pressure loss for ducts and hoods, it is important to be sure that the device pressure loss is given on the same basis as for the duct system; that is, all must be related to standard air or all must be related to actual conditions.

The relationship between pressure drop and air density for a change in operating conditions is

$$\Delta P_{act} = \Delta P_{std} \times \frac{\rho_{act}}{\rho_{std}} \qquad (8.15)$$

▶ ▶ ▶ ▶ ▶ ▶ ▶ ▶ ▶

Example 8.2

Calculate the actual pressure loss expected when moving 10,000 acfm of air (at 500 F and 1 atm) through 250 feet of straight duct (20-inch diameter, circular cross-section). Assume a flanged hood opening. The air must also pass through a control device with a rated pressure drop of 5.0 in. H_2O at 110 F and 1 atm.

Solution

From Figure 8.2, the intersection of the 10,000 cfm line and the 20-inch duct diameter line yields a linear velocity of about 4600 fpm, and a pressure drop of 1.3 inches of water per 100 feet of duct. Thus, for standard air,

$$\text{duct SP loss} = \frac{1.3 \text{ in. } H_2O}{100 \text{ ft}} \times 250 \text{ ft} = 3.2 \text{ in. } H_2O$$

$$\text{hood SP loss} = (1.0 + 0.5)\,VP$$

but

$$VP = \left(\frac{4600}{4005}\right)^2 = 1.3 \text{ in. } H_2O$$

so

$$\text{hood SP loss} = 1.5(1.3) = 2.0 \text{ in. } H_2O$$

$$\begin{array}{l} \text{control device loss} \\ \text{(for standard air)} \end{array} = 5.0 \text{ in. } H_2O \times \frac{\rho_{std}}{\rho_{act}}$$

$$= 5.0 \text{ in. } H_2O \times \frac{T_{act}}{T_{std}}$$

$$= 5.0 \text{ in. } H_2O \times \frac{570}{530}$$

$$= 5.4 \text{ in. } H_2O$$

total pressure loss $= 3.2 + 2.0 + 5.4$
(for standard air)

$$= 10.6 \text{ in. } H_2O$$

total pressure loss $= 10.6 \times \dfrac{\rho_{act}}{\rho_{std}}$
(actual air)

$$= 10.6 \times \frac{T_{std}}{T_{act}}$$

$$= 10.6 \times \frac{530}{960}$$

$$= 5.8 \text{ in. } H_2O$$

◀ ◀ ◀ ◀ ◀ ◀ ◀ ◀ ◀

Duct problems can arise in connection with either of the two main functions of ducts, which are (1) to carry the air from point A to point B, and (2) to provide proper resistance (in multiple-hood systems) to properly balance the total airflow among all hoods. However, in the path of greatest resistance (the critical path), the usual objective is to minimize the pressure loss so as to minimize fan power costs. The main duct, the fan entrance and exit ducts, and the final control device are always included in the path of greatest resistance. Only one hood and one branch duct will be included in the critical path, but sometimes these can be major contributors to the overall pressure loss. Resistance in the critical path can be reduced (and airflow increased) by eliminating duct elbows, by replacing tight elbows with wide-curvature ones or with elbows with turning vanes, and by supplying larger diameter ducts.

Whether airflow is increased by reducing the system resistance or by increasing the fan power, there is a unique point of operation for every fan–duct system. That point is defined by the intersection of the fan curve with the system resistance curve, as shown in Figure 8.6.

◆ 8.4 FANS ◆

Fans provide the energy needed to move air through hoods, ducts, and control equipment. Fans move air and provide a total pressure gain to overcome the pressure losses created when air flows through the ducting and control devices. Most fan rating tables give data in

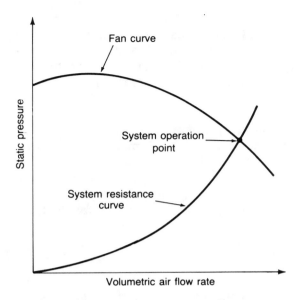

Figure 8.6
The system resistance curve and the fan operating curve.

terms of fan static pressure (FSP). The fan static pressure is the fan total pressure minus the fan outlet velocity pressure. The fan total pressure is simply the TP at the fan outlet minus the TP at the fan inlet. Thus, the fan static pressure is related to the duct system pressures as follows:

$$FSP = TP_{outlet} - TP_{inlet} - VP_{outlet} \qquad (8.16)$$

Equation (8.16) can also be written as

$$FSP = (SP_{outlet} + VP_{outlet}) - (SP_{inlet} + VP_{inlet}) - VP_{outlet}$$

or

$$FSP = SP_{outlet} - SP_{inlet} - VP_{inlet}$$

which is the usual form in which it is published. In terms of duct system pressure losses (without regard to sign), we write

$$FSP = \Delta P_{\text{suction side}} + \Delta P_{\text{discharge side}} - VP_{inlet} \qquad (8.17)$$

Equation (8.17) is the basic equation by which we select a fan to match with a desired duct system.

The two basic types of fans are (1) centrifugal or radial flow, and (2) propeller or axial flow (Danielson 1973). In a centrifugal fan, air enters at the eye of the rotor, turns at right angles, and is accelerated and compressed by centrifugal force into the discharge. In an axial flow fan, the air flows straight through the device along the axis of rotation. Airfoil blades pull the air in on the leading edge and discharge it from the trailing edge. The centrifugal force is converted into a pressure rise by stationary vanes. Table 8.5 lists the major types of fans. Figure 8.7 is a photograph showing a large fan wheel and a small fan wheel.

To select the proper fan for any given application, the three basic items of information that are required are air volumetric flow rate, fan static pressure increase to be supplied, and gas density at the fan. Other factors that are usually important in choosing the right fan are the types and concentrations of any contaminants (dusts, liquids, or combustible gases) present in the main airflow, the space available for installation, and the importance of noise as a limiting factor.

FAN CURVES

The performance of a fan is summarized by its "fan curves," which present quantitatively the relationships among airflow, static pressure delivered, brake horsepower, and mechanical efficiency. In general, centrifugal fans with backward inclined blades are desirable for their stable, efficient operation. However, backward inclined fans must operate in relatively dust-free environments; for dusty air, radial or straight blade fans are better. A typical fan curve for a backward inclined fan is presented in Figure 8.8.

FAN LAWS

For a given fan, as the speed of rotation is increased, both the airflow rate and the static pressure are increased. Similarly, for two *geometrically similar* fans at constant speeds of rotation, the larger fan produces more flow at a higher pressure. Of course, the higher performance means more power consumption in both of these cases. The quantitative relationships among speed of rotation, airflow rate, and static pressure are known as the **fan laws**, which are

$$\frac{Q_1}{N_1 D_1^3} = \frac{Q_2}{N_2 D_2^3} = \text{constant} \qquad (8.18a)$$

$$\frac{\text{FSP}_1}{\rho_1 N_1^2 D_1^2} = \frac{\text{FSP}_2}{\rho_2 N_2^2 D_2^2} = \text{constant} \qquad (8.18b)$$

$$\frac{\dot{w}_1}{\rho_1 N_1^3 D_1^5} = \frac{\dot{w}_2}{\rho_2 N_2^3 D_2^5} = \text{constant} \qquad (8.18c)$$

Table 8.5   Types of Fans

Centrifugal Fan Types	
Description	**Applications**
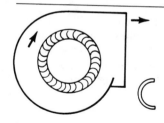 *Forward Curved:* The wheel's blades are small and curved forward in the direction of the wheel's rotation. This fan runs at a relatively low speed to move a given amount of air. This wheel type is most often called a squirrel-cage wheel.	Primarily for low-pressure heating, ventilating and air conditioning such as domestic furnaces, central station units and packaged air conditioning equipment.
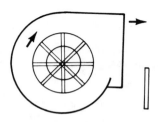 *Radial Blade:* This wheel is like a paddle wheel . . . with or without side rims. The blades are perpendicular to the direction of the wheel's rotation and the fan runs at a relatively medium speed to move a given amount of air.	The radial blade type is designed for material handling applications, features rugged construction and simple field repair. Also used for high-pressure industrial requirements.
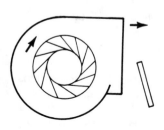 *Backward Inclined:* The wheel's blades are flat and lean away from the direction of the wheel's rotation. This fan runs at a relatively high speed to move a given amount of air. It is more efficient than the above-listed types.	General heating, ventilating and air conditioning systems. Used in many industrial applications where the airfoil blade might be subjected to erosion from light dust.
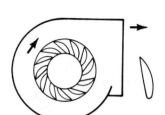 *Airfoil Blade:* Although not a "Basic Type," this is an important refinement of the backward inclined wheel design. It has the highest efficiency and runs at a slightly higher speed than the standard flat blade to move a given amount of air.	Most efficient of all centrifugals. Usually used in both larger HVAC systems and clean air industrial applications where the energy savings are significant. Can be made with special construction for dusty air.

Continued

Table 8.5   *(continued)*

Centrifugal Fan Types	
Description	**Applications**
Radial Tip: The wheel's blades are somewhat cupped in the direction of the wheel's rotation but the blade leans back so that its outside tip approaches a radial position. This fan runs at approximately the same speed as a backward inclined wheel to move a given amount of air.	This type is also designed for material handling or dirty or erosive applications and is more efficient than the radial blade.

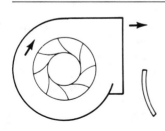

Axial Fan Types	
Propeller: Wheels usually have two or more single thickness blades in a simple ring enclosure. Efficiencies are generally low and use is limited to low pressure.	High-volume air moving applications such as air circulation within a space or ventilation through a wall without attached duct work.
Tubeaxial: The wheel is similar to the propeller type except it usually has more blades of a heavier design. The wheel is enclosed in a drum or tube to increase efficiency and pressure capability.	Ducted HVAC applications where air distribution on the downstream side is not critical. Industrial applications include drying ovens, paint spray booths, and fume exhaust systems.
Vaneaxial: Most efficient axial type fan. Uses straightening vanes to improve efficiency and pressure capability. Blades often have airfoil shapes and may be available with adjustable pitch. Pressure capabilities are medium to high.	General HVAC systems especially where straight-through flow and compactness is required. Good downstream air distribution. Used in many industrial applications.

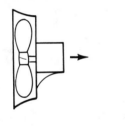

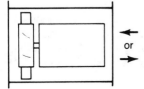

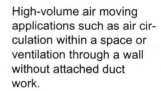

Continued

Table 8.5   *(continued)*

	Axial Fan Types	
	Description	**Applications**
	*Inline Centrifugal:* This type is actually a centrifugal fan, with airfoil or backward inclined wheel in a vaneaxial casing. Good efficiency but lower than a similar centrifugal type.	Used primarily for low-pressure return air systems in heating, ventilating and air conditioning applications. Has straight-through flow.

Adapted from the Chicago Blower Corporation, Glendale Heights, IL, 1978.

Figure 8.7
Centrifugal fan wheels. [This photo shows a large (120-in. diameter) fan wheel (on a 22-foot-long shaft) that has been rebuilt and is ready for shipment. In the same shipment is a small (22-in. diameter) rebuilt wheel.]
(Courtesy of Barron Industries, Leeds, Alabama.)

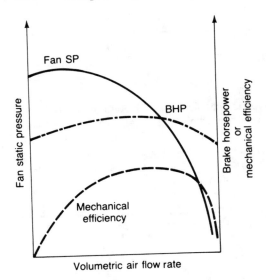

Figure 8.8
A typical fan curve for a backward inclined fan.

where

     Q = volumetric flow rate

     N = fan speed

     D = fan wheel diameter

     FSP = static pressure

     ρ = air density

     $\dot{w}$ = fan power (brake horsepower)

     1, 2 = subscripts referring to different operating points

Note that the fan laws apply to the same fan or to geometrically similar fans that are being operated similarly (at the same point of rating or at constant efficiency). As Crawford (1976) states, "It is certainly possible to operate the two fans, even though they are geometrically similar, at different conditions so that the fan laws do not hold. . . ." The fan laws are summarized in Table 8.6.

     Geometric similarity means that all dimensions of the two fans are in the same ratio to each other. To further clarify the statement of similarity of operation, **fan efficiency** can be defined as the power input into the fluid divided by the power input into the fan; that is,

$$\eta = \frac{k(\Delta P)Q}{\dot{w}} \tag{8.19}$$

where

     η = fractional efficiency

     k = units conversion factor

Table 8.6   Fan Law Relationships

Independent Variable	Parameters Held Constant	Fan Law
N	D, ρ	$Q \propto N$ $FSP \propto N^2$ $\dot{w} \propto N^3$
D	N, ρ	$Q \propto D^3$ $FSP \propto D^2$ $\dot{w} \propto D^5$
ρ	N, D	$Q = \text{constant}$ $FSP \propto \rho$ $\dot{w} \propto \rho$
D	FSP, ρ	$Q \propto D^2$ $N \propto 1/D$ $\dot{w} \propto D^2$

Note that the product of Eqs. (8.18a) and (8.18b) divided by Eq. (8.18c) yields the fan efficiency, which therefore must be constant for the fan laws to be valid. However, dividing Eq. (8.18b) by the square of Eq. (8.18a) and holding D constant (that is, dealing with only one fan) yields

$$FSP_1 = \rho_1 Q_1^2 \eta \qquad \textbf{(8.20)}$$

Thus, on SP versus Q coordinates, and for constant density, lines of constant efficiency are parabolas, as shown in Figure 8.9. In applying the fan laws, we can go from point A to point B or from point C to point D in Figure 8.9, but not from point A to D or B to C or A to C or B to D.

▶ ▶ ▶ ▶ ▶ ▶ ▶ ▶ ▶

Example 8.3

A fan turning at 1000 rpm supplies 800 ft³/min of air at an SP gain of 3.0 in. H₂O, and draws 1.5 hp. If the fan is speeded up to move 1500 ft³/min, calculate the new speed, the new SP gain, and the new power consumption.

Solution

$$N_2 = N_1\left(\frac{Q_2}{Q_1}\right) = 1000\left(\frac{1500}{800}\right) = 1875 \text{ rpm}$$

$$FSP_2 = FSP_1\left(\frac{N_2}{N_1}\right)^2 = 3.0(1.875)^2 = 10.5 \text{ in. H}_2\text{O}$$

$$\dot{w}_2 = \dot{w}_1\left(\frac{N_2}{N_1}\right)^3 = 1.5(1.875)^3 = 9.9 \text{ hp}$$

◀ ◀ ◀ ◀ ◀ ◀ ◀ ◀ ◀

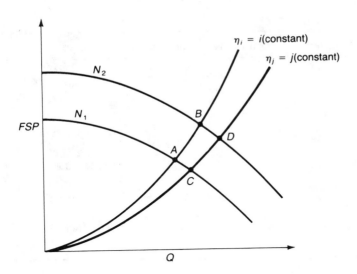

Figure 8.9
Fan diagram illustrating application of the fan laws.

▶ ▶ ▶ ▶ ▶ ▶ ▶ ▶ ▶

Example 8.4

To increase the air volumetric flow rate as stated in Example 8.3, a larger fan is suggested, to be operated at the same speed. If the wheel diameter of the existing fan is 20 inches, calculate the theoretical new diameter and the new power requirement.

Solution

$$D_2 = D_1\left(\frac{Q_2}{Q_1}\right)^{1/3} = 20\left(\frac{1500}{800}\right)^{1/3} = 24.7 \text{ in.}$$

$$\dot{w}_2 = \dot{w}_1\left(\frac{D_2}{D_1}\right)^5 = 1.5\left(\frac{24.7}{20}\right)^5 = 4.3 \text{ hp}$$

◀ ◀ ◀ ◀ ◀ ◀ ◀ ◀ ◀

▶ ▶ ▶ ▶ ▶ ▶ ▶ ▶ ▶

Example 8.5

A fan moves 10 m^3/s of standard air ($\rho = 1.183 \text{ kg/m}^3$) and delivers 900 N/m^2 SP when turning at 1000 rpm. If the density changes to 1.00 kg/m^3, estimate the new flow rate, the FSP, and the percent power decrease.

Solution

The flow rate will be unchanged at 10 m³/s.

$$\mathrm{FSP}_2 = \mathrm{FSP}_1\left(\frac{\rho_2}{\rho_1}\right) = 900(0.845) = 761 \text{ N/m}^2$$

$$\dot{w}_2 = \dot{w}_1\left(\frac{\rho_2}{\rho_1}\right) = \dot{w}_1(0.845)$$

$$\% \text{ decrease} = \frac{\dot{w}_1 - \dot{w}_2}{\dot{w}_1} \times 100 = 15.5\%$$

◄ ◄ ◄ ◄ ◄ ◄ ◄ ◄ ◄

FAN RATING TABLES

Every fan has a separate fan curve for each speed. Fan rating tables, supplied by fan manufacturers for homologous (geometrically similar) families of fans, cover the entire range of operating points for fans. One table is supplied for each size of fan, and each table gives data for Q, SP, N, and $\dot{w}$ at many points. Very large ranges of airflow rates and static pressures can be obtained within a geometrically similar family. For values between tabular entries, we use linear interpolation. Table 8.7 provides some fan rating tables adapted from the Chicago Blower Corporation (1978), and Example 8.6 illustrates their use.

▶ ▶ ▶ ▶ ▶ ▶ ▶ ▶ ▶

Example 8.6

From Table 8.7, select a fan to move 6000 cfm of standard air and deliver 7.5 in. H₂O. How fast will it turn and what brake horsepower will it draw?

Solution

The required cfm is too high for a 15 SISW (single inlet, single width) fan and too low for a 30 SISW fan. (*Note:* A complete set of tables would have closer sizes, and one of several sizes might be satisfactory.) From the 20 SISW table, using double linear interpolation, N = 2245 rpm and $\dot{w}$ = 10.0 bhp.

◄ ◄ ◄ ◄ ◄ ◄ ◄ ◄ ◄

FAN SELECTION

The basic type of fan is determined by the type of gas flow to be handled. Next, the fan size is chosen from the rating tables. Usually, fans in the middle of a rating table are near their peak efficiencies. If a design operating point is near the top or bottom of a table, you should

Table 8.7 Examples of Fan Rating Tables

15 SISW

WHEEL	OUTLET
16-3/16 in. diameter	14-5/8 × 13-1/16 in. inside
1.32 sq. ft. inside area	

MAXIMUM BHP = .212 $\left(\dfrac{RPM}{1000}\right)^3$

TIP SPEED, fpm = 4.24 × RPM

CLASS I – RPM 3302

CFM	OV FPM	1/4" SP RPM	1/4" SP BHP	1/2" SP RPM	1/2" SP BHP	3/4" SP RPM	3/4" SP BHP	1" SP RPM	1" SP BHP	1-1/4" SP RPM	1-1/4" SP BHP	1-1/2" SP RPM	1-1/2" SP BHP	1-3/4" SP RPM	1-3/4" SP BHP	2" SP RPM	2" SP BHP	2-1/2" SP RPM	2-1/2" SP BHP	3" SP RPM	3" SP BHP	3-1/2" SP RPM	3-1/2" SP BHP
660	500	559	.03																				
792	600	607	.04	752	.08																		
924	700	662	.06	787	.10																		
1056	800	719	.07	834	.12	942	.17																
1188	900	778	.09	887	.14	983	.20	1080	.26														
1320	1000	839	.11	942	.17	1031	.23	1118	.29	1205	.36	1290	.44										
1452	1100	902	.14	998	.20	1085	.27	1164	.33	1242	.40	1321	.48	1400	.56								
1584	1200	966	.17	1057	.24	1139	.31	1214	.38	1287	.45	1359	.53	1432	.61	1504	.70						
1716	1300	1031	.21	1117	.28	1196	.35	1268	.43	1336	.50	1402	.58	1469	.67	1536	.76	1668	.95				
1848	1400	1098	.25	1178	.33	1253	.40	1323	.48	1389	.56	1451	.64	1513	.73	1575	.82	1700	1.02				
1980	1500	1165	.30	1240	.38	1312	.46	1379	.54	1444	.63	1504	.72	1562	.80	1619	.90	1735	1.10	1851	1.31		
2112	1600	1234	.35	1304	.44	1373	.52	1437	.61	1499	.70	1558	.79	1613	.89	1667	.98	1775	1.18	1885	1.40	1994	1.64
2244	1700	1302	.41	1369	.50	1433	.59	1496	.69	1556	.78	1613	.88	1668	.98	1718	1.07	1820	1.28	1923	1.50	2025	1.74
2376	1800	1370	.47	1434	.57	1495	.67	1557	.77	1613	.87	1669	.97	1722	1.07	1773	1.17	1869	1.39	1966	1.61	2063	1.85
2508	1900	1439	.54	1500	.65	1559	.75	1617	.86	1673	.96	1725	1.07	1778	1.17	1828	1.28	1921	1.50	2012	1.73	2104	1.97
2640	2000	1508	.62	1566	.73	1623	.84	1678	.95	1733	1.06	1783	1.17	1834	1.28	1882	1.40	1976	1.63	2062	1.86	2150	2.11
2904	2200	1647	.81	1702	.93	1753	1.05	1803	1.17	1853	1.29	1904	1.41	1950	1.53	1996	1.65	2084	1.90	2169	2.15	2248	2.41
3168	2400	1787	1.02	1839	1.15	1855	1.28	1931	1.41	1978	1.55	2024	1.68	2070	1.82	2114	1.95	2197	2.21	2278	2.48	2356	2.76
3432	2600	1927	1.28	1975	1.42	2020	1.56	2063	1.70	2106	1.84	2148	1.99	2190	2.13	2233	2.28	2313	2.56	2390	2.85	2466	3.14
3696	2800	2071	1.58	2115	1.73	2156	1.88	2195	2.03	2236	2.18	2275	2.33	2315	2.49	2355	2.64	2432	2.96	2505	3.26	2578	3.57

CLASS II – RPM 3586

CFM	OV FPM	4" SP RPM	4" SP BHP	4-1/2" SP RPM	4-1/2" SP BHP	5" SP RPM	5" SP BHP	5-1/2" SP RPM	5-1/2" SP BHP	6" SP RPM	6" SP BHP	7" SP RPM	7" SP BHP	8" SP RPM	8" SP BHP	9" SP RPM	9" SP BHP	10" SP RPM	10" SP BHP	11" SP RPM	11" SP BHP	12" SP RPM	12" SP BHP
1980	1500	2070	1.78																			
2112	1600	2096	1.89	2199	2.13																	
2244	1700	2128	1.99	2225	2.25	2319	2.52															
2376	1800	2159	2.10	2256	2.37	2346	2.64	2435	2.92													
2508	1900	2195	2.23	2288	2.50	2378	2.78	2469	3.07	2549	3.36											
2640	2000	2235	2.36	2324	2.64	2410	2.92	2497	3.22	2578	3.52	2712	3.98									
2904	2200	2327	2.68	2406	2.96	2484	3.25	2564	3.55	2644	3.86	2737	4.13	2889	4.81							
3168	2400	2429	3.04	2502	3.32	2573	3.62	2645	3.93	2718	4.25	2799	4.51	2945	5.18	3085	5.90					
3432	2600	2537	3.45	2605	3.75	2671	4.05	2738	4.37	2805	4.69	2863	4.91	3007	5.62	3143	6.35	3273	7.10			
3696	2800	2647	3.89	2715	4.22	2778	4.54	2840	4.86	2902	5.19	2938	5.37	3072	6.08	3207	6.85	3335	7.63	3405	7.89	
3960	3000	2759	4.38	2825	4.72	2887	5.07	2949	5.42	3007	5.76	3026	5.89	3150	6.62	3275	7.39	3399	8.19	3456	8.42	
4224	3200	2874	4.93	2937	5.29	2998	5.65	3059	6.01	3116	6.39	3122	6.47	3237	7.22	3353	7.99	3471	8.82	3522	9.03	
4488	3400	2992	5.54	3052	5.90	3111	6.28	3169	6.66	3225	7.05	3226	7.12	3335	7.88	3442	8.68	3550	9.50	3586	9.65	3581	9.25
4752	3600	3113	6.19	3170	6.58	3225	6.97	3282	7.37	3338	7.78	3335	7.84	3437	8.62	3540	9.43					
5016	3800	3234	6.90	3291	7.32	3344	7.73	3397	8.14	3451	8.56	3444	8.60	3546	9.43						
5280	4000	3355	7.64	3412	8.11	3464	8.55	3518	8.99	3567	9.41	3555	9.42								

Use Class II fan in the shaded area.

20 SISW

WHEEL — 21-9/16 in. diameter — OUTLET — 19-7/16 × 17-3/8 in. inside — 2.34 sq. ft. inside area

MAXIMUM BHP = $.894\left(\dfrac{\text{RPM}}{1000}\right)^3$

TIP SPEED, fpm = 5.65 × RPM

CLASS I – RPM 2469 CLASS II – RPM 2895

Table (lower static pressures)

OV FPM	CFM	1/4" RPM	1/4" BHP	1/2" RPM	1/2" BHP	3/4" RPM	3/4" BHP	1" RPM	1" BHP	1-1/4" RPM	1-1/4" BHP	1-1/2" RPM	1-1/2" BHP	1-3/4" RPM	1-3/4" BHP	2" RPM	2" BHP	2-1/2" RPM	2-1/2" BHP	3" RPM	3" BHP	3-1/2" RPM	3-1/2" BHP
500	1170	419	.06																				
600	1404	455	.08	564	.15																		
700	1638	496	.10	590	.18																		
800	1872	539	.13	625	.21	707	.31																
900	2106	584	.17	665	.26	737	.35	810	.46														
1000	2340	629	.21	706	.31	774	.41	839	.52	904	.65	968	.78										
1100	2574	676	.26	748	.36	813	.48	873	.59	932	.72	991	.85	1050	1.00								
1200	2808	725	.31	793	.43	854	.55	911	.67	965	.80	1019	.94	1074	1.09	1128	1.25						
1300	3042	774	.37	838	.50	897	.63	951	.76	1002	.90	1052	1.04	1102	1.19	1152	1.35	1251	1.69				
1400	3276	823	.45	883	.58	940	.72	992	.86	1042	1.01	1088	1.15	1135	1.31	1181	1.47	1275	1.82				
1500	3510	873	.53	930	.68	984	.83	1035	.97	1083	1.12	1128	1.28	1171	1.44	1214	1.60	1302	1.96	1389	2.34		
1600	3744	925	.62	978	.78	1030	.94	1078	1.09	1125	1.25	1169	1.42	1210	1.58	1250	1.75	1331	2.11	1414	2.50	1495	2.91
1700	3978	976	.73	1027	.89	1075	1.06	1122	1.23	1167	1.39	1210	1.56	1251	1.74	1289	1.91	1365	2.28	1442	2.68	1519	3.09
1800	4212	1028	.84	1076	1.02	1121	1.19	1167	1.37	1210	1.55	1252	1.73	1292	1.91	1330	2.09	1402	2.47	1474	2.87	1547	3.30
1900	4446	1079	.97	1125	1.17	1169	1.34	1213	1.53	1254	1.72	1294	1.90	1333	2.09	1371	2.29	1441	2.67	1509	3.08	1578	3.51
2000	4680	1131	1.11	1175	1.31	1217	1.50	1258	1.69	1300	1.90	1337	2.09	1376	2.29	1412	2.49	1482	2.90	1547	3.31	1612	3.75
2200	5148	1235	1.44	1276	1.65	1314	1.86	1352	2.08	1390	2.29	1428	2.52	1463	2.73	1497	2.94	1563	3.38	1627	3.84	1686	4.29
2400	5616	1340	1.83	1379	2.06	1414	2.29	1449	2.52	1483	2.76	1518	2.99	1553	3.23	1585	3.47	1648	3.94	1708	4.42	1767	4.92
2600	6084	1445	2.28	1481	2.53	1515	2.78	1547	3.03	1580	3.28	1611	3.54	1643	3.79	1675	4.06	1735	4.56	1793	5.07	1849	5.60
2800	6552	1553	2.81	1586	3.08	1617	3.35	1646	3.61	1677	3.88	1706	4.16	1736	4.43	1766	4.70	1824	5.27	1879	5.80	1933	6.36

Table (higher static pressures) (shaded = Class II)

OV FPM	CFM	4" RPM	4" BHP	4-1/2" RPM	4-1/2" BHP	5" RPM	5" BHP	5-1/2" RPM	5-1/2" BHP	6" RPM	6" BHP	7" RPM	7" BHP	8" RPM	8" BHP	9" RPM	9" BHP	10" RPM	10" BHP	11" RPM	11" BHP	12" RPM	12" BHP
1500	3510	1552	3.17																				
1600	3744	1572	3.35	1649	3.79																		
1700	3978	1596	3.54	1669	4.01	1739	4.50																
1800	4212	1619	3.74	1692	4.22	1760	4.71	1826	5.20														
1900	4446	1647	3.97	1716	4.45	1784	4.95	1852	5.46	1912	5.99	2034	7.10										
2000	4680	1676	4.21	1743	4.70	1807	5.20	1872	5.73	1934	6.28	2053	7.36	2167	8.52								
2200	5148	1745	4.77	1804	5.26	1863	5.78	1923	6.31	1983	6.87	2099	8.03	2208	9.24	2314	10.46						
2400	5616	1822	5.41	1876	5.91	1929	6.45	1984	6.99	2038	7.56	2147	8.74	2255	10.01	2358	11.25						
2600	6084	1903	6.14	1954	6.67	2003	7.21	2054	7.77	2103	8.35	2204	9.56	2304	10.83	2405	12.19	2455	12.60				
2800	6552	1985	6.93	2036	7.51	2084	8.08	2130	8.65	2176	9.25	2269	10.49	2362	11.78	2456	13.15	**2501**	**13.54**	**2554**	**13.98**		
3000	7020	2069	7.80	2119	8.41	2165	9.03	2212	9.64	2255	10.25	2342	11.52	2428	12.85	**2515**	**14.22**	**2549**	**14.58**	**2594**	**14.98**		
3200	7488	2155	8.77	2203	9.41	2248	10.05	2294	10.70	2337	11.37	2419	12.68	**2501**	**14.03**	**2582**	**15.45**	**2603**	**15.69**	**2641**	**16.06**	**2686**	**16.45**
3400	7956	2244	9.85	2289	10.50	2333	11.18	2377	11.86	2419	12.55	**2501**	**13.95**	**2578**	**15.34**	**2655**	**16.78**	**2663**	**16.90**	**2689**	**17.17**	**2729**	**17.48**
3600	8424	2335	11.02	2378	11.72	2419	12.40	2462	13.12	**2503**	**13.84**	**2583**	**15.30**	**2659**	**16.78**	**2732**	**18.26**	**2731**	**18.29**	**2746**	**18.45**	**2776**	**18.76**
3800	8892	2426	12.28	2468	13.03	**2508**	**13.76**	**2548**	**14.48**	**2588**	**15.23**	**2666**	**16.75**	**2741**	**18.32**	**2814**	**19.87**	**2805**	**19.78**	**2807**	**19.83**	**2827**	**20.01**
4000	9360	**2516**	**13.59**	**2559**	**14.43**	**2598**	**15.21**	**2638**	**15.99**	**2675**	**16.74**	**2751**	**18.33**	**2824**	**19.93**	**2895**	**21.60**	**2882**	**21.43**	**2877**	**21.37**	**2865**	**21.44**

Use Class II fan in the shaded area.

The BHP shown does not include belt drive loss.

The performance shown is for fan with outlet duct.

Table 8.7   (continued)

WHEEL	OUTLET	inside area
32-3/8 in. diameter	29-3/16 × 26-1/8 in. inside	5.27 sq. ft. inside area

30 SISW

MAXIMUM BHP $= 6.01 \left(\dfrac{RPM}{1000}\right)^3$ TIP SPEED, fpm $= 8.48 \times RPM$

CLASS I RPM 1647   CLASS II RPM 1807   CLASS III RPM 2166

Static Pressure 1/2″ through 5″

CFM	OV FPM	1/2″ SP RPM	BHP	1″ SP RPM	BHP	1-1/2″ SP RPM	BHP	2″ SP RPM	BHP	2-1/2″ SP RPM	BHP	3″ SP RPM	BHP	3-1/2″ SP RPM	BHP	4″ SP RPM	BHP	4-1/2″ SP RPM	BHP	5″ SP RPM	BHP
5270	1000	459	.58	552	1.01	626	1.45	697	1.95	768	2.51										
5797	1100	482	.67	575	1.14	647	1.62	713	2.13	777	2.69	841	3.31								
6324	1200	507	.78	599	1.29	670	1.81	732	2.33	792	2.91	851	3.53	909	4.19						
6851	1300	536	.90	623	1.45	693	2.00	754	2.56	810	3.15	865	3.78	919	4.46	973	5.17				
7378	1400	565	1.04	646	1.62	717	2.22	777	2.82	831	3.42	882	4.07	934	4.76	984	5.48	1034	6.24	1085	7.05
7905	1500	595	1.19	668	1.79	741	2.44	800	3.08	853	3.73	902	4.39	951	5.08	999	5.82	1046	6.59	1092	7.40
8432	1600	625	1.36	692	1.98	765	2.68	824	3.36	877	4.05	925	4.74	971	5.45	1016	6.20	1061	6.99	1105	7.80
8959	1700	656	1.55	718	2.21	788	2.94	848	3.66	900	4.39	948	5.12	992	5.85	1035	6.61	1078	7.41	1121	8.24
9486	1800	687	1.76	746	2.45	810	3.19	872	3.98	924	4.74	971	5.51	1015	6.29	1057	7.07	1098	7.87	1138	8.72
10013	1900	719	1.98	775	2.72	833	3.46	895	4.31	948	5.11	995	5.92	1039	6.74	1080	7.56	1119	8.38	1157	9.23
10540	2000	751	2.23	805	3.00	858	3.77	918	4.64	972	5.51	1019	6.35	1062	7.21	1103	8.07	1142	8.93	1179	9.80
11594	2200	815	2.79	865	3.64	913	4.49	963	5.35	1019	6.34	1066	7.28	1110	8.21	1150	9.15	1189	10.10	1225	11.05
12648	2400	880	3.44	927	4.37	971	5.30	1015	6.22	1063	7.19	1113	8.28	1158	9.31	1198	10.33	1236	11.35	1272	12.38
13702	2600	945	4.20	989	5.20	1031	6.21	1071	7.21	1112	8.22	1158	9.30	1204	10.48	1246	11.61	1284	12.70	1319	13.81
14756	2800	1012	5.08	1052	6.15	1092	7.24	1130	8.32	1167	9.40	1206	10.49	1248	11.66	1292	12.94	1331	14.17	1367	15.35
15810	3000	1078	6.08	1116	7.23	1154	8.39	1190	9.55	1225	10.71	1260	11.86	1296	13.04	1336	14.30	1377	15.67	1415	17.00
16864	3200	1145	7.22	1181	8.43	1216	9.66	1250	10.91	1284	12.14	1316	13.38	1349	14.61	1384	15.87	1421	17.21	1460	18.67
17918	3400	1212	8.49	1246	9.78	1280	11.08	1312	12.40	1344	13.72	1375	15.03	1406	16.34	1437	17.65	1469	18.99	1504	20.40
18972	3600	1279	9.91	1312	11.28	1344	12.65	1375	14.04	1405	15.44	1435	16.84	1464	18.22	1493	19.61	1522	21.00	1553	22.41
20026	3800	1347	11.50	1378	12.93	1408	14.38	1438	15.85	1467	17.32	1495	18.79	1523	20.26	1550	21.73	1578	23.19	1606	24.66

Static Pressure 6″ through 15″

CFM	OV FPM	6″ SP RPM	BHP	7″ SP RPM	BHP	8″ SP RPM	BHP	9″ SP RPM	BHP	10″ SP RPM	BHP	11″ SP RPM	BHP	12″ SP RPM	BHP	13″ SP RPM	BHP	14″ SP RPM	BHP	15″ SP RPM	BHP
8959	1700	1203	9.99	1286	11.88	1374	14.42	1457	17.15	1535	20.07	1618	23.94	1682	26.46	1759	30.83	1818	33.56	1878	36.40
9486	1800	1218	10.50	1295	12.39	1382	14.99	1464	17.78	1554	21.54	1643	25.71	1701	28.22	1785	32.93	1838	35.65	1892	38.46
10013	1900	1234	11.04	1308	12.96	1394	15.63	1490	19.25	1584	23.27	1675	27.70	1730	30.28	1817	35.26	1868	38.05	1918	40.90
10540	2000	1252	11.63	1324	13.58	1426	17.05	1524	20.91	1619	25.20	1713	29.92	1765	32.54	1853	37.83	1902	40.67	1950	43.59
11594	2200	1294	12.95	1360	14.94	1463	18.66	1564	22.81	1661	27.40	1756	32.43	1805	35.09	1896	40.72	1941	43.59	1986	46.53
12648	2400	1340	14.44	1403	16.52	1507	20.51	1608	24.95	1707	29.82	1802	35.15	1850	37.92	1941	43.87	1985	46.82	2028	49.82
13702	2600	1387	16.03	1449	18.27	1554	22.53	1655	27.24	1753	32.40	1849	38.01	1896	40.94	1988	47.19	2031	50.29	2073	53.40
14567	2800	1434	17.72	1496	20.12	1601	24.66	1702	29.64	1800	35.09	1896	41.00	1943	44.09	2035	50.65	2078	53.92	2120	57.19
15810	3000	1482	19.54	1543	22.09	1648	26.90	1749	32.18	1848	37.92	1944	44.14	1990	47.39	2082	54.25	2125	57.69	2166	61.12
16864	3200	1529	21.48	1591	24.18	1696	29.28	1797	34.85	1895	40.90	1991	47.43	2038	50.83	2129	58.02				
17918	3400	1576	23.49	1639	26.41	1743	31.82	1845	37.70	1943	44.06	2039	50.92	2085	54.46						
18972	3600	1620	25.49	1686	28.75	1791	34.47	1892	40.67	1991	47.36	2087	54.55	2133	58.27						
20026	3800	1666	27.68	1731	31.09	1836	37.15	1938	43.69	2037	50.72	2133	58.24								
21080	4000	1716	30.20	1775	33.50	1880	39.84	1982	46.70	2081	54.05										
22134	4200	1769	32.94	1822	36.24	1926	42.81	2027	49.91	2126	57.54										
23188	4400	1826	35.88	1874	39.30	1976	46.16	2076	53.55												
24242	4600	1883	39.03	1929	42.57	2030	49.78														
25296	4800	1942	42.37	1985	46.07																

The RPM shown does not include belt drive loss.

The performance shown is for fan with outlet duct.

WHEEL	48 in. diameter	OUTLET	43-3/8 × 38-15/16 in. outside	11.58 sq. ft. inside area	MAXIMUM BHP = $51.0\left(\dfrac{RPM}{1000}\right)^3$	44½ SISW
CLASS I RPM 817		CLASS II RPM 1066		CLASS III RPM 1394	TIP SPEED, fpm = 12.57 × RPM	

CFM	OV FPM	1/2" SP RPM	BHP	1" SP RPM	BHP	1-1/2" SP RPM	BHP	2" SP RPM	BHP	2-1/2" SP RPM	BHP	3" SP RPM	BHP	3-1/2" SP RPM	BHP	4" SP RPM	BHP	4-1/2" SP RPM	BHP	5" SP RPM	BHP
11580	1000	292	1.3	356	2.2																
12738	1100	307	1.5	369	2.5	425	3.6														
13896	1200	324	1.7	383	2.8	433	3.9														
15054	1300	342	2.0	397	3.2	445	4.3	492	5.7												
16212	1400	360	2.3	411	3.5	459	4.8	502	6.1	548	7.7										
17370	1500	378	2.6	426	3.9	473	5.3	514	6.6	554	8.2	599	10.0								
18528	1600	397	3.0	442	4.4	487	5.9	527	7.3	565	8.8	604	10.5	647	12.5						
19686	1700	416	3.4	459	4.9	501	6.4	541	8.0	578	9.5	613	11.1	650	13.1	691	15.2				
20844	1800	434	3.8	476	5.4	516	7.0	555	8.7	591	10.3	625	11.9	658	13.7	694	15.8				
22002	1900	453	4.3	494	6.0	531	7.6	569	9.4	605	11.1	638	12.8	669	14.6	701	16.6	736	18.8		
23160	2000	472	4.8	513	6.6	548	8.3	584	10.2	619	12.0	652	13.8	682	15.5	712	17.5	743	19.7	775	22.0
25476	2200	510	6.0	550	8.0	582	9.9	615	11.8	648	13.9	680	15.9	709	17.9	737	19.8	764	21.8	792	24.0
27792	2400	549	7.4	587	9.6	619	11.6	648	13.7	678	15.9	708	18.1	737	20.4	765	22.5	791	24.6	816	26.7
30108	2600	588	9.0	624	11.3	656	13.6	683	15.8	710	18.1	738	20.5	766	22.9	793	25.4	819	27.7	844	30.0
32424	2800	628	10.9	662	13.4	693	15.9	720	18.3	745	20.7	770	23.1	796	25.7	822	28.3	847	31.0	872	33.5
34740	3000	668	13.0	700	15.6	730	18.3	757	21.0	781	23.5	804	26.1	828	28.7	852	31.5	876	34.3	900	37.1
37056	3200	709	15.4	739	18.2	768	21.1	794	23.9	817	26.7	840	29.4	861	32.1	884	35.0	906	37.9	929	40.9
39372	3400	750	18.2	778	21.0	805	24.1	831	27.1	855	30.1	876	33.0	897	35.9	917	38.8	938	41.9	959	45.0
41688	3600	790	21.2	817	24.2	843	27.4	868	30.7	892	33.9	913	37.0	933	40.0	952	43.1	972	46.2	992	49.4
44004	3800	831	24.6	857	27.7	882	31.0	906	34.5	929	37.9	950	41.2	970	44.5	989	47.7	1007	50.9	1025	54.2

CFM	OV FPM	6" SP RPM	BHP	7" SP RPM	BHP	8" SP RPM	BHP	9" SP RPM	BHP	10" SP RPM	BHP	11" SP RPM	BHP	12" SP RPM	BHP	13" SP RPM	BHP	14" SP RPM	BHP	15" SP RPM	BHP
25476	2200	850	29.0																		
27792	2400	866	31.4	919	36.9																
30108	2600	890	34.6	937	39.7	985	45.6	1037	51.9												
32424	2800	917	38.4	961	43.4	1003	48.9	1048	55.1	1095	61.8										
34740	3000	945	42.6	988	47.8	1028	53.1	1068	59.0	1109	65.5	1152	72.5								
37056	3200	973	46.9	1016	52.7	1055	58.2	1092	63.9	1130	70.1	1168	76.8	1198	79.9						
39372	3400	1002	51.4	1044	57.7	1083	63.8	1120	69.7	1155	75.7	1190	82.1	1207	84.0						
41688	3600	1032	56.1	1072	62.8	1111	69.5	1148	75.9	1182	82.2	1216	88.4	1226	89.0	1249	91.7				
44004	3800	1063	61.1	1101	68.2	1139	75.3	1176	82.4	1210	89.1	1243	95.7	1249	95.1	1262	96.5	1293	99.9		
46320	4000	1095	66.5	1132	73.9	1168	81.4	1204	88.9	1238	96.3	1271	103.4	1275	102.3	1283	102.3	1300	104.5		
48636	4200	1130	72.4	1163	79.9	1198	87.7	1233	95.6	1267	103.5	1299	111.2	1303	110.3	1307	109.2	1316	110.0	1340	112.8
50952	4400	1165	78.8	1197	86.5	1229	94.5	1262	102.7	1295	111.0	1328	119.2	1331	118.8	1334	117.2	1339	116.5	1351	118.1
53268	4600	1201	85.7	1231	93.6	1262	101.7	1293	110.1	1325	118.7	1356	127.4	1359	127.3	1361	126.0	1364	124.3	1370	124.4
55548	4800	1237	93.0	1267	101.2	1296	109.5	1325	118.1	1356	126.9	1386	135.9	1387	136.0	1389	135.2	1391	133.3	1393	131.8

The BHP shown does not include belt drive loss.   The performance shown is for fan with outlet duct.
(Adapted from the Chicago Blower Corporation, Glendale Heights, IL [Airfoil SQA Fans–Class I and II], 1978.)

consider a smaller or larger fan, respectively. If the design point is near the left or right margins of the table, consider a modified version of the fan type.

Fans are usually designed and rated for air at 70 F, 1 atm, 50% relative humidity, and an altitude at mean sea level. At these conditions, the air density is 0.075 lb_m/ft^3. If the fan is to be used at conditions other than these standard values (and most fans are), then corrections must be made for air density. As can be seen in Table 8.7, the *volumetric* flow rate does not change. Therefore, the fan design procedure is as follows (*Industrial Ventilation* 1972):

1. Use the actual air volumetric flow rate Q to design the system.

2. Calculate the SP losses as if standard air were being moved (duct friction loss charts are also based on standard air). Correct any control device pressure drops to standard air.

3. Select a fan from the rating tables using the actual Q and the standard air SP from Steps 1 and 2. The speed selected is the actual speed needed.

4. Calculate the actual SP delivered (and needed) and the actual BHP drawn by multiplying the values from the table by the ratio of the actual air density to the standard air density.

Note that some manufacturers supply density correction factor tables to correct for altitude and temperature; *the correction factors are often the inverse of the density ratio*. One can always tell if a density correction factor is to be used as a multiplier or a divisor; the actual FSP and $\dot{w}$ for a fan moving less dense air will always be less than those for a fan moving denser air. The brake horsepower required for moving high temperature (low density) air is significantly less than for standard air. When specifying a motor to operate the fan, the designer must specify a motor with enough horsepower to handle the cooler air present during plant start-ups.

Fan tests are run with well-designed fan inlet and outlet ducts to ensure that the air enters and leaves the fan with minimum turbulence and maximum SP regain. If the actual fan connections in the field are not good, substantial dynamic losses can occur.

Fan noise can sometimes be a problem. Generally, fan noise is caused by mechanical vibrations, by bearings, motors, or drive belts, or by air turbulence inside the fan (Sorg 1983). Mechanical noise can be muffled by acoustical linings in ducts, by turning vanes in duct elbows, or by insulation around the fan and/or motor. Also, the fan and motor can be located outside of noise-sensitive areas. Turbulence noise is proportional to air velocities; larger ducts can help to alleviate noise problems (and reduce SP losses). Also, using larger, slower fans rather than smaller, faster ones tends to reduce turbulence noise.

Example 8.7

Design the ducts and select a fan (from the tables given in this text) to move air through the system shown in Figure 8.10. The final control device (FCD) has a maximum SP loss of 5.0 inches of water with a flow rate of 15,000 scfm. Other data are shown in the figure.

Solution

The path of greatest resistance involves hood 2 and duct B. From Figure 8.2, the duct diameter is equal to 24 inches, and the friction loss for standard air is 0.53 in. H_2O/100 ft. The actual velocity in this size duct is 3180 fpm, which is close to the design velocity of 3200 fpm. We determine the fitting losses from Table 8.2. Note that $\rho_{act}/\rho_{std} = T_{std}/T_{act} = 530\ R/610\ R = 0.87$.

Acceleration loss	*Standard*	*Actual*
$SP_a = (3180/4005)^2 =$	0.63 in. H_2O	0.55 in. H_2O
Hood 2 loss	0.50 in. H_2O	0.44 in. H_2O

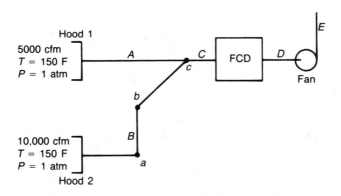

Hood and duct data

Hood 1 - SP loss = 0.8 in. H_2O (std. air)
Hood 2 - SP loss = 0.5 in. H_2O (std. air)

Duct A - length = 70 ft
Duct B - length = 80 ft
Duct C - length = 15 ft
Duct D - length = 10 ft
Duct E - length = 15 ft

Note: design duct velocity = 3200 fpm

Fitting data

Fitting

a   90° ell
b   45° ell
c   45° branch entry

Figure 8.10
Diagram for Example 8.7.

Duct B loss

actual length	=	80 ft
equivalent length (90° elbow) (20 × 2 ft) =		40 ft
equivalent length (45° elbow) (10 × 2 ft) =		20 ft
		140 ft

From Figure 8.2,

$$\text{SP loss} = \frac{0.53 \text{ in. } H_2O}{100 \text{ ft}} \times 140 \text{ ft} = 0.74 \text{ in. } H_2O \ (0.65 \text{ actual})$$

Entrance loss at point C

$$\text{SP loss} = 36 \text{ ft} \times \frac{0.53 \text{ in. } H_2O}{100 \text{ ft}} = 0.19 \text{ in. } H_2O \ (0.17 \text{ actual})$$

Duct C loss

From Figure 8.2 for Q = 15,000 cfm and diameter = 30 inches, SP = 0.38 in. H_2O/100 ft.

$$\text{SP} = \frac{0.38 \text{ in. } H_2O}{100 \text{ ft}} \times 15 \text{ ft} = 0.06 \text{ in. } H_2O \ (0.05 \text{ actual})$$

FCD (including entrance and exit losses)

$$= 5.0 \text{ in. } H_2O \text{ (std) } (4.35 \text{ actual})$$

Duct D $\dfrac{0.38 \text{ in. } H_2O}{100 \text{ ft}} \times 10 \text{ ft} = 0.04 \text{ in. } H_2O \ (0.03 \text{ actual})$

Duct E $\dfrac{0.38 \text{ in. } H_2O}{100 \text{ ft}} \times 15 \text{ ft} = \underline{0.06} \text{ in. } H_2O \ (0.05 \text{ actual})$

$$\text{Total SP Loss} = 7.2 \text{ in. } H_2O \text{ (std)}$$

$$(\text{loss up to point } C = 1.9 \text{ in. } H_2O)$$

Check duct A. From Figure 8.2 for 5000 cfm at 3200 fpm, diameter = 17 inches. Use 18″ duct with V = 2830 fpm and P = 0.60 in. H_2O/100 ft.

$$\text{SP}_a = (2830/4005)^2 = 0.50$$

$$\text{SP}_C = \frac{0.60 \text{ in. } H_2O}{100 \text{ ft}} \times 70 \text{ ft} = 0.42 \text{ in. } H_2O \ (0.37 \text{ actual})$$

$$\text{Hood } 1 = \underline{0.8} \text{ in. } H_2O$$

$$(\text{total to point } C) \ 1.7 \text{ in. } H_2O \text{ (std)}$$

Since duct A has a lower pressure drop of $1.9 - 1.7 = 0.2$ in. H_2O, we can (1) decrease the diameter of duct A and accept a higher gas velocity in duct A, (2) accept a higher volumetric flow rate in duct A, or (3) install a balancing damper in duct A (or some combination of the three).

Since the air is clean at the fan, we select a backward inclined blade fan. The fan is selected for 15,000 cfm at $7.2 - 0.6 = 6.6$ in. H_2O

FSP. We select a fan from the 30 SISW table. By interpolation, $N = 1490$ rpm and $\dot{w} = 19.8$ bhp with standard air. The actual BHP (after correcting for density) is expected to be 17 bhp. Therefore, we select a 20 or 25 hp motor, depending on the expected drive losses.

◀ ◀ ◀ ◀ ◀ ◀ ◀ ◀

◆ 8.5 Cooling Hot Airstreams ◆

Process exhaust gas is often hot and must be cooled before it is routed through control equipment. Depending on the cooling method, the volumetric flow rate of the cooled gas might be reduced, thus decreasing the required size of downstream control equipment, ducts, and fans. Thus, installing equipment to cool hot exhaust gases can make good economic sense, even if the heat is not recovered. With the current high cost of energy, recovering a portion of the heat is almost always cost-effective. A potential problem that can arise from cooling to a low temperature is the condensation of moisture, with consequent corrosion or plugging of downstream equipment. Options for cooling hot gases are (1) dilution with ambient air, (2) spraying water into the gas stream, and (3) heat exchange (with or without heat recovery).

Air Dilution

Air dilution is one of the easiest methods for cooling hot airstreams. However, unless the gas is very hot and the target cool temperature is substantially above the ambient temperature, the final air volume can be very large. Solving the material and energy balance equations (assuming the heat capacity for air is virtually constant over the temperature range of interest), and applying the ideal gas law leads to

$$Q_d = Q_e \left(\frac{T_e - T_f}{T_f - T_d} \right) \left| \frac{T_d}{T_e} \right|$$ (8.21)

where

Q_d = dilution airflow rate needed

Q_e = exhaust airflow rate to be cooled

T_e = temperature of the exhaust air

T_d = temperature of the dilution air

T_f = final temperature of the mixed stream

The absolute value signs $| \; |$ in Eq. (8.21) serve as a reminder that, at this point, T_d and T_e must be expressed as absolute temperatures.

Water Injection

Water injection cools very efficiently because of the high heat of vaporization of water. As the water that is sprayed into the mixing chamber evaporates, the water absorbs considerable heat from the air-stream, thus lowering the air temperature. The final volume of the cooled stream is considerably less than if air dilution were used, but the resulting humidity of the stream can be very high, which can cause problems in downstream units. The material and enthalpy balances are

$$\dot{M}_a + \dot{M}_w = \dot{M}_f \tag{8.22}$$

$$\dot{M}_a \left(h_{a_i} - h_{a_f} \right) = \dot{M}_w \left(h_{w_f} - h_{w_i} \right) \tag{8.23}$$

where

$\dot{M}$ = mass flow rates, lb_m/min

h = specific enthalpies, Btu/lb_m

a, w = subscripts indicating air and water, respectively

i, f = subscripts indicating initial and final states, respectively

Equations (8.22) and (8.23) are very similar to the equations that were used to develop Eq. (8.21), but the enthalpy change of the water in Eq. (8.23) is due not only to temperature change, but also to the latent heat of vaporization of water. Equation (8.23) can be solved directly if the enthalpies are tabulated for air and water vapor, or can be written as

$$\dot{M}_a C_{p_a} \left(T_a - T_f \right)$$
$$= \dot{M}_w \left[\Delta H_v + C_{p_{wv}} \left(T_f - T_v \right) + C_{p_{wl}} \left(T_v - T_w \right) \right] \tag{8.24}$$

where

C_{p_a} = average specific heat of air over the temperature range, Btu/lb_m-°F

ΔH_v = heat of vaporization of water, Btu/lb_m

$C_{p_{wv}}$ = specific heat of water vapor, Btu/lb_m-°F

$C_{p_{wl}}$ = specific heat of water liquid, Btu/lb_m-°F

T_v = temperature at which the water vaporizes, °F

Detailed correlations are available to predict the C_ps in Eq. (8.24) as functions of temperature, and some of these correlations are presented in Appendix B. However, for the accuracy of these calculations, average values can be used. The value of C_{p_a} ranges from 0.24 to 0.26 Btu/lb_m-°F over the range from 0 F to 2000 F, and the value of $C_{p_{wv}}$ ranges from 0.44 to 0.51 from 60 F to 1000 F. At 1 atm, ΔH_v ranges from 970 Btu/lb_m (at 212 F) to 1060 Btu/lb_m (at 60 F). Note that T_v is often well below 200 F, depending on the partial pressure of the water in the system. Since enthalpy is a thermodynamic state function, there is no difference in the total enthalpy change calculated by (1) assuming that the water is completely vaporized at T_w and that the water vapor is then heated to T_f, or by (2) assuming that the water is heated as a liquid from T_w to T_f and then vaporized. For simplicity, we will assume that all the vaporization occurs at the inlet water temperature. Thus, the third term inside the brackets of Eq. (8.24) becomes zero.

Equation (8.24) can be solved for the water injection rate required to achieve any desired T_f. Once the water injection rate is obtained, it can be converted to moles, then added to the number of moles of exhaust air. We can then calculate a final volumetric flow rate. Alternatively, a psychrometric chart can be used to estimate these values (see Appendix B).

HEAT EXCHANGE

The three main advantages of using a heat exchanger to cool the exhaust gases are (1) the final air volume is the smallest of the three cooling methods, (2) there is no increase in moisture content of the gases (although the relative humidity increases as the temperature decreases), and (3) heat energy ($) can be recovered. The two main disadvantages are (1) the capital cost of purchasing and installing the heat exchanger and (2) the operating and maintenance costs of running the heat exchanger.

An older method of cooling very hot gases is the use of large U-tubes that transfer heat to ambient air by convection and radiation (Danielson 1973). These coolers are basically large steel tubes (30 to 60 feet long and 1 to 3 feet in diameter), connected in series and parallel, through which the hot gases flow. With the tenfold or more increase in the cost of energy that occurred in the 1970s, recovery of heat from hot gas streams became much more attractive economically than before. Even though energy prices remained flat during the 1980s, heat recovery was still often justified economically. In the late 1990s, energy prices spiked up again and may increase substantially in the future. Heat exchangers allow the recovery of useful heat.

A popular type of heat exchanger is the shell-and-tube heat exchanger, an example of which is depicted in Figure 8.11. One fluid passes through the tubes while the other fluid flows countercurrently in the shell (the space outside the tubes). Detailed design of heat exchangers is a complicated process and is beyond the scope of this text. However, we will provide a basic summary of the design equations as follows (refer to Figure 8.11 for nomenclature):

$$\dot{H}_1 = \dot{M}_1 C_{p_1} \left(T_{1h} - T_{1c} \right) \tag{8.25}$$

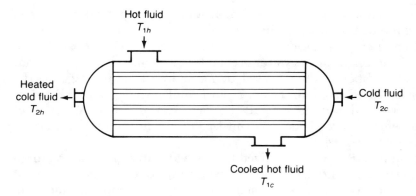

Figure 8.11
A shell-and-tube heat exchanger.

where $\dot{H}_1$ = rate of heat given off by the hot fluid (assuming no phase change), Btu/hr

$$\dot{H}_2 = \dot{M}_2 C_{p_2} \left(T_{2h} - T_{2c} \right)$$ (8.26)

where $\dot{H}_2$ = rate of heat absorbed by the cold fluid (no phase change), Btu/hr

$$\dot{H}_x = UA\Delta T_{\mathrm{LM}}$$ (8.27)

where

$\dot{H}_x$ = rate of heat exchanged, Btu/hr

U = overall heat transfer coefficient, Btu/hr-ft^2-°F

A = heat transfer area, ft^2

ΔT_{LM} = log mean temperature difference [defined in Eq. (8.28)], °F

$$\Delta T_{\mathrm{LM}} = \frac{(T_{1h} - T_{2h}) - (T_{1c} - T_{2c})}{\ln\left(\dfrac{T_{1h} - T_{2h}}{T_{1c} - T_{2c}}\right)}$$ (8.28)

For shell-and-tube heat exchangers with relatively thin-walled tubes and for relatively clean airstreams (negligible wall fouling), the overall heat transfer coefficient can be estimated by the following equation (McCabe et al. 1985):

$$U = \frac{1}{\dfrac{1}{h_o} + \dfrac{x_w}{k_m} + \dfrac{1}{h_i}}$$ (8.29)

where

h_o, h_i = individual heat transfer coefficients on the outside and the inside of the tube, respectively, Btu/hr-ft^2-°F

x_w = thickness of the tube wall, ft

k_m = thermal conductivity of the tube metal, Btu/hr-ft-°F

Usually, if only one stream is a gas, its individual heat transfer coefficient is so low ($1/h$ is so high) that U is approximately equal to h. If both streams are gases, then both h_o and h_i should be included. The term x_w/k_m is usually negligible when one or both of the streams is a gas.

Individual gas heat transfer coefficients depend on the type of gas, the flow regime, and the thermal and transport properties of the gas. Ganapathy (1976) has developed a graphical means of estimating individual gas heat transfer coefficients for various gas streams. Some individual coefficients for air or flue gases as estimated using Ganapathy's procedure are given in Table 8.8.

Table 8.8   Some Calculated Values of Individual Heat Transfer Coefficients

Case Stream	Service	Temperature, °F	h, Btu/hr-ft^2-°F
1. Air or Flue Gas	Outside of tubes. Staggered bundle of 1″ tubes. Mass flux G = 5000 lb$_m$/hr-ft^2	600	17
2. Air or Flue Gas	Same as above, but G = 500 lb$_m$/hr-ft^2	600	4.3
3. Air or Flue Gas	Same as Case 1	1000	22
4. Air or Flue Gas	Same as Case 2	1000	5.5
5. Nitrogen-Rich Gas with 18% H$_2$O Content	Inside 1.5″ tubes. Mass flux G = 100 lb$_m$/hr-ft^2	1400	12
6. Nitrogen-Rich Gas with 18% H$_2$O Content	Same as Case 5	800	9
7. Nitrogen-Rich Gas with 18% H$_2$O Content	Same as Case 5, except G = 500 lb$_m$/hr-ft^2	1400	40
8. Nitrogen-Rich Gas with 18% H$_2$O Content	Same as Case 7	800	32

The heat exchanged $\dot{H}_x$ is always equal to $\dot{H}_2$, but can be substantially less than $\dot{H}_1$ owing to heat losses from the exchanger to the surroundings. The normal preliminary design calculation procedure is as follows:

1. Use Eq. (8.25) to calculate $\dot{H}_1$.
2. Estimate any heat losses and calculate $\dot{H}_2$.
3. Use Eq. (8.26) to calculate T_{2h} if $\dot{M}_2$ is known; otherwise, estimate T_{2h} and calculate $\dot{M}_2$.
4. Estimate a value for U from Eq. (8.29) or from experience with similar heat exchangers.
5. Use Eq. (8.28) to calculate ΔT_{LM}.
6. Use Eq. (8.27) to calculate A.

Some general ranges for values of U (all in Btu/hr-ft^2-°F) are as follows: For air-to-air heat exchangers, U ranges from 0.2 to 5; for air-to-water, U ranges from 0.4 to 10; and for water-to-water, U ranges from 50 to 2000.

Example 8.8

Calculate the final volumetric flow rate of a 100,000 cfm exhaust gas stream cooled from 900 F to 300 F by (a) dilution with ambient air at 80 F, (b) spray injection of water at 70 F, and (c) heat exchange.

Solution

(a) From Eq. (8.21),

$$Q_d = 100,000\left(\frac{900-300}{300-80}\right)\left(\frac{540}{1360}\right)$$

$$= 108,000 \text{ cfm} \quad (\text{at } 80 \text{ F})$$

$$Q_f = 100,000\left(\frac{760}{1360}\right)+108,000\left(\frac{760}{540}\right)$$

$$= 208,000 \text{ cfm} \quad (\text{at } 300 \text{ F})$$

(b) Rearranging Eq. (8.24),

$$\dot{M}_w = \frac{\dot{M}_a C_{p_a}\left(T_a - T_f\right)}{\Delta H_v + C_{p_{wv}}\left(T_f - T_v\right)}$$

Assuming 1 atm pressure and a molecular weight of 29 for the exhaust stream,

$$\dot{M}_a = \frac{100,000 \text{ ft}^3}{\text{min}} \times \frac{(1 \text{ atm})(29 \text{ lb}_m/\text{lbmol})}{\left(0.730 \text{ atm-ft}^3/\text{lbmol-°R}\right)(1360 \text{ R})}$$

$$= 2920 \text{ lb}_m/\text{min}$$

$$\dot{M}_w = \frac{(2920)(0.25)(900-300)}{1054+0.45(300-70)}$$

$$= 378 \text{ lb}_m/\text{min}$$

$$Q_w = 45 \text{ gal/min}$$

Converting these flow rates to a molar basis, we obtain

molar flow rate of water = 378/18 = 21 lbmol/min

total molar flow rate of cooled gas = 2920/29 + 21

$$= 121.7 \text{ lbmol/min}$$

Finally, we use the ideal gas law to calculate the volumetric flow rate. Thus,

$$Q_f = 121.7 \frac{\text{lbmol}}{\text{min}} \times 0.730 \frac{\text{atm-ft}^3}{\text{lbmol-}^\circ\text{R}} \times \frac{760 \text{ R}}{1 \text{ atm}}$$

$$= 67,500 \text{ ft}^3/\text{min} \quad \text{(at 300 F)}$$

(c) For heat exchange, nothing is added to the airstream. The ideal gas law can be used directly.

$$Q_f = Q_e \left(\frac{760}{1360}\right) = 56,000 \text{ cfm} \quad \text{(at 300 F)}$$

◀ ◀ ◀ ◀ ◀ ◀ ◀ ◀ ◀

♦ 8.6 COSTS ♦

HOODS

The fabricated equipment costs of canopy hoods have been related to the metal plate area, the thickness of the metal, and the fabrication labor hours (Neveril et al. 1978). However, in most instances the fabrication of a hood is a customized application that is highly site specific. It is probably best to obtain quotes from local fabricators when estimating the costs of hoods.

DUCTS

The costs of ductwork (including straight ducts and various types of fittings) depend on the size of the ductwork, the materials of construction, and the metal thickness. Neveril (1978) and Vatavuk and Neveril (1980b) reported the costs of straight ducts and duct fittings graphically as functions of duct diameter and metal thickness. Those authors presented their cost-estimating graphs in 1977 dollars. For purposes of this text, two of those graphs have been updated to 1992 dollars using the Marshall-Swift Index; those updated graphs are presented in Figures 8.12 and 8.13. Of course, the 1992 costs obtained

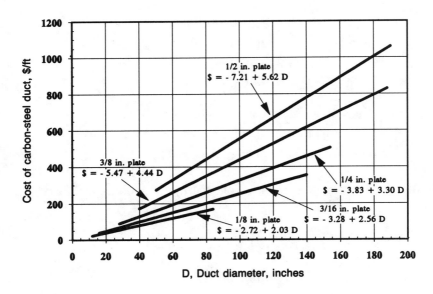

(a) Carbon-steel straight ducts

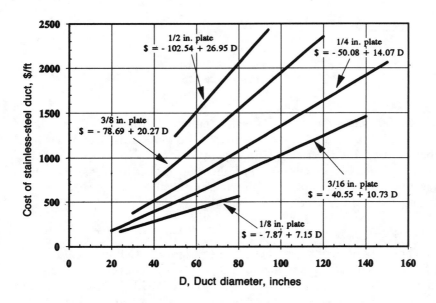

(b) Stainless-steel straight ducts

Figure 8.12
Fabricated costs (in 1992 dollars) of straight ducts of various wall thicknesses.

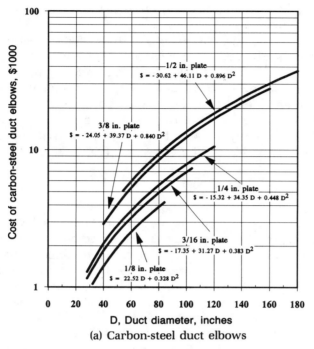

(a) Carbon-steel duct elbows

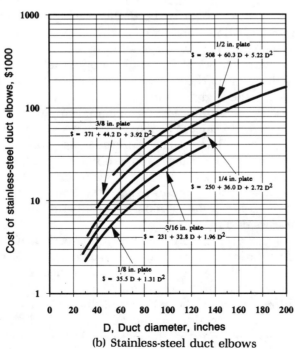

(b) Stainless-steel duct elbows

Figure 8.13
Fabricated costs (in 1992 dollars) of duct elbows (including flanges).

from these figures can be escalated to any desired year using the techniques discussed in Chapter 2.

The fabricated costs of straight ducts (in 1992 dollars/foot of length) are presented in Figure 8.12 (for carbon steel and stainless steel, separately) for various metal thicknesses. Figure 8.13 presents the costs of carbon- and stainless-steel elbows (with flanges included). According to Vatavuk and Neveril (1980b), the cost of tees is roughly one-third that of like-sized elbows, and the cost of a transition piece (reducer or expander) is roughly one-half that of an elbow that is the same diameter as the larger side of the transition piece.

FANS

The great majority of fans used in air pollution control (APC) applications are either backward inclined or radial tip fans. Backward inclined fans are loosely grouped into four classes based on fan outlet velocity and fan static pressure. Based on APC system requirements (for flow rate and pressure drop), most backward inclined fans selected will be Class II or III fans. Class I includes fans that cover the lowest fan outlet velocity and fan static pressure applications, and Class IV fans are simply those that "give performance greater than Class III." Class I and Class IV situations are unusual, and backward inclined fans for these cases must be specially ordered. Radial tip fans are not classified.

While it is very important to understand the principles of fan selection using the fan rating tables presented earlier, manufacturers have many fan sizes (wheel diameters) for each family of fans. It is not unusual to have four or more fan sizes that could supply the required flow rate and pressure drop. From these, the engineer must select the most efficient fan size—the one that will minimize operational costs.

Today most factory representatives have small computer programs that help the engineer to select the most efficient fan. Figures 8.14 and 8.15 plot the results of many runs* of one such program (supplied by the Chicago Blower Corporation), and can help the engineer to quickly choose the most efficient fan size for a given service. While these curves were prepared from computerized data from only one company, the curves are very useful for preparing cost estimates during the preliminary process design.

To use Figure 8.14 or 8.15, simply find the intersection of the desired flow rate and fan static pressure. Just as was done when selecting the fan, the flow rate should be the actual volume needed and the FSP is the APC system pressure drop based on moving standard density air. This point of intersection will be in a zone of the chart between two solid lines, which will correspond to the most efficient wheel diameter for that service. It is noted that the shaded line dividing Class II

---

*The authors acknowledge the work of Mr. Frank Marshall (a Ph.D. student at UCF) in developing this section on fan and motor costs.

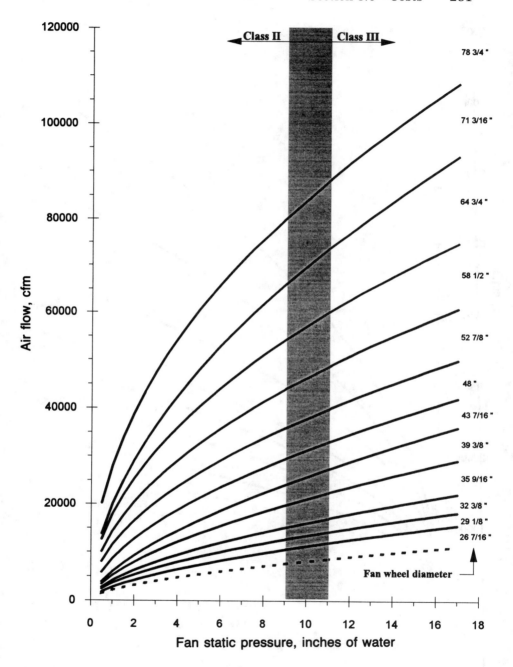

Figure 8.14
Most efficient fan size as a function of flow rate and fan static pressure—
backward inclined fans.

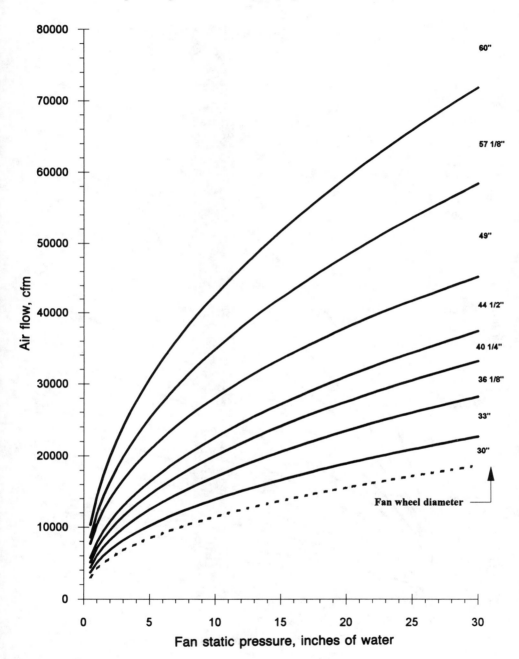

Figure 8. 15
Most efficient fan size as a function of flow rate and fan static pressure—
radial tip fans.

from Class III fans is not an exact line of demarcation. Once the most efficient fan size has been chosen, the cost can be estimated using Figure 8.16. The cost curves for fans that are presented in this figure were based on the wheel diameter of the most efficient fan size.

For example, consider a backward inclined fan that is to deliver 30,000 acfm of low-density air through a system pressure drop of 10 in. H_2O. If the FSP is 15 in. H_2O after correcting to standard density, then the fan cost is based on 30,000 cfm at 15 in. H_2O. From Figure 8.14, the fan is Class III with a wheel size of 39 3/8 inches. From Figure 8.16 the cost of that fan is about $4,000 (the equation given in the figure yields a cost of $3,885).

The costs in Figure 8.16 were based on carbon-steel fans in "low-temperature" service, and must be adjusted for higher temperatures. If the service is for temperatures between 250 F and 600 F, multiply the cost obtained from the graph by 1.03. If a stainless-steel fan is needed (for higher temperatures and/or corrosive service), multiply the carbon-steel cost by 2.5. High-pressure fans (those that deliver up to 70 in. H_2O) are available (mostly for use with high-energy venturi scrubbers) but are considerably more expensive than their lower-pressure counterparts.

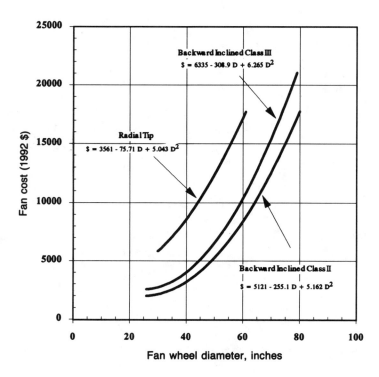

Figure 8.16
FOB costs (in 1992 dollars) for fans as a function of wheel diameter.

After determining the fan cost, the motor and motor starter costs must be estimated. Table 8.9 (for motors) and Table 8.10 (for motor starters) are used for these purposes. Since the efficiencies of the most efficiently sized fans are about 75–80% and losses in the belt drive and motor are about another 15–20%, the motor size (horsepower) needed can be estimated by the following equation:

$$M = (0.0001575 \times Q \times \text{FSP})/0.6 \qquad\qquad \textbf{(8.30)}$$

where

M = motor size, hp

Q = airflow rate, cfm

FSP = fan static pressure, in. H_2O

Table 8.9   FOB Costs for Electric Motors (in 1992 dollars)

Size of Motor		Cost of Motor by Type, $	
hp	rpm[a]	Enclosed	Explosion-Proof[b]
0.5	low	245	403
0.75	low	274	433
1	low	320	448
1.5	low	345	478
2	high	398	406
2	low	361	448
3	high	482	–
3	low	471	489
5	high	591	–
5	low	533	593
7.5	high	728	–
7.5	low	648	748
10	high	848	–
10	low	767	823
15	high	1152	–
15	low	1095	1223
20	high	1497	–
20	low	1260	1465
25	high	1801	–
25	low	1550	1880
30	high	2284	–
30	low	1959	2152
40	high	3028	–
40	low	2613	2892
50	high	3758	–
50	low	3362	3322
60	high	4667	–
60	low	4478	–

Table 8.9   *(continued)*

| Size of Motor | | Cost of Motor by Type, $ | |
hp	rpm[a]	Enclosed	Explosion-Proof[b]
75	high	5894	–
75	low	5233	–
100	high	7939	–
100	low	7526	–
125[c]	high	10,997	–
125	low	10,891	–
150	high	13,436	–
150	low	12,622	–
200	high	17,496	–
200	low	15,278	–
250	high	21,928	–
250	low	19,167	–
300	high	26,150	–
300	low	22,850	–
400	high	34,675	–
400	low	30,315	–
500	high	43,747	–
500	low	38,250	–

[a] Motor speeds listed as "high" or "low" rpm. Ranges of actual speeds are 3450 to 3575 rpm for high category, and 1705 to 1790 rpm for low category. Motor rpms should be matched to required fan rpms through belt or gear drives.
[b] Made of nonsparking materials.
[c] These sizes and up available in 460 volt operation only.

Adapted from the Chicago Blower Corporation, Glendale Heights, IL, 1992.

Table 8.10   FOB Costs for Electric Motor Starters (in 1992 dollars)

NEMA[a] Size	Motor Size, hp	Cost, $
00	0 – 3	174
0	5 – 7.5	216
1	7.5 – 10	246
1.75	10 – 15	384
2	15 – 20	486
3	20 – 40	810
4	40 – 60	1830
5	60 – 250	4317
–[b]	300	5000
–	400	7500
–	500	10,000

[a] NEMA refers to the National Electric Manufacturers' Association.
[b] These items are special-order, custom-built units. Costs are approximate.

Adapted from the Chicago Blower Corporation, Glendale Heights, IL, 1992.

Because motors only come in certain sizes, the calculated size M must be rounded to the nearest standard size (always round up unless M is only very slightly larger than a standard size). The standard sizes for electric motors are given along with cost data for motors in Table 8.9. Most fans are coupled to motors with a belt drive. The cost of the belt drive and the cost of the belts are small and add only about 5% to the cost of the fan plus motor plus starter.

COOLERS

Vatavuk and Neveril (1981a) provided cost-estimating graphs for quenchers, spray chambers, and air dilution systems. A spray chamber can control the water addition rate and provide any desired humidity in the gas stream, whereas a quencher is a simpler device that floods the gas stream with water and saturates it with water vapor while cooling it to the saturation temperature. The costs of spray chambers and quenchers from Vatavuk and Neveril (1981a) were updated using the Marshall-Swift Index and are shown in Figure 8.17(a) on the opposite page.

A dilution air cooling system consists basically of a duct tee, a damper, and a temperature controller. Although the capital costs of dilution air systems are significantly less than those of other cooling systems, the savings are usually more than offset by the higher fan, duct, and power costs incurred by handling the additional airflow. The costs of dilution air systems from Vatavuk and Neveril (1981a) were updated using the M-S Index and are shown in Figure 8.17(b).

Vatavuk and Neveril (1982) correlated heat exchanger costs with heat transfer area over a wide range of sizes. That correlation was updated to 1992 dollars using the Marshall-Swift Index and is presented as follows:

$$C = 53,742(A)^{-0.44} \exp\left(0.0672(\ln A)^2\right) \qquad \textbf{(8.31)}$$

where

C = cost (in 1992 dollars)

A = heat transfer area, ft^2

Equation (8.31) applies for heat exchangers with areas ranging from 200 to 50,000 ft^2.

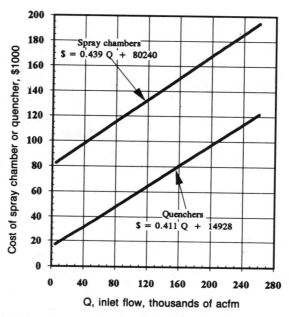

(a) Costs of carbon-steel quenchers and spray chambers
(does not include pumps or external piping)

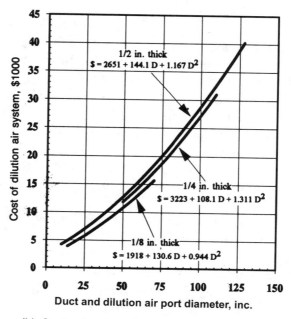

(b) Costs of carbon-steel dilution air systems
(for stainless steel, multiply cost by 3.5)

Figure 8.17
Fabricated costs (in 1992 dollars) of different systems for cooling hot exhausts.

♦ Problems ♦

8.1 A process in a plant is to be revised such that the airflow rate through a system will increase from 8000 cfm to 10,000 cfm. The existing fan runs at 1600 rpm and provides 6.0 in. H_2O SP. Estimate the new speed required for this fan and the new SP.

8.2 The existing fan in Problem 8.1 has a 15 hp motor; if fan plus drive losses are estimated to be 35%, will a new motor be needed?

8.3 If a process airflow rate were to be doubled, calculate the required size of the new fan (relative to the size of the old fan) if it is necessary to run it at the same rpm (because of motor/ drive considerations).

8.4 From the tables in this chapter, select a fan to move 12,000 cfm of standard air at 6.5 in. H_2O SP. Specify the size, speed, and BHP of the fan.

8.5 Derive the value of k in Eq. (8.19) for units of hp/in. H_2O-cfm.

8.6 From the tables in this chapter, select a fan to move 13,000 cfm of air at 500 F and 0.8 atm through a system ΔP of 4.4 in. H_2O. Specify the size, speed, and BHP of the fan.

8.7 Calculate the mechanical efficiency of a 30 SISW fan operating at 13,702 cfm and 6 in. H_2O.

8.8 Derive the value of k in Eq. (8.19) for units of kW/(Pa - m^3/s).

8.9 Calculate the pressure drop expected when moving 30,000 cfm of standard air through 250 feet of 30-in. diameter circular duct with three 90° elbows and two 45° elbows.

8.10 Design a 2:1 rectangular duct and fan system to pull 40,000 scfm of air carrying a medium-density dust through an FCD with a ΔP of 6.0 in. H_2O. Assume the duct system must extend 120 feet and has four 90° elbows.

8.11 Specify the dimensions of (a) a square duct and (b) a 3:1 rectangular duct that are flow-equivalent to the circular duct of Problem 8.9.

8.12 Derive Eq. (8.21) from the basic material and enthalpy balances. Show all your steps in detail.

8.13 Calculate the flow rate of dilution air (in cfm at 90 F) needed to cool 50,000 cfm of air from 1200 F to (a) 500 F, (b) 300 F, and (c) 150 F.

8.14 Calculate the flow rate of water injection (in gallons per minute) necessary to do the cooling for the three cases in Problem 8.13. Assume the water is available at 60 F.

8.15 Calculate the required area of an air-to-air heat exchanger to cool 20,000 cfm of air from 1200 F to 400 F. Assume ambient air at 80 F will be heated to 800 F by the exchange. Assume heat losses are 10% and that the overall heat transfer coefficient is 1.0 Btu/hr-ft^2-°F.

8.16 Estimate the fabricated cost (in 2002 dollars) of the duct in Problem 8.9.

8.17 Estimate the total installed cost (in 2002 dollars) of the system (excluding the FCD) described in Example 8.7. Assume that the TIC of the hoods is $22,000 in 2000 dollars.

8.18 Estimate the DEC (in 2002 dollars) of a backward inclined fan system designed to move 25,000 cfm of air at an FSP of 12 in. H$_2$O.

8.19 The design of a duct system requires an 18-in. diameter duct, but the building only has space for a 12-in. deep duct. How wide should the rectangular duct be in order to accommodate the same flow at the same pressure drop?

8.20 Develop a system resistance curve for a system with 400 feet of 16-in. circular duct, two 90° elbows, and three 45° elbows for flows between 3000 and 8000 scfm. Using Table 8.7, find the system operating point (flow rate and static pressure loss) for a 20 SISW fan operating at 2100 rpm. What is the fan efficiency at this operating point? Is this fan a good choice?

8.21 A 500-foot duct is required to carry 80,000 acfm (at T = 1500 F) of dust-free gases from an industrial oven to the stack. A heat exchanger is proposed to be located next to the oven in order to reduce the airstream temperature to 520 F. Determine the required duct size with and without the heat exchanger. Also, calculate the savings in the cost of the duct if 1/8-in. carbon steel is used to construct the duct (1/8-in. carbon steel is sufficient to carry gases at 520 F) instead of 3/16-in. stainless steel (which would be needed for the duct to operate at 1500 F).

8.22 A stream of ambient air at 80 F is being brought into a furnace at a rate of 25,000 acfm. If a heat exchanger can be added that preheats that stream of air from 80 F to 280 F but would incur 5.0 in. H$_2$O pressure drop, is it worth it? Assume the exchanger costs $100,000 and assume that maintenance and other fixed charges amount to $12,000/year. Assume that the existing fan can handle the increased pressure drop imposed by the addition of the exchanger. Assume that fuel is worth $3.00/million Btu, and that electricity costs $0.08/kWh.

8.23 If 20,000 acfm of exhaust gases are cooled from 1600 F to 400 F by spraying 70 F water into the airstream, estimate the water addition rate needed, in gal/min.

REFERENCES

Airfoil SQA Fans, Bulletin SQA-105, Chicago Blower Corporation, Glendale Heights, IL, April 1978.

ASHRAE Handbook—1985 Fundamentals Volume. Atlanta, GA: The American Society of Heating, Refrigerating and Air-Conditioning Engineers, 1985.

Crawford, M. *Air Pollution Control Theory.* New York: McGraw-Hill, 1976.

Danielson, J. A., Ed. *Air Pollution Engineering Manual* (2nd ed.), AP-40. Washington, D.C.: U.S. Environmental Protection Agency, May 1973.

Ganapathy, V. "Quick Estimation of Gas Heat Transfer Coefficients," *Chemical Engineering,* September 13, 1976, p. 199.

Hemeon, W. C. L. *Plant and Process Ventilation.* New York: Industrial Press, 1963.

Industrial Ventilation: A Manual of Recommended Practice (12th ed.). Lansing, MI: American Conference of Governmental Industrial Hygienists, 1972.

McCabe, W. L., Smith, J. C., and Harriott, P. *Unit Operations of Chemical Engineering* (4th ed.). New York: McGraw-Hill, 1985.

McDermott, H. J. *Handbook of Ventilation for Contaminant Control.* Ann Arbor, MI: Ann Arbor Science, 1976.

Menkel, B. E. "Basic Design of a Plating Exhaust System," *Plating and Surface Finishing,* 72(4), April 1985.

Neveril, R. B. *Capital and Operating Costs of Selected Air Pollution Control Systems,* EPA-450/5-80-002. Washington, D.C.: U.S. Environmental Protection Agency, December 1978.

Neveril, R. B., Price, J. U., and Engdahl, K. L. "Capital and Operating Costs of Selected Air Pollution Control Systems-III," *Journal of the Air Pollution Control Association,* 28(10), October 1978.

Sorg, G. R. "Fan Acoustic Basics: Part L" *Pollution Engineering,* Vol. XV, 3, March 1983.

Vatavuk, W. M., and Neveril, R. B. "Part III: Estimating the Size and Cost of Pollutant Capture Hoods," *Chemical Engineering,* December 1, 1980(a), p. 111.

_____. "Part IV: Estimating the Size and Cost of Ductwork," *Chemical Engineering,* December 29, 1980(b), p. 71.

_____. "Part V: Estimating the Size and Cost of Gas Conditioners," *Chemical Engineering,* January 26, 1981(a), p. 127.

_____. "Part VII: Estimating Costs of Fans and Accessories," *Chemical Engineering,* May 18, 1981(b), p. 171.

_____. "Part XII: Estimating the Size and Cost of Incinerators," *Chemical Engineering,* July 12, 1982, p. 129.

◆◆◆ CHAPTER 9

A PARTICULATE CONTROL PROBLEM

There is no substitute for hard work.

Thomas A. Edison

◆ 9.1 INTRODUCTION ◆

Air pollution engineers have many responsibilities. In addition to design work, these responsibilities can include monitoring, permitting, performance testing, maintenance scheduling, and operations trouble-shooting. Happily, engineers sometimes have the opportunity to participate in the design and construction of an air pollution control system from initial conception to start-up of the equipment. For many individuals, such design work is a most satisfying and rewarding experience, perhaps more so than any other part of their job. This chapter deals with the process design of an air pollution control system.

A **system** *is defined as one or more major devices connected together with appropriate ducting, fans, pumps, controls, and other auxiliary equipment, all working together to achieve the control objective.* The designing of systems is an art, and, as such, the talent for design must be developed through experience. In the solution of any one problem by different engineers, it is likely that many good yet different designs would be generated. In this chapter, we present a step-wise approach to one design problem to illustrate the procedures and thought processes. If so desired by the instructor, this problem can be used as a student design project concurrent with teaching the first eight chapters of this text. A four- to six-week period is suggested for completion of this problem, if assigned concurrently.

◆ 9.2 PROBLEM STATEMENT ◆

A mixture of natural talc and borax is obtained via surface mining techniques and is delivered in open-top trucks to the plant site. At the plant site it is crushed by large jaw crushers into pieces no larger than 10–15 cm. The material then goes to a roller mill to further reduce the size. After passing through the roller mill, the material is screened; material smaller than 1 cm is stored for drying, and larger material is recycled for further crushing.

The material that passes the screening step is conveyed to storage silos; from there, it goes to a large rotary kiln dryer where it is dried from 8% moisture to 0.2%. The moist solids feed rate to the dryer is 50,000 kg/hr. The kiln is fired with a low-sulfur fuel oil at a rate of 6.0 liters of fuel per 1000 kg of feed.

The dried product is passed through an air contact cooler, and is then stored before shipment. The exhaust gases leave the kiln at 550 F. The dusty gases exiting the kiln must be treated by appropriate control equipment before being exhausted to the atmosphere. A process flow schematic diagram is presented in Figure 9.1.

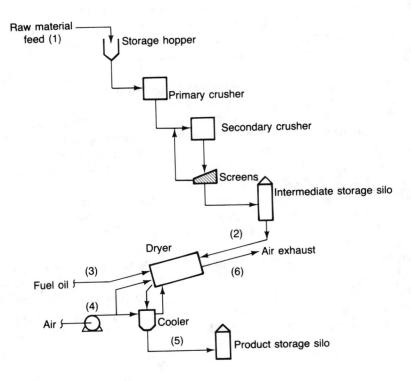

Figure 9.1
Process flow schematic diagram.

First Requirement

Complete the following material balance table and answer the questions that follow. Some additional data are given as follows:

fuel oil specific gravity = 0.90
fuel oil composition (weight percent): C = 88%; H = 11.7%; S = 0.3%
kiln particulate emissions = 5.0% of dry solids feed rate
combustion air supplied per kg of fuel = 18 kg/kg
auxiliary air supplied per kg of moist solids feed = 0.30 kg/kg
inlet air composition (mole percent): N_2 = 78%; O_2 = 21%; Ar = 1%

Material Balance Table
Component, thousand kg/hr

Stream Number	Dry Solids	N_2	O_2	CO_2	H_2O	S	Other	Total
1		0	0	0	4.17	0	0	52.0
2	46.0	0	0	0	4.00	0	0	50.0
5		0	0	0		0	0	
3	0	0	0	0	0			
4	0			0	0	0		
6						0		

Questions

1. Calculate the mass emission rate of particles from the dryer, in kg/hr.
2. Calculate the percentages of the raw feed that are lost in the crushing step and in the drying step.
3. Calculate the mass emission rate of SO_2 from the dryer, in kg/hr.
4. Calculate the concentration of particles emitted from the dryer, in µg/std m^3.
5. Calculate the concentration of SO_2 in the exhaust, in µg/std m^3 and in ppm.
6. If emission limits require that less than 30 kg/hr of particles be emitted, calculate the required efficiency of the collection system.
7. Calculate the volume percent of moisture in the exhaust gases.
8. Calculate the exhaust gas flow rate, in dry std m^3/hr and in acfm.
9. Calculate the volume percentages of CO_2 and O_2 in the exhaust gases on a dry basis.

◆ 9.3 OPTIONS FOR FINAL CONTROL ◆

Several proposals have been made to control particulate emissions, all of which involve a cyclone precleaner from which product can be recovered. After the cyclone precleaner, the gases could be routed to a baghouse, an electrostatic precipitator, or a wet scrubber.

Each of these final control devices can be designed to be very efficient; however, the major difference between the first two and the third one is that the wet scrubber produces a sludge that must be discarded. Disposal of such sludges is sometimes costly and/or difficult. Furthermore, the small amount of SO_2 in the exhaust can create a serious corrosion problem in a wet collection system. (*Note:* If it were required to remove SO_2, a wet scrubber that could remove both particles and SO_2 might be a good option.) Finally, if we use a dry collection method, we might be able to salvage some of the dust as additional product that can be sold. Based on these considerations, we shall, at this point, discard the option of wet scrubbing. In practice, the design engineer probably would investigate this option more fully before deciding whether to discard it or to include it as an option in the preliminary design and cost estimation. A proposed process flow diagram is presented in Figure 9.2.

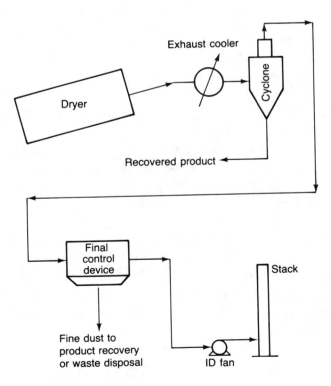

Figure 9.2
Proposed process flow diagram for exhaust control.

◆ 9.4 Major Items of Equipment ◆

Before we proceed to the design of the individual pieces of equipment, we will briefly consider each item shown in Figure 9.2. The cyclone precleaner is a good device for removing large particles, especially if very high efficiencies are not required. Cyclones can withstand high temperatures and erosive/corrosive conditions. As discussed in Chapter 4, the overall efficiency of collection for a cyclone depends on the dimensions of the cyclone, the properties of the particles (density and size distribution), the properties of the gas (viscosity), and the power input to the cyclone (in the form of gas pressure drop or gas velocity). Higher efficiencies usually require higher power inputs. Therefore, we will want to trade off efficiency versus pressure drop, recognizing that the particulate loading on the final control device (FCD) will increase as we decrease the efficiency of the cyclone.

An exhaust cooler will be necessary for either choice of FCD because fabric filter bags cannot withstand temperatures of 550 F, and because ESPs are smaller, less costly, and easier to operate at lower temperatures. Therefore, a cooler must be placed somewhere in front of the FCD. The reason we have placed it upstream of the cyclone is because both the volumetric flow rate and the viscosity of air decrease with a decrease in temperature. Therefore, the cyclone pressure drop is decreased without decreasing its efficiency. Furthermore, cooling the gas early in the system results in smaller (less expensive) ducts.

The precise placement of the exhaust gas cooler is not a clear-cut choice. In front of the cyclone, the cooler will accumulate dust faster than it would if it were behind the cyclone. Thus, it will require more frequent cleaning. This higher maintenance cost must be traded off against the reduced operating cost of the cyclone and reduced capital cost of the smaller ducts in making the final design decision.

The FCD will be either a baghouse or an ESP. The design of either device depends on additional data about the particulate distribution and loading in the gas stream entering the FCD (which will be different from that of the gas entering the cyclone). We will discuss the designs of both devices in more detail later in this section.

The induced draft (ID) fan is required to pull the gases through the system and to push them into the stack. We will place the fan at the clean end of the system to prevent erosion of the blades and subsequent maintenance problems.

The stack is necessary to ensure safe dispersal of the exhaust gases. In this problem, because the FCD is so efficient and because the SO_2 concentrations are low, we will assume that the detailed design of the stack is not critical. Stack design involves modeling of plume rise and plume dispersion (see Chapter 20). In this problem, we will simply use a stack height and diameter that are similar to other stacks in similar service.

The designing of the ductwork is sometimes postponed until late in the project. However, early consideration of the ductwork can result in significant savings in cost. In designing the ductwork, we try to minimize the overall cost by properly sizing the duct for the required gas flow and by laying out the equipment to minimize the required length of ducting. We will now address the design of the individual pieces of equipment.

THE COOLER

The cooler reduces the exhaust gas temperature. If we bring the gases to a temperature that is "too low," condensation will occur somewhere downstream (probably in the FCD). With some SO_2 (and likely some SO_3) present in the gas, such condensation could be disastrous. Even if there were no SO_3, condensation would moisten the particulate layer and perhaps foul the FCD (perhaps a sticky cake on the bags or perhaps bridging in the ESP hoppers). The *dew point* (the temperature at which liquid first condenses) for a stream of air that contains SO_3 is a function of the relative humidity and the SO_3 content of the stream, as shown by the following equation [developed by Verhoff and Banchero (1974) and presented in transposed form by Pierce (1977)]:

$$\frac{1000}{T_{DP}} = 1.7842 + 0.0269 \left(\log \overline{P}_{H_2O} \right) - 0.1029 \left(\log \overline{P}_{SO_3} \right)$$
$$+ 0.0329 \left(\log \overline{P}_{H_2O} \right) \left(\log \overline{P}_{SO_3} \right) \tag{9.1}$$

where

T_{DP} = dew point, °K

$\overline{P}_{H_2O}$ = partial pressure of water vapor, atm

$\overline{P}_{SO_3}$ = partial pressure of sulfur trioxide, atm

Second Requirement

1. Estimate the dew point of the exhaust gases assuming zero SO_3 content.

2. Estimate the dew point of the exhaust gases assuming 10 ppm SO_3.

Actually, the dew point will be different from either of those calculated in the second requirement because of the SO_2/SO_3 split. Assume that if we cool the exhaust gases to a temperature of 350 F, nowhere in the system will the dew point be encountered. (Calculations of heat losses from the ducts and the FCD would be necessary to check the validity of this assumption.)

The two practical alternatives for the cooler are (1) a shell-and-tube heat exchanger and (2) a water injection, direct contact cooler. If we use a heat exchanger, we could use the enthalpy of the exhaust gas to preheat the air coming into the dryer and thus save on fuel costs, or we could simply run water through the tubes, discarding the heat absorbed from the exhaust. The first choice would require greater surface area because air-to-air heat exchange is less efficient than air-to-water. Thus, a bigger, more costly heat exchanger would be required, but a fuel savings would result. A potentially significant problem with the heat exchanger is the possible fouling of the tubes by the dust in the exhaust gases. If we inject water directly to cool the gas, we increase the relative humidity; thus, we cannot go as low in temperature.

Third Requirement

1. Calculate the change in enthalpy when reducing the exhaust gas from 550 F to 350 F.
2. Use the following data to calculate the operating cost savings if we recover heat to reduce fuel consumption.

   fuel cost = $1.50/gal

   fuel LHV = 18,000 Btu/lb$_m$

   air-to-air heat exchanger efficiency: 80% of heat removed is transferred to incoming air (the rest is "lost")
3. With regard to item 2, calculate the exit temperature of the pre-heated air if it enters at 80 F.
4. Assuming an overall heat transfer coefficient of 3.0 Btu/hr-ft^2-°F, calculate the heat transfer area required for the air-to-air heat exchanger of item 2.
5. If we use direct contact, water injection cooling, calculate the flow rate of water injection (in lb$_m$/hr and in gal/min) required to cool the exhaust to 400 F.
6. If the heat exchanger costs $100,000 and the water injection chamber costs $30,000 (all other things being equal), is the heat exchanger worth the additional cost? Assume the plant operates 2500 hours per year. One simple way to answer this question is to calculate the payback period. If the payback period is less than a preset goal (say 2 years), the project is attractive.

Note that if we include the heat recovery project, it will reduce fuel consumption (and burner air) and change the overall material balance. The total available enthalpy in the exhaust gases will be reduced; thus, an iterative procedure will be required. However, in this problem, the major portion of the airflow (the auxiliary air) does not change because of the solids' drying requirements. Also, the clinker cooler (*clinker* is the dried product) requires some minimum amount of cool air. Therefore, we shall proceed without iterating on the original material balance.

THE CYCLONE

Normally, in a completely open design problem, we would integrate the cyclone design with that of the FCD, the ID fan, and so forth to try to minimize system cost while achieving the overall collection efficiency target. This would include varying the cyclone size, pressure drop, and efficiency, and observing the effect on the total system cost. For this problem, assume that we have determined that an overall efficiency of about 60–70% with a maximum pressure drop of 6 in. H_2O is "best."

Based on a plant test, we have obtained the following data for the particles exiting the dryer:

Size Range, μm	Weight Percent in Size Range
0–2	1.0
2–4	9.0
4–6	10.0
6–10	30.0
10–18	30.0
18–30	14.0
30–50	5.0
>50	1.0

Fourth Requirement

1. Is the particle size distribution log-normal?

2. Design two identical cyclones to operate in parallel to meet the previously stated objectives. The particles have a density of 100 lb_m/ft^3. The gas inlet velocity to the cyclones must be kept below 100 ft/sec. (The gas temperature is 350 F.)

3. Calculate the cut diameter, the overall efficiency, and the pressure drop for your cyclones.

4. For your final design, complete the following table:

Size Range, μm	$d_{p(avg)}$	$\dfrac{d_{p(avg)}}{d_{pc}}$	Percent Efficiency	Weight Percent in Size Range	Weight Percent Collected	Weight Percent Passed
0–2						
2–4						
4–6						
6–10						
10–18						
18–30						
30–50						
>50						

5. Calculate the mass flow rate of particles in the exit gas from the cyclones (in kg/hr) and the loading (in g/m^3 or gr/ft^3).

6. Calculate the new particle size distribution in the exit gas from the cyclones.

7. Estimate the total installed cost for these cyclones (in 2002 dollars).

FINAL CONTROL DEVICE—ALTERNATIVE I : BAGHOUSE

Baghouses are often considered to be the most efficient devices for particulate collection. The potential disadvantages of using a baghouse include the possibility of damage from hot or chemically corrosive gases, excessive pressure drop owing to excessive loading, excessive maintenance owing to frequent bag failures, blinding of the fabric owing to moist, sticky particles, and high cost relative to other alternatives. On the other hand, the advantages of baghouses include the ability to collect very small particles, the ability to collect a variety of different dusts equally well, and the highest collection efficiency of any pollution control device.

As noted in Chapter 6, the design of a baghouse to meet certain process requirements is not a rigid procedure. Much depends on the designer's experience and preferences. There are several basic methods for cleaning the dust from the bags (shaking, reverse airflow, or pulse-jet). In addition, we can choose to collect the dust on the inside or the outside of the bag surface. Deciding on the degree of compartmentalization and the amount of excess bag area is the designer's prerogative. Other important variables for which there are ranges of acceptable values are the type of fabric, the face velocity, and the pressure drop. In this problem, to limit your expenditure of excessive time and effort, we will make several of these choices for you.

We do not have the option to cool the gas to the temperature required for cotton, wool, or dacron fabrics (200–250 F). Teflon^R bags can withstand high temperatures, as can glass fiber bags. However, Teflon^R bags are very expensive, and glass fiber bags have poor abrasion resistance. Therefore, we will choose Nomex^R fabric bags, which can withstand temperatures up to 400 F.

Because the total volume of air to be filtered is not that large, and because a pulse-jet baghouse can operate with much higher air-to-cloth ratios, we will choose a pulse-jet baghouse over a reverse-air or shaker baghouse. This selection will give us much lower capital costs. Once we have chosen a pulse-jet baghouse, we must choose the smaller of the standard sizes for Nomex^R bags (6 in. × 8 ft bags instead of 11½ in. × 30 ft bags). We make this choice because the pulse of compressed air dissipates its energy quickly and longer bags are not cleaned as well as shorter bags.

Fifth Requirement

1. Prepare a preliminary process design of a pulse-jet baghouse. Assume the gas temperature has dropped to 300 F.

Include the following:

    a. Specify the number of bags if the air-to-cloth ratio is typical for talc.

    b. Determine the power for the pulse-jet air compressor if air is supplied at 100 psig and the volume supplied is 150 scfm (assume a compression efficiency of 50%).

    c. Sketch the layout of the bags in the baghouse and give rough overall dimensions.

2. Estimate the installed cost (in 2002 dollars) of the baghouse and the bags (including the purchase price of the bags). Include the compressor capital cost, but not the fan cost. Assume that the compressor, which is dedicated to the baghouse, costs $29,000 and that the bags and cages cost $35.00 and $10.00 each, respectively (all costs in 2002 dollars).

3. Assuming an average pressure drop (at operating temperature) across the baghouse of 4 in. H_2O, calculate the annual operating cost for the baghouse (include only depreciation of the baghouse system at 12% of the TIC, maintenance at 5%, power at 8 cents/kWh, taxes and insurance at 2%, and a complete change of bags every 4 years).

FINAL CONTROL DEVICE—ALTERNATIVE 2: ESP

An electrostatic precipitator is the other viable choice for the application described in this design problem. Recall the simple yet useful model described by the Deutsch equation:

$$\eta = 1 - e^{(-Aw/Q)} \tag{9.2}$$

From this equation, we can immediately recognize the key variables. The variables under our control in this problem are the surface area and (to a certain extent) the drift velocity. The effective drift velocity is primarily a function of the size and type of the particles. However, we can influence the drift velocity by manipulating variables such as electrode voltage or by conditioning the flue gas. Recall that particle resistivity is a function of gas temperature, humidity, and chemical composition.

It is good practice to conduct a plant test on the particles in the exhaust stream if there is an opportunity to do so. The engineers conducting the test should be careful to duplicate the expected actual inlet operating conditions as closely as possible. Certainly, they would want to cool the gas stream and use a cyclone preseparator to try to simulate the exhaust gas stream as it would exist just before entering the ESP. However, since pilot-scale tests can be expensive and time-consuming, the design engineer should make some preliminary calculations using "typical" values of the design parameters to determine whether an ESP is even feasible.

Assume that the initial calculations have been made, and they show that an ESP is feasible. Further assume that a pilot-scale test results in the following values of certain variables for the pilot-scale ESP:

> inlet loading = 5.0 gr/ft^3
>
> outlet loading = 0.10 gr/ft^3
>
> gas velocity = 5 ft/sec
>
> corona power = 12.0 W
>
> gas flow rate = 100 ft^3/min
>
> collection surface area = 25 ft^2

Sixth Requirement

1. Using the results of the plant test and following the guidelines in Chapter 5, design a full-size ESP for our application. At a minimum, you must:

   a. Calculate an effective drift velocity from the plant test data.

   b. Specify for the full-size precipitator:

      (1) The required collection efficiency

      (2) The total collection surface

      (3) The number of plates (specify in your design of the full-size ESP: height and length)

      (4) The spacing between the plates, and the gas linear velocity

      (5) The corona power required

      (6) The fan power required assuming a ΔP of 0.5 in. H$_2$O and a fan efficiency of 0.60

      (7) The number of high-tension sections

2. Estimate the installed cost of your ESP (in 2002 dollars).

3. Calculate the annual operating cost for the ESP. Include depreciation at 12% of the TIC, maintenance at 3%, taxes and insurance at 2%, and power at 8 cents/kWh. Ignore labor charges, overhead charges, and all other charges.

DUCTWORK

The ducting connects the various pieces of equipment, creating a closed system through which to carry the gas. Since purchasing and installing large ducts is expensive, and since trying to force gas to flow through small ducts by using a large pressure drop is also expensive, it behooves us to give proper consideration to the design of ducts. The design problem is twofold: First, we want to optimize the effective diameter of the duct (trade off installed capital costs versus pressure drop/power costs). Second, we want to arrange the large pieces of equipment to minimize the total length and the number of duct fittings and bends. We will address the first problem here, recognizing that the second problem is best left to common sense.

The pressure drop caused by transporting the gas through the ducts must be estimated before a fan can be specified. The Bernoulli equation [Eq. (8.3)], in conjunction with the familiar friction factor chart, is the basis for calculations involving fluid flow and pressure drop in circular channels. From this base, charts relating pressure loss to velocity for airflow in ducts have been developed (see Figures 8.2 and 8.3 in Chapter 8). The charts in Chapter 8 show the relationship of pressure drop and velocity for standard air in circular ducts of various diameters.

Rectangular ducts are often used (instead of circular ducts) because they are easier to fit into certain installations. A method of equating a rectangular duct to a circular duct was discussed in Chapter 8. The equivalence described is that of equal flow rates with equal pressure drop. Figure 8.5 can be used to find the equivalent rectangular duct given the diameter of the circular duct and either one dimension or the aspect ratio of the rectangular duct.

Seventh Requirement

1. Assuming design gas velocities in circular ducts of about (a) 3200, (b) 4000, and (c) 4500 ft/min, select standard diameters for the ducts (assume an average temperature of 300 F).

2. For a total equivalent length (including fittings) of 400 feet, calculate the duct pressure drop (in in. H_2O) for standard air and for flowing air for the three cases in item 1.

3. Calculate the equivalent square duct sizes for the three cases in item 1.

4. Assume installed duct costs of $80/ft, $100/ft, and $120/ft for the small, medium, and large circular ducts. Assume a power cost of 8 cents/kWh and a fan efficiency of 0.60. Considering only power costs and depreciation (at 12% of the TIC per year), which duct size has the lowest operating cost?

5. Assuming all three sizes will transport dust satisfactorily, if the largest duct did not have the lowest total cost, give two reasons why you might choose to install the largest duct anyway.

INDUCED DRAFT FAN

Selection of the best fan for the system involves matching the requirements of the system (total ΔP and volumetric flow rate) with the capabilities of a fan. The careful use of manufacturers' tables of data and/or fan curves is necessary to ensure proper selection Improper selection can result in poor performance of the system and/or costs that are higher than necessary. The objective of fan selection is to select a fan that not only moves the required amount of air through the system resistance (pressure drop), but also operates efficiently and stably. Since it is possible to slightly adjust a given fan's performance (for example, by changing the speed or by placing a damper in the system), an exact design is not critical. However, proper selection is important from both economic and operational viewpoints.

Eighth Requirement

1. Use the data in Table 9.1 to select a fan for your system. Assume that the pressure loss from the fan outlet through the stack is 1 in. H_2O (at flowing conditions—that is, at 300 F). If your design includes a heat exchanger (from the third requirement), assume a ΔP through the heat exchanger of 3.0 in. H_2O at an average operating temperature of 450 F.

2. Estimate the total installed cost of this fan (in 2002 dollars). include an allowance for a motor, a starter, and a belt drive.

STACK

In this problem, it is assumed that the exhaust gases are clean enough so that dispersion from a tall stack is unnecessary. Usually, a height of several meters is sufficient to act as the fan outlet duct. Also, an exit gas velocity of 45–70 ft/sec is adequate for this type of exhaust. Once an outlet velocity is known, it is easy to calculate the required stack diameter from the continuity equation. If the stack is, in fact, the outlet duct for the fan, and is mounted directly on the fan, we can obtain the outlet velocity from the fan's rating table.

♦ 9.5 SUMMARY ♦

A final step in any design project is the written summarization of the design. Even a preliminary process design should be summarized and a neat design package prepared for communicating the results of your work to management. A short cover letter and an executive summary of the design (including an overall cost estimate) is necessary, followed by capsulized (but detailed) sections on each major component of the system. Since we have discussed the items one at a time in this chapter, we will skip the detailed reporting. However, some benefits will be derived from writing an executive summary.

Final Requirement

Write a cover letter and a short executive summary of your design. Highlight the comparison of the ESP and the baghouse, and recommend one or the other of these alternatives. Submit the summary along with all of your solutions to the various requirements of this design project.

Table 9.1 Fan Rating Table for Eighth Requirement

CFM	14" SP		16" SP		18" SP		20" SP		22" SP		24" SP		26" SP		28" SP	
	RPM	BHP	RPM	BHP	RPM	BHP	RPM	BHP	RPM	BHP	RPM	BHP	RPM	BHP	RPM	BHP
10,000	995	35	1044	40	1093	45	1143	50	1192	55	1242	60	1292	66	1341	72
12,000	1007	43	1056	48	1105	54	1155	60	1204	66	1254	72	1303	78	1352	86
14,000	1018	50	1068	56	1118	63	1167	70	1216	77	1265	84	1314	93	1364	100
16,000	1029	57	1079	64	1130	72	1178	80	1227	88	1276	96	1324	105	1375	114
18,000	1039	64	1089	72	1141	81	1189	90	1238	99	1288	108	1336	117	1386	128
20,000	1049	71	1100	80	1150	90	1200	100	1250	110	1300	120	1349	129	1398	141
22,000	1060	78	1110	89	1160	100	1210	110	1260	120	1310	131	1360	142	1410	154
24,000	1078	86	1121	98	1171	110	1220	120	1271	131	1321	142	1371	154	1421	168
26,000	1096	93	1133	106	1182	119	1231	130	1281	142	1330	153	1381	166	1432	180
28,000	1114	100	1155	114	1197	128	1242	140	1290	153	1340	164	1392	178	1445	192

REFERENCES

Pierce, R. R. "Estimating Acid Dewpoints in Stack Gases," *Chemical Engineering,* April 11, 1977.

Verhoff, F. H., and Banchero, J. T. "Predicting Dew Points of Flue Gases," *Chemical Engineering Progress,* 70(8), 1974.

◆◆◆ CHAPTER 10

PROPERTIES OF
GASES AND VAPORS

Experts were still unsure what caused the massive gas leak. A sharp change in temperature, impurities inside the tank, or even a miniscule crack could have caused a rapid buildup in pressure. . . . The chemical has a very low boiling point so by the time it started leaking out, it had turned from liquid to gas. . . . The vapor cloud drifted slowly downwind over the densely populated shanty town. . . . The effects of the chemical on human beings resemble those of nerve gas.

Derived from several accounts of the tragic methyl isocyanate release in Bhopal, India, December 1984, in which over 2000 people died and over 50,000 were left injured.

◆ 10.1 INTRODUCTION ◆

In air pollution work we often encounter the terms gas and vapor. People sometimes use the terms interchangeably. In this text, we wish to make a distinction between gases and vapors, although we realize that under various conditions of temperature and pressure the differences are often slight.

One reason for making this distinction is that the behavior of gases can be accurately predicted using the ideal gas law, whereas concentrated vapors can show considerable deviation from ideal behavior (which, after all, is based on spherical, non-interacting, point-mass molecules). However, the main reason for making a distinction between gases and vapors is because different control techniques can be applied to vapors than can be applied to gases. In this chapter we will discuss, in general terms, some of the differences in the properties of pollutant gases and vapors, as well as some of the fundamental principles on which various pollution control processes are based. Specific design-oriented discussions of gas and vapor pollution control equipment are presented in Chapters 11–16.

Gases and vapors have some properties in common. Both are composed of widely separated, freely moving molecules. Both will expand to fill a larger, differently shaped container. Both exert pressure in all directions. In short, both are in the gaseous state of matter. The major difference lies in the internal energy of the molecules. A substance in the gaseous state is considered a true gas if it is far removed from the liquid state. Usually, this means that the temperature of the substance is above its critical point (the highest temperature at which it can be condensed). On the other hand, a vapor is a substance in the gaseous state that is not far from being a liquid. The vapor can exist as dispersed molecules at a temperature not far above or even below its dew point (the temperature at which pure vapors at atmospheric pressure will condense). A vapor can usually be adsorbed onto surfaces or condensed into a liquid relatively easily.

We recognize that the discussion in the preceding paragraph lacks precision. Therefore, we will provide some specific examples that might help you to distinguish between gases and vapors. In air, we consider oxygen, nitrogen, argon, and carbon dioxide to be gases, whereas we refer to the water content as water vapor. The pollutants SO_2, NO, NO_2, and CO are considered gases, whereas most volatile organic compounds (VOCs) are considered vapors (exceptions would be methane, ethane, ethylene, and other VOCs with low boiling points).

♦ 10.2 VAPOR PRESSURE ♦

Every liquid exerts a *vapor pressure*. **Vapor pressure** *is defined as the pressure exerted by a pure component vapor in equilibrium with a flat surface of its pure component liquid at a certain temperature.* Vapor pressure is a measure of the escaping tendency or volatility of the *liquid;* thus, we refer to it as the *liquid's* vapor pressure. For instance, at 150 F, *n*-decane (a component in kerosene) has a low vapor pressure (0.3 psi), whereas *i*-pentane (a component in gasoline) has a high vapor pressure (over 3 atm).

Vapor pressure increases rapidly with an increase in temperature. A common curve-fit equation for the vapor pressure/temperature relationship is the Antoine equation, which is

$$\log P_{v_i} = A_i + \frac{B_i}{C_i + T} \tag{10.1}$$

where

P_{v_i} = vapor pressure of pure liquid i

T = temperature

A_i, B_i, C_i = curve-fit constants for component i

The vapor pressures of many organic substances are given in *International Critical Tables of Numerical Data* (1930), and in Jordan (1954). Also, vapor pressure data are often simply presented in tables or graphs (see Appendix B).

The vapor pressure should not be confused with the **partial pressure** of a vapor in a **mixture** of gases. In general, the partial pressure can take on any value less than or equal to the vapor pressure, as long as no liquid is present. If the system is not at equilibrium, there are no theoretical ties between vapor pressure and partial pressure. When the system is at equilibrium with both vapor and liquid present, the vapor phase is said to be *saturated*, and the partial pressure is equal to the vapor pressure.

When a pure liquid is placed in contact with air in a closed space, some of the liquid will volatilize (evaporate) until vapor–liquid equilibrium is established. At equilibrium, the partial pressure of the vapor (and thus the gas-phase composition) is dependent on the liquid's vapor pressure; that is,

$$\overline{P}_i = y_i P = P_{v_i} \tag{10.2}$$

where

$\overline{P}_i$ = partial pressure of component i, atm

y_i = mole fraction of component i in the gas

P = total pressure, atm

Equation (10.2) is useful in conjunction with vapor pressure data or Eq. (10.1) to predict the temperature to which an airstream (with a given concentration of a VOC vapor) must be cooled to condense an appreciable quantity of the VOC. The following example problem illustrates the use of Eq. (10.2).

▶ ▶ ▶ ▶ ▶ ▶ ▶ ▶ ▶

Example 10.1

An airstream is presently at 1 atm and 160 F, and contains 40,000 ppm toluene. To what temperature must the air be cooled to remove two-thirds of the toluene vapor?

Solution

At 1 atm, a concentration of 40,000 ppm toluene corresponds to a partial pressure of 0.588 psi, which is below the liquid's vapor pressure at 160 F (see Appendix B, Figure B.1). However, as we reduce the temperature, the vapor pressure decreases, whereas the partial pressure remains the same. When the partial pressure equals the vapor pressure, condensation will occur. We need to find the temperature at which the vapor pressure of toluene equals one-third of the inlet partial pressure; that is,

$$P_{v_i} = \frac{1}{3}\overline{P}_i = 0.196 \text{ psi}$$

From Figure B.1 in Appendix B, when $P_{v_i} = 0.196$ psi, T is approximately 40 F.

Note that in the design of a heat exchanger to cool the air in Example 10.1, we must provide enough surface area to not only cool the air (remove sensible heat), but also to condense the vapors (remove latent heat). Usually, regular cooling water is not cool enough to allow the use of a condenser as a final control device (FCD). Chilled water or a refrigerant might be sufficiently cool to remove enough vapor from the air to meet the pollution control objective. However, note from Example 10.1 that even when cooled to 40 F, the exit airstream will still contain 13,333 ppm toluene.

In general, for nonideal, multicomponent mixtures of liquids and vapors, the following equation relates gas-phase composition and liquid-phase composition at equilibrium (Smith and Van Ness 1975):

$$\phi_i y_i P = \gamma_i x_i P_{v_i} \tag{10.3}$$

where

ϕ_i = vapor-phase activity coefficient for component i

γ_i = liquid-phase activity coefficient for component i

x_i = mole fraction of component i in the liquid

For ideal solutions (either gases or liquids), the activity coefficient is equal to 1.0. However, many mixtures of volatile organic liquids do not behave ideally, and γ_i (which is a strong function of composition and relative concentrations) can range from 1 up to 10 or higher.

The relative ratios of VOCs in the gas is a function of the liquid mixture of solvents in contact with the air, the airflow rate, and the molecular diffusivities of the VOC components in the air. Bishop et al. (1982) have developed a computer model to predict relative vapor ratios in room air exposed to liquid solvent mixtures (an important piece of information for industrial hygienists and others concerned with indoor air pollution). The primary equation of the model is

$$R_{ij} = y_i/y_j = \frac{\gamma_i x_i P_{v_i}}{\gamma_j x_j P_{v_j}} \left(\frac{\mathscr{D}_i}{\mathscr{D}_j}\right)^k \tag{10.4}$$

where

R_{ij} = vapor ratio

$\mathscr{D}_i$ = molecular diffusivity of component i in air, cm^2/s

k = an empirical, flow-dependent constant (often equal to about 0.5)

The calculation of the γ's for use in Eq. (10.4) is not trivial, and requires the use of a computer routine such as UNIFAC (Fredenslund, Jones, and Prausnitz 1975). Bishop et al. (1982) have shown that the computerized routine based on Eq. (10.4) gave far superior results when compared with the "approximate solution" recommended by the American Conference of Governmental Industrial Hygienists.

◆ 10.3 DIFFUSIVITIES ◆

The second law of thermodynamics explains that matter will diffuse spontaneously from a region of high concentration to one of low concentration. Fick's first law can be taken as a definition of the **diffusivity** (the proportionality constant between the rate of flux of matter and the concentration gradient). Thus,

$$\frac{\dot{M}}{A} = -\mathscr{D}\frac{dC}{dx} \qquad (10.5)$$

where

$\dot{M}$ = mass transfer rate, mol/s

A = area normal to the direction of diffusion, cm^2

$\mathscr{D}$ = diffusivity, cm^2/s

$\dfrac{dC}{dx}$ = concentration gradient, mol/cm^4

Since the concentration decreases in the direction of diffusion, a negative sign is necessary in Eq. (10.5).

The diffusivity of a particular substance depends on the substance itself and on the medium through which it is moving. Diffusivities are important in almost all applications of gas and vapor pollution control. In absorption, adsorption, catalytic incineration, and absorption coupled with chemical reaction, the gas flows past (and contacts) a solid or liquid as part of the process. According to the boundary layer theory of fluid flow, a laminar sublayer (or film) is formed even in the case of turbulent flow of a fluid past a fixed object or when two immiscible fluids flow past each other. The pollutant is carried about in the bulk gas (or liquid) by bulk flow and by turbulent eddies. However, the only way that the pollutant passes from one phase to the other is via molecular diffusion across the laminar film surrounding the interface. Diffusion is often the rate-limiting step in pollutant removal processes. Our qualitative description of diffusion will be expanded and quantified in Chapter 13 on absorption. Appendix B presents some data on gas diffusivities.

10.4 GAS-LIQUID AND GAS-SOLID EQUILIBRIA

SOLUBILITY

In the absorption process, once a gas molecule has diffused through the stagnant gas film it must be absorbed into the liquid. Normally, we assume that at an interface (which has no depth), local equilibrium is established virtually instantaneously. Once absorbed, the pollutant then diffuses through a stagnant liquid film into the bulk liquid. However, even though the *rate* of absorption is very fast and thus not a limitation, the *extent* of absorption (the solubility) is crucial to the overall objective of mass transfer.

Most air pollution control equipment works at or near atmospheric pressure and with relatively dilute solutions (both gas and liquid streams). For dilute solutions, the concentrations of the pollutant in the gas and the liquid are often linearly related. This relationship is known as Henry's law, which can be expressed as

$$\overline{P}_i = H_i x_i \qquad (10.6)$$

where H_i = Henry's law constant for pollutant i, atm/mole fraction.

Appendix B presents some data on Henry's law constants. Note that (1) all Henry's law constants have units, (2) Henry's law constants vary with temperature, and (3) as x_i increases, the relationship eventually becomes nonlinear; that is, at some value of x_i, Eq. (10.6) is no longer valid.

▶ ▶ ▶ ▶ ▶ ▶ ▶ ▶ ▶

Example 10.2

(a) Calculate the equilibrium partial pressure of CO in air that is in contact with water that is at 20 C and has a $1.0(10)^{-6}$ mole fraction of CO. (b) Calculate the mole fraction of H_2S in water that is in contact with air that is at 20 C and contains H_2S with a partial pressure of 0.05 atm.

Solution

(a) From Appendix B, $H_{CO} = 5.36(10)^4$ atm/mole fraction, so $\overline{P}_{CO} = 0.0536$ atm.

(b) From Appendix B, $H_{H_2S} = 4.83(10)^2$ atm/mole fraction. Therefore,

$$x_{H_2S} = \frac{0.05 \text{ atm}}{483 \text{ atm/mole fraction}} = 1.04(10)^{-4} \text{ mole fraction}$$

◀ ◀ ◀ ◀ ◀ ◀ ◀ ◀ ◀

Note that Eq. (10.6) could also be written as

$$x_i = H_i' \overline{P}_i \tag{10.7}$$

In Eq. (10.7), H_i' is obviously the inverse of H_i, but it still might be called the Henry's law constant by some references. It is important to know which defining formula is being used. In the form of Eq. (10.6), Henry's law constants are typically very large numbers. Furthermore, the largest numbers correspond to the least soluble gases.

ADSORPTION

Adsorption is another type of equilibrium that is important in air pollution work. Adsorption involves a gas–solid equilibrium that is quite similar in principle to the solubility equilibrium exhibited by gas–liquid systems. However, the equilibrium relationships are rarely linear. When air containing a pollutant vapor is contacted with certain solids and allowed sufficient time to equilibrate, some of the vapor molecules will have been attracted to and retained by the solid surface. The amount adsorbed on the solid depends on the type of vapor, the partial pressure of the vapor, the type of solid, the amount of surface area available for adsorption, and the temperature. A typical adsorption isotherm is shown in Figure 10.1.

Physical adsorption is exothermic, giving up an amount of heat similar to the heat of condensation. **Chemisorption** involves the breaking and re-forming of bonds, and is much more energetic than physical adsorption. The state of matter of the vapor molecules on the surface can be somewhere between liquid and vapor, and there can be either several layers or a single layer of molecules on the surface.

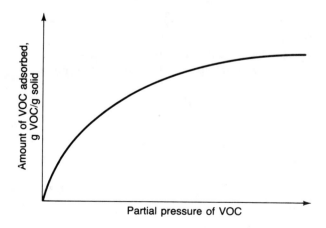

Figure 10.1
A typical adsorption isotherm.

Physical adsorption is reversible, and this characteristic is used as a means of separating VOCs from air. Several solids are effective as adsorbents; the principle solid used in air pollution control is activated carbon. Design of systems based on activated carbon adsorption is discussed in detail in Chapter 12.

♦ 10.5 CHEMICAL REACTIONS ♦

After absorption or adsorption occurs, a chemical reaction often takes place to permanently change the pollutant to a less harmful form. Classic examples are SO_2 scrubbing to form $CaSO_4$, and catalytic incineration of VOCs to form CO_2 and H_2O. Furthermore, homogeneous (one phase) reactions are quite important in pollutant formation and control. We will now briefly review two fundamental principles of chemical reactions—kinetics and thermodynamics.

KINETICS

For our purposes, the study of kinetics is the study of chemical reaction rates. A chemical reaction rate can be expressed as the rate of "disappearance" of reactants or the rate of "appearance" of products. In this text, we will use the symbol r_i to indicate the generation rate of component i. Since products are generated and reactants are "used up," r_i is positive if i is a product and negative if i is a reactant. However, the rate of reaction is always positive. Thus,

$$\text{reaction rate} = r_P = -r_R \tag{10.8}$$

where

   r_P = rate of generation of product P, mol/L-s

   r_R = rate of generation of reactant R, mol/L-s

Reaction rates have units of moles/(volume-time).

The rate of reaction is proportional to the concentrations of the reactants (frequency of collisions) and to the temperature (energy of collisions). Consider the reaction

$$R + S \rightarrow P + Q$$

The rate of reaction can be expressed as

$$r_P = kC_R^x C_S^y \tag{10.9}$$

where

   k = reaction rate constant (temperature dependent)

   C_R, C_S = concentrations of reactants, mol/L

   x, y = exponents

The reaction is said to be of order x in reactant R, of order y in reactant S, and of order $(x + y)$ overall. Often (but not always), x and y are simple whole numbers that are related to the stoichiometry. However, x and y usually must be determined experimentally.

The rate constant k is usually represented by an Arrhenius equation such as

$$k = Ae^{-E/RT} \tag{10.10}$$

where

A = frequency factor

E = activation energy

R = universal gas constant, in energy units

T = absolute temperature

▶ ▶ ▶ ▶ ▶ ▶ ▶ ▶ ▶

Example 10.3

For a first order reaction $R \rightarrow P$, values of the rate constant k are determined to be 10.0 s^{-1} and 5.00 s^{-1} at T = 700 K and 670 K, respectively. Estimate the values of A and E.

Solution

From Eq. (10.10) applied at T = 700 K and T = 670 K,

$$\frac{k_1}{k_2} = \frac{e^{-E/RT_1}}{e^{-E/RT_2}} = \exp\left[-\frac{E}{R}\left(\frac{1}{T_1} - \frac{1}{T_2}\right)\right]$$

$$2.0 = \exp\left[-\frac{E}{1.987}\left(\frac{1}{700} - \frac{1}{670}\right)\right]$$

$$\ln 2 = 3.22(10)^{-5} E$$

$$E = 21{,}530 \text{ cal/mol}$$

Therefore,

$$A = \frac{k}{e^{-E/RT}} = \frac{10}{e^{-21{,}530/1.987(700)}}$$

$$= 5.28(10)^7 \text{ s}^{-1}$$

◀ ◀ ◀ ◀ ◀ ◀ ◀ ◀ ◀

Reactions occur in reactors (of course), and two ideal reactor models are commonly used to describe real reactors. The first is the *continuous stirred tank reactor* (CSTR) model, which depicts a continuous flow through a tank in which the contents are rapidly mixed. The concentrations of all species are uniform throughout the tank and are the same as those in the outlet stream. A steady-state (accumulation rate = zero) material balance for component i in the CSTR (with constant volume V) yields

$$0 = Q_{in}C_{i_{in}} - Q_{out}C_{i_{out}} + r_i V \qquad (10.11)$$

where

Q = volumetric flow rate, L/s

V = volume of reactor, L

Eq. (10.11) can be solved algebraically for any one of the variables given the others.

The other ideal reactor model is the *plug flow reactor* (PFR) model, which depicts one-dimensional flow through a long tube. The velocity is constant at all radial positions in the tube, and axial dispersion is negligible. The steady-state material balance for component i in a plug flow reactor must be written for a differential slice of the reactor as follows:

$$0 = Q_V C_{i_V} - Q_{V+\Delta V}C_{i_{V+\Delta V}} + r_i \Delta V \qquad (10.12)$$

If the flow rate is constant and if r_i is not a function of position, then Eq. (10.12) becomes

$$\frac{dC_i}{r_i} = \frac{1}{Q}dV \qquad (10.13)$$

Both conditions on which Eq. (10.13) is predicated are satisfied for the case of an isothermal reacting system with a constant molar flow rate. However, for many actual gas-phase reactions, both the volumetric flow rate and the reaction rate can change with position in the reactor. In such cases, it is often convenient to express reaction rates and concentrations in terms of mole fraction and solve Eq. (10.12) numerically and iteratively.

▶ ▶ ▶ ▶ ▶ ▶ ▶ ▶ ▶

Example 10.4

The reaction of Example 10.3 is to occur isothermally at 640 K in (a) a CSTR or (b) a PFR. Calculate the required volume of each reactor to give 99% conversion of R to P when the volumetric flow rate is constant at 100 L/s.

Solution

(a) Equation (10.11) for the CSTR can be rearranged to give

$$V = \frac{-Q\left(C_{R\text{in}} - C_{R\text{out}}\right)}{r_R}$$

$$V = \frac{QC_{R\text{in}}\left(1 - C_{R\text{out}}/C_{R\text{in}}\right)}{-kC_{R\text{out}}}$$

$$V = \frac{100C_{R\text{in}}\left(1 - 0.01\right)}{k\left(0.01C_{R\text{in}}\right)}$$

$$V = \frac{9900}{k}$$

At 640 K,

$$k = 5.28(10)^7 e^{-21,530/(1.987)(640)} = 2.34 \text{ s}^{-1}$$

Therefore,

$$V = 4230 \text{ L (for a CSTR)}$$

(b) Starting with Eq. (10.13) for a PFR,

$$\int_{C_{R\text{in}}}^{C_{R\text{out}}} \frac{dC_R}{-kC_R} = \frac{1}{Q}\int_0^v dV$$

$$\frac{-1}{k}\ln\frac{C_{R\text{out}}}{C_{R\text{in}}} = \frac{V}{Q}$$

$$\frac{-1}{k}\ln\left(0.01\right) = \frac{V}{100}$$

$$V = \frac{100}{2.34}\ln 100$$

$$V = 197 \text{ L (for a PFR)}$$

Note the difference in the required volumes of the PFR and CSTR for the same degree of conversion.

THERMODYNAMICS

Many reactions are very energetic. Some reactions release large amounts of heat (exothermic), whereas others require considerable heat input (endothermic). One obvious effect of such heat release or absorption is that it changes the temperature of the reacting mix and thus affects the reaction rates. Another effect is that, for gases, a change in temperature affects the concentrations (in mol/L but not in mole fraction) and the volumetric flow rates.

Some of the most energetic reactions are the combustions of organics. The heat of reaction (heat of combustion) of just one kilogram of methane can raise the temperature of 50 kilograms of air by almost 900 Kelvin degrees. However, when using the values of the heat of combustion as reported in tables of data, keep in mind the difference between the *higher heating value* (HHV) and the *lower heating value* (LHV). The HHV is tabulated most often and includes the energy released by condensing the water formed in the combustion reaction. The LHV is the HHV minus the heat of condensation of the water. The LHV represents the net heat of combustion that is available to heat the gaseous combustion products in industrial combustion processes.

Another important aspect of thermodynamics is that of chemical equilibrium. Many chemical reactions do not go to 100% completion. An equilibrium that depends on temperature is established between reactants and products. Consider the oxidation of SO_2 to SO_3:

$$SO_2 + \tfrac{1}{2}O_2 \rightarrow SO_3 \qquad (10.14)$$

For this gas-phase reaction (assuming an ideal mixture), we can define an equilibrium constant K_p as

$$K_p = \frac{\overline{P}_{SO_3}}{\overline{P}_{SO_2}\left(\overline{P}_{O_2}\right)^{1/2}} \qquad (10.15)$$

A large numerical value of K_p indicates that an *equilibrium* mixture of SO_2, O_2, and SO_3 will be predominantly SO_3. The value of K_p is a strong function of temperature, as shown in Table 10.1.

Note that equilibrium concentrations are not often observed in industrial reactors. For instance, in the $SO_2 \rightarrow SO_3$ reaction, SO_3 formation and removal actually occurs via a complex multi-step mechanism, and produces concentrations of SO_3 that are quite different from those expected from equilibrium considerations.

Temperature, °K	K_p
295	$2.6(10)^{12}$
500	$2.6(10)^5$
1000	1.8
1500	0.038

Table 10.1 Equilibrium Constants for the Reaction $SO_2 + \tfrac{1}{2}O_2 \rightarrow SO_3$

Adapted from Wark and Warner, 1981.

Generally, the *equilibrium constants* for exothermic reactions decrease with increasing temperature. On the other hand, *kinetic* constants always increase with temperature. Therefore, it is possible for an optimum temperature to exist for the conversion of a given reactant to a given product. The optimum temperature will be high enough to allow the reactions to proceed rapidly, yet low enough to allow the equilibrium to achieve the desired degree of conversion.

◆ PROBLEMS ◆

10.1 Calculate the partial pressure and gas-phase mole fraction of NO in air that is in contact with water that has a mole fraction of $1.0(10)^{-6}$ NO. The temperature and pressure are 20 C and 1 atm.

10.2 Calculate the mole fraction of ethane in water at 10 C when the ethane concentration in air that is in equilibrium with the water is 10,000 ppm. The total pressure is 5.0 atm.

10.3 Calculate the partial pressure and gas-phase concentration (in ppm) of n-octane in air when the air is saturated at 100 F. The total pressure is 1 atm.

10.4 Calculate the partial pressure and gas-phase concentration (in ppm) of ethanol in air when air at 18 psi is bubbled through pure ethanol at 25 C. Assume the bubbler achieves 40% saturation.

10.5 To what temperature must air be cooled to remove 40% of its heptane vapor content? The airstream is presently at 150 C and 1 atm, and contains 50,000 ppm heptane.

10.6 If the airstream of Problem 10.5 were flowing at 50 m³/min, calculate the amount of money (in dollars per day) that could be saved by condensing 40% of the heptane. Assume that solvent (liquid) heptane has a specific gravity of 0.75 and sells for $1.00/liter.

10.7 Two reactions (I and II) have activation energies of 50,000 cal/mol and 30,000 cal/mol, respectively, and identical pre-exponential factors. Determine the ratio of the rate constants k_I/k_{II} at (a) 500 C and (b) 800 C.

10.8 For the reactions of Problem 10.7, determine the factor by which the rate constant k_I increases when the temperature increases from 500 C to 800 C. Also determine the factor for k_{II}.

10.9 Derive an equation and solve for the volume V of a CSTR, operated isothermally for the second-order reaction $2R \rightarrow P$, where $r_R = -kC_R^2$. Data are $k = 6.0$ L/mol-s, $C_{R_{in}} = 0.005$ mol/L, $C_{R_{out}} = 0.0005$ mol/L, and $Q = 100$ L/s.

10.10 Derive an equation and solve for the volume V of a PFR for the same data as in Problem 10.9.

10.11 For a first-order reaction, values of the rate constant k are determined experimentally to be 20 sec^{-1} at 1000 K and 10 sec^{-1} at 950 K. Estimate the values of A and E for the reaction.

10.12 A reactor processes air flowing at 100 actual cubic meters per min (acmm) at 50 C and 1 atm. The airstream is contaminated with a gaseous pollutant (800 ppm). The contaminant is destroyed in a 30.0 m^3 vessel via a first-order reaction. The rate constant for the reaction is $k = 15.0$ min^{-1}. The final concentration in the exit stream is 10 ppm. Does this reactor more closely approximate a CSTR or a PFR?

10.13 Calculate the partial pressure and gas-phase concentration (in ppm) of toluene in air when air at 8.0 pounds per square inch gage (psig) is bubbled through pure toluene at 24 C. Assume the bubbler achieves 60% saturation of the airstream.

10.14 A natural gas pipeline is routed under a river. A local group is concerned about a methane leak that might contaminate the water. Of course, most of the methane that might leak would bubble out of the water and disperse into the atmosphere. But, assuming that the methane were to "hang around" indefinitely in the air above the water at a partial pressure of 0.03 atm, estimate the concentration of methane in the water (mol fraction).

10.15 A CSTR is being fed a 50 L/min stream containing 0.08 mol/L of reactant A. The reaction proceeds according to $2A \rightarrow B$ with $-r_A = k\,[A]^3$, with $k = 2.5$ $L^2/(mol)^2$-min. The concentration of A exiting the reactor is 0.03 mol/L.

a. Calculate the volume of the reactor (L).

b. Given that B has a molecular weight of 75, and that the process runs 24 hours per day, calculate the production rate of B (kg/day).

REFERENCES

Bishop, E. C., Popendorf, W., Hanson, D., and Prausnitz, J. "Predicting Relative Vapor Ratios for Organic Solvent Mixtures," *Journal of the American Industrial Hygiene Association, 43,* September 1982.

Fredenslund, Aa., Jones, R. L., and Prausnitz, J. M. "Group-Contribution Estimation of Activity Coefficients in Nonideal Liquid Mixtures," *Journal of the American Institute of Chemical Engineers, 21,* 1975, p. 1086.

International Critical Tables of Numerical Data. New York: McGraw-Hill.

Jordan, T. E. *Vapor Pressure of Organic Compounds.* New York: Interscience 1954.

Smith, J. M., and Van Ness, H. C. *Introduction to Chemical Engineering Thermodynamics* (3rd ed.). New York: McGraw-Hill, 1975.

Wark, K., and Warner, C. F. *Air Pollution—Its Origin and Control* (2nd ed.). New York: Harper & Row, 1981.

CHAPTER 11

VOC INCINERATORS

One fire burns out another's burning.

W. *Shakespeare,* Romeo and Juliet, *c. 1596*

◆ 11.1 INTRODUCTION ◆

As stated in Chapter 1, volatile organic compounds (VOCs) make up a major class of air pollutants. This class includes not only pure hydrocarbons but also partially oxidized hydrocarbons (organic acids, aldehydes, ketones), as well as organics containing chlorine, sulfur, nitrogen, or other atoms in the molecule. Within this class there are hundreds of individual compounds, each with its own properties and characteristics. These VOCs are emitted from combustion processes, from many types of industrial operations, and from solvent evaporation, among other sources.

One method of pollution control that can be applied broadly to VOCs is incineration. In this chapter, we will concentrate on the problem of incineration of VOC vapors (as opposed to incineration of bulk liquids or solids). Note that vapor incinerators (also called **thermal oxidizers or afterburners**) can sometimes be used successfully for air polluted with small particles of combustible solids or liquids. Incineration can be used for odor control, to destroy a toxic compound, or to reduce the quantity of photochemically reactive VOCs released to the atmosphere. The VOC vapors might be in a concentrated stream (such as emergency relief gases in a petroleum refinery), or might be a dilute mixture in air (such as from a paint-drying oven). For large volume, intermittent (but concentrated) VOC streams, elevated flares are usually used. In the case of a dilute fume in air, the two methods for incineration are direct thermal oxidation and catalytic oxidation. A schematic diagram of an afterburner is presented in Figure 11.1.

The alternatives to incineration are recovery of the vapors (which can be achieved by recompression, condensation, or carbon adsorption)

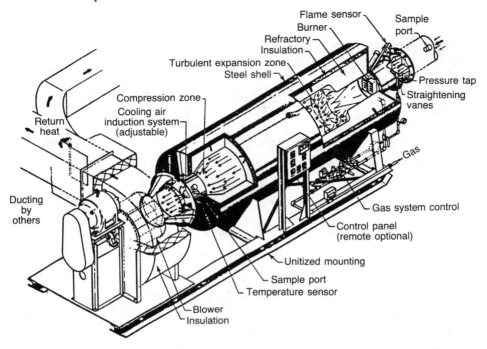

Figure 11.1
Sectional view of a direct-flame afterburner.
(Courtesy of KTI Gas Processors, Inc., Santa Ana, CA.)

or liquid absorption coupled with either recovery or chemical oxidation. The main advantage of incineration is its potential for very high efficiency. If held for a sufficient length of time at a sufficiently high temperature, organics can be oxidized to any desired degree of completeness. As an example, in 1984, the city Environmental Protection Board in Jacksonville, Florida, voted to require local paper mills to reduce their emissions of malodorous organic sulfur compounds. The proposal called for a reduction in concentration from 10,000 ppm to 5 ppm in certain exhaust gases. Only incineration could accomplish such a stringent control objective (99.95% destruction).

Quite often, there are several VOC sources within a manufacturing plant (such as printing presses or parts-painting stations), the emissions from which are gathered by several hoods and a common duct system and routed to a thermal oxidizer. It should be pointed out that the total reduction in VOC emissions from these sources depends not only on the destruction efficiency of the thermal oxidizer but also on the system's **capture efficiency.** The capture efficiency is defined as *the fraction of VOCs emitted from the processing point that is actually gathered by the side baffles, hoods or other capturing devices, and routed to the incinerator.* Emissions that are not captured are termed

fugitive emissions. Under recent U.S. EPA rules, the total VOC reduction efficiency is the product of the capture efficiency and the destruction efficiency. For example, even if one has a highly efficient afterburner that achieves, say, 99% destruction, if the capture system achieves only 80% capture efficiency, then the total VOC reduction efficiency is only 79.2%.

Both capture and destruction are now being considered by the EPA in regulating sources and in granting permits to operate. The actual degree of capture is difficult to measure, requiring a mass balance approach, but there is one easy case—that of 100% capture. As defined by the EPA, 100% capture is presumed to be achieved when the source is located inside a **total enclosure** (such as a room) and all airflow is *into* the enclosure except for exhaust points which are ducted to an afterburner. In order to qualify for the presumption of 100% capture, the total enclosure must meet the following criteria (McIlwee and Sharp 1991):

1. The sum of the areas of all openings (doors, windows, etc.) must be less than 5% of the sum of the enclosure's surface area (walls, floor, and ceiling).

2. Air must flow inward at all openings with an average face velocity of at least 200 ft/min.

3. All sources emitting VOCs inside the enclosure must be "distant" from any openings (at least 4 equivalent diameters).

4. All exhaust streams must be directed to a thermal oxidizer or other final control device.

5. All windows and doors not counted in the 5% of area rule must be closed during normal operations.

The main disadvantage of incineration is the high fuel cost. Also, some of the products of combustion of certain pollutants are themselves pollutants. For example, when a chlorinated hydrocarbon is burned, HCl or Cl_2 or both will be emitted. Depending on the amounts of these by-product pollutants, additional controls might be required.

♦ 11.2 THEORY ♦

OXIDATION CHEMISTRY

Basic chemical reaction theory was presented in Chapter 10. In this section, we will extend that theory to the specific case of the oxidation of VOC vapors in air. For simplicity, consider only the case of a premixed dilute stream of a pure hydrocarbon (HC) in air. The stoichiometry of complete combustion is

$$C_xH_y + (b)O_2 + 3.76(b)N_2 \rightarrow xCO_2 + \left(\frac{y}{2}\right)H_2O + 3.76(b)N_2 \quad \textbf{(11.1)}$$

where

C_xH_y = the general formula for any hydrocarbon

$b = x + (y/4)$, the stoichiometric number of moles of oxygen required per mole of C_xH_y

3.76 = the number of moles of nitrogen present in air for every mole of oxygen

In Eq. (11.1), we have included the nitrogen as a reminder that when combustion occurs using air, much additional gas (the nitrogen) is always present. For simplicity, in future equations we will not include nitrogen explicitly. Note that the formation of nitrogen oxides is not accounted for in Eq. (11.1). Furthermore, if sulfur or chlorine (two common impurities) were present in the VOC, sulfur oxides or HCl gas, respectively, would be formed.

The kinetics of oxidation reactions are as important as the stoichiometry. The actual detailed mechanisms of combustion are complex, and do not occur in a single step as might be inferred from Eq. (11.1). The combustion of methane, the simplest hydrocarbon, involves a branching chain reaction, the main reactions of which are depicted in Figure 11.2 (Glassman 1977). For higher order HCs, the mechanisms are further complicated by the larger number of possible intermediates.

Because of the necessity to simplify kinetic models for air pollution design work, several authors (Cooper, Alley, and Overcamp 1982; Hemsath and Susey 1974; Lee, Hansen, and Macauley 1979) have taken the approach of developing "global models." A global model ignores many of the detailed steps that are required in mechanistic models, and ties the kinetics to the main stable reactants and products. Since carbon monoxide is a very stable intermediate, the simplest global model for the oxidation of HC is a two-step model as follows:

$$C_xH_y + \left(\frac{x}{2} + \frac{y}{4}\right)O_2 \rightarrow xCO + \left(\frac{y}{2}\right)H_2O \tag{11.2}$$

$$xCO + \left(\frac{x}{2}\right)O_2 \rightarrow xCO_2 \tag{11.3}$$

Again, the reactions do not actually occur in single steps according to Eqs. (11.2) and (11.3).

Several authors have shown the importance of atoms and radicals such as O, H, OH, CH_3, CH_2, and HO_2 in combustion. Recent research has shown that thermal oxidation of organics can be enhanced by injection of H_2O_2 or O_3 into the hot gases downstream of the flame zone. These particular compounds break apart into radicals even at temperatures as low as 500 C, and speed up the oxidation of the VOCs (Cooper et al. 1991; Clausen et al. 1992). Detailed mechanistic modeling of these reactions is very complex and time consuming. Therefore, global modeling has been pursued, and Eqs. (11.2) and (11.3) have

$$CH_4 + O_2 \rightarrow CH_3 + HO_2 \quad \text{chain initiation}$$

$$\left.\begin{array}{l} CH_3 + O_2 \rightarrow CH_2 + OH \\ OH + CH_4 \rightarrow H_2O + CH_3 \\ OH + CH_2O \rightarrow H_2O + CHO \end{array}\right\} \text{chain propagation}$$

$$CH_2 + O_2 \rightarrow HO_2 + HCO \quad \text{chain branching}$$

$$\left.\begin{array}{l} HCO + O_2 \rightarrow CO + HO_2 \\ HO_2 + CH_4 \rightarrow H_2O_2 + CH_3 \\ HO_2 + CH_2O \rightarrow H_2O_2 + HCO \end{array}\right\} \text{chain propagation}$$

$$\left.\begin{array}{l} OH \rightarrow wall \\ CH_2O \rightarrow wall \end{array}\right\} \text{chain termination}$$

Figure 11.2
Major reactions involved in the oxidation of methane.
(Adapted from Glassman, 1977, as portrayed in Barnes et al., 1978.)

been successfully used to establish a global model of the overall kinetics of VOC oxidation in vapor incinerators. Using Eqs. (11.2) and (11.3), a global kinetic model that is of the first order in each reactant results in the rate equations.

$$r_{HC} = -k_1[HC][O_2] \tag{11.4}$$

and

$$r_{CO} = xk_1[HC][O_2] - k_2[CO][O_2] \tag{11.5}$$

where

r_i = rate of formation of component i, mol/L-s

[ ] = concentration, mol/L

HC = generic symbol for any hydrocarbon

k = rate constant, s^{-1} or L/mol-s (as appropriate)

In the presence of excess oxygen, the rate equations reduce to

$$r_{HC} = -k_1[HC] \tag{11.6}$$

$$r_{CO} = xk_1[HC] - k_2[CO] \tag{11.7}$$

In a typical afterburner, the oxygen mole fraction is 0.1–0.15 and the HC mole fraction is 0.001, so Eqs. (11.6) and (11.7) are applicable. A third equation could be written for the generation of CO_2, or CO_2 production could be calculated from a material balance on carbon. The generation rate equation for CO_2 is

$$r_{CO2} = k_2[CO] \tag{11.8}$$

Equations (11.6), (11.7), and (11.8) represent a special case of a general set of consecutive first-order irreversible reactions. Levenspiel (1962) discussed the first-order series reactions represented by

$$A \xrightarrow{k_1} R \xrightarrow{k_2} S \tag{11.9}$$

and presented solutions for the concentrations of all components as functions of the dimensionless reaction time $(k_1 t)$ and for various ratios of k_2/k_1. Figure 11.3 is a reproduction of Levenspiel's graphical representation of the 1:1:1 stoichiometry series reactions represented by Eq. (11.9). Note the large difference in the concentration of the intermediate R depending on the value of the ratio of k_2/k_1. Such behavior is representative of the VOC $\rightarrow$ CO $\rightarrow$ CO_2 system; CO can be present in large or small concentrations in an afterburner depending on the temperature of operation and the residence time. Figure 11.4, reproduced from a study by Hemsath and Susey (1974), illustrates the formation of CO in a VOC incinerator. Such formation can be an important factor in choosing the design temperature for an incinerator.

THE THREE TS

The importance of the three Ts of incineration—temperature, time, and turbulence—has been recognized for many years. In 1973, Danielson suggested that for good destruction, afterburners should be

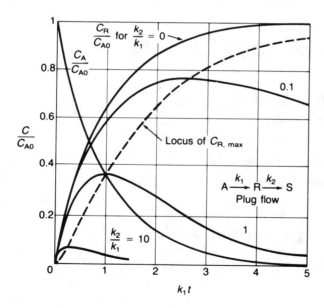

Figure 11.3

Concentration/residence time curves in a plug flow reactor for a 1:1:1 series reaction.

Note: C_i = concentration of i, mol/L. (Adapted from Levenspiel, 1962.)

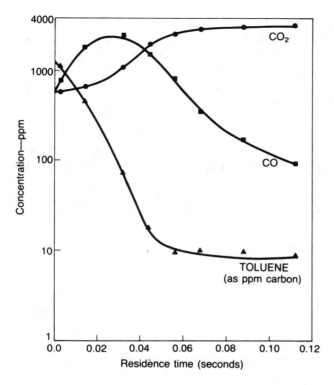

Figure 11.4
Concentration/residence time curves in a plug flow reactor for toluene, CO, and CO_2 at 1525 F.
(Adapted from Hemsath and Susey, 1974.)

designed for temperatures of 1000–1500 F, for residence times of 0.3–0.5 sec, and for flow velocities (to promote turbulent mixing) of 20–40 ft/sec. More recently, these general guidelines have been modified to include higher temperatures (1200–2000 F) and longer residence times (0.2–2.0 sec) in order to promote more complete destruction of the VOCs (Buonicore and Davis 1992).

In recent years, it has become standard practice to specify an afterburner as part of a hazardous waste incinerator (HWI) facility. U.S. EPA regulations for HWIs require a minimum of 99.99% destruction and removal efficiency (DRE) of the principal organic hazardous constituents (POHCs) being burned. In these cases, the final operating permits issued by the EPA almost always require a minimum of 2.0 seconds residence time in the afterburner (as well as specifying a certain minimum temperature, the value of which depends on the compounds being burned, but which is often 1800 F or higher). Hospital (biomedical) waste incinerators often utilize afterburners as well, and state or local regulations often specify a minimum temperature and residence time combination for these facilities. For example, the regulatory requirement of 1600 F and 1.0 sec is not uncommon for an afterburner on a hospital waste incinerator.

Since the kinetic constants in Eqs. (11.6) and (11.7) increase exponentially with temperature, VOC destruction rates are very sensitive to temperature. However, since the reactions in an afterburner proceed at a finite rate, sufficient time must be provided at the desired temperature to allow the reactions to reach the desired degree of completion. Turbulence ensures sufficient mixing of oxygen and VOCs during the process.

In a mathematical sense, the three Ts are related to three characteristic times—a chemical time, a residence time, and a mixing time—given by the following equations:

$$\tau_c = 1/k \tag{11.10}$$

$$\tau_r = V/Q = L/u \tag{11.11}$$

$$\tau_m = L^2/D_e \tag{11.12}$$

where

τ_c, τ_r, τ_m = chemical, residence, and mixing times, respectively, s

V = volume of the reaction zone, m^3

Q = volumetric flow rate (at the temperature in the afterburner), m^3/s

L = length of the reaction zone, m

u = gas velocity in the afterburner, m/s

D_e = effective (turbulent) diffusion coefficient, m^2/s

The ratio of the mixing time to the residence time is called the *Peclet number,* Pe, and the ratio of the residence time to the chemical time is known as the *Damkohler number,* Da. Barnes, Putnam, and Barrett (1978) have noted that if Pe is large and Da is small, then mixing is the rate-controlling process in the afterburner. If Pe is small and Da is large, then the chemical kinetics are rate-controlling. At the temperatures of most afterburners, as long as a reasonable flow velocity is maintained, the mixing processes will not be the limiting factor. However, as the temperature is raised, the chemical time decreases rapidly, and at some point the overall rate is limited by the mixing processes in the device.

PREDICTING VOC KINETICS

Although kinetics are important to the proper design of an afterburner, kinetic data are scarce and are difficult and costly to obtain by pilot studies. Because of the lack of detailed data, past methods for determining the design or operating temperature of an incinerator were very rough at best. Ross (1977) summarized the older methods by suggesting that the design temperature be set "several hundred degrees (F) above the VOC autoignition temperature." The **autoignition temperature** *is the temperature at which combustible mixtures of the VOC in air will ignite without an external source* (that is, without spark or flame). Some autoignition temperatures are presented in Table 11.1. Note that excessively high design temperatures will result in very high estimated costs for purchasing and operating a VOC incinerator. Thus, the prediction of temperatures higher than actually needed might preclude further consideration of the use of an afterburner when, in fact, it could be a feasible alternative.

More precise methods for the quantitative prediction of kinetic data and/or design temperatures have been proposed by several authors. Lee and coworkers, in two studies (Lee, Hansen, and Macauley 1979; Lee, Morgan, Hansen, and Whipple 1982), conducted experiments on several VOCs and proposed a purely statistical model to predict the temperatures required to give various levels of destruction in an isothermal plug flow afterburner. (See Chapter 10 for a discussion of plug flow reactors.) Their model depends on a number of properties of the VOC, the most important of which are the autoignition temperature, the residence time, and the ratio of hydrogen to carbon atoms in the molecule. Their method gives excellent correlation coefficients, and the standard deviation of predicted temperatures is about 20 °F. Two of their equations are as follows:

Table 11.1   Autoignition Temperatures of Selected Organics in Air

Substance	Autoignition Temperature, °F	Substance	Autoignition Temperature, °F
Acetone	1000	Ethylene Dichloride	775
Acrolein	453	Hexane	820
Acrylonitrile	898	Hydrogen	1076
Ammonia	1200	Hydrogen Cyanide	1000
Benzene	1075	Hydrogen Sulfide	500
n-Butane	896	Isobutane	950
1-Butene	723	Methane	999
n-Butyl Alcohol	693	Methanol	878
Carbon Monoxide	1205	Methyl Chloride	1170
Chlorobenzene	1245	Methyl Ethyl Ketone	960
Cyclohexane	514	Phenol	1319
Ethane	986	Propane	871
Ethanol	799	Propylene	851
Ethyl Acetate	907	Styrene	915
Ethylbenzene	870	Toluene	1026
Ethyl Chloride	965	Vinyl Chloride	882
Ethylene	842	Xylene	924

$$T_{99.9} = 594 - 12.2W_1 + 117.0W_2 + 71.6W_3 + 80.2W_4 + 0.592W_5 \\ - 20.2W_6 - 420.3W_7 + 87.1W_8 - 66.8W_9 + 62.8W_{10} - 75.3W_{11} \quad \text{(11.13)}$$

$$T_{99} = 577 - 10.0W_1 + 110.2W_2 + 67.1W_3 + 72.6W_4 + 0.586W_5 \\ - 23.4W_6 - 430.9W_7 + 85.2W_8 - 82.2W_9 + 65.5W_{10} - 76.1W_{11} \quad \text{(11.14)}$$

where

$T_{99.9}$ = temperature for 99.9% destruction efficiency, °F

T_{99} = temperature for 99% destruction efficiency, °F

W_1 = number of carbon atoms

W_2 = aromatic compound flag (0 = no, 1 = yes)

W_3 = C=C (double bond) flag—not counting the aromatic ring—
(0 = no, 1 = yes)

W_4 = number of nitrogen atoms

W_5 = autoignition temperature, °F

W_6 = number of oxygen atoms

W_7 = number of sulfur atoms

W_8 = hydrogen/carbon ratio

W_9 = allyl (2-propenyl) compound flag (0 = no, 1 = yes)

W_{10} = carbon-double-bond–chlorine interaction (0 = no, 1 = yes)

W_{11} = natural logarithm of residence time (sec)

Cooper, Alley, and Overcamp (1982) combined collision theory with empirical data and proposed a method for predicting an "effective" first-order rate constant k for hydrocarbon incineration over the range from 940 to 1140 K. Their method depends on the molecular weight and the type of the HC. Once k is found, the design temperature can be obtained. Recall that the rate constant k can be written as

$$k = Ae^{-E/RT} \quad \text{(11.15)}$$

where

E = activation energy, cal/mol

A = pre-exponential factor, s^{-1}

R = ideal gas law constant, 1.987 cal/mol-°K

T = absolute temperature, °K

The pre-exponential factor can be given by

$$A = \frac{Z'Sy_{O_2}P}{R'} \quad \text{(11.16)}$$

where

Z' = collision rate factor

S = steric factor

y_{O_2} = mole fraction oxygen in the afterburner

P = absolute pressure, atm

R' = gas constant, 0.08205 L-atm/mol-°K

The **steric factor** S in Eq. (11.16) (a factor to account for the fact that some collisions are not effective in producing reactions because of molecular geometry) can be calculated from

$$S = \frac{16}{MW}$$ (11.17)

where MW = molecular weight of the HC

The collision rate factor Z' can be estimated from Figure 11.5 for three classes of compounds. The pre-exponential factor A can then be calculated for an estimated mole fraction of oxygen in the afterburner. The activation energy E (in kcal/mol) is correlated with molecular weight as shown in Figure 11.6, the equation for which is:

$$E = -0.00966(MW) + 46.1$$ (11.18)

Once A and E have been estimated, k can be calculated for any desired temperature. In an isothermal plug flow reactor (PFR), the HC destruction efficiency, the rate constant, and the residence time are interdependent, and are related as

$$\eta = 1 - \frac{[\text{HC}]_{\text{out}}}{[\text{HC}]_{\text{in}}} = 1 - e^{-k\tau_r}$$ (11.19)

where η = HC destruction efficiency

Although this method is lengthier than the method of Lee et al., it does result in kinetic constants. Thus, in addition to being useful for the ideal case of an isothermal PFR, the approach described in the preceding discussion can also be used in the design of a nonisothermal afterburner, which is more representative of actual conditions.

The destruction of the VOC will often occur quickly relative to CO destruction. In some instances, CO destruction can "control" the design. CO production has been observed in afterburners by several investigators. The kinetics of CO destruction have been studied by many individuals (Dryer and Glassman 1973; Howard, Williams, and Fine 1973; Williams, Hottel, and Morgan 1969).

The following expression for CO oxidation was published by Howard et al. (1973) after review of numerous experimental studies; it is valid over a wide range of temperatures (840–2360 K).

destruction rate of CO = $1.3(10)^{14}e^{-30,000/RT} \{O_2\}^{1/2} \{H_2O\}^{1/2} \{CO\}$

(11.20)

where $\{\ \}$ indicates concentration in mol/cm^3

A kinetic model for only the destruction of CO, as given by Eq. (11.20), can be combined with the VOC kinetic model (which produces CO) to build a complete global model for the processes occurring in an afterburner. If CO destruction is much slower than VOC destruction, then higher temperatures are needed, simply to prevent excess CO emissions. In such a case, precise estimation of the VOC kinetics becomes less crucial.

▶ ▶ ▶ ▶ ▶ ▶ ▶ ▶ ▶

Example 11.1

Estimate the temperature required in an isothermal plug flow incinerator with a residence time of 0.5 sec to give 99.5% destruction of toluene. Use the three methods discussed in this section.

Solution

(a)

$$\text{autoignition temp} + 300\ F = 1026 + 300 = 1326\ F$$

(b) Method of Lee et al.:

From Eqs. (11.13) and (11.14),

$$T_{99.9} = 594 - 12.2(7) + 117 + 0 + 0 + 0.592(1026) - 0$$
$$- 0 + 87.1(1.14) - 0 + 0 - 75.3(\ln 0.5)$$
$$= 1386\ F$$

$$T_{99} = 577 - 10.0(7) + 110.2 + 0 + 0 + 0.586(1026) - 0$$
$$- 0 + 85.2(1.14) - 0 + 0 - 76.1(\ln 0.5)$$
$$= 1369\ F$$

$T_{99.5}$ will be between T_{99} and $T_{99.9}$. Since this method is approximate, a linear average is satisfactory. Thus,

$$T_{99.5} = 1378\ F$$

(c) Method of Cooper et al.

First, we rearrange Eq. (11.19) and calculate the required value of k.

$$k = \frac{-\ln(1 - 0.995)}{0.5} = 10.6\ \text{s}^{-1}$$

From Eq. (11.18), we calculate E as

$$E = -0.00966(92) + 46.1 = 45.2\ \text{kcal/mol}$$

We calculate S from Eq. (11.17), and estimate Z' from Figure 11.5. Thus,

$$S = \frac{16}{92} = 0.174$$

$$Z' = 2.85(10)^{11}$$

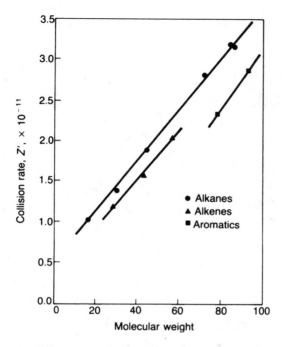

Figure 11.5
Collision rate factor for various hydrocarbons.

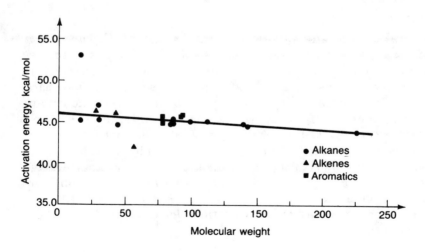

Figure 11.6
Activation energies for hydrocarbon incineration as a function of molecular weight.

For an assumed oxygen mole fraction of 0.15 and a pressure of 1 atm, we calculate A from Eq. (11.16) as

$$A = \frac{2.85(10)^{11}(0.174)(0.15)(1.0)}{0.08205} = 9.07(10)^{10} \text{ s}^{-1}$$

Finally, we rearrange Eq. (11.15) to solve for T, knowing k, A, and E. Thus,

$$T = \frac{-E}{R} \frac{1}{\ln(k/A)}$$

$$= \frac{-45,200}{1.987} \frac{1}{\ln\left[10.6/9.07(10)^{10}\right]}$$

$$= 995 \text{ K} = 1331 \text{ F}$$

◄ ◄ ◄ ◄ ◄ ◄ ◄ ◄ ◄

NONISOTHERMAL NATURE OF VOC OXIDATION

When completely burned, even small concentrations of VOCs in air can cause substantial increases in the temperature of the gas stream. At the same time, heat losses from the afterburner can be significant and can cause substantial decreases in gas temperature. Ideally, we should account for both of these phenomena in the design process; particular methods will be described in Section 11.3.

CATALYTIC OXIDATION

A **catalyst** *is an element or a compound that speeds up a reaction without undergoing permanent change itself.* Typically, gaseous molecules diffuse to and adsorb onto the surface of the catalyst, which is where the reaction takes place. Product gases desorb and diffuse back into the bulk gas stream. The detailed mechanisms of the reaction are not known. Perhaps the catalyst weakens the internal VOC bond energies, or perhaps it alters the VOC molecular geometry. However, because the mechanism of the reaction is changed by the catalyst, the reaction proceeds much faster and/or at much lower temperatures with the use of a catalyst than with direct thermal incineration. Further discussion of the theory of catalysis is beyond the scope of this text. However, some design considerations for catalytic incinerators are given in the following section.

◆ 11.3 DESIGN CONSIDERATIONS ◆

THERMAL OXIDIZERS

The process design of a VOC thermal oxidizer or afterburner involves specifying a temperature of operation along with a desired residence time, and then sizing the device to achieve the desired residence time and temperature with the proper flow velocity. Selection of the proper piece of equipment depends on such factors as mode of operation (continuous or intermittent), oxygen content, and the concentration of the VOC (Hemsath and Susey 1972). Proper selection and proper sizing are very important when trying to minimize the overall cost of the incineration option. For this reason, it is desirable to keep the volume of the stream to be treated as low as possible. However, most insurance regulations limit the maximum VOC concentration in such streams to 25% of the *lower explosive limit* (LEL) of the VOC. Even so, many process streams encountered in industry have concentrations of 5% or less of the LEL. If the process stream could be concentrated from 5% up to 25% of the LEL (for instance, by reducing the flow rate of dilution air), the total volume to be incinerated would drop by 80%! Some values of the LEL are presented in Table 11.2.

As mentioned previously, thermal oxidizers play a key role in the incineration of hazardous wastes. Current regulations require 99.99% DRE of POHCs. However, for a mixture of wastes, determining which

Table 11.2 Lower Explosive Limits (LELs) for Selected Organics

Organic	LEL (Percent by Volume in Air)
Acetone	2.15
Benzene	1.4
n-Butane	1.9
n-Butanol	1.7
Cyclohexane	1.3
Ethane	3.2
Ethanol	3.3
Ethyl Acetate	2.2
Heptane	1.0
Hexane	1.3
Isobutane	1.8
Isopropanol	2.5
Methane	5.0
Methanol	6.0
Methyl Acetate	4.1
Methyl Ethyl Ketone	1.8
Propane	2.4
Toluene	1.3
Xylene	1.0

POHCs will "control" the overall DRE of the mixture is not trivial. The minimum effective temperature in a thermal oxidizer is a function of the POHC that is the most difficult to destroy. Dellinger and others have developed for the U.S. EPA a ranking of gas-phase thermal stability of numerous POHCs. The ranking order is related to, but is not identical with, the order of the compounds' heats of combustion per unit mass (Taylor et al. 1990). Selection of the critical POHC in a mixed waste should be based on this ranking. A partial listing is given in Table 11.3.

During the design process, material and energy balances are performed on the device to calculate the flow rate of fuel gas required to raise the temperature of air at a given flow rate to the specified temperature. The energy effects of the chemical reactions of the pollutants and the heat transfer through the walls of the reactor should be accounted for, but they are sometimes ignored. Nonideal flow patterns and nonideal mixing result in radial profiles of concentration and temperature that theoretically should be taken into account as well. However, at best, this last part of the design process becomes mathematically difficult, and usually is not worth the extra effort required unless extremely high efficiencies will be required (as with a highly toxic vapor).

Thus, we are left with two approaches to the afterburner design problem. The traditional approach, which assumes an isothermal plug flow reactor, is mathematically very easy and is amenable to hand calculation, but results in estimates that must be considered approximate

Table 11.3 Thermal Stability Ranking for Selected POHCs

POHC	T_{99}, °F	ΔH_c, kcal/g
Benzene	1150	10.03
Naphthalene	1070	9.62
Chlorobenzene	990	6.60
Acrylonitrile	985	5.57
Chloromethane	950	3.25
Toluene	895	10.14
Trichloroethene	865	1.74
Pyridine	785	7.83
Ethyl Cyanide	770	4.57
Acetyl Chloride	765	2.77
Isobutyl Alcohol	715	7.62
1,1,2,2-Tetrachloroethane	690	1.39
Methyl Ethyl Ketone	650	8.07
Tetrachloromethane	645	0.24
Trichloromethane	625	0.75

Notes: Higher thermal stabilities indicated by higher temperatures for 99% destruction (T_{99}'s). All T_{99}'s reported in this table were experimentally determined. ΔH_c = heat of combustion.

Adapted from Taylor et al., 1990.

at best. The second approach, which allows for nonisothermal operation, usually requires the use of a simple computer program to handle
the large number of calculations, but is not conceptually difficult. We
will now discuss the first approach in detail. We conclude this section
with an outline of the procedures for the more accurate approach.

Material and Energy Balances.   A labeled schematic diagram
of a vapor incinerator is presented in Figure 11.7. In a typical design
problem, information about the relationships among time, temperature, and efficiency of destruction for the vapor to be incinerated is not
available. Pilot-scale studies, experimental data from the literature,
prediction techniques (discussed earlier in this chapter), or regulations enable us to specify both the design temperature and the residence time. We will now illustrate methods for calculating the fuel
flow rate and for sizing the afterburner to achieve the desired residence time and temperature. *(Note:* A variation of this problem is to
calculate the expected temperature given a specific flow rate of fuel
gas and contaminated airstream.)

The steady-state overall material balance reduces to

$$0 = \dot{M}_G + \dot{M}_{PA} + \dot{M}_{BA} - \dot{M}_E \qquad (11.21)$$

where $\dot{M}$ denotes the mass flow rates (in kg/min or lb_m/min), and the
subscripts refer to the streams identified in Figure 11.7. The
steady-state enthalpy balance (other forms of energy are insignificant)
is as follows:

$$0 = \dot{M}_{PA}\, h_{PA} + \dot{M}_G\, h_G + \dot{M}_{BA}\, h_{BA} - \dot{M}_E\, h_E + \dot{M}_G\, (\Delta H_c)_G$$

$$+ \sum \dot{M}_{VOC_i}\, (\Delta H_c)_{VOC_i}\, X_i - q_L \qquad (11.22)$$

where

h = specific enthalpy, kJ/kg or Btu/lb_m

ΔH_c = net heat of combustion (lower heating value), kJ/kg or Btu/lb_m

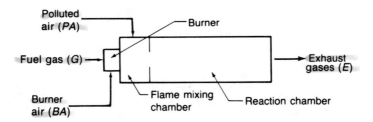

Figure 11.7
Schematic diagram of a vapor incinerator.

X_i = fractional conversion of VOC_i

q_L = rate of heat loss from the incinerator, kJ/min or Btu/min

In the simple analysis, we either ignore q_L or consider all losses to be represented as a simple percentage of the heat input. We assume all heat losses to occur at the front of the combustion chamber and assume the reaction to proceed isothermally at the outlet temperature. The heat effects of the VOC reactions are often ignored, but we will consider them. At a concentration of 1000 ppm, the heat released by VOC oxidation is roughly 10% of the heat supplied by the fuel. In the simple approach we will assume that the VOC oxidation is completed at the front end of the reactor. We can account for the heat losses by decreasing the rated heating value of the fuel gas.

Actually, the temperature profile in the device will vary considerably from the isothermal case. Depending on the inlet VOC concentration, the inlet temperature, the heat transfer coefficient, and the surface/volume ratio of the device, the temperature might first increase (owing to heat released by the VOC oxidation) and then decrease (owing to heat losses through the walls). The comparison of the temperature profiles (actual conditions versus the simple model) is shown in Figure 11.8.

If we assume that the enthalpy functions of all streams are similar to those for pure air, we greatly decrease our need for new data. For many afterburner systems, that assumption is not a bad one. Equation (11.22) then reduces to

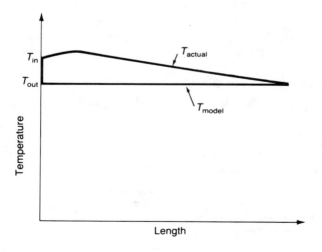

Figure 11.8
Comparison of the isothermal model and actual temperature profiles in an afterburner (with the same input of fuel for both cases).

$$0 = \dot{M}_{PA}\, h_{T_{PA}} + \dot{M}_G\, h_{T_G} + \dot{M}_{BA}\, h_{T_{BA}} - \dot{M}_E\, h_{T_E} + \dot{M}_G$$

$$\times (\Delta H_c)_G\,(1 - f_L) + \sum \dot{M}_{\text{VOC}_i}\,(\Delta H_c)_{\text{VOC}_i}\, X_i\,(1 - f_L) \qquad \textbf{(11.23)}$$

where

f_L = fractional heat loss

h_{Ti} = enthalpy of air at temperature T_i, kJ/kg or Btu/lb$_{\text{m}}$

Substituting Eq. (11.21) into Eq. (11.23) and solving for the mass flow rate of the fuel gas, we obtain

$$\dot{M}_G = \frac{\dot{M}_{PA}\left(h_{T_E} - h_{T_{PA}}\right) + \dot{M}_{BA}\left(h_{T_E} - h_{T_{BA}}\right) - \sum \dot{M}_{\text{VOC}_i}\,(\Delta H_c)_{\text{VOC}_i}\, X_i\,(1 - f_L)}{(\Delta H_c)_G\,(1 - f_L) - \left(h_{T_E} - h_{T_G}\right)}$$

$$\textbf{(11.24)}$$

The burner is often supplied outside ambient air drawn in with the fuel gas in a preset ratio R_B, as determined by the burner manufacturer. Thus, we let $\dot{M}_{BA} = R_B \dot{M}_G$, and substitute that relationship into Eq. (11.21). If we can also assume that $T_{BA} = T_G$, then an equation equivalent to Eq. (11.24) results as follows:

$$\dot{M}_G = \frac{\dot{M}_{PA}\left(h_{T_E} - h_{T_{PA}}\right) - \sum \dot{M}_{\text{VOC}_i}\,(\Delta H_c)_{\text{VOC}_i}\, X_i\,(1 - f_L)}{(\Delta H_c)_G\,(1 - f_L) - \left(R_B + 1\right)\left(h_{T_E} - h_{T_{BA}}\right)} \qquad \textbf{(11.25)}$$

Note that in Eqs. (11.24) and (11.25), we have expressed all enthalpies as enthalpy differences. A numerical value of enthalpy is referenced to a given temperature (and the reference temperature varies with different tables of data). When two enthalpies are subtracted, the reference temperature drops out. All items on the right side of Eqs. (11.24) and (11.25) are known or can be found in tables or charts of data (see Appendix B), because the outlet temperature is set at the start of the design. The following example problem serves to illustrate this procedure.

▶ ▶ ▶ ▶ ▶ ▶ ▶ ▶ ▶

Example 11.2

Calculate the mass flow rate of methane required for an afterburner to treat 2465 acfm of polluted air. The air enters at 200 F, and the desired exhaust temperature is 1350 F. It is estimated that the burner will bring in 200 scfm of outside air. The fuel gas enters at 80 F, and the burner air enters at 80 F. The lower heating value (LHV) of methane is 21,560 Btu/lb$_{\text{m}}$. Assume 10% overall heat loss. Also, ignore any benefits of the oxidation of the pollutants.

Solution

From Table B.2 in Appendix B, the density of the inlet polluted air is 0.060 lb_m/ft^3. Therefore,

$$\dot{M}_{PA} = 2465 \text{ acfm} \frac{0.060 \text{ lb}_m}{\text{acf}} = 148 \frac{\text{lb}_m}{\text{min}}$$

$$\dot{M}_{BA} = 200 \text{ scfm} (0.074) \frac{\text{lb}_m}{\text{scf}} = 14.8 \frac{\text{lb}_m}{\text{min}}$$

From Table B.7 in Appendix B, the enthalpies (all in Btu/lb_m) are as follows:

$$h_{T_E} = 328 \quad h_{T_{BA}} = 4.8 \quad h_{T_{PA}} = 33.6 \quad h_{T_G} = 4.8$$

Substituting these data into Eq. (11.24) and solving for $\dot{M}_G$, we obtain

$$\dot{M}_G = \frac{148(328 - 33.6) + 14.8(328 - 4.8)}{21,560(0.9) - (328 - 4.8)} = 2.53 \frac{\text{lb}_m}{\text{min}}$$

◀ ◀ ◀ ◀ ◀ ◀ ◀ ◀ ◀

Sizing the Device. After performing the heat and material balances, the rest of the preliminary process design is simple. Turbulent flow is required in an afterburner to ensure adequate mixing and to approach the condition of plug flow. Therefore, a linear velocity of 20–40 ft/sec is recommended (calculated based on the throat diameter). The average linear velocity throughout the main body of the device should be about 10–20 ft/s. Furthermore, sufficient residence time must be provided to allow the reactions to go to completion. Time and temperature are not independent, but the residence time is often set based on prior experience in the preliminary design stage. Usually, residence times of 0.4–0.9 sec are sufficient. As mentioned earlier, a residence time of 2.0 sec or longer is required for a hazardous waste incinerator, and 1.0 sec or longer is often mandated for a biomedical waste incinerator. The pressure drop in a thermal oxidizer is low (without considering any heat recovery exchangers); it is almost always less than 4.0 in. H_2O for the basic afterburner.

The applicable equations for sizing the afterburner (from our knowledge of the total mass flow rate of exhaust gas at the design temperature, the gas linear velocity, and the residence time) are presented below. The length of the reaction chamber is given by

$$L = u\tau_r \tag{11.26}$$

where all terms have been defined previously.

The volumetric flow rate of the exhaust gas is

$$Q_E = \frac{\dot{M}_E \, RT_E}{P\,(\text{MW})_E} \tag{11.27}$$

The diameter of the reaction chamber is

$$D = \sqrt{\frac{4Q_E}{\pi u}} \tag{11.28}$$

▶ ▶ ▶ ▶ ▶ ▶ ▶ ▶ ▶

Example 11.3

Specify the length and diameter of the afterburner of Example 11.2, given that the design velocity is 15 ft/sec, and the desired residence time is 1.0 sec.

Solution

$$L = 15(1.0) = 15 \text{ ft}$$

$$\dot{M}_E = 148 + 14.6 + 2.5 = 165 \text{ lb}_\text{m}/\text{min}$$

Assuming the exhaust gases have a molecular weight of 28.0, we can use the ideal gas law to obtain Q:

$$Q = \frac{(165)(0.730)(1810)}{1.0(28.0)} = 7790 \frac{\text{ft}^3}{\text{min}}$$

$$D = \sqrt{\frac{4(7790)}{\pi(15)(60)}} = 3.32 \text{ ft} = 40 \text{ in.}$$

◀ ◀ ◀ ◀ ◀ ◀ ◀ ◀ ◀

In the more accurate approach to afterburner design, we start with Eqs. (11.21) and (11.22) written for a small slice of the afterburner. The small slice is treated as a CSTR in which a portion of the VOC is combusted. The fraction combusted depends on the temperature, the kinetic constants, and the residence time in the CSTR. We assume an inlet temperature, and we have kinetic expressions [Eqs. (11.6) and (11.7)] for both the VOC and CO. Because of heat losses and gains, the temperature of the gas exiting the CSTR is different from the temperature of the entering gas. The heat loss depends on the heat transfer coefficient, the wall surface area, and the gas temperature itself. Likewise, the heat gain depends on the extent of reaction of the VOC, which depends on the rate constant, which depends on the temperature. Thus, an iterative calculation procedure is necessary to calculate the outlet temperature and the VOC and CO concentrations. Once the procedure has converged for one CSTR, the calculations are repeated for each succeeding CSTR downstream until the incinerator outlet is reached. If the outlet VOC and CO concentrations are not those desired, the process is repeated starting with a new assumed inlet temperature. A logic flowchart for this calculation procedure is shown in Figure 11.9.

CATALYTIC OXIDIZERS

Catalytic oxidizers can reduce the required temperature by hundreds of degrees and can save considerable amounts of space for equipment as compared with thermal oxidizers. In most cases, the gases are heated by a small auxiliary burner (as with a thermal oxidizer but to a

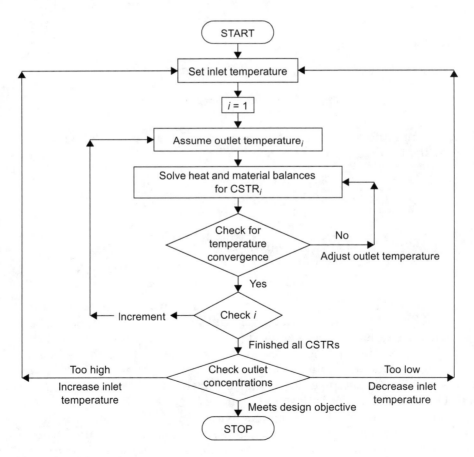

Figure 11.9
Logic flowchart for nonisothermal afterburner calculations.

much lower temperature). The gases are then passed directly through the catalyst bed, which is contained in the same unit as the burner.

In air pollution control work, the catalyst is usually a noble metal such as palladium or platinum (other metals are used, including Cr, Mn, Cu, Co, and Ni) deposited on an alumina support in a configuration to give minimum pressure drop. The pressure drop consideration is often critical for incinerator designs. A honeycomb arrangement typically results in a pressure drop of 0.05–0.5 in. H_2O/inch of bed depth, whereas a packed bed of 1/8-inch diameter pellets results in pressure drops of 1.0–10 in. H_2O/inch of bed (Snape 1977). A picture of various honeycomb catalysts is presented in Figure 11.10. In addition to exhibiting good reactivity and low pressure drop, the catalyst must be able to resist attrition (crumbling, breakage, or other mechanical wear), withstand high temperature excursions, and last a reasonable length of time in service.

Figure 11.10
Honeycomb catalysts.
(Courtesy of Corning Glass Works, Corning, NY.)

The overall rate of catalytic oxidation depends on both the rate of mass transfer (diffusion of the VOC to the surface of the catalyst) and the rate of the chemical oxidation reactions on the catalyst. At low temperatures (below 500 F), the chemical rate usually controls the rate of the process, whereas at higher temperatures, mass transfer is the limiting factor. Hawthorn (1974) discusses the theoretical basis for such behavior in detail. **Catalyst activity** refers to the degree to which a chemical reaction rate is increased compared with the same reaction without the catalyst. Catalysts can also be very *selective* (more active for some compounds, less active for others). Such activity and selectivity translates into lower operating temperatures required for the desired percentage of destruction. Table 11.4 presents inlet air temperatures required for 90% destruction of some VOCs using several commercial catalysts. Figure 11.11 presents some typical temperature/performance curves for a platinum-on-alumina catalyst.

The overall rate of catalytic oxidation is most often limited by mass transfer rather than by reaction kinetics. Thus, the design of catalytic units reduces to specification of the proper length of bed to permit sufficient residence time (*based on mass transfer rates*) to achieve the desired degree of VOC destruction. The design procedures have been reviewed by Retallick (1981). His basic equation is

$$\frac{[\text{VOC}]_L}{[\text{VOC}]_0} = e^{-L/L_m} \qquad (11.29)$$

Table 11.4   Some Temperatures Used for Catalytic Incineration

Compound[1]	Reported Temperatures,[2] °F
H_2	68, 250
CO	300, 500, 600
n-pentane	590
n-heptane	480, 570, 580
n-decane	500
Benzene	440, 480, 500, 575
Toluene	460, 480, 575
Methyl ethyl ketone	540, 570, 660
Methyl isobutyl ketone	540, 570, 660

[1] All compounds were premixed in air at 10% of their lower explosive limits (LELs).
[2] The reported temperatures are those required to give 90% destruction of the VOC. The temperatures were reported by various investigators using various commercial catalysts.

Adapted from Hawthorn, 1974.

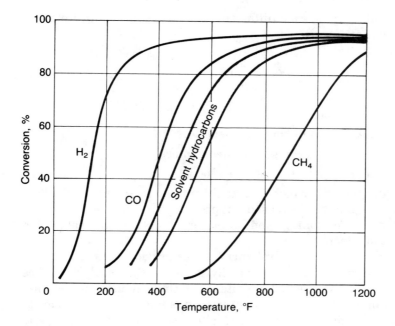

Figure 11.11
Typical conversion efficiencies for various hydrocarbons at various temperatures for a platinum-on-alumina catalyst.
(Adapted from Hawthorn, 1974.)

where

$[VOC]_L$ = concentration of the VOC (reactant) at length L

$[VOC]_o$ = inlet concentration of the VOC

L = length of the catalyst bed

L_m = length of one mass transfer unit

Retallick gives equations for calculating L_m under both turbulent and laminar flow conditions. Because of the small size of the individual channels, even at velocities of 40 ft/sec, the Reynolds number (based on channel diameter) is usually below 1000, so the flow is considered to be laminar. Thus,

$$L_m = \frac{ud^2}{17.6\mathscr{D}}$$
(11.30)

where

u = linear velocity in the channel, m/s

d = effective diameter of the channel, m

$\mathscr{D}$ = diffusivity, m^2/s

For turbulent flow when diffusion effects are significant, Retallick (1981) notes that

$$L_m = \frac{2}{fa}(Sc)^{2/3} \tag{11.31}$$

where

f = Fanning friction factor, dimensionless

Sc = Schmidt number ($\mu/\rho \mathscr{D}$), dimensionless

a = surface area per unit volume of bed, m^{-1}

The calculated length (which is usually from 2 to 10 inches for 99% conversion) should be increased to provide a design safety factor against catalyst poisoning and blinding; doubling the calculated length is often adequate.

The total amount of catalyst surface area provided is a critical factor in the design. Values are typically given as the catalyst surface area per unit gas volumetric flow rate, or as the exhaust gas flow rate per unit volume of catalyst. This latter form is known as *space velocity* because it has units of time^{-1}. Using the former ratio, Wark and Warner (1981) suggested a range of 0.2–0.5 ft^2 of catalyst surface area per scfm of waste gas. In terms of space velocity, Snape (1977) recommends 50,000–100,000 hr^{-1} for honeycomb supports, and 30,000 hr^{-1} for pelletized supports. Hawthorn (1974) reports that space velocities of 500–2000 min^{-1} are typically used to achieve 85–95% destruction of VOCs. Note that the inverse of space velocity is space time, which is similar to residence time. Comparing typical space times in catalytic oxidizers (0.03 to 0.1 s) to typical residence times in thermal oxidizers (0.3 to 1.0 s), we can see that the time required for oxidation is an order of magnitude longer in thermal systems than in catalytic systems.

HEAT RECOVERY

Heat recovery is another important consideration for afterburner design. Obviously, since fuel gas is expensive, it is desirable to recover heat from a vapor incinerator. At an energy cost of $0.40/therm (1 therm = 10^5 Btu), recovering useful heat equal to just 50% of the enthalpy released when 10,000 cfm of air is cooled from 1400 F to 400 F (recovering 260 Btu/lb$_m$ of air) results in a savings of $317 per day.

A very common method of recovering some of the energy in an incinerator's exhaust is by installing a heat exchanger. Heat exchangers that are used to preheat the incoming waste gas as shown in Figure 11.12(a) (primary heat recovery) are called *recuperators*. In recuperators, energy recoveries of 40–60% are common, and recoveries of 80% are often practical (Mueller 1977). When the hot exhaust gases are used to preheat the incoming VOC-in-air stream, the thermodynamic properties of the two streams are so similar that the energy recovery can be approximated by a simple equation, as follows:

$$E = \frac{\Delta T \text{ recovered}}{\Delta T \text{ available}} \times 100$$

$$= \frac{T_2 - T_1}{T_3 - T_1} \times 100$$

(11.32)

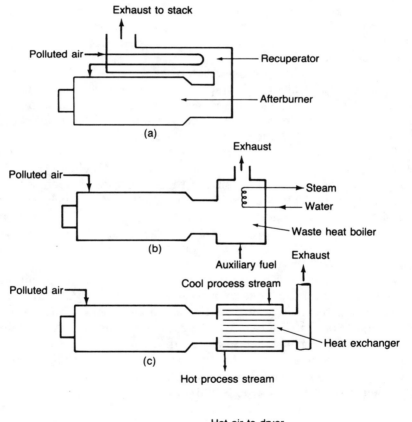

Figure 11.12
Various processing schemes for heat recovery from an afterburner.

where

E = percent heat recovery

$T_{1,2,3}$ = temperature of the VOC-in-air stream prior to preheat, temperature of the stream after preheat but before incineration, and temperature of the hot afterburner exhaust, respectively

Preheating the waste gas results in direct savings by reducing the fuel gas flow rate. However, for air-to-air heat exchange with low heat-transfer coefficients, conventional heat exchangers are large and can be very expensive. In addition, for higher percentages of energy recovery, more gas-metal contact is required; thus, pressure drop is increased. As a rule of thumb, recuperators with heat recoveries of 30, 50, and 70% will have pressure drops of 4, 8, and 15 in. H_2O, respectively (Vatavuk 1990).

If there are other process needs for heat, more effective use of the hot exhaust is possible. For example, the exhaust gas can be routed to a waste heat boiler to generate process steam, used to preheat a liquid process stream, or blended with fresh air to provide a hot air drying stream. Figure 11.12 depicts these applications.

There are many types of heat recovery equipment, and several variations of heat recovery processing schemes; Mueller (1977) has provided a good review of several of these schemes. Remember that heat recovery projects must be justified economically such that the cost of the heat exchanger is paid back by net energy savings (fuel gas savings less increased fan power). Recovery of much of the heat is almost always justifiable, but recovery of the "last 10%" of the heat energy is usually very expensive.

In the past twenty years or so an alternative to the traditional shell-and-tube heat exchanger has evolved. Known as *regenerative* heat recovery, this technology utilizes two chambers packed with ceramic heat transfer media. The two chambers are joined by an insulated duct with a small burner installed in it. The duct serves as the thermal oxidizer, and the burner is small because so much of the heat is recovered that very little additional fuel must be burned to achieve the desired operating temperature. The external ducts and valves are arranged so that the cool incoming waste gas stream can enter from either side, and the hot exhaust gases can exit in either direction (see Figure 11.13).

The flow through the packed beds is alternated as follows. First, the polluted air comes in through Bed 1, which is hot. As the cool air flows through the hot bed, the air becomes heated and the bed cools off. Meanwhile, the hot exhaust gases are flowing out through Bed 2, which is cool. Again the gases exchange heat with the ceramic media, this time heating the media and cooling the gases. After a period of time, when Bed 1 has become cool and Bed 2 hot, the valves are switched and the direction of flow is reversed. The cycle repeats itself.

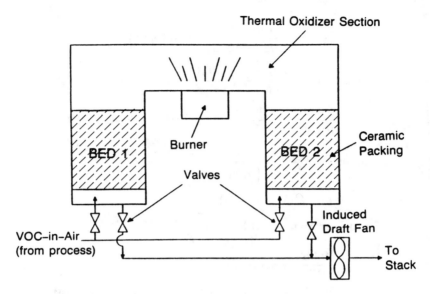

Figure 11.13
Thermal oxidizer with regenerative heat recovery.

Regenerative heating is capable of 80–95% heat recovery. Vertical flow-through beds are preferred because they provide more uniform gas flow distribution (especially when packing settles), and because the bottom face of the packing is always cool, avoiding the need to support hot packing. Pressure drops for these systems range from 15 in. H_2O (for 80% heat recovery) to 20 in. H_2O (for 95% recovery).

▶ ▶ ▶ ▶ ▶ ▶ ▶ ▶ ▶

Example 11.4

Calculate the enthalpy change for 500 m^3/min of air at 1 atm and 700 C being cooled to 200 C. If fuel gas (methane) costs \$6.50/GJ, and if electricity costs \$0.08/kWh, calculate the daily savings from recovering the heat content of this stream by reducing the temperature to 200 C. Assume that the increased pressure drop due to the heat exchanger is 5 in. H_2O. The existing fan can handle the ΔP but is only 60% efficient. If a heat exchanger to do the job has a TIC of \$100,000 and the appropriate capital recovery factor is 0.18, is the heat exchanger justified? Assume the system operates 250 days per year. If another heat exchanger could be purchased for \$50,000 to reduce the temperature an additional 50 C with an additional pressure drop of 3 in. H_2O, is the second heat exchanger justified?

Solution

From Figure B.2 in Appendix B, h_{700} = 750 kJ/kg and h_{200} = 205 kJ/kg, so Δh = 545 kJ/kg. The density of air at 700 C is

$$\rho = \frac{P(MW)}{RT} = \frac{1.0(29)}{(0.08205)973} = 0.363 \text{ kg/m}^3$$

$$\text{fuel savings} = \frac{500 \text{ m}^3}{\text{min}} \times \frac{0.363 \text{ kg}}{\text{m}^3} \times \frac{545 \text{ kJ}}{\text{kg}}$$

$$\times \frac{\$6.50}{10^6 \text{ kJ}} \times \frac{1440 \text{ min}}{\text{day}} = \$926/\text{day} = \$232,000/\text{yr}$$

$$\text{fan power} = \frac{1}{0.60} \times \frac{0.00415 \text{ kW}}{(\text{m}^3/\text{min})(\text{in. H}_2\text{O})} \times \frac{500 \text{ m}^3}{\text{min}}$$

$$\times 5 \text{ in. H}_2\text{O} = 17.3 \text{ kW}$$

$$\text{fan power cost} = 17.3 \text{ kW} \times \frac{250 \text{ day}}{\text{yr}} \times \frac{24 \text{ hr}}{\text{day}}$$

$$\times \frac{\$0.08}{\text{kWh}} = \$8300/\text{year}$$

$$\text{total annual cost} = \$100,000(0.18) + \$8300 = \$26,300$$

For the first exchanger, the annual cost = $26,300/yr and the annual savings = $232,000/yr, so the project is justified.

For the second exchanger, an additional reduction of 50 C will result in a savings of $76/day. The annual cost = ($9,000 + $5,000) = $14,000 and the annual fuel savings = $19,100/yr, so the second exchanger may or may not be justified.

FLARES

At the start of this chapter, we mentioned the problem of infrequent but large-volume releases of a VOC in a virtually pure form. Systems for safely venting process equipment during process upsets or emergencies are called *emergency relief systems*. Typically, such systems have many safety valves tied into one collection system. They are designed with very large pipes so that large volumes can be handled at very low pressures. The lines lead to a water seal drum and to a flare stack. The flare burns the VOC at a safe height above the process area. A schematic diagram of an emergency flare system is shown in Figure 11.14, and a detail drawing of a flare tip is shown in Figure 11.15.

Flare tips use steam to create turbulent mixing of air and the VOC at the top and to provide some cooling of the flare tip and stack. The VOC is ignited at the top by a continuous pilot. With the proper steam flow, smokeless operation can be maintained under almost all conditions of VOC flow.

In the late 1970s, some questions were raised about the emissions of unburned VOCs from flares. A possibility existed that regulators

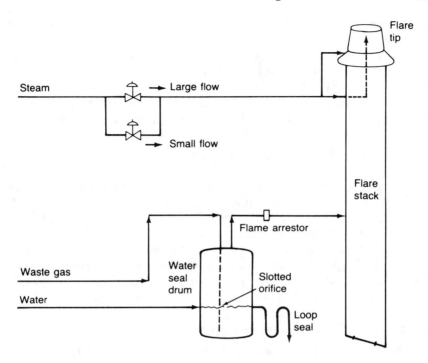

Figure 11.14
Schematic diagram of an emergency relief flare system.

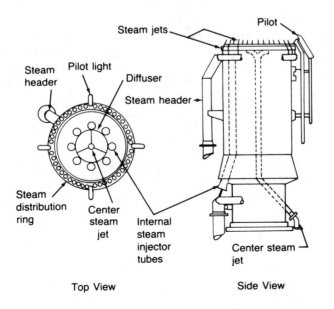

Top View   Side View

Figure 11.15
Detail of a flare tip.

would require "better control" (such as the use of thermal oxidizers or vapor recovery systems) even though the primary purpose of flares was for safe release of VOCs under emergency situations and not for pollution control. However, tests (Keller and Noble 1983) showed that flares provide good VOC destruction efficiencies. In some extensive testing, efficiencies of over 99.5% were observed for smokeless burning, and of 98.5% or more for all other types of operating conditions.

◆ 11.4 Costs ◆

The capital costs for vapor incinerators can be very high because of the high temperatures they must be able to withstand. However, the major contributor to annual operating cost is the fuel cost. Heat recovery can reduce the net fuel cost considerably; some people have claimed that in some extreme cases, an incinerator can become a payout item (Elnicki 1977). However, in most cases, even with good heat recovery and integration into an existing process plant, incinerators will represent a net cost addition to the annual cost of operating the plant.

Purchased Equipment Costs

Most thermal and catalytic incinerators treating waste gas flows of up to about 20,000 scfm are sold as package units. That is, the burner, the combustion chamber, the heat exchanger, the fan and motor, on-board instruments and controls, and a short exhaust stack are all factory assembled, skid-mounted, and sold as one unit (see Figure 11.1, at the beginning of this chapter).

Costs for package unit *thermal incinerators* (with a nominal operating temperature of 1500–1700 F) were reported (van der Vaart et al. 1996) to follow an equation of the form

$$P = aQ^b \qquad (11.33)$$

where

P = manufacturer's f.o.b. price, in 1988 dollars

Q = total gas flow rate exiting the incinerator, reported in scfm (valid for $1000 \leq Q \leq 50,000$ scfm)

a, b = curve fit constants that depend on heat recovery efficiency (see Table 11.5)

Table 11.5 Constants for Thermal Incinerator Cost Equation

Heat Recovery Efficiency	Constants	
%	a	b
35	$13,150	0.261
50	$17,060	0.250
70	$21,340	0.250

Recall that the delivered equipment cost (DEC) is about 14–18% higher than the manufacturer's f.o.b. price owing to freight, taxes, and process instrumentation and controls.

Equation 11.33 and Table 11.5 apply to incinerators with recuperative type heat recovery exchangers. For incinerators designed with the newer regenerative type of heat recovery, the following equation should be used.

$$P = \$220,000 + 11.6 \times Q \qquad \text{(11.34)}$$

where P and Q have the same meanings as in Eq. (11.33), except that (11.34) is valid for $10,000 \le Q \le 100,000$ scfm.

Costs for package unit *catalytic incinerators* (not including the catalyst costs) were also reported by van der Vaart and associates (1996). They were fit to a similar equation as (11.33):

$$P = aQ^b \qquad \text{(11.35)}$$

where

P = manufacturer's f.o.b. price, in 1988 dollars

Q = total exhaust gas flow rate, scfm (valid for $2000 \le Q \le 50,000$ scfm)

a, b = curve fit constants that depend on heat recovery efficiency (see Table 11.6)

Because catalytic units operate at much lower temperatures, Eq. (11.35) is valid for equipment designed for 600 F gases approaching the catalyst bed.

The initial price of the catalyst was not included in Eq. (11.35) because the useful life of the catalyst (typically 3–8 years) is less than that of the incinerator system (10–15 years). The initial cost of a catalyst should be treated as a separate capital investment (similar to bags in a baghouse). The annualized catalyst cost is given by Vatavuk (1990) as

$$C_c = \left(P_c + P_l\right)\text{CRF} \qquad \text{(11.36)}$$

where

C_c = catalyst replacement cost, \$

P_c = initial price of the catalyst (including taxes, etc.), \$

P_l = labor costs for replacing the catalyst, \$

CRF = capital recovery factor (depends on useful life and interest rates)

Vatavuk (1990) reports that the initial prices of catalysts are about \$3000/cubic foot for precious metal catalysts and about \$600/cubic foot for common metal catalysts (both in 1988 dollars). The labor cost for replacing the catalyst is small, usually less than 10% of the price of the catalyst.

Heat Recovery Efficiency	Constants		**Table 11.6**
%	a	b	Constants for Fixed- Bed or Structured
35	3623	0.419	Catalyst Catalytic
50	1215	0.558	Incinerator Cost
70	1443	0.553	Equation

As described in Chapter 2, the total installed cost (TIC) can be estimated based on the manufacturer's f.o.b. price (P). For all incinerators, the delivered equipment cost (DEC) is about 1.15 times P. For small package incinerators (Q less than 20,000 scfm), installation costs are minimal (just connect the utilities), so the TIC is estimated roughly as 1.25 times the DEC. For custom installations or larger units, TIC is about 1.6 times DEC as indicated in Chapter 2.

HEAT EXCHANGERS

In some cases, a retrofit project for an existing incinerator may be proposed to add a heat exchanger or to replace an inefficient heat exchanger with a more efficient one. In such cases, it is helpful to be able to estimate the cost of a stand-alone heat exchanger. A cost equation based on a variety of designs and sizes of heat exchangers has been derived by Vatavuk and Neveril (1982). Equation (11.37) is their equation (updated to 1992 dollars) for estimating the cost based on heat exchange area.

$$C = 53,752 A^{-0.44} \exp\left[0.0672(\ln A)^2 \right] \qquad (11.37)$$

where

C = f.o.b. cost (in 1992 dollars)

A = heat exchange area, ft^2

Equation (11.37) applies over the range of heat transfer area of 200 to 50,000 ft^2. The heat transfer area A can be estimated using the methods described in Chapter 8. Knowing the desired percentage of heat recovery (that is, the desired heat transfer rate) and the appropriate values for the overall heat transfer coefficient U and the log mean temperature difference ΔT_{LM}, we can calculate A directly from Eq. (8.27).

▶ ▶ ▶ ▶ ▶ ▶ ▶ ▶ ▶

Example 11.5

Estimate the total installed cost (in 2002 dollars) of a package incinerator to treat 13,000 scfm. The design temperature is 1500 F, and the design residence time is 1.0 sec. A recuperator with 60% heat recovery is desired.

Solution

Using Eq. (11.33) for a 50%—and then a 70%—recuperator, we get

$P_{50} = \$17{,}060 \, (13{,}000)^{0.250} = \$182{,}165$

$P_{70} = \$21{,}340 \, (13{,}000)^{0.408} = \$227{,}866$

A simple linear average will suffice for this estimate, so

$P_{60} = \$205{,}016$

Allowing for instrumentation, sales tax, and freight

DEC $= 1.15 \, P_{60} = \$235{,}768$

and then a modest amount for a foundation, ductwork, and connecting to plant utilities.

Because this is a package unit,

TIC $= 1.25 \times$ DEC $= \$294{,}710$

Finally, to adjust from 1988 dollars to 2002 dollars, we use the VAPCI from Chapter 2 plus 2% per year for 2001 and 2002.

TIC $= \$294{,}710 \, [(110/100) \times 1.04] = \$337{,}000$ (rounded)

◆ PROBLEMS ◆

11.1 Calculate the final temperature of the exhaust formed when 1.00 lb_m/min of methane at 70 F is combusted adiabatically (no heat loss) with 50.0 lb_m/min of air at 200 F. Recalculate the final temperature if there are 10% heat losses.

11.2 The desired gas temperature exiting the flame mixing chamber and entering the reaction chamber of a thermal incinerator is 750 C. The polluted airflow rate is 500 m^3/min at 77 C. The fuel gas (methane) is available at 20 C and the burner draws in supplemental air (at 25 C) at the rate of 14 kg air/kg fuel gas. Calculate the required flow rate of methane. Ignore heat losses and heat generation from pollutant combustion. Give your answer in kg/min and in m^3/min.

11.3 Recalculate the required flow rate of methane for the afterburner of Problem 11.2 if the exit temperature of the exhaust gases from the outlet must be 750 C, the heat losses are 12%, and the pollutant (1000 ppm toluene) is 96% combusted in the device.

11.4 Using the method of Lee et al., predict the temperature required in an isothermal plug flow incinerator to reduce the xylene level in a waste gas from 1000 ppm to 10 ppm. Assume a residence time of 0.7 sec.

11.5 Rework Problem 11.4 using the method of Cooper et al.

11.6 Size an afterburner and specify the flow rate of methane needed to control 8000 cfm of air from a paint-drying oven. The oven exhaust gases are available at 250 F. Assume that the design temperature and residence time have been established as 1300 F and 0.6 sec, respectively. The type of burner used in the process requires outside air at a mass ratio of 12:1 (air to fuel gas).

11.7 Size an afterburner and specify the flow rate of methane needed to control 500 m³/min of air from a rendering plant. The exhaust gases are available at 200 C, and must be raised to 700 C and held for 0.8 sec to reduce emissions of odorous compounds. The burner requires ambient air at a mass ratio of 14:1 (air to fuel gas).

11.8 By how much could the flow rate of fuel gas of Example 11.2 be reduced if a recuperator were used to preheat the waste gas by recovering the enthalpy released when reducing the exhaust gas temperature from 1350 F to 650 F? Determine the preheated waste gas temperature as it enters the afterburner.

11.9 Rework Problem 11.8 if the exhaust temperature is reduced to 400 F instead of 650 F.

11.10 A catalytic incinerator is investigated to replace the thermal incinerator of Example 11.2. Because of the lower required temperature, the required fuel gas flow rate drops to 1.44 lb_m/min. At a cost of $4.00/MBtu for fuel, calculate the maximum justifiable cost of the catalytic system. Assume that the costs of the shells of both systems are the same, so the only additional cost is that of the catalyst itself. Assume the catalyst must be replaced every 5 years, and that the incinerator operates 24 hours per day, 300 days per year.

11.11 Recommend the volume of catalyst and the length of a catalyst bed required for 99.5% control of hexane in air. The airflow rate is 10,000 cfm (at 30 ft/sec), and the channels in the particular honeycomb catalyst that will be used have an effective diameter of 0.05 in. The diffusivity of hexane in air is 0.075 cm²/sec. Assume that a space velocity of 50,000 hr^{-1} is sufficient, and that flow in the channels is laminar.

11.12 Compare the estimates of TIC (in 2000 dollars) for a thermal oxidizer designed to treat 30,000 scfm of contaminated air at 1700 F if it is equipped with either (a) 50% recuperative or (b) 95% regenerative heat recovery.

11.13 Estimate the TIC (in 2000 dollars, excluding catalyst) of a catalytic incinerator with 35% heat recovery. The 20,000 scfm solvent-in-air stream is at 600 F when it flows into the device.

11.14 Consider an existing thermal oxidizer that treats 20,000 scfm of contaminated air at 1600 F and is equipped with a 35% heat recuperator. It is desired to add a second gas-to-gas heat exchanger to recover an additional 35% of the heat content of the exhaust. Assuming an overall heat transfer coefficient of 4.0 Btu/hr-ft^2-°F and a ΔT_{LM} of 200 F, estimate the cost of the heat exchanger (in 2000 dollars).

11.15 Calculate the mass ratio of air to fuel if methane is completely combusted with the stoichiometric amount of air. Calculate the same ratio if twice the stoichiometric amount of air is used. Calculate the mole fraction of oxygen left in the exhaust gases for the second case.

11.16 Calculate the length and diameter of the reaction zone of an afterburner that treats 2000 acfm of polluted air (measured at 250 F). The gases in the afterburner are at a temperature of 1400 F for a residence time of 0.50 sec. Assume the burner will bring in 10 acfm of methane and 350 acfm of outside air (both measured at 80 F and 1 atm). The gas velocity in the afterburner is 20 ft/sec.

11.17 An isothermal plug flow afterburner with a residence time of 0.75 sec is being designed to provide 99.9% destruction of hexane ($C_6 H_{14}$). Predict the required temperature using the three methods outlined in this chapter.

11.18 A candle is being burned with a small amount of air in a perfectly insulated chamber. The candle wax has a heating value of 15,000 Btu/lb. You, being a typical college student, start burning the candle at both ends, and the rate of wax combustion increases to 0.1 lb/hour. Room air at 60 F enters the chamber in a ratio of 50 lb of air per lb of candle wax consumed.

  a. Draw a labeled diagram and derive mass and energy balances for this situation using your own symbols.

  b. Calculate the temperature of the gases coming out of the chamber, in degrees F.

  c. Comment on your calculated temperature relative to your experience with real candles. Is your calculated temperature reasonable? Explain any differences.

11.19 How much methane fuel is needed if you want to burn it along with 100 kg of air to achieve a final temperature of the combustion gases of 1000 C? The air and methane both start at room temperature. Give your answer in kg of methane and in standard cubic meters of methane. Assume adiabatic combustion.

11.20 Starting from mass and energy balances, develop the equations to calculate the mass of methane fuel needed if you want to burn it along with X lb of air to achieve a final temperature of the combustion gases of T degrees F. Assume that X lb of air is more than enough to achieve complete combustion of methane to CO_2 and H_2O, and that no pollutants are formed. Also assume adiabatic combustion, and that the air and methane both start at 60 F. Now assume that X is 30 lb, and T is 1800 F. Calculate the mass of methane you need. Give your answer in lb of methane.

11.21 Assume that the high-temperature oxidation of an organic solvent in excess air can be modeled as an irreversible first-order reaction ($r = -k\ C_{\text{solvent}}$). Now, consider a mixture of the solvent in air coming from a paint-baking oven that flows at the rate of 20 m³/s (at T = 150 C and 1 atm). The stream flows into a thermal oxidizer (total reactor volume is 60 m³) that you are modeling as two equal-volume CSTRs in series. In the first CSTR the temperature is 750 C, but the temperature drops to 700 C in the second CSTR. At 750 C, the reaction rate constant k is 25 sec^{-1}, and k follows the Arrhenius model with an activation energy of 45,000 cal/gmol-K.

Calculate the final concentration of solvent (in ppm) given that the inlet concentration of solvent in air is 500 ppm by volume. You may ignore any change in gas volumetric flow due to the change in moles by reaction, but you must account for changes in flow rate due to temperature.

REFERENCES

Barnes, R. H., Putnam, A. A., and Barrett, R. E. *Chemical Aspects of Afterburner Systems*, a topical report to the U.S. Environmental Protection Agency (Contract No. 68-02-2629), Battelle, Columbus, OH, August 15, 1978.

Buonicore, A. J. and Davis, W. T., Eds. *Air Pollution Engineering Manual,* Air and Waste Management Association. New York: Van Nostrand Reinhold, 1992.

Clausen, C. A., Cooper, C. D., Hewett, M., and Martinez, A. "Enhancement of Organic Vapor Incineration Using Ozone," *Journal of Hazardous Materials, 31,* 1992, pp. 75–98.

Cooper, C. D., Alley, F. C., and Overcamp, T. J. "Hydrocarbon Vapor Incineration Kinetics," *Environmental Progress, 1*(2), May 1982.

Cooper, C. D., Clausen, C. A., Tomlin, D., Hewett, M., and Martinez, A. "Enhancement of Organic Vapor Incineration using Hydrogen Peroxide," *Journal of Hazardous Materials, 27,* 1991, pp. 273–285.

Danielson, J. A., Ed. *Air Pollution Engineering Manual* (2nd ed.), AP-40. Washington, DC: U.S. Environmental Protection Agency, 1973.

Dryer, F. L., and Glassman, I. "High-Temperature Oxidation of CO and CH$_4$" *14th Symposium (International) on Combustion.* Pittsburgh, PA: The Combustion Institute, 1973, p. 987.

Elnicki, W. "Heat Recovery Economics," Chapter 18 in *Air Pollution Control and Design Handbook,* P. N. Cheremisinoff and R. A. Young, Eds. New York: Dekker, 1977.

Glassman, L. *Combustion.* New York: Academic Press, 1977.

Hawthorn, R. D. "Afterburner Catalysts—Effects of Heat and Mass Transfer between Gas and Catalyst Surface," *American Institute of Chemical Engineers Symposium Series, 70*(137), 1974.

Hemsath, K. H., and Susey, P. E. *Fume Incineration in Theory and Practice*, a paper presented at the 71st National Meeting of the American Institute of Chemical Engineers, Dallas, TX, February 20–23, 1972.

_____. "Fume Incineration Kinetics and Its Applications" *American Institute of Chemical Engineers Symposium Series, 70*(137), 1974, p. 439.

Howard, J. B., Williams, G. E., and Fine, D. H. "Kinetics of Carbon Monoxide Oxidation in Postflame Gases," *14th Symposium (International) on Combustion.* Pittsburgh, PA: The Combustion Institute, 1973, p. 975.

Keller, M., and Noble, R. "RACT for VOC—A Burning Issue," Pollution Engineering, July 1983.

Lee, K. C., Hansen, J. L., and Macauley, D. C. *Predictive Model of the Time-Temperature Requirements for Thermal Destruction of Dilute Organic Vapors,* a paper presented at the 72nd Annual Meeting of the Air Pollution Control Association, Cincinnati, OH, June 24–29, 1979.

Lee, K. C., Morgan N., Hansen, J. L., and Whipple, G. M. *Revised Model for the Prediction of the Time-Temperature Requirements for Thermal Destruction of Dilute Organic Vapors and Its Usage for Predicting Compound Destructibility,* a paper presented at the 75th Annual Meeting of the Air Pollution Control Association, New Orleans, LA, June 20–25, 1982.

Levenspiel, O. *Chemical Reaction Engineering.* New York: Wiley, 1962.

McIlwee, R. J., and Sharp, R. C. "The Basics of VOC Capture Systems." Ontario, CA: Smith Engineering Co., 1991.

Mueller, J. H. "Heat Recovery," Chapter 19 in *Air Pollution Control and Design Handbook,* P. N. Cheremisinoff and R. A. Young, Eds. New York: Dekker, 1977.

Retallick, W. B. "Design of Transfer-Limited Catalytic Incinerators," *Chemical Engineering*, January 12, 1981, p. 123.

Ross, R. D. "Thermal Incineration," Chapter 17 in *Air Pollution Control and Design Handbook,* P. N. Cheremisinoff and R. A. Young, Eds. New York: Dekker, 1977.

Snape, T. H. "Catalytic Incineration," Chapter 21 in *Air Pollution Control and Design Handbook,* P. N. Cheremisinoff and R. A. Young, Eds. New York: Dekker, 1977.

Taylor, P. H., Delligner, B., and Lee, C. C. "Development of a Thermal Stability Based Ranking of Hazardous Organic Compound Incinerability," *Environmental Science and Technology, 24*(3), 1990.

van der Vaart, D. R., Spivey, J. J., Vatavuk, W. M., and Wehe, W. M. "Thermal and Catalytic Incinerators," Chapter 3 in *OAQPS Control Cost Manual* (5th ed.). EPA 453/B-96-001, U.S. Environmental Protection Agency, Research Triangle Park, NC, 1996.

Vatavuk, W. M. Estimating Costs of Air Pollution Control. Chelsea, MI: Lewis Publishers, 1990.

Vatavuk, W. M., and Neveril, R. B. "Part XII: Estimate the Size and Cost of Incinerators," *Chemical Engineering,* July 12, 1982, p. 129.

Wark, K., and Warner, C. E. *Air Pollution—Its Origin and Control* (2nd ed.). New York: Harper & Row, 1981.

Williams, G. C., Hottel, H. C., and Morgan, A. C. "The Combustion of Methane in a Jet-Mixed Reactor," *12th Symposium (International) on Combustion.* Pittsburgh, PA: The Combustion Institute, 1969, p. 913.

CHAPTER 12

GAS ADSORPTION

A number of processes in food technology are associated with odor-ous effluents which can be readily controlled by adsorption. . . . [Examples are] poultry and meat processing, food canning, food cooking, fat and scrap rendering, fish dehydration. . . . Many [chemical manufacturing] processes involve losses to atmosphere of small amounts of objectionable major product, by-product, solvent, or plasticizer. . . . When solvent losses are involved, cycling recovery systems are advantageously used.

Amos Turk, 1962

♦ 12.1 INTRODUCTION ♦

The removal of low-concentration gases and vapors from an exhaust stream by the adherence of these materials to the surface of porous solids is an example of a practical application of adsorption. With the proper selection of the adsorbing solid (adsorbent) and the contact time between the solid and the vapor-laden exhaust stream, very high removal efficiencies are possible. In addition, the process can be designed to provide economical recovery of the adsorbed vapor (adsorbate). Gas adsorption is used for industrial applications such as odor control; the recovery of volatile solvents such as benzene, ethanol, trichloroethylene, freon, and so forth; and the drying of process gas streams. However, in this chapter, we will concentrate on the control of volatile organic compounds (VOCs) in fixed- and fluidized-bed systems.

Before reviewing the basic theory and design of adsorption systems, we will first describe the components and the operation of a typical fixed-bed carbon system as shown in Figure 12.1. The system in Figure 12.1 uses two horizontal cylindrical vessels in which a bed of granular activated carbon (the adsorbent) is supported on a heavy screen. Note that the valve arrangement permits each bed to be isolated. The air-vapor mixture from a plant process enters the main

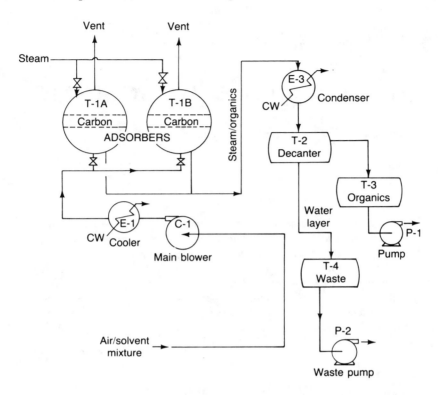

Figure 12.1
Simplified flow sheet for a fixed-bed carbon solvent recovery system.

blower and is passed through a cooler. The reason for cooling is that the amount of VOC that can be adsorbed per unit mass of carbon increases as the temperature decreases. The cooled gas stream then passes upward through the adsorbent bed, where the vapor is removed and the remaining air is either vented or returned to the source process. A vapor monitor in the vent stack detects bed saturation by monitoring the vapor concentration in the exhaust air. When the monitor detects that a preset maximum vapor concentration has been exceeded (breakthrough), the air-vapor stream is automatically switched to the idle bed, which has been regenerated and cooled.

The saturated (expended) bed is then regenerated by direct contact with low-pressure steam, which is admitted at the top of the bed. The adsorbed vapor is displaced from the carbon, and the mixture of steam and vapor is condensed and collected in the decanter for initial separation. If the condensed vapor is sufficiently insoluble in water, decanting is usually sufficient; otherwise, additional separation (for example, by distillation—see Figure 2.3 on p. 76) might be required.

♦ 12.2 ADSORPTION THEORY ♦

PHYSICAL AND CHEMICAL ADSORPTION

The two distinct adsorption mechanisms that are recognized are *physical adsorption* and *chemisorption*. **Physical adsorption,** also referred to as *van der Waals adsorption*, involves a weak bonding of gas molecules to the solid. The bond energy is similar to the attraction forces between molecules in a liquid. The adsorption process is exothermic, and the heat of adsorption is usually slightly higher than the heat of vaporization of the adsorbed material. The forces holding the gas molecules to the solid are easily overcome by either the application of heat or the reduction of pressure; either of these methods can be used to regenerate (clean) the adsorbent.

Chemisorption involves an actual chemical bonding by reaction of the adsorbate with the adsorbing solid. Heats of chemisorption are of roughly the same magnitude as heats of reaction, and chemisorption is not easily reversible. The oxidation of SO_2 to SO_3 on activated carbon is an example of chemisorption. Activated carbon and alumina can act as reaction catalysts with a number of gaseous mixtures; this fact must be considered when designing control and recovery systems. Except in some very specialized applications, if an adsorbate is chemically adsorbed to a significant extent, recovery of this material by an adsorption process is not feasible.

ADSORPTION ISOTHERMS

Generally, the capacity of an adsorbent to adsorb a particular adsorbate is directly proportional to the molecular weight and inversely proportional to the vapor pressure of the adsorbate. The capacity of an adsorbent for a specific gas or vapor can be presented as an isotherm, as shown in Figure 12.2. A point on an isotherm represents the mass of adsorbate per unit mass of adsorbent under equilibrium conditions at the indicated temperature and gas-phase concentration. The isotherms shown in Figure 12.2 are typical for many organic solvents on activated carbon. However, isotherms for other gas–solid systems can assume other shapes, including S-curves and curves with distinct flat sections.

One of the best mathematical models for describing adsorption equilibrium is the **Langmuir isotherm**, which is given by

$$\frac{\overline{P}}{a} = \frac{1}{k_1} + \frac{k_2}{k_1}\overline{P} \tag{12.1}$$

where

$\overline{P}$ = partial pressure of the adsorbate

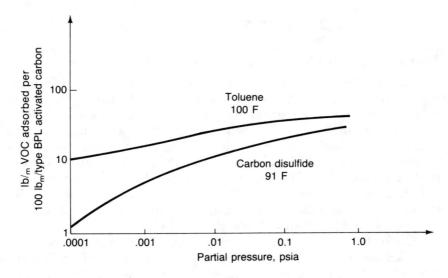

Figure 12.2
Adsorption isotherms for activated carbon.

a = mass of adsorbate adsorbed per unit mass of adsorbent

k_1, k_2 = constants

Equation (12.1) is based on the assumptions that (1) the adsorbed phase is a unimolecular layer, and (2) at equilibrium, the rate of adsorption is equal to the rate of desorption from the surface.

It can be shown that the rate of contact of gas molecules with the adsorbent is proportional to the partial pressure $\overline{P}$ of the adsorbate. However, actual contact with the surface is limited to an area $(1-f)$ not already occupied by adsorbate molecules, where f is the occupied fraction of the total solid surface. The rate of adsorption can then be found by

$$r_a = C_a \overline{P}(1-f) \tag{12.2}$$

where

r_a = rate of adsorption

C_a = constant

Conversely, the rate of loss of molecules from the surface is proportional to the fraction of the surface occupied. Thus,

$$r_d = C_d f \tag{12.3}$$

where

r_d = rate of loss by desorption

C_d = constant

At equilibrium, the rate of adsorption r_a is equal to the rate of desorption r_d, and the fraction of surface covered f is

$$f = \frac{C_a \overline{P}}{C_a \overline{P} + C_d} \tag{12.4}$$

Since we assume a unimolecular coverage, the mass of adsorbate per unit mass of adsorbent a is also proportional to the fraction of surface covered. Thus,

$$a = C'_a f \tag{12.5}$$

where C'_a = constant

Combining Eqs. (12.4) and (12.5), we obtain

$$a = \frac{\left(C_a C'_a / C_d\right) \overline{P}}{\left(C_a / C_d\right) \overline{P} + 1} \tag{12.6}$$

Since C_a, C_d, and C'_a are constants, Eq. (12.6) can also be written as

$$a = \frac{k_1 \overline{P}}{k_2 \overline{P} + 1} \tag{12.7}$$

where k_1, k_2 = constants

Equation (12.7) can be rearranged to give the Langmuir equation. In the Langmuir form, Eq. (12.1) is a straight line when plotted on rectangular coordinates with $\overline{P}/a$ as the ordinate and $\overline{P}$ as the abscissa.

At very low adsorbate partial pressure, $k_2 \overline{P}$ is approximately equal to zero, and Eq. (12.7) becomes

$$a = k_1 \overline{P} \tag{12.8}$$

At high partial pressure,

$$a = \frac{k_1 \overline{P}}{k_2 \overline{P}} = \frac{k_1}{k_2} \tag{12.9}$$

Hence, over a narrow intermediate range of $\overline{P}$,

$$a = k\left(\overline{P}\right)^n \tag{12.10}$$

where

k = constant

n = constant (with a value between 0 and 1)

Equation (12.10) is the form of Eq. (12.7) that is called the *Freundlich equation*.

ADSORPTION POTENTIAL

Goldman and Polanyi (1928) used the concept of *adsorption potential* to develop a single plot for the effect of temperature on adsorbent capacity. **Adsorption potential** *is defined as the change in free energy accompanying the compression of one mole of vapor from the equilibrium partial pressure $\overline{P}$ to the saturated vapor pressure P_v at the temperature of adsorption T.*

$$\Delta G_{ads} = RT \ln\left(\frac{P_v}{\overline{P}}\right) \qquad (12.11)$$

where

ΔG_{ads} = change in free energy of adsorption, cal/gmol

T = adsorption temperature, °K

P_v = vapor pressure at temperature T (in the same units as $\overline{P}$)

Dubinin (1947) found that when similar gases were adsorbed on the same adsorbent, the adsorption potentials were very nearly equal when the amount adsorbed was determined based on the product of the number of moles adsorbed multiplied by the molal volume. Thus,

$$\left[\frac{RT}{V'} 2.303 \log\left(\frac{P_v}{\overline{P}}\right)\right]_i = \left[\frac{RT}{V'} 2.303 \log\left(\frac{P_v}{\overline{P}}\right)\right]_j \qquad (12.12)$$

where

V' = specific molal volume, cm^3/gmol

i, j = subscripts denoting different gases

Grant and Manes (1966) used the work of Dubinin as modified by Lewis et al. (1950) to prepare plots of adsorption potential for hydrocarbons and reduced-sulfur gases. These plots are shown in Figure 12.3. Note that fugacities (thermodynamic potentials) were used in place of pressures to calculate the adsorption potentials. At the low partial pressures encountered in air pollution control, it is satisfactory to substitute $\overline{P}$ for $\overline{f}$ and P_v for f_v to facilitate the use of the Grant plot.

EXPERIMENTAL DETERMINATION OF ISOTHERMS

Adsorption isotherms can be obtained by several experimental methods, both static and dynamic. In one common procedure, a small bucket containing a few milligrams of adsorbent is suspended in a cell maintained at constant temperature and adsorbate partial pressure (Sloan 1974). An inert diluent gas (typically helium) is used to dilute the adsorbate. The adsorbent bucket is suspended from the balance arm of an electrobalance capable of measurements accurate to a few micrograms. The combined weight of the bucket and adsorbent is

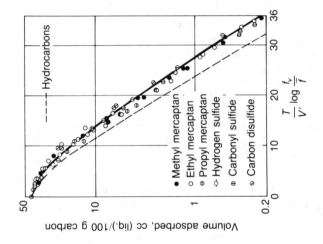

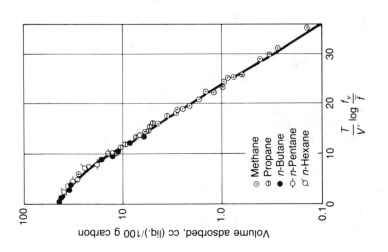

Figure 12.3

Adsorption potential plots for hydrocarbon and reduced-sulfur gases.
(Adapted from Grant and Manes, 1966.)

recorded continually as saturation is reached. This method provides very reliable data, but is time consuming.

Adsorption and desorption rates can be determined dynamically with a gas chromatograph (Lambert 1970). The adsorbent is packed in the chromatographic column, and a pulse of adsorbate is injected into the carrier gas stream. Isotherm data are calculated from the resulting chromatogram. For a detailed description of the procedure, refer to Sloan (1974).

◆ 12.3 PHYSICAL PROPERTIES ◆ OF ADSORBENTS

The adsorbents used for air pollution control include activated carbon, alumina, bauxite, and silica gel. Activated carbon is by far the most frequently used adsorbent, and has virtually displaced all other materials in solvent recovery systems.

ACTIVATED CARBON

The term *activated* as applied to adsorbent materials refers to the increased internal and external surface area imparted by special treatment processes. Any carbonaceous material can be converted to activated carbon. Coconut shells, bones, wood, coal, petroleum coke, lignin, and lignite all serve as raw materials for activated carbon. However, most industrial grade carbon is made from bituminous coal.

Activated carbon is manufactured by first dehydrating and carbonizing the carbonaceous raw material. Activation is completed during a controlled oxidation step in which the carbonized material is heated in the presence of an oxidizing gas. For certain carbons, the dehydration can be accomplished by using chemical agents. The ideal raw material has a porous structure that provides a uniform pore distribution and high adsorptive capacity when activated. Activated carbon is tailored for specific end use by both raw material selection and control of the activation process. Table 12.1 lists some properties of various adsorbents used to treat polluted air. Table 12.2 lists some properties of various activated carbons. As noted in Tables 12.1 and 12.2, carbons for gas-phase application have a surface area in the range from 800 to 1200 m^2/g. Most of the pore volume is distributed over a narrow range of pore diameters, usually ranging from 4 to 30 angstroms (1 Å = 0.0001 μm).

Several standardized tests are used to characterize gas-phase activated carbon. The *carbon tetrachloride number*—an important measure of the capacity—is defined as the g CCl_4/100 g carbon after the carbon is saturated in a flowing stream of CCl_4. Similarly, the *iodine number* (the g I_2/100 g carbon) is another measure of adsorptive capacity. The surface area is calculated from information from an isotherm and from the average area occupied by a single adsorbate molecule.

Carubba (1984) has described a rapid method for pore volume determination based on n-butane adsorption.

Carbon molecular "sieves" offer promise as an important addition in the design of adsorption systems. Molecular sieves differ from activated carbon in that the raw material is normally a polymeric material that produces uniform pore sizes of about 5 angstroms. Carbon sieves can be used to separate liquids with low boiling points.

To minimize pressure drop in fixed beds, granular or pelletized carbon is used. Typically, the particle size of granular carbon is about

Table 12.1 Physical Properties of Several Adsorbents

Composition	Internal Porosity, %	External Void Fraction, %	Bulk Dry Density, lb_m/ft^3	Surface Area, m^2/g
Acid-Treated Clay	30	40	35–55	100–300
Activated Alumina and Bauxite	30–40	40–50	45–55	200–300
Aluminosilicate "Sieves"	45–55	35	41–44	600–700
Bone Char	50–55	18–20	40	100
Carbons	55–75	35–40	10–30	600–1400
Fuller's Earth	50–55	40	30–40	130–250
Iron Oxide	22	37	90	20
Magnesia	75	45	25	200
Silica Gel	70	40	25	320

Adapted from Standen, 1963(a) and (b).

Table 12.2 Characteristics of Activated Carbons from Various Raw Materials

Raw Material	I_2 Number	Molecular Number	CCl_4 Number	Butane Pore Volume, cc/g	Application
Lignite	550	490	34	0.23	Liquid Phase
Bituminous Coal	900/1000	200/250	60	0.45	Vapor/Liquid Phase
Petroleum Acid– Sludge Coke	1150	180	59	0.46	Vapor Phase
Coconut	1350	185	63	0.49	Vapor Phase
Subbituminous Coal	1050	230	67	0.48	Vapor Phase
Wood	1230	470	76	0.57	Vapor/Liquid Phase

Adapted from Carubba et al., 1984.

0.05 in., which corresponds to a screen size between 10 and 20 mesh on the U.S. sieve scale. Activated carbon must be sufficiently hard to minimize attrition in the bed and the resultant buildup of fines (small particles of carbon that can restrict gas flow through the bed). To determine the carbon hardness, the carbon is vibrated on a 14-mesh screen with steel balls for 30 minutes. The hardness is the weight percent remaining on the screen, which should be at least 50% for satisfactory service in fixed beds (Turk 1968).

OTHER ADSORBENTS

Activated alumina is prepared by heating alumina trihydrate to 400 C in an airstream. The average bulk density of activated alumina ranges from 45 to 55 lb_m/ft^3, with a surface area of 300 m^2/g. The major adsorption application of activated alumina is in the drying of gas streams, but it is sometimes used in solvent recovery.

Silica gel is another adsorbent that is used primarily in gas drying applications. Silica gel is prepared by treating sodium silicate with a strong acid, followed by washing to remove excess electrolyte. The washed solid is dried and activated by heating. The average pore size and surface area of the final product are greatly affected by the final drying steps. The three main types of silica gel range in density from 0.15 to 0.70 g/mL. Average pore diameters range from 25 to 200 Å, with the denser product having the smaller pore diameter (Standen 1963b). Commercial grades of adsorbent gel have been available for over 50 years; hence, there is a great deal of adsorption application data in the literature.

♦ 12.4 FIXED-BED ADSORPTION SYSTEMS ♦

BREAKTHROUGH CURVES AND
THEIR RELATIONSHIP TO SYSTEM DESIGN

The dynamics of fixed-bed adsorption are shown graphically in Figure 12.4. In the upper part of the figure, a fixed bed is shown at three different times representing the interval from the initial admission of adsorbate to the time at which a significant concentration of adsorbate breaks through the bed. The plot in the lower half of the figure depicts the effluent concentration from the bed as a function of the volume of effluent, or time of operation. This type of plot is called a *breakthrough curve,* and provides valuable information on the adsorption rate in the bed.

During depletion of the bed, an active *adsorption zone* (AZ) moves through the bed. Behind the AZ, the adsorbent is saturated, whereas in front of the AZ, the bed is virtually free of adsorbate. The length (or height) of the zone is a function of the rate of transfer of adsorbate from the gas to the adsorbent. A shallow AZ indicates good adsorbent utilization and is represented by a steep breakthrough curve. Con-

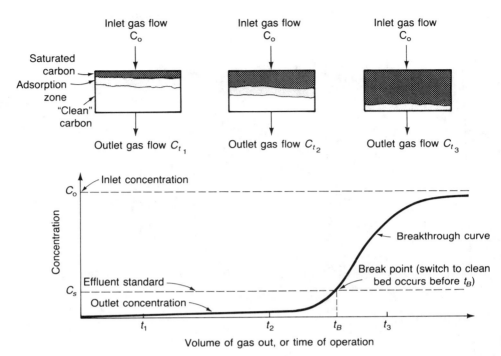

Figure 12.4
The adsorption wave and breakthrough curve.

versely, a wide or deep AZ denotes poor bed utilization and is indicated by a gradual slope on the breakthrough curve. The length of the AZ determines the minimum depth of the adsorbent bed.

Under actual plant operating conditions, bed capacity will seldom exceed 30 to 40% of that indicated by an equilibrium isotherm. Factors that contribute to bed capacity loss are shown graphically in Figure 12.5. Curve A represents the bed adsorbate concentration indicated by an isotherm for the bed operating temperature. In this case, the isotherm predicts a concentration of 30 g adsorbate/100 g carbon throughout the bed. A significant loss occurs because the bed must be taken out of service at or before the time that the leading edge of the AZ exits the bed. Since the adsorbent in the AZ is not saturated, the bed capacity is reduced by the area bounded by curve B. The heat of adsorption liberated as the AZ passes through the bed lowers the capacity as shown by curve C. Moisture reduces the bed capacity by two mechanisms: moisture in the influent gas displaces adsorbate (shown by curve D), and residual moisture remaining in the bed after regeneration further occupies some additional bed volume (shown by curve E). All of these losses are difficult to predict accurately. Pilot-scale breakthrough curves must be determined under conditions closely approximating plant operating conditions. Several mathematical procedures

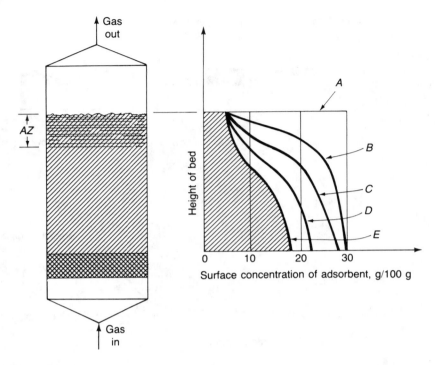

Figure 12.5
Factors affecting carbon capacity: *A*—theoretical saturation under equilib-
rium conditions; *B*—loss due to adsorption zone; *C*—loss due to heat wave;
D—loss due to moisture in gas; *E*—loss due to residual moisture on carbon.

for calculating breakthrough curves have been reported (Treybal 1968;
Wark, Warner, and Davis 1998). However, each of these procedures
requires isotherm and mass transfer rate data, which will not normally
be available for a specific control application. Hence, system design is
based primarily on previous plant experience and pilot-scale studies.

PRESSURE DROP ACROSS FIXED BEDS

Figure 12.6 is a plot of pressure drop versus superficial bed velocity for several particle sizes. The fan energy usage resulting from pressure drop is a significant part of the overall operating cost of a fixed-
bed system. Optimum bed velocity increases with bed depth over the
normal operating range of 50–100 ft/min.

In the absence of measured pressure loss data, the pressure drop
across the bed can be estimated with the following equation (Ergun 1952):

$$\frac{\Delta P g_c \varepsilon^3 d_p \rho_g}{D(1-\varepsilon)G^2} = \frac{150(1-\varepsilon)\mu}{d_p G} + 1.75 \qquad \textbf{(12.13)}$$

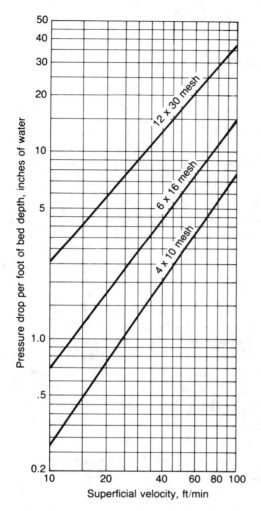

Figure 12.6
Pressure drop across a carbon bed as a function of the superficial bed velocity and carbon mesh size.
(Courtesy of Pittsburgh Carbon Company, Pittsburgh, PA.)

where

ΔP = pressure drop, lb_f/ft^2

g_c = gravitational constant, 4.17×10^8 lb_m-ft/lb_f-hr^2

ε = void fraction, ft^3 voids/ft^3 packed bed

d_p = particle diameter, ft

ρ_g = gas density, lb_m/ft^3

D = bed depth, ft

G = gas superficial mass flux, $lb_m/hr\text{-}ft^2$

μ = gas viscosity, $lb_m/hr\text{-}ft$

A simpler, empirical equation published by the Union Carbide Corporation is as follows:

$$\Delta P = 0.37D\left(\frac{V}{100}\right)^{1.56} \qquad (12.14)$$

where

ΔP = bed pressure drop, in. H_2O

D = bed depth, in.

V = superficial gas velocity, ft/min

Equation (12.14) is valid for velocities of 60–140 ft/min and bed depths of 5–50 inches, and for 4×6 mesh sized carbon.

▶ ▶ ▶ ▶ ▶ ▶ ▶ ▶ ▶

Example 12.1

An activated-carbon bed that is 12 ft × 6 ft × 2 ft deep is used in a benzene recovery system. The system is on-line for one hour and is then regenerated for one hour. The influent gas stream contains 5000 ppm benzene (by volume) at 1.0 atm and 100 F. The operating capacity of the bed is 10 lb_m benzene per 100 lb_m carbon. The physical properties of the carbon are as follows: bulk density = 30 lb_m/ft^3, void fraction = 0.40, and particle size = 4×10 mesh (0.011 ft). Determine the pressure drop across the bed from (a) Eq. (12.13), (b) Eq. (12.14), and (c) Figure 12.6.

Solution

(a) To use Eq. (12.13), we will assume that the gas stream behaves ideally. We need to find the mass flux of the gas stream.

$$\text{total mass of carbon in bed} = (12\text{ ft})\,(6\text{ ft})\,(2\text{ ft})\left(30\,\frac{lb_m}{ft^3}\right) = 4320\ lb_m$$

$$lb_m\text{ benzene adsorbed/hr} = 4320\left(\frac{10}{100}\right) = 432\ lb_m\text{benzene}$$

$$\text{MW benzene} = 78.1\ lb_m/lbmol$$

$$\text{volume of benzene vapor adsorbed/min} = 432\,\frac{lb_m}{hr} \times \frac{lbmol}{78.1\ lb_m}$$

$$\times \frac{hr}{60\text{ min}} \times 0.73\,\frac{atm\text{-}ft^3}{lbmol\text{-}°R} \times \frac{560\ °R}{1\text{ atm}} = 37.7\,\frac{ft^3}{min}$$

$$\text{total gas flow rate} = 37.7\,\frac{ft^3}{min} \times \frac{10^6}{5000\text{ ppm}} = 7540\text{ acfm}$$

$$\text{superficial gas velocity} = \frac{7540}{12(6)} = 104\text{ ft/min}$$

Assume that the influent gas has the properties of air (that is, $MW = 29$).

$$\rho_g = \frac{P(MW)}{RT} = \frac{(1.0)(2.9)}{(0.73)(560)} = 0.071\frac{\text{lb}_\text{m}}{\text{ft}^3}$$

$$G = 7540\frac{\text{ft}^3}{\text{min}} \times 60\frac{\text{min}}{\text{hr}} \times 0.071\frac{\text{lb}_\text{m}}{\text{ft}^3} \times \frac{1}{12(6)} = 446\frac{\text{lb}_\text{m}}{\text{hr-ft}^2}$$

From Appendix B,

$$\mu_g = 0.047 \frac{\text{lb}_\text{m}}{\text{hr-ft}^2}$$

Substituting the preceding values into Eq. (12.13), we obtain

$$\frac{\Delta P(4.17)(10)^8(0.4)^3(0.011)(0.071)}{2(1-0.4)(446)^2}$$

$$= \frac{150(1-0.4)(0.047)}{(0.011)(446)} + 1.75$$

$$0.087\Delta P = 0.862 + 1.75 = 2.61$$

$$\Delta P = 29.9\frac{\text{lb}_\text{f}}{\text{ft}^2}$$

$$\Delta P(\text{in. H}_2\text{O}) = \frac{29.9}{144} \times \frac{406.8 \text{ in. H}_2\text{O}}{14.7 \text{ psi}}$$

$$= 5.7 \text{ in. H}_2\text{O}$$

(b) From Eq. (12.14),

$$\Delta P = 0.37 \times 24 \times \left(\frac{104}{100}\right)^{1.56} = 9.4 \text{ in. H}_2\text{O}$$

(c) From Figure 12.6,

$$\Delta P = 7.8 \text{ in. H}_2\text{O/ft} \times 2 \text{ ft} = 15.6 \text{ in. H}_2\text{O}$$

Note the range of these estimates.

ADSORBENT REGENERATION

Material reversibly adsorbed on carbon or other adsorbents can be removed by

1. Contact with a hot inert gas (air can be used if the adsorbate is noncombustible)

2. Contact with low-pressure steam

3. Pressure reduction over the bed [also referred to as *pressure swing adsorption* (PSA)]

Methods 1 and 2 are essentially the same. However, for most solvents, steam regeneration is much more effective than inert gas regeneration, which is used only in those cases where water would contaminate the recovered solvent either by chemical reaction or by formation of a mixture that is difficult to separate. Regeneration by pressure reduction is not often economical in recovery or pollution control adsorption systems, but has been used in the recovery of gasoline vapors from tank-truck loading operations (see Figure 12.7).

Most fixed-bed carbon systems are designed for steam regeneration. The quantity of steam required for regeneration is a function of carbon loading, ease of adsorbate removal, and bed geometry. The steam flow rate during regeneration can be specified as pounds of steam per pound of recovered solvent or as pounds of steam per pound of carbon. A properly designed system should require no more than 1–4 pounds of steam per pound of recovered solvent or 0.2–0.4 pounds of steam per pound of carbon. If the latter quantity is the largest (which is usually the case for low carbon loading), then that quantity should be specified.

Low-pressure saturated steam provides rapid heating of the adsorber vessel and carbon bed. Less than 10% of the total heat input is used in the desorption process. Approximately 70% of the heat exits as steam and approximately 20% heats the vessel and carbon. Efficient use of regenerating steam has the added benefit of minimizing the overhead condenser size.

The method used to contact the carbon bed with steam is an important design consideration. Rapid bed heat-up is desired, but care must be taken not to "cook" the contaminants in the bed (cooking permanently lowers the adsorbent capacity). One of the effects of cooking is the formation of polymers with low volatility. Steam flow into the bed is usually in the direction opposite the inlet gas flow so that contaminants are not driven farther into the bed.

SAFETY CONSIDERATIONS

Most industrial solvents are flammable, with lower explosive limits (LELs) in the range from 1.0 to 2.0% by volume. Safety and insurance regulations specify that inlet vapor concentrations must not exceed 25% of the LEL. For most control or solvent recovery applications, this requirement does not present a problem. To minimize the size and total cost of a system, it is desirable to operate the system with concentrations as close as possible to 25% of the LEL.

Although the possibility is extremely remote, there have been instances of bed ignition. The most probable cause for bed ignition is the buildup of contaminants that react exothermically to produce

Figure 12.7
A carbon adsorption system at a gasoline terminal. (In this system, carbon adsorbs gasoline vapors displaced from tank trucks during loading operations. The carbon is regenerated by vacuum, and the vapors are absorbed back into liquid gasoline.)

localized ignition temperatures. Thermocouples are mounted just downstream of the bed to detect temperature surges. If the temperature exceeds a predetermined maximum temperature, the bed is flooded with water or blanketed with an inert gas. The former method is commonly used, but there is some concern that, in the worst case, contacting an incandescent bed with water might produce an amount of hydrogen sufficient to create a potentially explosive mixture.

12.5 DESIGN OF FIXED-BED CARBON ADSORPTION SYSTEMS

The design and operation of a carbon adsorption (CA) system involves heat and mass transfer, fluid dynamics, process control, and chemical analysis. Each of these fundamental areas of engineering must be carefully considered and integrated into a working, economical system. Although engineers most often design from basic fundamentals, in CA design, the mass transfer aspects are usually based on pilot-plant and empirical data.

It has been shown that the shape of the breakthrough curve and hence the length of the adsorption zone are functions of the volatility of the adsorbate and the operating conditions of the adsorbent bed. Nevertheless, it is possible to make some generalizations which can be used in a preliminary design. In typical solvent recovery operations where adsorbate volatilities may be *moderate* (for VOCs with 8–12 carbons) to *high* (for C_4–C_7 molecules), the AZ length will vary from 0.5 to 1.5 feet.

The carbon bed is contained in a steel vessel shaped like a large cylindrical drum positioned on its side, with a semispherical head on each end. The bed itself is shaped as a rectangular prism one to two feet tall, with a width slightly less than the diameter of the vessel, and with a length equal to that of the vessel minus the heads. The carbon is supported on a screen positioned inside the vessel such that the gas must flow upward or downward through the large rectangular face, and cannot flow around it. The actual dimensions of the bed typically are such that the length of the bed (L) is equal to twice the width (W). The following three requirements must be met.

1. The bed must contain a sufficient mass of adsorbent to provide a reasonable bed cycle time.

2. The superficial bed velocity must be high enough to provide satisfactory mass transfer rates, but low enough to allow a reasonable pressure drop.

3. The minimum bed depth must be greater than the length of one adsorption zone.

An important consideration that arises in the selection of pollution control equipment is whether to design the unit or to purchase a package unit from a supplier. Some generalizations can be made regarding this decision. If the CA system capacity is less than 20,000 cfm, purchase of a package unit from a reputable company will usually save money. Systems larger than 20,000 cfm are usually less expensive if designed in-house. Each case must be carefully considered. A proposed package system design requires a thorough evaluation before final acceptance. In dealing with the supplier of a package system, there is a two-way exchange of design information. Opgrande and associates (1979) reported that in the case of one CA project, the following information was required by the supplier:

1. Purpose of the system

2. Type of solvent

3. Whether the solvent was to be reused

4. Type of operation

5. Method of operation (continuous, batch, or other)

6. Gas stream composition, temperature, pressure, and humidity

7. Price of solvent

8. Gas stream flow rates

9. Cooling water availability, pressure, and temperature

10. Electrical power availability, voltage cycles, and so forth

11. Steam availability, pressure, and so forth

After receiving the information, the suppliers submitted proposals specifying the following:

1. Operational details of the system

2. Materials of construction

3. Cycle time

4. Operating conditions for the unit (such as gas stream flow rate, gas composition, temperature, pressure, and so forth)

5. Vessel size

6. Carbon supply

7. Recovery capacities

8. Installation site size requirement

9. System total operating weight

10. Electrical requirements

11. Steam requirements

12. Cooling water requirements

13. Instrument air requirements

14. Instrumentation package description

15. Cost and delivery, terms, and so forth

16. Pollution control guarantee

Example 12.2 illustrates the preliminary design procedures for a typical CA application.

Example 12.2

Prepare a preliminary design for a carbon adsorption (CA) system (including a fan) to control a stream of solvent-laden air from a plastics extruder local exhaust system. The exhaust stream temperature is 95 F, and it contains 1880 ppm of n-pentane (n-C_5). The plant engineer has provided the following information:

1. Other gaseous contaminants: none
2. Particulate matter contaminants: plant fugitive dust only
3. Flow rate: 5500 acfm (continuous)
4. Extruder exhaust pressure (intake for the new fan): –4.5 in. H_2O

Solution

We use Figure 12.3 and a capacity factor (percent of equilibrium adsorption capacity utilized during operation) to estimate the amount of carbon required (in pounds). We calculate the value of $[(T/V') \log (P_v/\overline{P})]$. Since n-pentane (in this case) is a pure component and the pressure is near atmospheric, we can calculate $\overline{P}$ from

$$\overline{P} = y_{\text{pentane}} \times 14.7 \text{ psia}$$

$$= \frac{1880 \text{ ppm}}{1,000,000} \times 14.7$$

$$= 0.0276 \text{ psia}$$

The mass flow rate of n-pentane to be adsorbed in one hour is:

$$\dot{M} = \frac{1 \text{ atm} \times 0.00188 \times 5500 \text{ cfm}}{0.73 \dfrac{\text{atm-ft}^3}{\text{lbmol-}^\circ\text{R}} \times 555 \ ^\circ\text{R}} \times \frac{72 \text{ lb}}{\text{lbmol}} \times \frac{60 \text{ min}}{\text{hr}}$$

$$\dot{M} = 110 \frac{\text{lb}}{\text{hr}} \ n\text{-pentane}$$

From Appendix B, P_v is about 16 at T = 95 F. The specific molal volume can be calculated from the density of liquid n-pentane and molecular weight as follows:

$$V' = \frac{72 \text{ g/gmol}}{0.64 \text{ g/cm}^3} = 112 \frac{\text{cm}^3}{\text{gmol}}$$

We can now solve for the abscissa of Figure 12.3 using pressures in place of fugacities.

$$\frac{T}{V'} \log \frac{P_v}{\overline{P}} = \frac{308 \text{ K}}{112 \text{ cc/gmol}} \log \frac{16}{0.0276} = 7.6$$

From Figure 12.3, the volume adsorbed is about 18 cc liquid/100 g carbon. Therefore, the theoretical equilibrium adsorption capacity is given by

$$\frac{18 \text{ cc}}{100 \text{ g C}} \times \frac{1 \text{ gmol}}{112 \text{ cc}} \times \frac{72 \text{ g}}{\text{gmol}} = \frac{11.6 \text{ g } n\text{-C}_5}{100 \text{ g C}}$$

Since the operating or dynamic capacity is 25–50% of the isotherm value, and since n-C_5 is fairly volatile, we choose a capacity factor of 30%. (For a final design, the capacity factor would be determined from experimental data.) Therefore,

$$\text{carbon capacity for design} = 0.3(11.6) = 3.5 \frac{\text{lb}_m n\text{-}C_5}{100 \ \text{lb}_m C}$$

We allow about one hour for regeneration and cooling. Thus, assuming 100% efficiency, a two-bed system will require a sufficient amount of carbon in each bed to adsorb 110 lb_m n-C_5. This gives a minimum of:

$$\frac{\text{lb}_m C}{\text{bed}} = 110 \frac{\text{lb}_m \ n\text{-}C_5}{\text{hr}} \times \frac{100 \ \text{lb}_m C}{3.5 \ \text{lb}_m \ n\text{-}C_5} = 3143 \frac{\text{lb}_m C}{\text{bed}}$$

Rounding this value to 3200 lb_m and using a bulk carbon density of 30 lb_m C/ft^3,

$$\text{bed volume} = \frac{3200 \ \text{lb}_m C}{30 \ \text{lb}_m C/\text{ft}^3} = 106.7 \ \text{ft}^3$$

Since n-C_5 is volatile, we use a minimum bed depth of 2 ft in our design. Assuming a rectangular bed with $L = 2W$,

$$\text{area of bed} = \frac{106.7}{2.0} = 53.4 \ \text{ft}^2$$

$$W^2 = \frac{53.4 \ \text{ft}^2}{2}$$

Therefore, $W = 5.2$ ft and $L = 10.4$ ft. Rounding these values, we will use bed dimensions of 5.25 ft $\times$ 10.5 ft.

Next, we check the superficial gas velocity.

$$V = \frac{5500 \ \text{acfm}}{(5.25)(10.5)} = 100 \ \frac{\text{ft}}{\text{min}}$$

This velocity is on the high end of the acceptable range. We will use this value in our preliminary design, but consideration should be given to buying some extra carbon and specifying a slightly longer and wider bed. (If a lower velocity is desired, and the bed depth must be kept at 2 feet, then the amount of carbon must be increased.) However, using our original dimensions:

mass of carbon/bed = 10.5 ft $\times$ 5.25 ft $\times$ 2.0 ft $\times$ 30 lb/ft^3 = 3308 lb carbon

total carbon to be purchased (two beds) = 6616 lb carbon

(Note that if we made the bed 12 feet long by 6 feet wide, then the carbon needed for two beds would be 8640 lb.)

The anticipated run time before needing regeneration for a bed with 3308 lb of carbon is:

$$\frac{3.5 \text{ lb } n\text{-C}_5}{100 \text{ lb C}} \times \frac{1 \text{ hour}}{110 \text{ lb } n\text{-C}_5} \times 3308 \text{ lb C} = 1.05 \text{ hr } (63.2 \text{ min})$$

(Note that if we had used a 12-ft by 6-ft bed with 4320 lb of carbon in it, we could have extended the run time to 82.5 minutes.)

Steam Requirement

Based on n-C$_5$ recovery, the steam requirement is

$$\text{steam required} = \frac{4 \text{ lb}_m \text{ stm}}{\text{lb}_m \text{ } n\text{-C}_5} \times 110 \frac{\text{lb}_m \text{ } n\text{-C}_5}{\text{hr}} \times 1.05 \text{ hr} = 440 \frac{\text{lb}_m \text{ stm}}{\text{regen}}$$

Based on the weight of carbon, the steam requirement is

$$\text{steam required} = 0.3 \frac{\text{lb}_m \text{ stm}}{\text{lb}_m \text{ C}} \times 3308 = 992 \frac{\text{lb}_m \text{ stm}}{\text{regen}}$$

Assume that the regeneration takes 45 minutes, and the rest of the time is for bed cooling; thus the required steam flow *rate* is 992 lb/0.75 hr, or 1323 lb/hr. Since the n-C$_5$ mass flow rate (carbon loading) is low, we will specify the higher of the two calculated steam rates, or 1323 lb$_m$ stm/hr. If plant steam is not available, a package boiler must be specified.

Bed Pressure Drop

For this application, assume that we will use either a 4 × 10 or 6 × 16 mesh carbon. From Figure 12.6 we obtain

$$\Delta P_{4 \times 10} = 7.5 \text{ in. H}_2\text{O/ft} \times 2 \text{ ft} = 15.0 \text{ in. H}_2\text{O}$$

or $\quad \Delta P_{6 \times 16} = 15.0 \text{ in. H}_2\text{O/ft} \times 2 \text{ ft} = 30.0 \text{ in. H}_2\text{O}$

Because of the high ΔP, we should consider adding excess carbon and making the bed larger, to reduce the superficial velocity and thus the ΔP. For now, assume the 4 × 10 mesh carbon.

Blower Horsepower Requirement

Although there is no significant process particle contamination, the local exhaust system will be picking up plant fugitive dust, so a fabric filter or guard chamber should be installed before the carbon adsorption system. We assume a filter ΔP of 3 in. H$_2$O for preliminary sizing and allow a pressure drop of 2 in. H$_2$O for the inlet and exhaust piping. Thus, the fan must provide a total ΔP of:

$$\Delta P = 3.0 + 2.0 + 15.0 - (-4.5) = 24.5 \text{ in. H}_2\text{O}$$

If we assume that the blower efficiency is 60%, then

$$\text{blower hp} = \frac{0.0001575 \dfrac{\text{hp}}{\text{cfm-in. H}_2\text{O}} \times 5500 \text{ acfm} \times 24.5 \text{ in. H}_2\text{O}}{0.60}$$

$$= 35.4 \text{ hp}$$

Preliminary Specification Summary

   adsorber bed size: 5.25 ft wide × 10.5 ft long × 2 ft deep

   mass of carbon per bed = 3308 lb_m

   steam required = 992 lb_m/regen × 24 regen/day = 23,800 lb/day

   fan hp = 35.4 hp (buy a 40 hp motor)

◀ ◀ ◀ ◀ ◀ ◀ ◀ ◀ ◀

◆   12.6 ECONOMICS OF FIXED-BED   ◆
ADSORPTION SYSTEMS

CAPITAL COSTS ESTIMATES

   Adsorber costs are somewhat more difficult to estimate than similar costs of other control systems due to the variety of adsorber applications, which entail increased complexity in design procedures. In 2000, the delivered cost of a carbon-steel packaged adsorber unit, complete with fan, instrumentation, overhead condenser, cooler, decanter, and with a flue gas throughput of 5000 to 10,000 acfm was in the neighborhood of $40 to $50 per acfm. As described earlier in Chapter 2, this would give an order of magnitude estimate for conceptual planning.

   Vatavuk (1996) analyzed vendor cost data for several different adsorber installations and developed a correlation that gives a ratio (R_c) of the *total* adsorber equipment cost to the combined cost of the adsorber *vessels* and the *contained carbon*. Vatavuk determined that R_c was related to the flue gas flow rate (Q) as follows:

$$R_c = 6.98Q^{-0.133} \tag{12.15}$$

where Q = flue gas flow, acmf ($4{,}000 \le Q \le 500{,}000$ acfm)

Process vessel cost is often estimated by calculating the weight of the vessel and multiplying this by the cost per pound of fabricated material of construction such as carbon-steel or one of the more corrosion-resistant alloys.

   Recall that the adsorber vessel is simply a large cylindrical drum positioned on its side. Based on several design rules for carbon adsorber vessels that restrict superficial bed velocities and bed depths, it can be shown that the cost of a single adsorber vessel can be related to the surface area of the vessel as shown below:

$$C_v = 313 \ S^{0.778} \tag{12.16}$$

where

C_v = f.o.b. vessel cost in 2000 dollars (fabricated of 304 stainless steel)

S = surface area including heads, ft^2 ($97 \leq S \leq 2{,}110 \ ft^2$)

As illustrated in Example 12.2, the geometry of the carbon bed is a rectangular prism, the size of which is determined by the quantity of adsorbate to be removed, the capacity of the carbon to adsorb this solvent, and the time required to regenerate an exhausted bed. For horizontal beds, recall that the length of the bed (L) typically is equal to twice the width (W). This means that the cylindrical vessel housing the bed must have a diameter sufficient to accommodate the bed supports as well as the depth of the bed. For beds with a depth of 2 to 3 feet, the diameter of the vessel must be at least 1 to 2 feet larger than the width of the bed.

Using the above relationships, the total adsorber system equipment cost C_A is found from the following:

$$C_A = R_c \left[C_c + N_T \times C_v \right] \tag{12.17}$$

where

C_c = cost of carbon contained in the system, year 2000 dollars

N_T = number of vessels required

The cost of activated carbon, C_c, depends on the quality and granule size, but is typically in the range of $2.00 to $2.50 per pound, in the year 2000. The total number of vessels, N_T, must provide for the vessels on-line adsorbing, and for those off-line being regenerated or cooling down.

For moderate to highly corrosive flue gas treatment, it may be necessary to use costly corrosion-resistant alloys to fabricate the adsorber vessels. The relative costs of several of these materials is shown below.

Material	Relative Cost
304 Stainless Steel	1.0
316 Stainless Steel	1.3
Monel-400	2.3
Nickel-200	3.2
Titanium	4.5

ADSORBER OPERATING COSTS

The major utility costs associated with adsorber operation are steam, electricity, and cooling-water costs. Cooling-water requirements range from 5 to 7 gallons of water per pound of steam. Electric power costs can be based on calculated blower horsepower or estimated from an average horsepower requirement of 6–8 hp/1000 scfm of gas throughput.

The disposal of the wastewater from the decanter should be considered to be an additional utility cost. This cost may or may not be significant, depending on the solubility and treatability of the recovered material.

▶ ▶ ▶ ▶ ▶ ▶ ▶ ▶ ▶

Example 12.3

Estimate the year 2000 total installed cost of the carbon adsorption system described in Example 12.2. Use data from this chapter and from Chapter 2 as appropriate.

Solution

Total mass of carbon is

$$3308 \text{ lb/bed times two beds} = 6616 \text{ lbs}$$
$$C_c = 6616 \text{ lb} \times \$2.50/\text{lb} = \$16{,}540$$

Since the bed depth is 2 feet, we add one foot to the bed width to estimate the vessel diameter. In this case, $D = 6.25$ feet. The cylinder length is still 10.5 feet. Thus the surface area, S, of the vessel (including the two semispherical heads) is:

$$S = \pi D L + 2 \ \pi D^2/2$$
$$S = 3.14159 \times 6.25 \times 10.5 + 2 \times 3.14159 \times (6.25)^{2}/2$$
$$S = 328.9 \text{ ft}^2$$

The cost of *one* vessel is:

$$C_V = 313 \ S^{ \ 0.778}$$
$$C_V = 313 \ (328.9)^{0.778}$$
$$C_V = \$28{,}433$$

Find R_c using Eq.(12.15),

$$R_c = 6.98 \ Q^{-0.133} \quad \text{In this case } Q = 5500 \text{ acfm and } R_c = 2.220$$

Now find C_A using Eq. (12.17),

$$C_A = 2.22 \ [\ \$16{,}540 + 2 \times \$28{,}433 \ ] = \$162{,}960$$

This is the system cost, f.o.b. vendor. Using the information in Chapter 2, Table 2.3:

$$DEC = 1.18 \times \$162{,}960$$
$$DEC = \$192{,}293$$
$$TIC = 1.75 \times DEC$$
$$TIC = \$336{,}000 \text{ (rounded)}$$

◀ ◀ ◀ ◀ ◀ ◀ ◀ ◀ ◀

Example 12.4

Estimate the annual operating cost for the adsorber described in Example 12.2. Assume that regenerating steam is supplied by the main plant boiler, and labor requirements are 2 person-hours per shift based on 350 operating days per year. Recovered n-pentane is worth 18 cents per pound. The carbon will be replaced every 5 years at a cost of $2.50 per pound. Labor costs $17.50/hr, electricity costs $0.08/kWh, steam costs $8.00/1000 lb_m and cooling water costs $0.12/1000 gal.

Solution

For this problem, we use the operating cost factors given in Table 2.4 on p. 94.

Direct operating costs

Labor: 2,100 hrs @ $17.50/hr	$36,750
Supervision: (15% of labor) =	5512
Maintenance: 5% of total installed cost (TIC) =	16,800

Utilities:

Electricity

$$35.4 \text{ hp} \times 0.746 \frac{\text{kW}}{\text{hp}} \times 24 \frac{\text{hr}}{\text{day}} \times 350 \frac{\text{days}}{\text{yr}} \times \frac{\$0.08}{\text{kWh}} = \quad 17,746$$

Steam

$$992 \frac{\text{lb}_\text{m} \text{ stm}}{\text{regen}} \times 24 \frac{\text{regen}}{\text{day}} \times 350 \frac{\text{day}}{\text{yr}} \times \frac{\$8.00}{1000 \text{ lb}_\text{m} \text{ stm}} = \quad 66,662$$

Cooling water

$$992 \frac{\text{lb}_\text{m} \text{ stm}}{\text{regen}} \times \frac{7 \text{ gal}}{\text{lb}_\text{m} \text{ stm}} \times 24 \times 350 \times \frac{\$0.12}{1000 \text{ gal}} = \quad 7000$$

Carbon replacement $\dfrac{6{,}616 \text{ lb}_\text{m}\text{C}}{5 \text{ years}} \times \dfrac{\$2.50}{\text{lb}_\text{m}\text{C}} =$ \qquad 3308

Indirect operating costs

Labor overhead: 60% of total labor	
$\quad = 0.6(5512 + 36{,}750) =$	25,357
Taxes: 1% of TIC =	3360
Insurance: 1% of TIC =	3360
Depreciation: 10% of TIC =	33,600
$\quad$ Total operating cost (rounded) =	$219,500

Credit for recovered n-pentane (assume 90% recovery):

$$\text{Credit} = 110\frac{\text{lb}_\text{m}\,n\text{-C}_5}{\text{hr}} \times 24 \times 350 \times 0.9 \times \frac{\$0.18}{\text{lb}_\text{m}} = \quad \$149{,}700$$

Net annual operating cost (rounded) $\qquad\qquad\qquad$ $\$70{,}000$

Note that this figure does not include costs for waste disposal.

$$\blacktriangleleft\ \blacktriangleleft\ \blacktriangleleft\ \blacktriangleleft\ \blacktriangleleft\ \blacktriangleleft\ \blacktriangleleft\ \blacktriangleleft\ \blacktriangleleft$$

◆ 12.7 FLUIDIZED-BED ADSORBERS ◆

The cyclic heating and cooling of the adsorbent necessary in fixed-bed adsorbers results in high steam requirements. Other problems associated with fixed-bed systems are poor steam and gas distribution and complex valving requirements.

In an effort to alleviate these problems, a number of fluidized or moving-bed adsorbers have been built. The obvious advantages of a fluidized system are excellent gas–solid contact and the absence of cyclic heating. Fluidized systems have been built that use activated alumina and activated carbon. Several of these units were designed specifically for SO_2 adsorption.

The Reinluft and alkalized alumina processes are two examples of early fluidized-bed SO_2 adsorbers (Teller 1967). Both of these processes not only adsorbed SO_2, but also converted the adsorbed gas to SO_3. These processes were intended primarily for large throughput volumes and major application in coal-fired power plants. A major operating problem associated with these fluidized-bed adsorbers was rapid attrition of the adsorbent and the resultant production of fines.

A fully commercialized fluidized-bed adsorber is the Purasiv HR unit shown schematically in Figure 12.8. In the Purasiv HR unit, adsorption and regeneration occur continuously in separate zones within the adsorber vessel. Inlet gas enters the fluidized top section, flows upward through that section, and exits at the top of the unit. The carbon flows downward continuously to the desorption section where it is heated indirectly and purged by nitrogen. The purge gas and desorbed solvent pass through a condenser and a separator. Noncondensable gases from the separator are passed back through a secondary adsorber section located between the primary adsorber and the desorbing section.

Regenerated carbon is cooled by heat exchange and is then lifted by air back to the top of the vessel. Make-up nitrogen is supplied to the intake of the nitrogen circulating blower. In one system handling 4500 acfm of an exhaust containing mixed ketones, a flow rate of 5 scfm of make-up nitrogen was required (Wohler 1979). Steam usage was 1.1 pound of steam per pound of solvent recovered.

Successful operation of the Purasiv process has depended in part on the development of a hard, spherical activated carbon that can

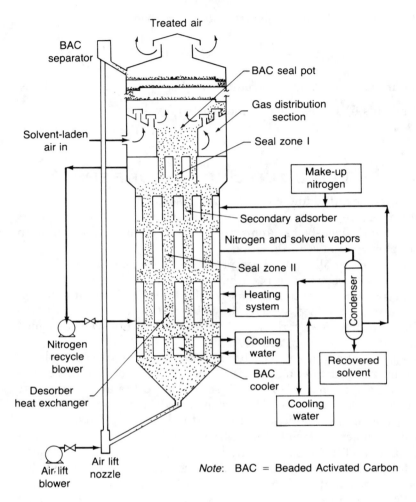

Figure 12.8
Schematic diagram of a Purasiv HR fluidized adsorber.
(Courtesy of the Union Carbide Corporation, Danbury, CT.)

withstand abrasion during fluidization. The beaded activated carbon (BAC) used in Purasiv units is produced from a molten pitch by a process developed in Japan.

Capital and operating cost data on the Purasiv HR system are limited. For systems up to 20,000 scfm, the capital cost is higher than that of fixed-bed systems. However, utility savings should provide a reasonable payout on the additional capital investment. For application with water-soluble solvents, the Purasiv system has the advantage of not requiring a costly separation step to produce a reusable solvent. Considering these advantages, fluidized-bed adsorption should become an important process in air pollution control.

♦ PROBLEMS ♦

12.1 Using the adsorption potential plot for hydrocarbons in Figure 12.3, calculate the amount of hexane adsorbed in pounds of hexane per 100 pounds of carbon at 80 F. The specific gravity of liquid hexane is 0.66. The concentration of hexane in air is 1000 ppm, and the air is at 1.0 atm.

12.2 Select points on the carbon disulfide isotherm shown in Figure 12.2 and determine whether the Langmuir or Freundlich equation is more applicable.

12.3 At a temperature of 91 F and a carbon disulfide partial pressure of 0.01 psia, calculate the adsorption capacity (in lb_m CS_2/100 lb_m C) given by (a) Figure 12.2 and (b) the generalized correlation in Figure 12.3. At 91 F, the vapor pressure of liquid CS_2 is 9.15 psia, and the specific gravity is 1.26.

12.4 A two-bed carbon adsorption system is to be designed to handle 8000 acfm of air containing 700 ppm of hexane. Laboratory studies indicate that carbon can adsorb 8 lb_m hexane per 100 lb_m carbon under the conditions at which the system will be operated. If the system is to operate at 90 F and 1.0 atm at a removal efficiency of 99%, determine the approximate dimensions of each carbon bed. Bed regeneration and cooling will require one hour.

12.5 A fixed-bed carbon adsorption system consisting of two 3-foot-deep beds containing 6 × 16 mesh carbon operates at a superficial gas velocity of 60 ft/min. Specify the horsepower of the fan motor for the 15,000-cfm system if the total pressure drop across the system excluding the pressure drop across the bed is 5 in. H_2O.

12.6 Estimate the total amount of steam required per day (in pounds) to regenerate the beds of the system described in Problem 12.4.

12.7 Estimate the total installed cost (in 2002 dollars) of a package carbon adsorption system that processes 15,000 acfm and contains 4000 lb_m of carbon. Each 2000 lb bed is contained in a carbon-steel vessel that is 7 ft in diameter and 14 ft long (excluding heads).

12.8 If the carbon in the adsorption system described in Problem 12.7 is replaced every six years and labor requirements are 4 person-hours per shift, what is the annual operating cost for the system? Use the same unit costs and cost factors as given in Example 12.4. Assume that the fan draws 25 hp, and that there are 12 regenerations per day, and each regeneration uses 600 lb steam.

12.9 You are assigned the task of process design for a fixed-bed carbon adsorber system. You are given the following data: the inlet airstream flows at 5000 acfm (at 95 F and 1 atm) and contains 0.2% n-pentane by volume; the proper operating carbon capacity for n-pentane is 3.5 kg n-pentane per 100 kg carbon. What is the minimum amount (in kg) of carbon needed for each adsorber bed if you decide to have the system operate for 1 hour between bed regenerations?

12.10 Methyl mercaptan (CH_3SH) is catalyzed to dimethyl disulfide (CH_3SSCH_3) when adsorbed onto carbon. Look up the chemical and physical properties of these two compounds and sketch a flow chart showing how a surplus carbon adsorption system might be modified to control a source which is presently emitting a dilute stream of CH_3SH in air.

12.11 A project to design a carbon adsorber to treat a stream of contaminated air with a flow rate of 4900 acfm is in progress at the manufacturing plant where you have just been hired. The previous engineer calculated that 140 cubic feet of carbon (4 × 6 mesh) was needed and set the superficial bed velocity at 87.5 ft/min, because she wanted to use an existing fan. Calculate the pressure drop that she anticipated for this carbon bed.

12.12 Engineer Smith is trying to save operating costs on an existing incinerator that has a 50% efficient recuperator on it already. It treats 20,000 acfm of air containing a low concentration of a toxic and odorous compound, whose only effective control is incineration. His associate, Engineer Brown, suggests that he pass the contaminated air through a carbon adsorption system prior to incineration to double the concentration of contaminant (and effectively halve the volume of air fed to the incinerator). Assuming that the compound is not too volatile and does not react with carbon, sketch a flow sheet of the system Brown has in mind. Roughly estimate the potential savings in fuel costs ($5.00/million Btu) resulting from the decrease in the volume of air fed to the afterburner. If the TIC of the carbon system is $800,000, is this a good idea?

12.13 You are designing a carbon adsorption system for a dilute hexane-in-air stream flowing at 5500 acfm. You need 110 cubic feet of carbon (4 × 6 mesh), and the minimum bed depth must be 1 ft. Calculate the bed pressure drops for bed depths of 1.0, 1.5, and 2.0 ft.

12.14 Estimate the daily steam requirements to regenerate the carbon for the system described in Example 12.1.

12.15 Estimate how much carbon (in pounds) must be purchased for a typical 2-bed carbon adsorption system to treat 5000 cfm of air at

95 F and 1 atm. The air contains 600 ppm of styrene. A run time of at least 2 hours is desired before switching beds. Pilot plant tests have shown that styrene adsorbs well (steep breakthrough curve, 6-inch active adsorption zone), and that the working capacity is 12 g styrene/100 g carbon. The carbon has a bulk density of 30 lb/cubic foot. Making other appropriate assumptions as needed, also estimate the bed dimensions (in feet).

REFERENCES

Carubba, R. V., Urbanic, J. E., Wagner, N. J., and Zanitch, R. H. "Perspectives of Activated Carbon," *Adsorption and Ion Exchange, American Institute of Chemical Engineers Symposium Series*, 80, 1984, p. 76.

Dubinin, M. M., and Raduschkevich, L. V. *Comptes Rendus de L'Academie des Sciences, 55,* 1947, p. 327.

Ergun, S. *Chemical Engineering Progress, 48,* 1952, pp. 87–89.

Goldman, F., and Polanyi, M. *Zeitschrift für Physikalische Chemie, 132,* 1928, p. 321.

Grant, R. J., and Manes, M. "Adsorption of Binary Hydrocarbon Gas Mixtures on Activated Carbon," *Industrial and Engineering Chemistry Fundamentals, 5,* 1966, p. 490.

Lambert, F. L. *Investigation of the Dynamic Parameters of the Adsorption of Hydrocarbons on Activated Carbon,* Ph.D. Thesis, Clemson University, Clemson, SC, 1970.

Lewis, W. K., Gilliland, E. R., Chertow, B., and Cadogan, W. P. "Pure Gas Isotherms," *Industrial and Engineering Chemistry, 42,* 1950, p. 1326.

Opgrande, J. L., Bobratz, C. J., and Hoab, G. R. "Carbon Adsorption Recovery Units for a Waste Gas Containing Toluene," *Proceedings of the Seminar on VOC Emission Control,* Atlanta, GA, September 18–19, 1979.

Sloan, E. D. *Nonideality of Binary Adsorbed Mixtures of Benzene and Freon–11 on Graphitized Carbon,* Ph.D. Thesis, Clemson University, Clemson, SC, 1974.

Standen, A., Ed. "Adsorption," *Kirk-Othmer Encyclopedia of Chemical Technology* (2nd ed.). New York: Interscience Publishers, 1963(a), p. 460.

———. "Amorphous Silica," *Kirk-Othmer Encyclopedia of Chemical Technology* (2nd ed.). New York: Interscience Publishers, 1963(b), p. 62.

Teller, A. J. "Recovery of Sulfur Oxides from Stack Gases," *Proceedings of the MECAR Symposium on New Development in Air Pollution Control,* New York, October 23, 1967.

Treybal, R. E. *Mass Transfer Operations* (2nd ed.). New York: McGraw-Hill, 1968.

Turk, A. "Source Control by Gas-Solid Adsorption," *Air Pollution,* A. Stern, Ed. New York: Academic Press, 1962, p. 500.

Vatavuk, W. M. *Estimating Costs of Air Pollution Control.* Chelsea, MI: Lewis Publishers, 1990.

———. *OAQPS Control Cost Manual* (5th ed.). EPA 453/B-96-001, U.S. Environmental Protection Agency, Research Triangle Park, NC, 1996.

Wark, K., Warner, C. F., and Davis, W. T. *Air Pollution* (3rd ed.). Menlo Park, CA: Addison-Wesley, 1998, p. 308.

Wohler, R. "Fluidized Bed Carbon Adsorption," *Proceedings of the Seminar on VOC Emission Control,* Atlanta, GA, September 18–19, 1979.

$\blacklozenge\blacklozenge\blacklozenge$ CHAPTER 13

GAS ABSORPTION

There are four major factors to be established in the design of any plant involving the diffusional operations: the number of ideal stages or their equivalent, . . . the length [height] of the required device, . . . the cross-sectional area of the equipment, . . . and the energy requirements.

Robert E. Treybal, 1968

$\blacklozenge$ 13.1 INTRODUCTION $\blacklozenge$

In air pollution control work, **absorption** refers to the selective transfer of material from a gas to a contacting liquid. The separation principle involved is the preferential solubility of a gaseous component in the liquid. In a majority of pollution control applications, the contacting liquid is water, and the process is sometimes referred to as *scrubbing* or *washing*. Frequently encountered examples of the application of gas absorption in pollution control include

1. Removal and recovery of ammonia in fertilizer manufacture

2. Removal of hydrogen fluoride from glass furnace exhaust

3. Control of sulfur dioxide from combustion sources

4. Recovery of water-soluble solvents such as acetone and methyl alcohol

5. Control of odorous gases from rendering plants

Gas absorption involves the diffusion of material from a gas through a gas–liquid interface and ultimate dispersion in the liquid. Dispersion or solution of the absorbed material in the liquid may be accompanied by chemical reaction.

Both molecular and turbulent (eddy) mass transfer are present in the absorption process. Eddy diffusion is many times faster than the molecular mechanism, and is maximized by designing absorption

393

equipment to operate at high levels of turbulence. Even so, the major resistance to mass transfer between the liquid and the gas is due to the laminar layers at the phase boundary. We will develop the theory to quantitatively describe mass transfer in Section 13.2.

GAS ABSORPTION EQUIPMENT

Gas absorption is usually carried out in packed towers such as shown in Figure 13.1. The gas stream enters the bottom of the column and passes upward through a wetted packed bed. The liquid enters the top of the column and is uniformly distributed over the column packing. The column packing can have any of a number of commercially available geometric shapes designed to give maximum gas–liquid contact and low gas-phase pressure drop. The requirements of a satisfactory packing are

1. High wetted area per unit volume
2. Minimal weight
3. Sufficient chemical resistance
4. Low liquid holdup
5. Low pressure drop
6. Low cost

Several types of packings are shown in Figure 13.2. Packings are available in a number of materials, including ceramics, glass, metal, and various plastics.

♦ 13.2 ABSORPTION TOWER DESIGN ♦

Since the majority of absorption operations are carried out in packed towers, we will emphasize the design of this equipment. The three major components of the process design of a packed tower are (1) performing the tower material balances to determine a liquid circulation rate, (2) calculating a packed height to provide the desired degree of separation, and (3) determining a column diameter sufficient to handle the required liquid and gas flow rates. Because mass transfer theory underlies all of the design equations, we will discuss it first.

MASS TRANSFER THEORY

Mass transfer from the gas to the liquid occurs across the gas–liquid interface provided by the wetted surface of the tower packing. We will develop a design procedure based on the two-film theory, which visualizes all resistance to transfer across the interface as resulting from a laminar film on each side of the interface. The model for the two-film theory is shown schematically in Figure 13.3. We assume that the gas and the liquid are in equilibrium at the interface.

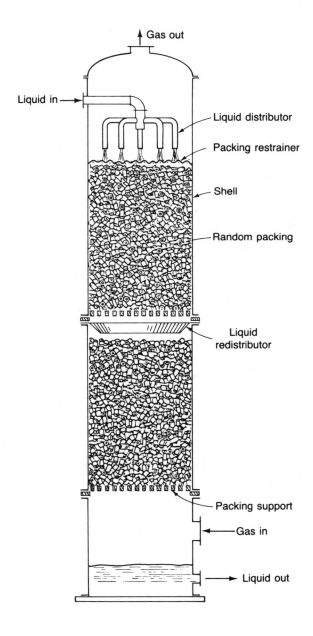

Figure 13.1
Schematic diagram of a packed gas absorption tower.
(Adapted from Treybal, 1968.)

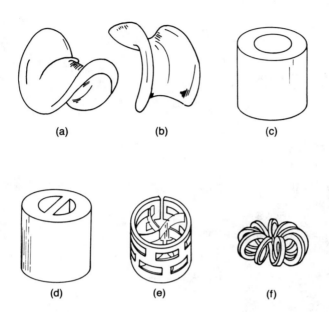

Figure 13.2
Typical tower packings: (a) Berl saddle; (b) Intalox saddle; (c) Raschig ring; (d) Lessing ring; (e) Pall ring; (f) Tellerette.

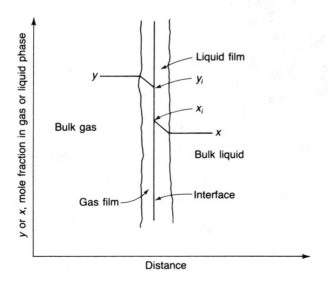

Figure 13.3
Schematic diagram of concentration gradients across gas and liquid films.

The rate of mass transfer of component A in a gas mixture of A and B by both molecular diffusion and phase bulk motion can be determined from

$$\frac{N_A}{A} = -D_M \frac{dy}{db} + y \frac{(N_A + N_B)}{A} \qquad (13.1)$$

where

N_A = transfer rate of component A, lbmol/hr

N_B = transfer rate of component B, lbmol/hr

D_M = molal diffusivity, lbmol/hr-ft

y = mole fraction of component A in the gas

b = distance in the direction of diffusion, ft

A = area through which mass is transferred perpendicular to the path of diffusion, ft^2

A similar expression can be written for the liquid in terms of x, the mole fraction of component A in the liquid. The second term on the right side of Eq. (13.1) accounts for mass transfer owing to the bulk movement of the fluid toward the interface.

Equation (13.1) can be integrated for the special case of component A diffusing through a stagnant film of component B. This case is applicable to the absorption of a soluble gas from a nonsoluble gas such as air when the contacting liquid is water. For example, assume that the gas in Figure 13.3 is a mixture of ammonia and air and that the liquid is water. Since ammonia is soluble in water, it diffuses toward and through the interface and is absorbed. Air must diffuse away from the interface to maintain a constant molar density in the gas. The bulk motion of the gas must be toward the interface at a velocity such that to a stationary observer the air molecules appear to be stationary. For this special case, the diffusion rate of component B relative to the interface is zero, and Eq. (13.1) becomes

$$\frac{N_A}{A} = -D_M \frac{dy}{db} + y \frac{N_A}{A} \qquad (13.2)$$

Rearranging Eq. (13.2), we obtain

$$-\frac{1}{N_A} \int_y^{y_i} \frac{dy}{1-y} = \frac{1}{D_M A} \int_0^{B_T} db \qquad (13.3)$$

where

y_i = y at the interface

B_T = thickness of the gas film, ft

Integrating Eq. (13.3) and rearranging the result gives the mass transfer rate of component A per unit area, which is referred to as the *flux of component A.*

$$\frac{N_A}{A} = \frac{D_M}{B_T} \ln\left(\frac{1-y_i}{1-y}\right) \tag{13.4}$$

The molal diffusivity can be replaced by $\mathscr{D}\rho_m$, where $\mathscr{D}$ is the volumetric diffusivity in ft^2/hr or cm^2/s, and ρ_m is the molar density of the gas in lbmol/ft^3 or gmol/cm^3. Gas-phase volumetric diffusivities have been predicted from the kinetic theory of gases, but are usually estimated from semi-empirical equations such as those presented by Treybal (1968).

In a majority of applications, gas absorption equipment is operated in the turbulent regime, and Eq. (13.4) must be modified to include eddy or turbulent transport. Thus,

$$\frac{N_A}{A} = \frac{(\mathscr{D}+E)}{B_T} \rho_m \ln\left(\frac{1-y_i}{1-y}\right) \tag{13.5}$$

where E = eddy diffusivity, ft^2/hr or cm^2/s

In actual practice, $\mathscr{D}$, E, and B_T cannot be determined independently and are combined into a single mass transfer coefficient defined by

$$\frac{k_y}{\phi} = \frac{N_A}{A(y-y_i)} \tag{13.6}$$

where

k_y = Drew-Colburn gas film coefficient, $\dfrac{\text{lb mol}}{\text{hr-ft}^2\text{-}\Delta y}$

ϕ = correction factor for phase bulk motion or drift

In Eq. (13.6), the drift correction factor can be obtained from

$$\phi = (1-y)_{\text{LM}} = \frac{(1-y_i)-(1-y)}{\ln\left(\dfrac{1-y_i}{1-y}\right)}$$

where $(1-y)_{\text{LM}}$ = log mean concentration gradient

Combining Eqs. (13.5) and (13.6) and the expression for ϕ gives

$$\frac{N_A}{A} = \frac{k_y(y-y_i)}{\phi} = \frac{k_y(y-y_i)}{(1-y)_{\text{LM}}} \tag{13.7}$$

A similar expression can be written for the liquid film as

$$\frac{N_A}{A} = k_x(x_i - x) \tag{13.8}$$

The drift correction factor for the liquid film is very nearly equal to 1.0, and so is omitted. The drift correction factor also can be omitted from Eq. (13.7) when y is less than 0.05, which is usually the case in pollution control applications.

MASS TRANSFER IN PACKED TOWERS

An Overview. Before we develop the detailed equations for tower design, we will briefly discuss the procedures involved in the process design of an absorption tower. The simplified approach to tower design applies only under certain conditions, but is mathematically easier than the detailed approach and is often applicable to pollution control problems in which gas- and liquid-phase concentrations (of pollutant) are very low (less than 5%). A quantitative illustration of the simplified approach can be found in Example 13.3 on p. 417.

First we must perform an overall tower material balance to determine the liquid circulation rate. Referring to Figure 13.4, we usually

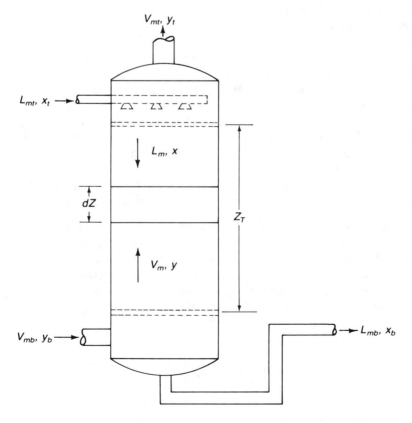

Figure 13.4
Schematic diagram of a packed tower showing the design variables.

will know the gas molar flow rate V_m and composition y at the bottom and top of the tower, and the liquid composition x at the top of the tower. The amount of pollutant in the liquid at the bottom of the tower is determined by its removal rate from the gas, but the liquid *composition* at the bottom of the tower depends on the liquid circulation rate.

We calculate the liquid circulation rate as a multiple (ranging from 1.5 to 3.0) of a minimum liquid rate that we can determine graphically or algebraically from a plot of the equilibrium curve and the tower top and bottom operating points. The line connecting the tower top operating point (x_t, y_t) to the intersection of the horizontal projection of the tower bottom gas composition y_b with the equilibrium curve is called the *minimum slope operating line*. The slope of this line is the ratio of the solute-free liquid (minimum) flow rate to the solute-free gas flow rate. Since we know the gas flow rate, we can easily calculate the minimum liquid flow rate. The actual liquid flow rate is simply a multiple of the minimum liquid flow rate.

Once the actual liquid flow rate is established, we can determine the height of the packed section by a graphical integration procedure. The graphical procedure is illustrated for the general case in Example 13.1, and for the simplified case in Example 13.3.

Finally, we determine the tower diameter by using a graphical correlation for tower flooding and pressure drop. We first determine the gas flux G_y at which flooding would occur in the tower, and we select an actual gas flux equal to 40–70% of the flooding flux. From the actual gas flux and the actual gas flow rate, we can determine the tower diameter.

We presented the preceding overview to assist you in understanding the general equations that we will develop in the following discussion. Despite the apparent complexities of the equations, the procedures are the same as those outlined above. We also believe that careful review of Examples 13.1, 13.2, and 13.3 will significantly enhance your comprehension of absorption tower theory and design.

The General Case (the Detailed Approach). Consider the steady-state absorption of component A from a nonabsorbing gas B by a nonvolatile liquid in the tower, as was shown in Figure 13.4. Letting y equal the mole fraction of A in the gas, and x equal the mole fraction in the liquid, the overall and component A balances over the differential section dZ can be written (without regard to sign) as

$$dL_m = dV_m \tag{13.9}$$

and

$$d(L_m x) = d(V_m y) = dN_A \tag{13.10}$$

where

    L_m = molar flow rate of the liquid, mol/hr

    V_m = molar flow rate of the vapor, mol/hr

Since only component A is transferred between phases (and assuming the liquid is nonvolatile),

$$d\left(L_m x\right) = dL_m = dN_A \text{ and } d\left(V_m y\right) = dV_m = dN_A \quad \textbf{(13.11)}$$

Writing a component balance from the bottom of the tower, b, to any point in the tower, we obtain

$$L_m x + V_{mb} y_b = L_{mb} x_b + V_m y \quad \textbf{(13.12)}$$

Letting L'_m and V'_m equal the molar flow rates of solute-free liquid and solute-free vapor,

$$V_m = \frac{V'_m}{1-y} \text{ and } L_m = \frac{L'_m}{1-x} \quad \textbf{(13.13)}$$

Substituting these expressions for V_m and L_m into Eq. (13.12), we obtain

$$V'_m\left(\frac{y_b}{1-y_b} - \frac{y}{1-y}\right) = L'_m\left(\frac{x_b}{1-x_b} - \frac{x}{1-x}\right) \quad \textbf{(13.14)}$$

or

$$\frac{y}{1-y} = \frac{L'_m}{V'_m}\left(\frac{x}{1-x}\right) + \left[\frac{y_b}{1-y_b} - \frac{L'_m}{V'_m}\left(\frac{x_b}{1-x_b}\right)\right] \quad \textbf{(13.15)}$$

When plotted on an x–y diagram as in Figure 13.5, Eq. (13.14) or Eq. (13.15) depicts the operating line for an absorption tower. The operating line shows the relationship between the bulk liquid-phase and gas-phase concentrations and the phase flow rates at any point in the packed section. Even though, in the general case, the operating lines are not straight, the minimum slope operating line is defined as the straight line connecting points (x_t, y_t) and (x_b^*, y_b), and has slope $(L'_{m\min}/V'_m)$.

An expression for the rate of transfer per unit area of gas–liquid contact as a function of the phase flow rates can be developed from Eqs. (13.7) through (13.10) as follows:

$$dN_A = d\left(V_m y\right) = k_y \frac{(y - y_i)}{\phi} dA = d\left(L_m x\right) = k_x\left(x_i - x\right) dA \quad \textbf{(13.16)}$$

The area for mass transfer is a function of the tower packing, and the usual practice is to report the available area per cubic foot of packing a. Again referring to Figure 13.4, the area for mass transfer in the differential tower section is

$$dA = aS\, dZ \quad \textbf{(13.17)}$$

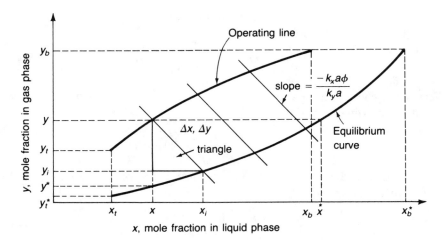

Figure 13.5
Graphical construction of the $(\Delta x, \Delta y)$ triangle, and representation of the overall gas-phase driving force.

where

S = cross-sectional area of the tower, ft^2

a = available area for mass transfer, ft^2/ft^3 of packing

Combining Eqs. (13.16) and (13.17) and substituting $V_m \, dy/(1-y)$ for $d(V_m y)$ results in

$$k_y a \frac{(y - y_i)}{\phi} dZ = \frac{V_m}{S} \frac{dy}{1-y} \qquad (13.18)$$

An equivalent equation for the liquid film is

$$k_x a (x_i - x) dZ = \frac{L_m}{S} \frac{dx}{1-x} \qquad (13.19)$$

Equation (13.18) or Eq. (13.19) must be integrated to find Z_T, the height of the packed section. To perform the integration, we must develop a relationship between the phase bulk concentrations, y and x, and the interfacial concentrations, y_i and x_i. We can develop this relationship using Eq. (13.16), remembering that the rate of transfer through the gas film must equal the rate of transfer through the liquid film. Thus, rearranging Eq. (13.16) gives

$$\frac{y - y_i}{x - x_i} = -\phi \frac{k_x a}{k_y a} \qquad (13.20)$$

On a plot of x versus y, Eq. (13.20) is a straight line, with a slope of $-\phi k_x a / k_y a$, connecting a point (x, y) on the operating line with the corresponding point (x_i, y_i) on the equilibrium curve, as shown in Figure 13.5. Equation (13.18) can be rearranged and integrated between the two ends of the tower as follows:

$$\left(\frac{k_y a}{G_{my}}\right)_{avg} \int_0^{Z_T} dZ = Z_T \left(\frac{k_y a}{G_{my}}\right)_{avg} = \int_{y_t}^{y_b} \frac{\phi \, dy}{(1-y)(y-y_i)} \qquad \textbf{(13.21)}$$

where

$G_{my} = V_m / S$ = molar flux of gas through the tower, lbmol/hr-ft^2

$\left(\dfrac{k_y a}{G_{my}}\right)_{avg}$ = average value of $\left(\dfrac{k_y a}{G_{my}}\right)_b$ and $\left(\dfrac{k_y a}{G_{my}}\right)_t$

Z_T = height of the packed section

b, t = bottom and top of tower, respectively

The right-side integral in Eq. (13.21) can be graphically integrated in the following manner (referring to Figure 13.5):

1. Plot the solute equilibrium curve y_i versus x_i at the tower operating temperature and pressure, which are assumed to be constant over the packed section.

2. Determine L_{min} from Eq. (13.14) by substituting x_b^* for x_b (where x_b^* is the liquid concentration in equilibrium with y_b) and substituting x_t and y_t for x and y.

3. Plot the actual operating line using Eq. (13.14) or Eq. (13.15). The actual molar liquid flow rate is set at some multiple (1.5 to 3.0) of the minimum rate L_{min}.

4. Calculate the value of $-\phi k_x a / k_y a$. Starting at y_t on the operating line, draw a series of parallel lines with slope $-\phi k_x a / k_y a$ at equal intervals between y_t and y_b (usually, six to ten lines are sufficient). The intersections of these lines with the equilibrium curve provide the values of x_i and y_i corresponding to x and y.

5. Prepare a table with the following headings:

y	$1-y$	y_i	$1-yi$	ϕ	$\phi/(1-y)(y-y_i)$

6. Determine the value of $\int_{y_t}^{y_b} \phi \, dy/(1-y)(y-y_i)$ by plotting $\phi/(1-y)$ $(y-y_i)$ versus y and determining the area under the curve from y_t to y_b.

7. Find Z_T from Eq. (13.21).

To base Z_T on the liquid film, steps 5, 6, and 7 are changed as follows:

5. The table headings become

x	$1-x$	x_i	$x_i - x$	$1/(1-x)(x_i - x)$

6. Plot $1/(1-x)(x_i - x)$ versus x and the area under the curve between x_t and x_b is the value of $\int_{x_t}^{x_b} dx/(1-x)(x_i - x)$.

7. Calculate Z_T as

$$Z_T = \frac{\text{area under curve}}{\left(\dfrac{k_x a}{G_{mx}}\right)_{avg}}$$

where $G_{mx} = L_m/S$ = molar flux of liquid through the tower, lbmol/hr-ft^2

Overall Mass Transfer Coefficients.   The procedure for determining Z_T can be simplified by introducing the concept of the overall gas- or liquid-phase transfer coefficients, defined by the following equations:

$$N_A = \frac{K_y aS(y - y^*)dZ}{\phi} = K_x aS(x^* - x)dZ \qquad \textbf{(13.22)}$$

where

K_y = overall gas-phase coefficient, lbmol/hr-ft^2-Δy

K_x = overall liquid-phase coefficient, lbmol/hr-ft^2-Δx

y^* = mole fraction of solute in the gas that would be in equilibrium with the bulk liquid (composition x)

x^* = mole fraction of solute in the liquid that would be in equilibrium with the bulk gas (composition y).

The relationships between K_y, K_x, k_x, and k_y are

$$\frac{1}{K_y} = \frac{1}{k_y} + \frac{m}{k_x} \qquad \textbf{(13.23)}$$

$$\frac{1}{K_x} = \frac{1}{k_x} + \frac{1}{mk_y} \qquad \textbf{(13.24)}$$

where m = slope of the equilibrium curve

The driving forces to be used with the overall coefficients are $y - y^*$ or $x^* - x$, as was shown in Figure 13.5.

The Transfer Unit Concept.   If we rewrite Eq. (13.21) as

$$Z_T = \left(\int_{y_t}^{y_b} \frac{\phi\, dy}{(1-y)(y - y_i)}\right)\left(\frac{G_{my}}{k_y a}\right)_{avg} \qquad \textbf{(13.25)}$$

we can see that the packed height is the product of two terms. The first term is a measure of the difficulty of the overall separation, and the second term is a measure of the mass-transfer efficiency of the packing under operating conditions. Thus, we can rewrite Eq. (13.25) as

$$Z_T = \left[N_{ty} \right]\left[H_y \right] \qquad (13.26)$$

where

N_{ty} = number of transfer units based on the individual driving force

H_y = height of a transfer unit (HTU) based on the individual gas film coefficient, ft

The number of transfer units and the corresponding HTUs can be defined in terms of the four different driving forces shown in Figure 13.5 and their corresponding mass-transfer coefficients. The overall and individual HTUs are related by

$$H_{oy} \frac{G_{mx}}{m G_{my}} = H_x + \frac{G_{mx} Hy}{m G_{my}} = H_{ox} \qquad (13.27)$$

where

H_x, H_y = HTUs based on individual liquid and gas film coefficients

H_{ox}, H_{oy} = HTUs based on overall liquid and gas film coefficients

If the overall HTU (for instance, H_{oy}) is used, it can be estimated by $(G_{my}/k_y a)_{\text{avg}}$ for use in Eq. (13.25). Some advantages of using the HTU concept are

1. We can expect less variation in the value of HTUs over the length of packing than when using overall transfer coefficients. For almost all gas-liquid systems, HTUs range from 0.3 to 9.0 ft, but for any given system, the HTU will be nearly constant throughout the tower.

2. The concept of mass transfer efficiency in terms of length of packing is easy to perceive.

3. Experimental mass transfer data for packed towers are often correlated as HTUs.

Transfer Coefficients and HTUs for Absorber Design. For effective tower design, an accurate value for either a mass transfer coefficient or an HTU must be determined. There has been a great deal of research done in developing correlations for predicting mass transfer coefficients and HTUs but there is a significant disparity in the work reported by various investigators. There are major problems in extrapolating transfer coefficients obtained on pilot-scale equipment to full-size installations. Following are some causes of these problems.

1. Difficulty in determining accurate interfacial areas in pilot-scale equipment

2. Difficulty in correcting for entrance effects in pilot-scale equipment

3. Inability to maintain constant operating conditions in small towers

Determination of the interfacial area can be avoided by developing correlations for $k_y a$ and $k_x a$ in which the area for gas-liquid contact is included in the mass transfer coefficient. These correlations are usually of the following form:

$$k_y a \text{ or } k_x a = bG_{my}^r G_{mx}^s \qquad (13.28)$$

where b, r, s = constants for a specific packing

Following are two equations of this type presented by Sherwood and Holloway (1940) and Hensel and Treybal (1952).

$$\frac{k_x a}{\mathscr{D}_x} = \alpha \left(\frac{G_x}{\mu_x} \right)^{1-\eta} \left(\frac{\mu_x}{\rho_x \mathscr{D}_x} \right)^{0.5} \qquad (13.29)$$

where

$\mathscr{D}_x$ = diffusivity of solute in liquid, ft^2/hr

α = constant from Table 13.1 (varies with packing)

μ_x = liquid viscosity, lb_m/hr-ft

ρ_x = liquid density lb_m/ft^3

η = constant from Table 13.1 (varies with packing)

and

$$k_y a = bG_y^r G_x^s \left(\frac{\mu_g}{\rho_g \mathscr{D}_g} \right)^{-2/3} \qquad (13.30)$$

where G_x, G_y = mass flux of liquid or gas, $\text{lb}_m/\text{hr-ft}^2$

Values for $b, r,$ and s for the absorption of ammonia in water are given in Table 13.2.

Type of Packing	α	η
⅜-in. Raschig rings	550	0.46
½-in. Raschig rings	280	0.35
1-in. Raschig rings	100	0.22
1½-in. Raschig rings	90	0.22
2-in. Raschig rings	80	0.22
½-in. Berl saddles	150	0.28
1-in. Berl saddles	170	0.28
1½-in. Berl saddles	160	0.28

Table 13.1 Constants to Be Used with Equation (13.29) (random packed unless otherwise noted)

Adapted from Badger and Banchero, 1955.

Type of Packing	b	r	s
½-in. Raschig rings	0.0065	0.90	0.39
1-in. Raschig rings	0.036	0.77	0.20
1½-in. Raschig rings	0.0142	0.72	0.38
1-in. Berl saddles	0.0085	0.75	0.40

Table 13.2 Constants to Be Used with Equation (13.30) for the Absorption of Ammonia in Water

Adapted from Badger and Banchero, 1955.

Correlations for H_x and H_y are usually presented as

$$H_x \text{ or } H_y = f\left(\text{Re}, \frac{G_x}{G_y}, \frac{\mu}{\rho \mathscr{D}}\right) \qquad (13.31)$$

where Re = Reynolds number of the liquid or gas, dimensionless

Some values of H_{oy} are presented in Table 13.3. Values of $K_x a$, $K_y a$, H_{oy}, and H_{ox} for specific applications are available from several references but must be selected with care to ensure that they apply to your specific case (Coulson and Richardson 1968; Perry 1963; Treybal 1968).

ABSORPTION AND CHEMICAL REACTION

From Eq. (13.22) we can see that the mass transfer rate increases as the equilibrium vapor pressure or back pressure decreases. For this reason, the absorbing liquid should have a large capacity for the solute gas. Liquid film resistance can be minimized by circulating fresh liquid through the absorber, but in most cases this approach is not economical.

The equilibrium vapor pressure can be decreased almost to zero by adding a chemical reactant to the absorbing liquid, which in effect ties up or chemically changes the solute gas.

Some examples of the use of chemically enhanced absorption are

1. Absorption of acid gases in alkaline solutions

2. Absorption of odorous gases in oxidizing solutions

3. Absorption of CO_2 and H_2S in amine solutions

Table 13.3   Height of a Transfer Unit H_{oy} Measured in a Pilot-Scale Tower

Raschig Rings Size (in.)	G_y $lb_m/hr\text{-}ft^2$	H_{oy} ($G_x = 500$ $lb_m/hr\text{-}ft^2$)	($G_x = 1500$ $lb_m/hr\text{-}ft^2$)
3/8	200	1.2	0.75
	600	2.0	1.05
1	200	1.3	0.71
	600	2.1	1.1
2	200	2.0	1.1
	600	3.4	1.9

Adapted from Perry, 1963.

The overall effect of the chemical reaction on the mass transfer rate is difficult to predict. The rate and order of the chemical reaction as well as the reversibility of the reaction are important factors. Refer to Sherwood, Pigford, and Wilke (1975) for a detailed discussion of simultaneous absorption and chemical reaction. Chapter 15 discusses an important application of chemically enhanced absorption—limestone scrubbing of flue gases to remove sulfur dioxide.

FLOODING, PRESSURE DROP, AND ALLOWABLE GAS AND LIQUID RATES IN PACKED TOWERS

The third major step in designing a packed tower is to calculate the diameter required to handle the gas flow to be treated. The diameter of a packed tower is determined by selecting a cross-sectional area that will provide gas and liquid mass velocities sufficient for good interfacial contact. At a constant G_x, increasing G_y will increase interfacial contact up to a point at which the flow of gas interferes with the downward flow of liquid. As the gas flow rate increases beyond this point (called the *flooding point*), a condition of flooding occurs in which a liquid layer develops in the tower and liquid accumulates until it is forced out the top of the tower. A tower must be operated below the flooding point but at a sufficiently high velocity to maintain good gas–liquid contact. There is a trade-off between high power costs at high gas velocities and high tower capital costs at low gas velocities. It is customary to operate towers at 40–70% of the flooding gas velocity as determined by the following procedure.

The flooding velocity is estimated using Figure 13.6, which is a logarithmic plot of

$$\frac{G_y^2 F_p \left(\mu_x\right)^{0.1}}{g_c \left(\rho_x - \rho_y\right)\rho_y} \quad \text{versus} \quad \frac{G_x}{G_y} \sqrt{\frac{\rho_y}{\rho_x - \rho_y}}$$

where

F_p = packing factor from Table 13.4, ft^{-1} (see page 420)

ρ_x = liquid density, lb$_m$/ft^3

ρ_y = gas density, lb$_m$/ft^3

μ_x = liquid viscosity, cp

g_c = proportionality constant, 32.17 ft-lb$_m$ /s^2 -lb$_f$

G_x = liquid mass flux, lb$_m$/s-ft^2

G_y = gas mass flux, lb$_m$/s-ft^2

To use Figure 13.6, we calculate the abscissa value from previously determined values of L_m and V_m (converted to mass flows). Note that G_x/G_y can be obtained from L_m/V_m because the cross-sectional area S divides out. In most pollution control applications, ρ_y is calculated

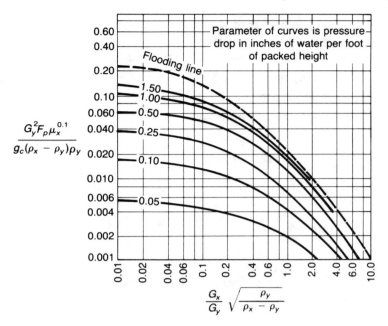

Figure 13.6
Generalized correlation for flooding and pressure drop in packed towers.
(Adapted from McCabe et al., 1985.)

assuming ideal gas behavior and ρ_x is taken as that for water (or the absorbing solution) at the average tower operating temperature. We determine a value of the ordinate at the intersection of the abscissa value with the flooding curve. From the ordinate value we obtain G_y at flooding, and the tower diameter is then based on $G_{y(\text{flooding})}/2$. We then use this new value of G_y to calculate a new ordinate value, which is then used to find the pressure drop per foot of packing. The preceding discussion is illustrated in Example 13.1.

The pressure drop for the tower can be calculated easily using Figure 13.6. First we calculate the actual G_y and G_x for the tower and then we calculate values for the abscissa and the ordinate for use in Figure 13.6. From those values, the intersection on the figure defines the actual operation of the tower, and the pressure drop per foot of packed height can be read directly from the chart. The total tower pressure drop is simply the pressure drop per foot times the height of the packed section in feet.

▶ ▶ ▶ ▶ ▶ ▶ ▶ ▶ ▶

Example 13.1

A 10,000 acfm exhaust from a heat treating process contains 25 mole percent ammonia, and the average exhaust conditions are 115 F and 1.0 atm. Estimate the height of packing, the tower diameter, and the tower pressure drop for an absorption tower that will provide 95% ammonia removal based on the following design parameters:

Average tower temperature is 86 F

Tower pressure is 1.0 atm

Pure water is used as absorbing liquid

Water rate is 1.5 times minimum

Packing is 1.0-in. ceramic Raschig rings

$k_y a$ is $15 \dfrac{\text{lbmol}}{\text{hr-ft}^3\text{-}\Delta y}$

$k_x a$ is $60 \dfrac{\text{lbmol}}{\text{hr-ft}^3\text{-}\Delta x}$

Solution

Since fresh (not recirculated) water is used, $y_b = 0.25$ and $x_t = 0$. Find outlet ammonia concentration y_t based on 1 mole of entering gas; that is,

moles of air entering = 0.75

moles of NH_3 entering = 0.25

moles of NH_3 leaving = $0.05 \times 0.25 = 0.0125$

$$y_t = \frac{0.0125}{0.75 + 0.0125} = 0.0164$$

Next plot the equilibrium curve for ammonia-water system at 1.0 atm and 86 F. Use the ammonia solubility data given in Appendix Table B.4. This table gives the concentration of NH_3 as lb NH_3/100 $lb_m H_2O$ versus the partial pressure of NH_3 in mm Hg. To convert to mole fractions, use the following equations:

$$y_e \text{ or } y_i = \overline{P}_A / 760$$

$$x_e \text{ or } x_i = \frac{C_A / 17}{C_A / 17 + 100/18}$$

where

$\overline{P}_A$ = partial pressure of NH_3, mm Hg

C_A = concentration of NH_3, in the water, g NH_3/100 g H_2O

Choose C_A from 2 to 15 g/100 g and read the corresponding $\overline{P}_A$ from the 30 C column. The following table results:

C_A, $\dfrac{\text{g NH}_3}{100 \text{ g H}_2\text{O}}$	x_e	$\overline{P}_A$, mm Hg	y_e
2	0.0207	19.3	0.0254
3	0.0308	29.6	0.0389
4	0.0406	40.1	0.0528
5	0.0503	51.0	0.0671
7.5	0.0736	79.7	0.105
10	0.0957	110	0.145
15	0.1371	179	0.236

These x–y data are plotted in Figure 13.7 (bottom curve).

Next, find the minimum water rate. From the plot where $y_b = y_e = 0.25$, then $x = x_b^* = 0.14$. Then, on the basis of 1 mole of gas entering $V'_m = 0.75$ mole. Now, substituting the values $x_t = 0$, $y_t = 0.0164$, $y_b = 0.25$, and $x = x_b^* = 0.14$ into Eq. (13.14) we get

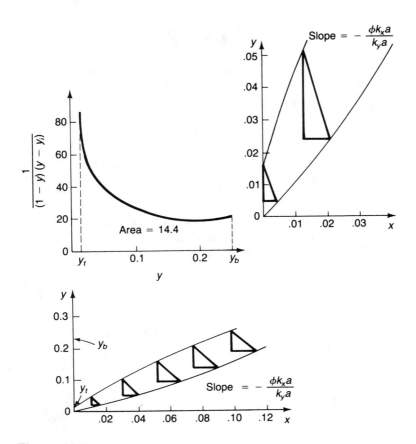

Figure 13.7
Graphs used in solution of Example 13.1.

$$L'_{m\min}\left(\frac{0}{1-0}-\frac{0.140}{1-0.140}\right)=0.75\left(\frac{0.0164}{1-0.0164}-\frac{0.25}{1-0.25}\right)$$

$$L'_{m\min}\,(-0.163)=0.75(0.01667-0.3333)=-0.2375$$

$$L'_{m\min}=1.457\text{ mol H}_2\text{O/mol gas}$$

The actual water rate is $1.5\times1.457=2.186$ mol H_2O/mol gas. The total molar flow rate of gas is

$$10{,}000\frac{\text{ft}^3}{\text{min}}\times\frac{1\text{ lbmol}}{359\text{ scf}}\times\frac{492}{575}=23.8\frac{\text{lbmols}}{\text{min}}\text{ or 1430 lbmol/hr}$$

Now solve for (x,y) points to plot the actual operating line from Eq. (13.14) rearranged as

$$2.186\frac{x}{1-x}=0.75\left(\frac{y}{1-y}-\frac{0.0164}{1-0.0164}\right)$$

$$\frac{x}{1-x}=0.343\left(\frac{y}{1-y}-0.0167\right)$$

y	1 – y	y/(1 – y)	y/(1 – y) – 0.0167	x/(1 – x)	x
0.025	0.975	0.0256	0.0089	0.0031	0.0031
0.05	0.95	0.0526	0.0359	0.0123	0.0122
0.10	0.90	0.111	0.094	0.0322	0.0312
0.15	0.85	0.177	0.160	0.0549	0.0520
0.20	0.80	0.250	0.233	0.0799	0.0740
0.25	0.75	0.333	0.316	0.108	0.0975

Use values of x and y calculated to plot the operating line. Now calculate $-\phi(k_xa/k_ya)$ and draw $\Delta x\,\Delta y$ triangles. Assume $\phi=1.0$, and then $-\phi(k_xa/k_ya)=-4$. To graphically integrate $\phi\,dy/[(1-y)(y-y_i)]$, we prepare the following table, using $\Delta x\,\Delta y$ triangles:

y	y_i	1 – y	$y-y_i$	$\dfrac{1}{(1-y)(y-y_i)}$
0.0164	0.0042	0.984	0.0122	83.30
0.052	0.023	0.950	0.0270	38.99
0.10	0.056	0.900	0.0440	25.25
0.15	0.097	0.850	0.0530	22.19
0.20	0.138	0.800	0.0620	20.16
0.25	0.186	0.750	0.0640	20.83

To evaluate $\int \phi \, dy/[(1-y)(y-y_i)]$, y is plotted as the abscissa and $1/(1-y)(y-y_i)$ as the ordinate. The area under the resulting curve between $y = 0.0164$ and $y = 0.25$ is the value of the integral. See Figure 13.7 (top left). This area under the curve is 14.4. To calculate Z_T from here, we need a value of G_{my} that depends on the tower area.

Next, we find the tower diameter, using Figure 13.6. First we calculate the loading parameter, $(G_x/G_y) \times \sqrt{\rho_y/(\rho_x - \rho_y)}$, as follows:

$$\text{molecular weight of entering gas} = 0.75 \times 29.0 + 0.25 \times 17.0$$
$$= 26.0 \text{ lb}_m/\text{mol}$$

$$\rho_x = 62.15$$

$$\rho_y = \frac{PMW}{RT} = \frac{14.7 \times 26.0}{10.73 \times 546} = 0.0652 \frac{\text{lb}_m}{\text{ft}^3}$$

$$\frac{G_x}{G_y} \sqrt{\frac{\rho_y}{\rho_x - \rho_y}} = \frac{2.186 \times 18}{1.0 \times 26} \sqrt{\frac{0.0652}{62.15 - 0.0652}} = 0.0490$$

From Figure 13.6, at the intersection of the abscissa value (which equals 0.0490) with the flooding curve, we see that

$$\frac{G_y^2 F_p \mu_x^{0.1}}{g_c (\rho_x - \rho_y) \rho_y} = 0.195$$

Thus, at flooding,

$$G_y = \left(\frac{0.195 \times 32.174 \times 0.0652 \times 62.08}{155 \times 0.80^{0.1}} \right)^{1/2} = 0.409 \frac{\text{lb}_m}{\text{ft}^2\text{-sec}}$$

The total gas flow entering the column, then, is

$$1430 \frac{\text{mol}}{\text{hr}} \times \frac{1 \text{ hr}}{3600 \text{ sec}} \times \frac{26.0 \text{ lb}_m}{\text{mol}} = 10.32 \frac{\text{lb}_m}{\text{sec}}$$

If we operate at 50% of flooding, then

$$G_y(\text{actual}) = \frac{0.409}{2} = 0.205 \frac{\text{lb}_m}{\text{sec-ft}^2}$$

Thus, the tower cross-section is

$$10.32 \frac{\text{lb}_m}{\text{sec}} \times \frac{\text{sec-ft}^2}{0.205 \text{ lb}_m} = 50.34 \text{ ft}^2$$

The diameter of the tower is

$$\left(\frac{50.34 \times 4}{\pi} \right)^{1/2} = 8.01 \text{ ft}$$

which can be rounded to 8.0 ft.

Now we calculate $(k_y a / G_{my})_{avg}$, as follows:

$$\left(G_{my}\right)_b = 1430 \frac{\text{lbmol}}{\text{hr}} \times \frac{1}{50.34 \text{ ft}^2} = 28.41 \frac{\text{lbmol}}{\text{hr-ft}^2}$$

$$\left(G_{my}\right)_t = \frac{1430 - (0.95)(0.25)(1430)}{50.34} = 21.66 \frac{\text{lbmol}}{\text{hr-ft}^2}$$

$$\left(G_{my}\right)_{avg} = 25.04$$

$$\left(\frac{k_y a}{G_{my}}\right)_{avg} = 0.599 \text{ ft}^{-1}$$

From Eq. (13.21), we see that

$$Z_T = \frac{14.4}{0.599 \text{ ft}^{-1}} = 24.0 \text{ ft}$$

which is the height of the packed section.

Finally, we calculate the ΔP for the tower. The abscissa remains the same at 0.049; the ordinate is now

$$\frac{(0.205)^2 (155)(0.80)^{0.1}}{32.2(62.2 - 0.065)(0.065)} = 0.049$$

From Figure 13.6, we read $\Delta P / Z$ as 0.45 in. H_2O/ft, and the tower ΔP is

$$0.45 \ \frac{\text{in. } H_2O}{\text{ft}} \times 24.0 \text{ ft} = 10.8 \ \text{ or } \ 11 \text{ in. } H_2O$$

◀ ◀ ◀ ◀ ◀ ◀ ◀ ◀ ◀

▶ ▶ ▶ ▶ ▶ ▶ ▶ ▶ ▶

Example 13.2

Rework Example 13.1, but base the calculation of the packed height on the overall gas-phase mass transfer coefficient.

Solution

First calculate the value of $K_y a$ from Eq. (13.23):

$$\frac{1}{K_y a} = \frac{1}{k_y a} + \frac{m}{k_x a}$$

From Figure 13.8, we see that

$$m \approx \frac{0.12 - 0.048}{0.08 - 0.04} = 1.8$$

$$\frac{1}{K_y a} = \frac{1}{15.0} + \frac{1.8}{60} = 0.0967$$

$$K_y a = 10.34 \frac{\text{lbmol}}{\text{hr-ft}^3\text{-}\Delta y}$$

Refer to Eq. (13.22). The driving force to be used with $K_y a$ is $y - y^*$, and the integral to evaluate graphically is $\phi \, dy/[(1 - y)(y - y^*)]$. Referring to Figure 13.8, we see that the $\Delta x \, \Delta y$ triangles are now replaced with vertical lines representing the $y - y^*$ driving forces at the same intervals between y_t and y_b. A table can be prepared as follows:

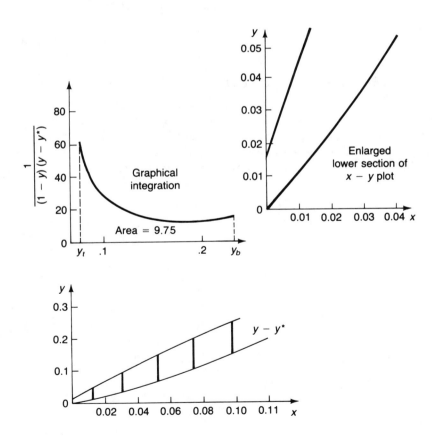

Figure 13.8
Graphs used in solution of Example 13.2.

y	y*	1 – y	y – y*	$\dfrac{1}{(1-y)(y-y^*)}$
0.0164	0.0000	0.984	0.0164	61.97
0.052	0.0150	0.948	0.0370	28.51
0.100	0.0300	0.900	0.0700	15.87
0.150	0.0680	0.850	0.0820	14.34
0.200	0.110	0.800	0.0900	13.89
0.250	0.156	0.750	0.0940	14.18

If the first and last columns are plotted as abscissa and ordinate, the area under the resulting curve from $y = 0.0164$ to $y = 0.250$ will be 9.75. Then

$$\frac{1}{2}\left[\left(\frac{G_{my}}{K_y a}\right)_t + \left(\frac{G_{my}}{K_y a}\right)_b\right] = \frac{1}{2}\left(\frac{21.66}{10.34} + \frac{28.41}{10.34}\right) = 2.42 \text{ ft}$$

$$Z_T = 2.42 \times 9.75 = 23.6 \text{ ft}$$

DESIGN SIMPLIFICATION FOR LEAN GAS APPLICATIONS

In many air pollution control applications, the exhaust gas to be treated by absorption will have a solute concentration below 5 mole percent. Referring to Figure 13.5, we can see that, over narrow concentration ranges, the operating and equilibrium curves can be assumed to be linear. In this case, both y and y^* are linear with x, and

$$\left(\frac{K_y a}{G_{my}}\right) Z_T = \int_{y_t}^{y_b} \frac{dy}{(1-y)(y - y^*)}$$

can be simplified to

$$\left(\frac{K_y a}{G_{my}}\right) Z_T = \frac{y_b - y_t}{\Delta y_{LM}} \tag{13.32}$$

where

$$\Delta y_{LM} = (y - y^*)_{LM} = \frac{\left(y_b - y_b^*\right) - \left(y_t - y_t^*\right)}{\ln\left(\dfrac{y_b - y_b^*}{y_t - y_t^*}\right)}$$

▶ ▶ ▶ ▶ ▶ ▶ ▶ ▶ ▶

Example 13.3

Rework Example 13.1. Keep all conditions the same, except let the inlet gas concentration be 5 mole percent ammonia.

Solution

The NH_3 mole fraction in the inlet air stream is $y_b = 0.05$. Based on 1 mole of gas inlet, the air in the exhaust equals 0.95 moles. Then

$$y_t = \frac{(0.05)(0.05)}{0.95 + (0.05)(0.05)} = 0.00262$$

Assume that the equilibrium curve is linear from $y = 0.0$ to $y = 0.05$ (see Figure 13.9, lower line). From this curve, we see that $x_b^* = 0.0380$. Now find the minimum water rate, L'_{min}; note that $V'_m = 0.95$. Then

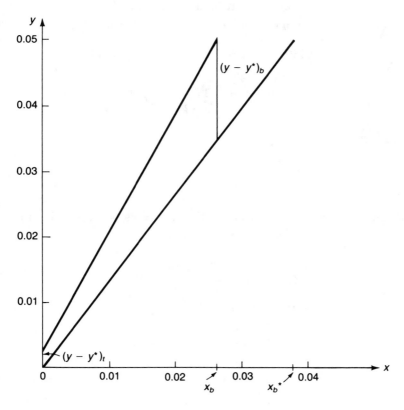

Figure 13.9
Graphical construction for Example 13.3.

$$L'_{m\min}\left(\frac{-0.0380}{1-0.0380}\right) = 0.95\left(\frac{0.00262}{1-0.00262} - \frac{0.05}{1-0.05}\right)$$

$$L'_{m\min}(-0.0406) = 0.95(0.00263 - 0.0526) = -0.0475$$

$$L'_{m\min} = 1.170 \text{ mol } H_2O/\text{mol gas}$$

$$L'_{m\text{act}} = 1.5 \times 1.170 = 1.755 \text{ mol } H_2O/\text{mol gas}$$

This is the operating line (see Figure 13.9, upper line).
We calculate that

$$x_b = \frac{(0.95)(0.05)}{1.755 + (0.95)(0.05)} = 0.0264$$

At this point we should note that G_y at the top of the packed bed will be larger than that calculated in Example 13.1 because there is more air in the feed stream. Thus, the cross-sectional area of the bed should be recalculated. However, to simplify the solution we will arbitrarily increase the bed diameter to 9.0 ft. (63.6 ft^2) and assume the additional area will keep the bed operation at 50% of flooding.

Based on a bed area of 63.6 ft^2, calculate $(G_{my})_t$ and $(G_{my})_b$

$$\left(G_{my}\right)_t = \frac{1430 - (0.95)(0.05)1430}{63.6} = 21.4\frac{\text{lbmol}}{\text{hr-ft}^2}$$

$$\left(G_{my}\right)_b = \frac{1430 \text{ lbmol}}{\text{hr}} \times \frac{1}{63.6 \text{ ft}^2} = 22.5\frac{\text{lbmol}}{\text{hr-ft}^2}$$

Calculate K_ya from

$$\frac{1}{K_ya} = \frac{1}{k_ya} + \frac{m}{k_xa}$$

k_ya and k_xa are given in Example 13.1 and m = slope of equilibrium curve from $y = 0.0$ to $y = 0.05$

$$\frac{1}{K_ya} = \frac{1}{15} + \frac{1.30}{60} = 0.088$$

$$K_ya = 11.3\frac{\text{lbmol}}{\text{hr-ft}^2 - \Delta y}$$

$$\left(G_{my}\right)_{\text{avg}} = \frac{21.4 + 22.5}{2} = 22.0\frac{\text{lbmol}}{\text{hr-ft}^2}$$

$$\frac{G_{my}}{K_ya} = \frac{22.0}{11.3} = 1.95 \text{ ft}$$

$$\frac{\left(y_b - y\,^*_b\right)\left(y_t - y\,^*_t\right)}{\ln\left|\dfrac{y_b - y\,^*_b}{y_t - y\,^*_t}\right|} = \frac{(0.05 - 0.035) - (0.00262 - 0)}{\ln\left[\dfrac{(0.05 - 0.035)}{(0.00262)}\right]}$$

$$= \frac{0.0124}{\ln\left[\dfrac{0.0150}{0.00262}\right]} = 0.0071$$

$$Z_T = \frac{(y_b - y_t)\left(G_{my}\right)}{\Delta y_{\text{LM}}\ K_y a} = \frac{(0.05 - 0.00262)}{0.0071}1.95$$

$$= 13.0 \text{ ft}$$

◀ ◀ ◀ ◀ ◀ ◀ ◀ ◀ ◀

◆ 13.3 ESTIMATING THE COST ◆ OF ABSORPTION TOWERS

Preliminary cost estimates for process vessels including packed towers are prepared by estimating an empty tower cost and adding to this estimate the cost of fittings and packing. The simplest procedure for estimating empty tower cost is to estimate the mass of the empty tower and then multiply this mass by an appropriate cost per unit mass of construction material. Carbon and stainless steels are used in a majority of applications, but in very corrosive service reinforced plastic or ceramic materials may be used.

Tower mass is for the most part a function of geometry and operating pressure. Tower wall thickness may be found using the following equation:

$$t = \frac{Pr_i}{SE_j - 0.6P} + C \qquad\qquad \textbf{(13.33)}$$

where

P = maximum operating pressure, psig

r_i = internal radius of shell, in.

t = minimum wall thickness, in.

S = allowable working stress, psi

E_j = joint efficiency, 0.8 to 1.0

C = corrosion allowance, in. (use 1/8 in. for noncorrosive service)

Table 13.4   Tower Packing Characteristics

Type	Material	Nominal Size, in.	Bulk Density,† lbm/ft³	Total Area,† ft²/ft³	Porosity ε	Packing Factors‡	
						F_p	f_p
Berl saddles	Ceramic	½	54	142	0.62	240	§1.58
		1	45	76	0.68	110	§1.36
		1½	40	46	0.71	65	§1.07
Intalox saddles	Ceramic	½	46	190	0.71	200	2.27
		1	42	78	0.73	92	1.54
		1½	39	59	0.76	52	1.18
		2	38	36	0.76	40	1.0
		3	36	28	0.79	22	0.64
Raschig rings	Ceramic	½	55	112	0.64	580	§1.52
		1	42	58	0.74	155	§1.36
		1½	43	37	0.73	95	1.0
		2	41	28	0.74	65	§0.92
Pall rings	Steel	1	30	63	0.94	48	1.54
		1½	24	39	0.95	28	1.36
		2	22	31	0.96	20	1.09
	Polypropylene	1	5.5	63	0.90	52	1.36
		1½	4.8	39	0.91	40	1.18

† Bulk density and total area are given per unit volume of column.
‡ Factor F_p is a pressure-drop factor and f_p a relative mass-transfer coefficient.
§ Based on NH_3–H_2O data; other factors based on CO_2–NaOH data.

Adapted from McCabe et al., 1985.

To ensure structural integrity, one should not specify wall thicknesses less than 3/16 in. Peters and Timmerhaus (1991) present an equation to determine the thickness of ellipsoidal heads for pressure vessels. In a majority of low-pressure applications, it is sufficiently accurate to increase the shell mass by 10% as an allowance for the heads. A further 10% may be added to account for flanges and internals. Packing cost is based on total packed volume and cost per unit volume of packing. Figure 13.10 and Table 13.5 provide tower and packing cost data.

The costs of absorption towers increased by about 15% from 1990 to 2000. The cost estimate resulting from Figure 13.10 should be increased by 15% to obtain an estimate in 2000-year dollars. These are f.o.b. costs and a factor of about 18% should be added to estimate the delivered equipment cost. If we apply the factors for scrubbers given in Chapter 2 (Table 2.3) to absorption towers, then another 97% should be added to cover both direct and indirect installation costs to get an estimate for the TIC.

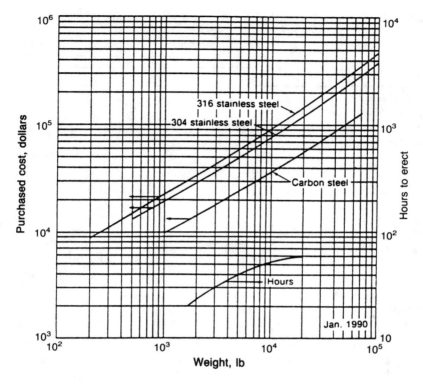

Figure 13.10
Purchased cost of towers in 1990 \$ (cost includes shell, heads, and skirt).
(Peters, M. S., and Timmerhaus, K. D., *Plant Design and Economics for Chemical Engineers*, © 1991 McGraw-Hill, Inc. Used by permission.)

ABSORPTION TOWER OPERATING COSTS

The major operating cost in absorption processes where water is the absorbent is the blower power cost. If the scrubbing water is not recirculated (that is, if fresh water is used), annual water costs will also be significant. Liquid disposal cost will depend on reuse potential, which may be quite good if the solute is a nontoxic acid or base. Acidic or alkaline absorber effluents may often be used for neutralization of other liquid wastes.

▶ ▶ ▶ ▶ ▶ ▶ ▶ ▶ ▶

Example 13.4

Estimate the 2002 installed cost of the absorption tower sized in Example 13.1.

Solution
Since the absorber operates at atmospheric pressure, Eq. (13.33) does not apply (that is, $P = 0$ psig). Thus, we may specify a wall thick-

Table 13.5 Purchased Cost of Tower Packings in January 1990 in $/ft^3
f.o.b. Factory*

	Size, in.			
	1	**1½**	**2**	**3**
Raschig rings:				
Chemical porcelain	12.8	10.3	9.4	7.8
Carbon steel	36.5	23.9	20.5	16.8
Stainless steel	155	117	87.8	—
Carbon	52	46.2	33.9	31.0
Intalox saddles:				
Chemical stoneware	17.6	13.0	11.8	10.7
Chemical porcelain	18.8	14.1	12.9	11.8
Polypropylene	21.2	—	13.1	7.0
Berl saddles:				
Chemical stoneware	27.0	21.0	—	—
Chemical porcelain	33.5	21.5	15.6	—
Pall rings:				
Carbon steel	29.3	19.9	18.2	—
Stainless steel	131	99.0	86.2	—
Polypropylene	21.2	14.4	13.1	

*The costs of packing for absorption towers increased by about 15% from 1990 to 2000. The estimate of cost resulting from Table 13.5 should be increased by 15% to obtain an estimate in 2000-year dollars.

Peters, M. S., and Timmerhaus, K. D., *Plant Design and Economics for Chemical Engineers*, © 1991 McGraw-Hill, Inc. Used by permission.

ness of 3/16 in. plus 1/8 in. corrosion allowance, giving us a total wall thickness of 5/16 in. (0.026 ft). The packed height of the tower is 24.0 ft. To this we must add 5 ft for vapor distribution and sump at the bottom of the tower and 4 ft for a liquid distributor and entrainment separator at the top. Thus, the total height of the tower shell is 33.0 ft.

We may then proceed as follows:

Volume of metal in shell is $\pi D Z_T t = 3.1416 \times 8.0 \times 33.0 \times 0.026$

Vol_m is 21.56 ft^3

Density of carbon steel is 489 lb$_m$/ft^3

Mass of tower is $489 \times 21.56 = 10,540$ lb$_m$

Allowing 10% for heads and 10% for fittings, the total mass is then 12,650 lb.

From Figure 13.10, we see that the cost in 1990 was about $40,000. We then add 18% for freight and 97% of that subtotal for installation, which brings the total installed cost to $93,000.

The volume of packing is

$$\frac{\pi D^2}{4} \times 24.0 = 1206 \text{ ft}^3$$

From Table 13.5, we see that 1-in. ceramic rings cost $12.80/ft^3 in 1990. If we add 8% for freight and taxes, or $1.02/ft^3, then the cost of packing is 13.82×1206 ft^3 = $16,667. Thus, to the total tower cost (installed) in 1990 of $93,000, we add $16,667, and we have $109,667. To get the cost in 2002 we increase the total by 18% (extrapolating the 15% increase during 1990 to 2000 for two more years).

$$\$109,667 \times 1.18 = \$130,000 \text{ (rounded)}$$

◀ ◀ ◀ ◀ ◀ ◀ ◀ ◀

◆ 13.4 STRIPPING OPERATIONS ◆

While we have concentrated on the application of absorption processes to the removal of pollutants from a gas stream, absorption theory and technology is frequently applied in the remediation of polluted surface and ground waters. In this latter application, referred to as *air stripping*, contaminated water is contacted with air in a packed bed or spray chamber and the contaminant is *stripped* from the water by the air stream. The exhaust air from this process may require treatment by absorption or incineration depending on the amount and toxicity of the contaminant removed.

The design of an air stripping column is similar to that for an absorber, and in most cases Eq. (13.32) may be used to find the packed height. Since the mass transfer driving force in a stripping column is in the reverse direction of that in an absorption column, $(y - y^*)$ will be negative. If the equilibrium curve and operating line are plotted for a stripper as in Figure 13.9, the operating line will be below the equilibrium curve. Typical air stripper operating parameters for the removal of halogenated organics are given by Ram, Christman, and Cantor (1990).

◆ PROBLEMS ◆

13.1 A 1-ft diameter packed column is used to scrub a soluble gas ($MW = 22$) from an air-gas mixture. Pure water enters the top of the column at 1000 lb_m/hr. The entering gas stream contains 5% soluble gas and 95% air. Ninety-five percent of the soluble gas is removed. Both the operating line and equilibrium curve may be assumed to be straight. The equation for the equilibrium curve is $y = 1.2x$, where x, y = mole fractions. The entering gas mixture flow rate is 800 lb_m/hr. The column operates at 30 C and 1 atm, and

$$k_y a = 12.0 \text{ lbmol/hr-ft}^3\text{-}\Delta y$$
$$k_x a = 8.0 \text{ lbmol/hr-ft}^3\text{-}\Delta x$$
$$\phi = 1.0$$

Calculate or find:

a. Concentration of the soluble gas in the effluent liquid if the column is operated at minimum liquid flow rate

b. Concentration of soluble gas in the liquid at a point in column where $y = 0.02$

c. $K_y a$

d. Height of packed section, Z_T

e. H_{oy}

f. Whether column is in danger of flooding if it is packed with ½-in. ceramic Raschig rings

13.2 An absorption column is to be designed to recover 99.5% of the NH_3 from a feed stream, which is supplied at 72 F and an NH_3 partial pressure of 10 mm Hg. The feed stream contains 2100 lb_m/hr of air. The column operates at 1.0 atm, is supplied with pure water at 72 F that flows at 1.5 times the minimum water rate. H_{oy} for the operation is 1.9 ft.

a. Find the minimum water rate, in lb_m/hr.

b. Find the height of the packing required if $\phi = 1.0$.

13.3 A gas stream containing a valuable hydrocarbon ($MW = 44$) is scrubbed with a nonvolatile oil ($MW = 300$, specific gravity = 0.90) in a tower packed with 1-inch Raschig rings. The entering gas analyzes 20 mole percent hydrocarbon with the remainder being an inert gas ($MW = 29$). The gas stream enters the column at 5000 lb_m/hr-ft^2, and hydrocarbon-free oil enters the top at 10,000 lb_m/hr-ft^2. The column is 4 ft in diameter. Ninety-five percent of the valuable hydrocarbon is to be recovered.

Use the equilibrium relationship and the equations for $k_y a$ and $k_x a$ given below. Estimate the height of packing required.

x_e	0.1	0.2	0.25	0.3	0.4	0.45	0.5
y_e	0.01	0.027	0.041	0.06	0.122	0.163	0.2

$$k_y a = 0.05 \, G_x^{0.75}$$
$$k_x a = 0.025 G_y^{0.6} G_x^{0.2}$$

13.4 An air-ammonia mixture, containing 1.5 volume percent NH_3, at 95 F and 1.0 atm is scrubbed in a packed tower with pure water at 95 F. The outlet NH_3 concentration must be no more than 0.1% to meet EPA regulations.

a. What volume of this gas in acfm can be processed in a 2-ft diameter column packed with 1-in. ceramic Raschig rings if the gas-to-liquid mass flow ratio is 1.0?

b. What height of packing is necessary if H_{oy} is 1.85 ft?

13.5 A 300 cfm stream of an air-SO_2 mixture containing 16.2% SO_2 by volume at 68 F and 1.0 atm is to be scrubbed with water in a countercurrent packed tower for the purpose of recovering 95% of the SO_2. The tower operates at constant temperature. Calculate the minimum water rate in lb_m/hr.

Equilibrium data

Liquid concentration, lb_m SO_2/100 lb_m H_2O	0.5	1.0	2.0	3.0	5.0
Partial pressure of SO_2, mm Hg	26	59	123	191	336

13.6 Derive Eq. (13.4) from Eq. (13.2). Show all intermediate steps.

13.7 A 2-ft diameter column packed with 1-in. ceramic Raschig rings was operated with air and water and found to flood at an air-flow rate of 356 cfm. Air entered the column at 80 F and 1.0 atm, and the column operated at these same conditions. What was the flow rate of liquid to the column, in gallons per minute?

13.8 A pollution control project at the XYZ Synthetic Fuels plant involves aeration of the liquid effluent from a crude oil washing process. In this process, crude oil is washed in a stirred tank and the oil is decanted off. The remaining wash water is neutralized with NH_4OH and sent to an aeration basin where air is sparged in at the rate of 3 scf/gal wash water. Air samples taken above the aeration basin show an average NH_3 content of 3000 ppm by volume. Engineer Smith thinks it might be economical to enclose the aeration basin and recover the NH_3 by absorption with water. The water from the absorption tower would then be used as the wash water in the washing process. Consider the following operating data:

Crude oil flow rate is 20,000 barrels/day, specific gravity is 0.83

Ambient temperature is 75 F

Tower packing consists of Raschig rings

Recovery of NH_3 is 90%

Wash water rate is 50 gallons/barrel of crude oil

Water rate in absorber is 1.25 times minimum

Power cost is 7¢ per kWh

Fan/motor efficiency is 75%

NH_3 is worth 7.5¢/lb$_m$

H_{oy} is 2.5 ft.

Using Figure 13.9 for equilibrium data, do the following:

a. Estimate the dimensions of the absorber.

b. Estimate the motor horsepower required to drive the fan.

c. Compute the value of the recovered NH_3.

d. Draw a sketch of your system showing flow rates.

e. Explain whether or not Smith has a good idea.

13.9 In the solution to Example 13.3, the diameter of the absorber was arbitrarily assumed to be 9.0 ft. Calculate the actual tower diameter necessary to operate at 50% of flooding.

13.10 You have a stream of air containing methanol at 5% by volume. The stream is flowing at 2000 actual ft^3/min at 1 atm pressure and 110 F. The regulations require 94% removal of the methanol, and you decide to try a water absorption column, using pure water as the absorbing liquid. The tower will be packed with 1-inch ceramic Berl saddles. Equilibrium data for the water-methanol system at 95 F (the temperature at which the tower will operate) are given below.

x, mol fraction	y, mol fraction
0.020	0.021
0.030	0.033
0.040	0.045
0.050	0.056
0.060	0.068

a. *Neatly* sketch a material balance diagram of the tower.

b. Calculate the concentration of methanol in the exit gases (ppm, dry basis).

c. Plot an equilibrium curve, then draw the minimum slope operating line on the equilibrium diagram, and then calculate a reasonable water inflow rate to accomplish the absorption required.

d. Calculate the concentration of methanol in the water being discharged from the tower (in mole fraction).

e. Estimate the diameter of the tower, based on bottom operating conditions. Assume that the density and viscosity of the bottom liquid are the same as those for pure water. Make other appropriate assumptions as needed.

REFERENCES

Badger, W. L., and Banchero, J. T. *Introduction to Chemical Engineering.* New York: McGraw-Hill, 1955.

Coulson, J. M., and Richardson, J. F. *Chemical Engineering*, Vol. II. New York: Pergamon Press, 1968.

Hensel, S. L., and Treybal, R. E. *Chemical Engineering Progress, 48,* 1952, pp. 362–370.

McCabe, W. L., and Smith, J. C. *Unit Operations of Chemical Engineering.* New York: McGraw-Hill, 1976.

McCabe, W. L., Smith, J. C., and Harriott, D. *Unit Operations of Chemical Engineering* (4th ed.). New York: McGraw-Hill, 1985.

Perry, J. H., Ed. *Chemical Engineers' Handbook* (3rd ed.). New York: McGraw-Hill, 1963.

Peters, M. S., and Timmerhaus, K. D. *Plant Design and Economics for Chemical Engineers.* McGraw-Hill, 1991.

Ram, N. M., Christman, R. F., and Cantor, K. P. *Significance and Treatment of Volatile Organic Compounds in Water.* Chelsea, MI: Lewis Publishers, 1990.

Sherwood, T. K., and Holloway, F. A. *Transactions of the American Institute of Chemical Engineers, 36,* 1940, p. 21.

Sherwood, T. K., Pigford, R. L., and Wilke, C. R. *Mass Transfer.* New York: McGraw-Hill, 1975.

Treybal, R. E. *Mass Transfer Operations.* New York: McGraw-Hill, 1968.

BIOLOGICAL CONTROL OF VOCS AND ODORS

Biofiltration . . . converts pollutants to harmless products, requires no fuel, utilizes inexpensive off-the-shelf components, generates no hazardous by-products, and is ultimately a low-cost alternative.... Biofiltration is simple and inexpensive when done right, but the knowledge to "do it right" is at the frontiers of science and engineering.

Joseph Devinny, Marc Deshusses, Todd Webster, 1999

◆ 14.1 INTRODUCTION ◆

Control of volatile organic compounds and odors by biological systems, a practice that emerged in Europe in the early 1980s, gained widespread acceptance in the United States during the 1990s. The fundamental principal of biological air pollution control is that gaseous pollutants are utilized by microbes as a food or energy source, and are destroyed in the process, being converted into innocuous metabolic end products (CO_2 and H_2O). The process requires careful attention to design and operation to ensure good contact between the contaminated air and the liquid phase or the biofilm containing the microbes, and to ensure that the microbial population is sustained and maintained in a healthy state. However, for the right applications, and with good attention to detail, these biological control systems can function well for years with high removal efficiencies and low operating costs.

The basic process involves contacting the contaminated airstream with a circulating liquid or a wetted bed that contains a healthy population of biomass. There are three main types of applications, as shown in Table 14.1; the differences are in how the water and biomass move through the system. Whether the water phase is flowing (as in a scrubber) or is stationary (as in a biofilter), the microbial population

429

Table 14.1 Types of Biological Air Pollution Control Systems

Application	Biomass	Water Phase
Biofilter	Fixed	Fixed
Biotrickling Filter	Fixed	Flowing
Bioscrubber	Flowing (suspended)	Flowing

lives in the water phase. The most popular of these applications seems to be biofiltration, and the rest of this chapter will be devoted to biofilters. A schematic diagram showing the cross-section of a biofilter is presented in Figure 14.1.

The most successful uses of biofilters are for highly soluble organic compounds with low molecular weight, such as alcohols, aldehydes, and ketones, and for a variety of odorous gases, including H_2S and NH_3. However, other types of compounds (e.g., styrene, and gasoline components such as benzene and toluene) have been successfully treated. Deshusses and Johnson (2000) found that maximum removal performance in a biofilter tended to follow the sequence: alcohols (best) → esters → ketones → aromatics → alkanes (worst). This sequence also tends to be in order of decreasing solubility in water (increasing Henry's coefficients).

Key concerns in the design and operation of biofilters include: (1) identifying the concentration and type of contaminants in the airstream, (2) finding the correct microbial population, (3) selecting a compatible medium, (4) maintaining adequate moisture, (5) sizing the bed to provide adequate residence time for the given airflow rate, and (6) controlling pH, nutrient levels, and temperature in the bed. This technology is best suited for high volumetric flow rate airstreams with low concentrations of pollutants. The capital costs are very reasonable, and the operating costs tend to be quite low. Figure 14.2 compares various technologies for VOC control, indicating the general ranges of operation for which they are best suited.

Biofilters have several advantages and disadvantages compared with other VOC control devices.

Advantages:

1. Effective removal of compounds
2. Little or no by-product pollutants produced (including CO and NO_x)
3. Uncomplicated installations
4. Low costs

Disadvantages:

1. Less suited to high concentration streams
2. Large area needed for installation
3. Careful attention to moisture control required
4. May become clogged by particulate matter and/or biomass growth

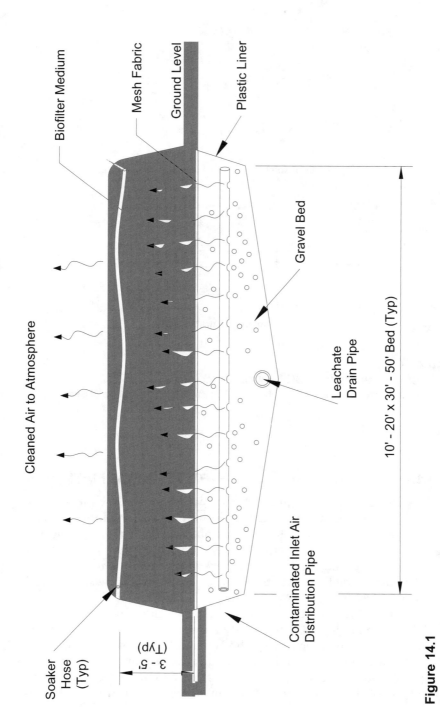

Figure 14.1
Schematic diagram showing the cross-section of a typical open-bed biofilter.

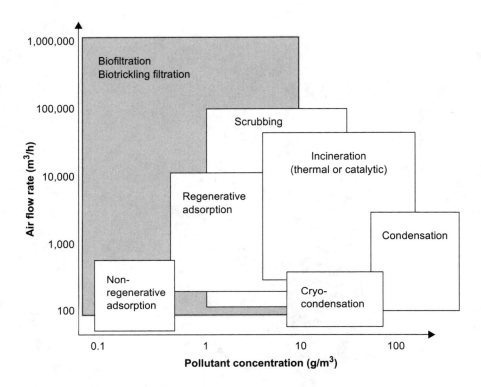

Figure 14.2
Ranges of suitability for VOC control technologies.
(As updated by Deshusses and Cox, 2002, based on Kosteltz, Finkelstein, and Sears, 1996.)

◆ 14.2 Theory and Descriptive Information ◆

The theoretical underpinnings of biofilters involve the mechanisms of absorption and mass transport of a gaseous pollutant from a carrier gas (air) into and through a liquid (water), and the mechanisms of absorption and metabolic degradation of that pollutant by bacteria. An excellent discussion of these mechanisms is provided in a book devoted to biofiltration by Devinny et al. (1999), while a brief summary is presented here.

First, the pollutant gas must be absorbed into the liquid film in which the biomass grows. The extent of absorption depends on the pollutant solubility (as indicated by the Henry's law constant), while the rate of mass transfer depends on the physical parameters of the system (surface area, turbulence, concentration driving force, and so on). In most cases, the mass transfer rate in a biofilter is controlled by dif-

fusion through the water film because the water is essentially stationary, and diffusion through a water film is much slower than through
an air film.

For purposes of this chapter, it is convenient to re-write Henry's
law as follows:

$$C_{i,G} = H_{i,D} \times C_{i,L} \qquad \text{(14.1)}$$

where

$C_{i,G}$ = concentration of pollutant i in air, g/m^3

$C_{i,L}$ = concentration of pollutant i in water, g/m^3

$H_{i,D}$ = Henry's constant for species i, dimensionless, although the
units really are (g/m^3 in air)/(g/m^3 in water)

▶ ▶ ▶ ▶ ▶ ▶ ▶ ▶ ▶

Example 14.1

(a) Convert the Henry's constant for H_2S at 1 atm and 30 C (see
Appendix B) from the units given to units convenient for biofilter calculations using Eq. (14.1).

(b) Use this value to calculate the equilibrium concentration of
H_2S that would exist in water if the concentration in air is maintained
at 150 ppm at 1 atm and 30 C.

Solution

(a) From Table B.3, the appropriate Henry's constant is 0.0609 ×
10^4 atm/mole fraction. This large number is a result of the units used
in Table B.3; the concentration in water will be very much less than
1.0 in units of mole fraction, even for fairly soluble gases. It is difficult
to work with this number (609 atm/mole fraction) because our brain
tells us that the partial pressure of H_2S in air *must* be less than the
total pressure (which is 1 atm). So it is suggested that the conversion
proceed in steps as follows:

First, rearrange Henry's Law as it appears in Table B.3:

$$H_i = \frac{\overline{P}_i}{x_i}$$

Divide both sides by total pressure to convert partial pressure to
mole fraction:

$$\frac{H_i}{P_T} = \frac{y_i}{x_i}$$

Multiply by molecular weight of i in the numerator and denominator
to convert mole fractions (essentially moles i per mole air or water) of
component i to mass of i:

$$\frac{H_i}{P_T} = \frac{\text{mass } i\big/\text{mole air}}{\text{mass } i\big/\text{mole water}}$$

Convert mole air to volume of air using the ideal gas law, and convert mole water to volume of water (1 mole liquid water occupies 0.0180 L).

$$\frac{H_i}{P_T} = \frac{\text{mass } i\big/\text{L air} \times 0.08206(T)\big/P_T}{\text{mass } i\big/\text{L water} \times 0.0180}$$

Convert L to m^3 in top and bottom, and recognize the definitions of $C_{i,G}$ and $C_{i,L}$.

$$\frac{H_i}{P_T} = \frac{1000\, C_{i,G} \times 0.08206(T)\big/P_T}{1000\, C_{i,L} \times 0.0180}$$

Cancel P_T on both sides and combine terms

$$H_i = \frac{C_{i,G}}{C_{i,L}} \times 4.559T \quad \text{(Keep in mind that } T \text{ is in deg K)}$$

Note that $C_{i,G}/C_{i,L}$ is just $H_{i,D}$. Rearrange to solve for $H_{i,D}$.

$$H_{i,D} = \frac{H_i}{4.559T} = \frac{0.2194 H_i}{T}$$

This is the general equation. Now, substitute for this particular problem to obtain a numerical value of $H_{i,D}$

$$H_{i,D} = \frac{0.2194 \times 0.0609(10)^4}{(273 + 30)} = 0.441$$

(b) To solve part (b), first convert the air concentration of 150 ppm H$_2$S to mass/volume units using Eq. (1.9) adjusted to 30 C:

$$C_{i,G} = \frac{1000 \times 150 \times 34}{24.86}$$
$$= 2.051 \times 10^5 \,\mu\text{g/m}^3 \text{ or } 0.2051 \text{ g/m}^3$$

Rearranging Eq. (14.1) and solving,

$$C_{i,L} = 0.2051 / 0.441$$
$$= 0.465 \text{ g/m}^3 \text{ or } 0.465 \text{ mg/L in the water}$$

As the contaminant is being absorbed into the liquid, it also diffuses toward the surface of the biofilm or the medium on which it is growing. Operators try to produce little or no liquid drainage from the biofilter (leachate), so the liquid phase is nearly stationary. Thus, diffusion through the liquid film is the rate-controlling mass transfer mechanism. Once the contaminant reaches the surface of the medium or the biofilm, it must be adsorbed. Two models for adsorption, Freundlich and Langmuir (see also Chapter 12), commonly are used to describe that equilibrium. The basic equations are presented below.

$$\text{Freundlich:} \qquad C_{\text{ads}} = k\, C_L{}^{1/n} \qquad (14.2)$$

where

C_{ads} = concentration of pollutant on the solid phase, mg/g

C_L = concentration of pollutant in the liquid phase, mg/m^3

k, n = empirical constants

$$\text{Langmuir:} \qquad C_{\text{ads}} = \frac{C_{\text{max}}\, C_L}{K_L + C_L} \qquad (14.3)$$

where

C_{max} = maximum concentration that can be attained on the solid, mg/g

K_L = empirical constant

After being adsorbed on the biofilm, the pollutant must then be absorbed into the cells of the bacteria, and metabolized. Finally, the end products (CO_2 mainly) must be expelled from the cells, and must diffuse outward through the liquid film. The metabolic reactions usually are described with Monod (variable-order) kinetics, which can approach first-order or zero-order kinetics (at low or high concentrations, respectively), as can be seen by inspection of the following equation.

$$r_i = \frac{k_i C_{L,i}}{C_{L,i} + K_i} \qquad (14.4)$$

where

r_i = rate at which pollutant i is consumed, g/min-m^3 (in the liquid phase)

k_i = maximum degradation rate of substance i, g/min-m^3

K_i = Monod constant for substance i, g/m^3

A mass balance can be written about a small volume within a biofilter that includes the gas, liquid, and solid phases. Assuming the gas transport processes are fast, and assuming that the biodegradation reactions are much slower than the adsorption rate, the steady-state mass balance results in a differential equation that balances the rate of diffusion of pollutant across the liquid film against the reaction consuming the pollutant.

$$D_i \frac{d^2 C_{L,i}}{dx^2} - r_i = 0 \qquad (14.5)$$

where D_i = diffusion coefficient of pollutant i in water, m^2/min

In order to solve Eq. (14.5), boundary conditions are needed. One boundary condition is provided by Henry's Law at the gas-liquid interface, and the other is that the slope of the concentration profile in the liquid approaches zero at the liquid-solid interface. With those boundary conditions, and assuming first-order kinetics for the term r_i, Heinsohn and Kabel (1999) present a solution that allows calculation of the theoretical removal efficiency of a biofilter. Their equations are presented below, but first recall that the removal efficiency of a biofilter is related to inlet and outlet concentrations in the gas:

$$\eta = (C_i - C_e)/C_i \qquad (14.6)$$

where

η = removal efficiency, fraction

C_i and C_e = inlet and outlet concentrations in air, respectively, g/m^3

According to Heinsohn and Kabel, the removal efficiency is a function of several biofilter parameters, that is:

$$\eta = 1 - \exp\left[-ZK/(H_i U_G)\right] \qquad (14.7)$$

where

Z = height of packed bed (medium), m

K = reaction unit (defined below), min^{-1}

H_i = Henry's constant (from Eq 14.1)

U_G = superficial gas velocity (face velocity), m/min

The reaction unit K is given by:

$$K = (a/\delta) D_i \, \phi \tanh \phi \qquad (14.8)$$

where

a = the ratio of packing surface area to bed volume, m^{-1}

δ = thickness of liquid film, m

ϕ = Thiele number for first-order kinetics, defined below

$$\phi = \delta (k_1/D_i)^{0.5} \qquad (14.9)$$

where k_1 = pseudo first-order rate constant, min^{-1}

Note that k_1 is obtained from combining the other parameters in Eq. (14.4) when $C_{L,i}$ is low, or can be obtained by experiments at low concentrations.

Use of the theoretical approach to design biofilters is limited because of the many assumptions needed to produce Eqs. (14.7) and (14.8). However, if it can be assumed that H_i and K do not change (i.e., $K/H_i = K_o$), then Eq. (14.7) becomes:

$$\eta = 1 - \exp\left(-K_o\ Z/U_G\right) \tag{14.10}$$

If some pilot data are available (that is, if η, Z, and U are known at one or more points), examination of Eq. (14.10) for a fixed value of K_o and for various values of Z and U may guide us in designing a biofilter that achieves a good removal efficiency.

Example 14.2

A pilot-plant biofilter to control styrene was built and operated to obtain data as follows: at superficial gas velocities of 0.5, 1.0, and 1.5 m/min, removal efficiencies of 63%, 39%, and 28%, respectively, were observed. The bed depth was 0.5 m. Develop a set of graphs that show how different efficiencies might be obtained by varying bed depth and gas velocity in the full-scale biofilter.

Solution

Solving Eq. (14.10) for K_o we get

$$K_o = \frac{\ln\left(1 - \eta\right)}{-Z/U_G}$$

Plugging in the measured values for each of the three data points and averaging yields a value for K_o of 0.99 min^{-1}. Assuming this value will remain constant, a spreadsheet was developed to calculate the efficiency at three different bed depths and numerous gas velocities. The results were plotted and appear in Figure 14.3.

Some authors believe that the rate of diffusion of the pollutant into the liquid film limits the overall rate, while others believe that the reaction is rate limiting. Still others suggest that the rate-limiting step can be the diffusion and desorption of end products out of the liquid film. In reality, since both gas and liquid concentrations vary throughout the bed, any or all of these conditions might limit the overall rate at different points within the biofilter.

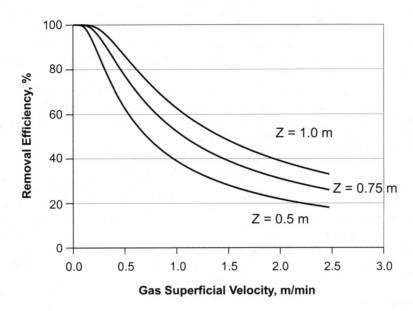

Figure 14.3
Solution to Example 14.2.

In order to design a biofilter, or to analyze its performance, the reader must be familiar with several terms that are in common use with regard to biofiltration, as defined by the following equations and discussion.

$$EBRT = V_f/Q \qquad (14.11)$$

where

$EBRT$ = empty bed residence time, min

V_f = volume of empty filter bed, m^3

Q = volumetric airflow rate, m^3/min

$$\tau = EBRT \times \theta \qquad (14.12)$$

where

τ = true residence time, min

θ = bed porosity (= volume of voids/volume of empty bed)

$$SL = Q/A \qquad (14.13)$$

where

SL = surface loading, m/min (also equal to U_G)

A = surface area of empty bed normal to the airflow rate, m^2

$$VL = Q/V_f \qquad (14.14)$$

where VL = volume loading, min^{-1} (note: $VL = 1/EBRT$)

In addition to surface loading (SL) and volume loading (VL), the terms *mass surface loading* and *mass volume loading* are also used, where these mass loadings are just SL and VL, respectively, multiplied by the concentration of the pollutant in the inlet gas stream, C_i (in units of g/m^3). That is,

$$ML_S = SL \times C_i \text{ and } ML_V = VL \times C_i \qquad (14.15)$$

Note that most authors report data with a time unit of hours instead of minutes, so that the more commonly seen units of ML_S and ML_V are g/m^2-hr, and g/m^3-hr, respectively.

The *elimination capacity* of a biofilter is

$$EC = \frac{Q \times (C_i - C_e)}{V_f} \qquad (14.16)$$

where EC = elimination capacity, g/m^3-hr

Note that EC also is calculated from $VL \times (C_i - C_e)$ or $(C_i - C_e)/EBRT$. A simple manipulation results in yet another way to calculate elimination capacity:

$$EC = ML_V \times \eta \qquad (14.17)$$

Elimination capacity is often a better performance measure than removal efficiency because it normalizes the bed volume and flow rates. However, it does vary with the input mass volume loading as shown in Example 14.3. The elimination capacity tracks linearly with the mass volume loading at low loading rates. That is, the efficiency stays constant as the mass volume loading increases, up to a point. The point at which the elimination capacity begins to level out (where the efficiency begins to drop off) as ML_V continues to increase is called the *critical load*.

Normally, it takes two to three months of laboratory or pilot-scale work to develop biofilter data for one pollutant species. Deshusses and Johnson (2000) developed a protocol for rapid (less than 48 hours) determination of elimination capacity and critical load, and applied their technique to various VOCs to establish data for these parameters. Table 14.2 presents their data, and gives the range of ECs observed by others.

▶ ▶ ▶ ▶ ▶ ▶ ▶ ▶ ▶

Example 14.3

A mixture of BTEX (benzene, toluene, and xylene) vapors in air is to be controlled with a compost biofilter. A pilot plant with bed dimensions of 4 m × 1 m × 0.75 m (deep) was operated with results as shown

Table 14.2 Maximum Elimination Capacity and Critical Load for Selected Compounds

| Compound | From Deshusses & Johnson[a] | | From Others[b] |
	EC_{max} g/m³-hr	Cr. Load g/m³-hr	EC_{max} g/m³-hr
Methanol	135–150	32–34	30–65, 100–120
Ethanol	148–150	78–80	20–40, 18–40, 90–130
Butanol	140	80–85	24–26, 70–76
Acetone	65–70	21–23	40–45, 100–150
MEK	30–35	20–22	22–43, 120
MIBK	40–50	13–15	25–30
Ethyl acetate	140–240	175–180	79–96, 170–200, 150–250, 280–350
Butyl acetate	32–34	28–32	40
Benzene	7–8	1	2–5, 23, 31–47
Toluene	8–20	6–8	5–18, 10–40, 20–25, 23–32, 45–55
Hexane	3–8	1	1.5, 2.1, 2.5
Isopentane	7–8	1–2	18–28, 2–3.5 (n-pentane)

[a] From Deshusses and Johnson experimental results (2000).
[b] Data from other authors as reported by Deshusses and Johnson (2000); different studies reported different ranges.

Table and data adapted from Deshusses and Johnson, 2000.

below. Other pilot-plant data include: gas flow = 5 m³/min, T = 30 C, P = 1 atm, ΔP = 4 cm of water. The average molecular weight of the BTEX mixture is 92.

Inlet BTEX conc., ppm	Removal Efficiency, %
25	95
37.5	95
50	95
75	90
100	75
125	65
150	53
175	47
200	40

Prepare a graph showing the elimination capacity as a function of the mass loading rate. What are the maximum efficiency and the maximum elimination capacity of the system? What is the critical load?

Solution

First, calculate the *EBRT*, surface loading, and volume loading.

$EBRT$ = (4 m × 1 m × 0.75 m)/5 m³/min = 0.60 min

SL = 5 m³/min/(4 m × 1 m) = 1.25 m/min

$VL = 1/EBRT = 1.667$ min⁻¹

Next, for each data point, calculate the inlet concentration (in g/m³) using the ideal gas law, and the mass volume loading and elimination capacity, using Eqs. (14.15) and (14.17).

Conc., ppm	Conc., g/m³	Effic., %	ML_V, g/m³-hr	EC, g/m³-hr
25	0.0925	95%	9.25	8.79
37.5	0.1388	95%	13.88	13.18
50	0.1850	95%	18.50	17.58
75	0.2776	90%	27.76	24.98
100	0.3701	75%	37.01	27.76
125	0.4626	65%	46.26	30.07
150	0.5551	53%	55.51	29.42
175	0.6476	47%	64.76	30.44
200	0.7401	40%	74.01	29.61

By inspection of the table, the maximum efficiency is 95%.

Plot the last two columns of the above table to create a graph as shown in Figure 14.4. From inspection of the graph, the maximum elimination capacity is 30 g/m³-hr. The critical load appears to be about 20 g/m³-hr.

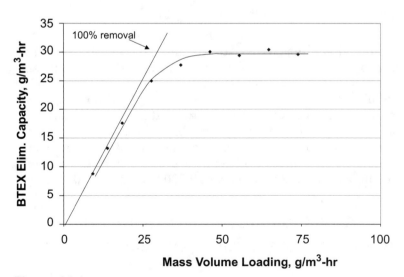

Figure 14.4
BTEX elimination capacity vs. mass volume loading from Example 14.3.

◀ ◀ ◀ ◀ ◀ ◀ ◀ ◀ ◀

The contaminant moves through the biofilter at a much slower effective velocity than the air, owing to its spending much of the time adsorbed onto the solid phase or absorbed in the liquid phase. If, for the moment, we ignore the (slow) biological reactions degrading the contaminant, a pseudo-steady-state mass balance can be written as follows:

$$U_{avg} A M_{tot} = U_G A M_G + U_L A M_L \qquad (14.18)$$

where

U = superficial velocity of contaminant: through the biofilter (avg), in the gas phase (G), or in the liquid phases (L), m/min

A = cross-sectional area of biofilter bed, m^2

M_{tot} = total mass of contaminant / total volume of biofilter, g/m^3

M_G, M_L = mass concentration of contaminant in gas (G) or liquid (L) phase, respectively, g/m^3

Equation (14.18) can be solved for the average velocity, noting that movement through the biofilter in the liquid phase is zero.

$$U_{avg} = (U_G M_G + 0)/M_{tot} \qquad (14.19)$$

Substituting $M_G + M_L$ for M_{tot}, Eq. (14.19) becomes

$$U_{avg} = \frac{U_G M_G}{M_G + M_L} = U_G/R \qquad (14.20)$$

where R = retardation ratio = $(M_G + M_L)/M_G$

The retardation ratio is equal to the equilibrium partition coefficient (a gas-liquid equilibrium constant similar to Henry's constant) plus 1, but can be measured more easily than the partition coefficient in an active biofilter. R is measured dynamically by injecting a spike of the contaminant in the inlet gas to an operating biofilter, and measuring the time for the peak to emerge in the exit gas (as in a gas chromatograph). The size of R can tell us something about how greatly the contaminant is partitioned in the biofilter; larger values allow more time in contact with biomass and tend to produce greater removal. Values of R can vary from 2 or 3 to over 10,000 (Devinny et al. 1999).

◆ **14.3 KEY CONSIDERATIONS IN THE** ◆
DESIGN AND OPERATION OF BIOFILTERS

DESIGN

Choosing the proper medium, sizing the filter properly, and including a good means for humidifying the air will provide for a stable biomass ecology, a good removal efficiency, and a reasonable pressure drop. The *medium* is the solid porous material that is used to support the biomass; it retains a thin film of water on both its external and internal surfaces. Natural media include peat, compost, bark, and wood chips; synthetic media that have been used include activated carbon and polyurethane foams. A good medium fulfills several or all of the characteristics discussed in the following paragraphs.

The medium itself should have a high content of inorganic nutrients. The nutrients nitrogen, phosphorus, and potassium are essential to biomass growth. Although these can be added continuously to an operating biofilter, having a reservoir of them in the medium is a big advantage, and protects against upsets in the nutrient feed system. If the gas feed to the biofilter is subject to shut down for hours or days, the medium should provide a source of organic carbon to sustain the biomass. The medium should be lightweight and bulky, yet have good structural properties (it should not collapse or compact under its own weight). The medium should provide much surface area for biomass growth, but with plenty of void spaces to allow air to flow through the biofilter without much pressure drop. The medium should be able to retain moisture on its surface; the ideal amount is such that 40 to 80% of the total weight of wetted media should be water. It should be suitable for bacterial attachment, and have a high sorptive capacity.

The pressure drop in the biofilter is an important economic consideration when large volumes of air are being treated. Pressure drop through a porous bed was discussed earlier in Chapters 6 and 12. All that needs to be added here is that total pressure drop in a biofilter will increase linearly as the bed height and as the square of the superficial gas velocity. Bulking agents are added to the medium to increase the porosity of the bed; bulking agents include wood chips, bark, polystyrene beads, and even shredded tires.

Although there are a number of case studies in the literature to guide one in selecting a medium, a pilot study is always a good idea to ensure that the medium is suitable for the specific situation being considered. In the pilot stage, several media can be tested and the best one selected. The pilot study also allows for testing different bed depths and gas velocities. Problems with plugging of the bed may not show up until after six months or more of operation, so it is important to run the pilot study for a long duration (Deshusses 2001). For a given gas volumetric flow rate to be treated, a larger gas velocity results in a smaller biofilter, and lower capital costs. Of course, the operating costs (including those due to pressure drop) must also be considered.

Biofilters are often installed in multiple parallel units to allow for continued operation of the plant while one biofilter is down for maintenance. Further redundancy is obtained by designing below the critical load. Thus, when one biofilter is removed from service, the others can continue to function at reasonably high efficiency. Also, the possibility of plant expansion must be taken into account when making the final decision on sizing the biofilters.

Example 14.4

You are making a preliminary design for a full-scale biofilter to treat 20,000 cfm of air (at 30 C and 1 atm) contaminated with 80 ppm of BTEX vapors. As one of the first steps in the process, calculate the volume of the bed and the *EBRT* needed for this biofilter. Use the pilot-plant data given in Example 14.3. To meet a local regulatory limit, the exhaust rate of BTEX compounds cannot exceed 40 pounds per day. The process operates 24 hr/day.

Solution

First convert the inlet flow and concentration to a mass flow of BTEX using the ideal gas law.

$$\dot{M} = \frac{PQMW}{RT}$$

$$= \frac{1 \text{ atm} \times \left(80 \times 10^{-6} \times 20{,}000 \text{ ft}^3/\text{min}\right) \times 92 \text{ lb/lbmol}}{0.73 \dfrac{\text{atm-ft}^3}{\text{lbmol-}^\circ\text{R}} \times 545 \text{ }^\circ\text{R}}$$

$$\dot{M} = 0.370 \text{ lb/min or } 532 \text{ lb/day}$$

The required efficiency is

$$\frac{532 - 40}{532} = 0.925 \text{ or } 92.5\%$$

Note that if we design for the maximum elimination capacity of 30 g/m³-hr, the best efficiency will be only 65%; this will not meet the required efficiency. Based on the data from the previous example problem, and to be conservative, we decide to design for a mass volume loading of about 18.5 g/m³-hr (the highest loading that produced an efficiency of 95% in the pilot plant; the corresponding elimination capacity is 17.6 g/m³-hr). Thus, we can solve for V_f from the following relationship:

$$ML_V = C_i \times Q/V_f = C_i/EBRT$$

We must be sure to convert units appropriately. Thus,

$$ML_V = 18.5 \text{ g/m}^3\text{-hr} \times \frac{1 \text{ m}^3}{35.32 \text{ ft}^3} \times \frac{1 \text{ lb}}{454 \text{ g}}$$

$$= 1.16 \times 10^{-3} \text{ lb/ft}^3\text{-hr}$$

The inlet concentration is most easily obtained from numbers already calculated:

$$C_i = \frac{532 \text{ lb/day}}{20,000 \text{ ft}^3/\text{min} \times 1440 \text{ min/day}}$$

$$C_i = 1.85 \times 10^{-5} \text{ lb/ft}^3$$

So, $V_f = \dfrac{C_i \times Q}{ML_V}$

$$V_f = \frac{1.85 \times 10^{-5} \text{ lb/ft}^3 \times 20,000 \text{ ft}^3/\text{min} \times 60 \text{ min/hr}}{\left(1.16 \times 10^{-3} \text{ lb/ft}^3\text{-hr}\right)}$$

$$= 19,140 \text{ ft}^3$$

To ensure consistency with the pilot results, we should design the full-scale bed with similar values of *SL* and *EBRT* as were used in the pilot tests. The pilot results were:

$SL = 1.25$ m/min = 4.1 ft/min and *EBRT* = 0.6 min

In the current design, the value of *EBRT* is

$EBRT = 19,140 \text{ ft}^3/20,000 \text{ cfm} = 0.96 \text{ min}$

which is sufficient. In other cases, we could easily adjust the *EBRT*, if needed, by changing the bed depth at fixed bed length and width. We can use the same *SL* as in the pilot plant (*SL* = 4.1 ft/min) to solve for the bed face area.

$$A = Q/SL$$

$$= \frac{20,000 \text{ ft}^3/\text{min}}{4.1 \text{ ft/min}}$$

$$A = 4878 \text{ ft}^2$$

If we specify a bed (or parallel beds) that has (have) a total length of 100 feet and width of 50 feet, then the face area will be 5000 ft². Further, if we specify a 4-foot depth, then the final *EBRT* can be calculated from

$EBRT = V_f/Q = (5000 \times 4) \text{ ft}^3 /20,000 \text{ ft}^3/\text{min} = 1.0 \text{ min}$

The final ML_V and *EC* can also be calculated based on the final dimensions chosen:

$$ML_V = C_i/EBRT \times 60$$
$$= (1.85 \times 10^{-5} \text{ lb/ft}^3/1.0 \text{ min}) \times 60 \text{ min/hr}$$
$$= 1.11 \times 10^{-3} \text{ lb/ft}^3\text{-hr}$$
$$EC = 0.95 \times ML_V = 1.05 \times 10^{-3} \text{ lb/ft}^3\text{-hr}$$

◀ ◀ ◀ ◀ ◀ ◀ ◀ ◀ ◀

One key consideration in the design of any biofilter is whether to design the unit with the air flowing upward or downward through the biofilter (see Figure 14.5). Both approaches have been used, although for an open top biofilter, only upflow is possible. According to Devinny et al. (1999), downflow has proven superior, but there are advantages and disadvantages to each (see Table 14.3).

OPERATING CONSIDERATIONS

Although many things must be considered to ensure the safe and efficient operation of biofilters, three parameters stand out as being critical—moisture content (both the bed and the air), temperature, and pH. The oxidation of organic compounds generates heat, and that amount of heat can significantly affect both the temperature and the humidity of the air as it moves through the biofilter. For example, the complete oxidation of 1 gram of ethanol in 1 m^3 of 20 C air theoretically generates enough heat to raise the temperature of that air by 18 degrees (to 38 C), which simultaneously reduces the relative humidity of the air, resulting in evaporation of liquid water from the bed.

Within the filter bed, it is crucial to maintain a liquid water film on all surfaces of the medium. This thin film of water not only provides for mass transfer of the contaminant from the air, but also supports the life of the biomass. If the air within the bed falls below about 98% relative humidity, water begins evaporating from the thin films,

Table 14.3   Comparison of Upflow and Downflow Biofilters

Upflow

Advantages
1. Salts, acids, and particulates can be easily washed out of the system, since these substances tend to accumulate near the bed entrance.
2. Upflow design allows for open top biofilters (generally, lower cost).

Disadvantages
1. Area of surface irrigation (top of bed) and area of highest rate of biological activity (bottom of bed) are different (although soaker hoses placed in the bottom third of the bed can help overcome this).
2. Since air and water move countercurrently, there may be local areas of water hold-up and air channeling through dry spots.

Downflow

Advantages
1. Water addition and highest rate of biological activity occur at same place in bed.
2. Reduces likelihood of dry spots.

Disadvantages
1. Salts, acids, and particles move down into the bed.
2. Requires complete enclosure, which usually results in higher capital costs.

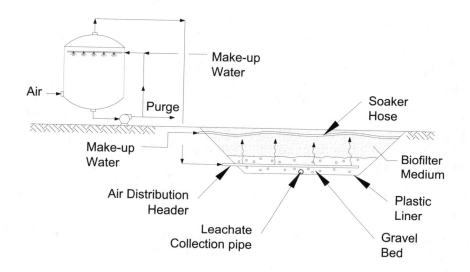

Figure 14.5 (a)   Upflow Biofilter

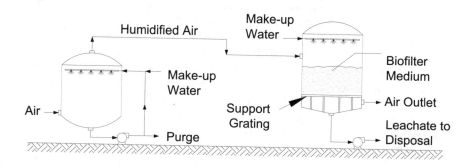

Figure 14.5 (b)   Downflow Biofilter

Figure 14.5
Schematic diagrams of (a) upflow and (b) downflow biofilters.

diminishing the functioning of the biomass. The best way to keep the air at high humidity is by saturating the air with water before it flows into the biofilter. Because evaporation of water absorbs heat, a relatively dry airstream will be cooled significantly during the humidification process. In practice, the actual temperature of the air exiting a well-designed chamber will be within one to three degrees of the theoretical adiabatic saturation temperature.

Humidification is often accomplished by forcing the air through a water-soaked packed tower, or through a spray chamber. In addition, some water is typically added to the bed through soaker hoses to help make up for the water that is lost by evaporation due to the heat released by the oxidation reactions, and/or by any leachate that may escape from the bottom of the bed. Soaker hoses may be placed at different depths throughout the bed.

▶ ▶ ▶ ▶ ▶ ▶ ▶ ▶ ▶

Example 14.5

A biofilter is to be designed to treat air containing a mixture of ethanol and propanol emissions from a pharmaceutical plant. Estimate the daily amount of water required to completely humidify an airflow of 65,000 cfm at 1 atm and 140 F, and 25% relative humidity. Also, estimate the final temperature of the air after humidification.

Solution

From the psychrometric chart (Figure B.3) in Appendix B, air at 140 F and 25% humidity contains 0.032 lb of water per pound of dry air. If we follow an adiabatic saturation line on the chart up and to the left to the 100% saturation line, we can immediately read the final temperature as 99.5 F. (In practice, the actual temperature of this air as it leaves a spray-tower saturator likely will be 101 to 103 F.)

From the right-hand y-axis, we read the final absolute humidity of the exiting saturated air as 0.042 lb water/lb dry air. This difference in inlet and outlet humidity (0.01 lb water/lb dry air) is the amount of water that must be supplied for every pound of dry air to fully saturate the inlet air before it enters the biofilter.

Also from the chart, we can estimate the humid volume of the inlet air as 15.9 ft³/lb of dry air. Using this humid volume, we estimate the dry air mass flow rate:

$$\dot{M}_{air} = \frac{65,000 \text{ cfm}}{15.9 \text{ ft}^3/\text{lb dry air}}$$

$$= \frac{4088 \text{ lb of dry air}}{\text{min}}$$

The water required is:

$$\dot{M}_{water} = \frac{4088 \text{ lb of dry air}}{\text{min}} \times \frac{0.01 \text{ lb water}}{\text{lb dry air}}$$

$$= 40.9 \text{ lb/min} \quad \text{(or a water supply rate of almost 5 gal/min)}$$

◀ ◀ ◀ ◀ ◀ ◀ ◀ ◀ ◀

Humidification was covered in Chapter 7, but additional information is presented here regarding the sizing of a humidifier tower. The design of a system to saturate a large volume of air in a short time is not as simple as just calculating the amount of water needed. Several types of contact vessels are possible, including packed towers, spray towers, venturi-type devices, and others. There are additional considerations if the gases entering the tower are very hot.

Whether the water-air contact is accomplished by spraying the water in a fine mist into an open tower, or by flowing the water down through a packed tower while the air flows upward, the objective is intimate contact of the air and water. Typically, much more water than is needed for evaporation is pushed through the tower because most of the water does not evaporate on the first contact. In fact, a rule of thumb is that the water flowing into the tower is 7 to 15 times the amount to be evaporated. The unevaporated water is collected from the bottom and is recycled except perhaps for a small purge stream, which is disposed, or perhaps treated and disposed.

A spray tower can be designed for a superficial gas velocity of 5 to 15 ft/s without causing undue entrainment of water drops. However, in a packed tower, the gas velocity is lower, usually in the range of 2 to 5 ft/s, to maintain a reasonable pressure drop. The tower is often a cylindrical vessel whose height is 2 to 3 times its diameter. Depending upon the particulate matter in the incoming gas and the dissolved solids content of the water, a small water purge stream may need to be discharged continuously from the tower. It is worth repeating that soaker hoses are often placed inside the bed to provide additional wetting of the media to compensate for incomplete humidification and/or temperature increases due to microbial reactions.

▶ ▶ ▶ ▶ ▶ ▶ ▶ ▶ ▶

Example 14.6

Size a spray tower(s) to humidify the air of Example 14.5.

Solution

The inflow air is 65,000 cfm at 140 F and 1 atm. From Example 14.5, we need to evaporate about 41 lb/min of water. Based on a 15:1 recirculation ratio, we will size the recirculation pump for 615 lb/min of water (74 gpm) through the tower.

Using the airflow rate, we estimate the tower area and diameter from:

$$A = Q / V$$

$$= \frac{65,000 \text{ cfm}}{(15 \text{ ft/s} \times 60 \text{ s/min})} = 72 \text{ ft}^2$$

$$D = (4A / \pi)^{0.5} = 9.6 \text{ ft}$$

Finally, the tower height is estimated from

$$Z = 2.5 \, D$$

$$= 24 \text{ ft}$$

◀ ◀ ◀ ◀ ◀ ◀ ◀ ◀ ◀

Aside from its effect on relative humidity, temperature is important because most microorganisms have a limited range of temperatures in which they can exist. Within that range, both mass transfer rates and metabolic reaction rates are faster at warmer temperatures. However, gas solubility in water tends to decrease with increasing temperature. In most biofilters, only a passive temperature control is practiced—that is, in cold climates, the biofilter is closed and well-insulated to prevent excessive cooling, and in warm climates, excess heat is removed via the evaporation of water within the bed.

Microorganisms are greatly affected by pH excursions. It should be noted that acetic acid is a common intermediate of biodegradation of organic compounds, and this in itself can lower the pH of the bed and disrupt operations. If the biofilter is treating H_2S or other compounds containing sulfur, significant amounts of sulfates (which hydrolyze to sulfuric acid) are produced. In this case, some leachate is desirable in order to remove the excess sulfates. During operation, the pH should be monitored; it can be adjusted if needed by adding dilute solutions of mild caustic through the soaker hoses.

Keep in mind that the biofilter functions by microorganisms consuming the contaminant as a food and/or energy source. If the bacteria grow too rapidly, this can produce significant amounts of new biomass, and may clog the filter. This condition can be discovered by regularly monitoring the pressure drop. Ideally, we would like the biofilter to operate with no net growth of biomass, but rather to just sustain the established microbial population. In that case, all of the organic contaminant is converted to CO_2 and H_2O and energy for the biomass. If biomass growth becomes too rapid, then the nutrient supply can be reduced. As a last resort, excessive biomass growth can be controlled by stressing the biomass (adding salts, or acids, for example).

CASE STUDY

Biofilters were used to control odors at the Orange County (Florida) South Water Reclamation Facility (SWRF)—a 20 mgd wastewater treatment plant located in southwestern Orange County. Odors had been noticed at various locations outside the plant boundaries, and a new tourist attraction had been announced for a large parcel of land adjacent to the SWRF. The county decided to take action.

An engineering study was conducted to determine the sources of the odors and the extent of control needed to ensure an acceptably low probability of future odor complaints (as indicated by dispersion modeling). Numerous odorous compounds have been found in industrial or municipal wastewaters (see Table 14.4). However, in municipal wastewater, H_2S is usually present in high concentrations, and often is the main compound responsible for odor complaints. Thus, there are many biofilter applications designed to remove H_2S at wastewater treatment plants. Indeed, in the case of the SWRF, H_2S was deemed to be the culprit.

The main source of H_2S was identified as fugitive odorous air emissions from the grit chamber/flow splitter (GCFS) building. H_2S was being released from the wastewater as it flowed through the grit chamber and flow splitter building, and these emissions were not being captured effectively. Furthermore, the air that was being captured was not being scrubbed efficiently by existing chlorine-caustic scrubbers. Given the high rate of fugitive emissions from the existing GCFS, and the inconsistent performance of the chlorine-caustic scrubbers, the county decided to improve the air collection system, and to route the air to three upflow biofilters in parallel.

The dimensions of each biofilter are identical: each filter measures approximately 25 feet long by 10 feet wide. The 3.5-foot depth of medium is supported on a top-graded, bottom-sloped gravel bed, which is underlain by a high-density polyethylene liner. With these dimensions, the bed cross-sectional area and empty bed volume are 250 ft^2 and 875 ft^3, respectively. A 12-inch diameter PVC pipe brings the air to the biofilters, where it connects to two 10-inch header pipes per bed. The air distribution pipes are parallel, 6-inch diameter, perforated PVC pipes, spaced 2.5 feet apart and installed within the gravel bed. For biofilters 2 and 3, soaker hoses were installed at two different levels within the medium (6 inches and 1.5 feet below the top of the bed), and were spaced every two feet. Soaker hoses were not placed in biofilter 1. A leachate drain pipe (6-inch diameter) was installed along the bottom central axis of each bottom-sloped gravel bed. Leachate is pumped to the headworks of the SWRF and mixes with the incoming wastewater for treatment. Unit 3 was initially covered with polyethylene to test this effect on its operation, but the cover was removed after only a few weeks.

Table 14.4 Odorous Compounds in Wastewater

Compound name	Formula	Molecular weight	Odor description	Odor threshold, ppm (v/v)
Acetaldehyde	CH_3CHO	44	Pungent, fruity	0.067
Acetic acid	CH_3COOH	60	Sour	1.0
Allyl mercaptan	CH_2CHCH_2SH	74	Disagreeable, garlic	0.0001
Ammonia	NH_3	17	Pungent, irritating	47
Amyl mercaptan	$CH_3(CH_2)_4SH$	104	Unpleasant, putrid	0.0003
Benzyl mercaptan	$C_6H_5CH_2SH$	124	Unpleasant, strong	0.0002
n-Butyl amine	$CH_3(CH_2)NH_2$	73	Sour, ammonia	0.080
Crotyl mercaptan	$CH_3(CH)_2CH_2SH$	88	Skunk-like	0.00003
Dibutyl amine	$(C_4H_9)_2NH$	129	Fishy	0.016
Diisopropyl amine	$(C_3H_7)_2NH$	101	Fishy	0.13
Dimethyl amine	$(CH_3)_2NH$	45	Putrid, fishy	0.34
Dimethyl disulfide	$(CH_3)_2S_2$	94	Decayed vegetables	0.0001
Dimethyl sulfide	$(CH_3)_2S$	62	Decayed cabbage	0.001
Diphenyl sulfide	$(C_6H_5)_2S$	186	Unpleasant	0.0001
Ethyl amine	$C_2H_5NH_2$	45	Ammonia-like	0.27
Ethyl mercaptan	C_2H_5SH	62	Decayed cabbage	0.0003
Hydrogen sulfide	H_2S	34	Rotten eggs	0.0005
Indole	$C_6H_4(CH)_2NH$	117	Fecal, nauseating	0.0001
Methyl amine	CH_3NH_2	31	Putrid, fishy	4.7
Methyl mercaptan	CH_3SH	48	Rotten cabbage	0.0005
Phenyl mercaptan	C_6H_5SH	110	Putrid, garlic	0.0003
Propyl mercaptan	C_3H_7SH	76	Unpleasant	0.0005
Pyridine	C_5H_5N	79	Pungent, irritating	0.66
Skatole	C_9H_9N	131	Fecal, nauseating	0.001
Thiocresol	$CH_3C_6H_4SH$	124	Skunky, irritating	0.0001
Thiophenol	C_6H_5SH	110	Putrid, garlic-like	0.00006
Trimethyl amine	$(CH_3)_3N$	59	Pungent, fishy	0.0004

Compiled from various sources.

The medium selected for use was a custom blend of locally available materials: wood chips (66%), screened bark mulch (16%), leaf compost (16%), and a small percentage of oyster shells (2%) that the county had previously used on other biofilter installations. Construction of the biofilters was completed in July 1998. In August 1998, the three units were balanced and tested to evaluate performance. They have been running ever since.

The airflows to the biofilters are as follows: about 2100 cfm to biofilter 1 (this is the outlet from an existing wet scrubber), and about 1200 cfm each to units 2 and 3 (gas taken directly from the GCFS). Biofilter 1 thus has a surface loading of 8.3 ft/min, and an empty bed residence time (*EBRT*) of 26 seconds. Biofilters 2 and 3 each have a surface loading of 4.8 ft/min, and an *EBRT* of 44 seconds. The airflows were initially balanced using a hand-held pitot tube, and are not routinely checked. The bed pressure drops routinely range between 2 and 3 inches of water.

To evaluate performance, the units were sampled for inlet and outlet H_2S concentrations weekly from August 1998 through January 2000. The concentration of H_2S to and from each filter was sampled at one location in the inlet duct, and at three locations on each bed that were spaced over the top surface of the bed. The H_2S concentration was measured using a hand-held H_2S meter, and sampling time was approximately 40 seconds per location. The gas exiting through the (open) top surface of the bed was sampled by placing a three-sided pyramidal-shaped funnel on the top surface of the bed (to isolate the biofilter exit gas from ambient air), and then the gas exiting from the top 4-inch-long exit spout was analyzed using a hand-held H_2S meter.

The weekly performance data were averaged by month and are summarized in Table 14.5. Even during the start-up phase, units 1 and 2 provided better than 99% removal of hydrogen sulfide. At first, unit 3 (with the cover in place) provided only about 75% removal. It was observed that the cover was tight fitting in some areas, and may have forced the air to flow at high velocity through small portions of the bed. After the cover was removed in late August, unit 3 performed as well as unit 2.

COST ESTIMATION

Since biofiltration is still a fairly new technology in the United States, information on costs is not as readily available as with many other technologies. However, several authors have reported capital costs for different installations, and these have been summarized by Devinny et al. (1999). In general, open top biofilters are less expensive than closed units. Also, the unit costs tend to be smaller for larger systems. For example, Devinny et al. (1999) report that smaller systems (about 100 m^3 of filter bed volume) can cost as much as $1000 to

$3000/m^3$ of bed volume ($28–$99/ft^3$), whereas larger systems (3000 m^3) range from about $300 to $1000/m^3$ ($9–$28/ft^3$). The cost of a specific system is influenced strongly by the volume of the bed, the type of medium, the amount of site preparation work required, the biodegradability of the contaminants, the humidifier tower, the size of the blower and pumps, the degree of instrumentation and computer control included, and the engineering and construction fees.

Mukhopadhyay and Moretti (1993) developed cost correlations for biofilters that depended only upon bed volume, as shown below:

$$C = 51,315 \ V^{0.49} \quad V < 200 \ m^3 \qquad (14.21)$$

and

$$C = \$231,254 + 2285 \ V \quad V \geq 200 \ m^3 \qquad (14.22)$$

Table 14.5   Monthly Average Performance of Biofilters at Orange County SWRF

Monthly Average H₂S Concentrations (ppm) and Removal Efficiencies[a]

	Biofilter 1			Biofilters 2 & 3 Combined		
Month	Inlet[b]	Outlet	Efficiency[c]	Inlet	Outlet	Efficiency
Aug-98	13.40	0.127	99.1	39.00	0.34	99.2
Sep-98	12.84	0.145	98.9	41.80	0.19	99.6
Oct-98	2.70	0.025	99.1	37.75	0.38	99.1
Nov-98	0.538	0.038	93.0	35.50	0.10	99.7
Dec-98	0.085	0.015	81.9	39.00	0.21	99.5
Jan-99	0.116	0.012	89.7	41.50	0.17	99.6
Feb-99	14.25	0.390	97.3	46.50	0.21	99.6
Mar-99	9.56	0.088	99.1	42.80	0.16	99.6
Apr-99	11.85	0.058	99.5	47.50	0.28	99.4
May-99	8.39	0.164	98.0	46.25	0.35	99.3
Jun-99	0.040	0.012	68.7	37.00	0.11	99.7
Jul-99	5.44	0.028	99.5	39.50	0.14	99.6
Aug-99	13.18	0.078	99.4	37.50	0.19	99.5
Sep-99	4.00	0.036	99.1	31.20	0.17	99.5
Oct-99	8.18	0.046	99.4	43.25	0.16	99.6
Nov-99	13.68	0.070	99.5	29.25	0.17	99.4
Dec-99	17.78	0.077	99.6	26.60	0.34	98.7
Average			99.04[d]			99.45

[a] Monthly average values are the average of 4 weekly readings (outlet is average from 3 locations from each biofilter).

[b] For any spot reading > 50.0 ppm (the upper limit of the meter), 50.0 ppm was used to calculate the average.

[c] Efficiency for biofilter 1 drops only when the inlet concentration drops to very low values.

[d] The months where inlet concentration was < 1.0 ppm were ignored in calculating the average efficiency.

Cooper et al., 2001.

where

C = equipment cost (f.o.b. supplier, in 1993 dollars)

V = bed volume, m^3

Once the equipment cost is estimated from the appropriate equation above, it can be converted into a total installed cost (TIC) estimate in the same manner as with other equipment. In this case, Eq. (14.23) gives the desired estimate.

$$TIC = 1.18\, C \times 1.6 \qquad (14.23)$$

The TIC can be escalated to present time using standard methods.

The direct or variable operating costs of a biofilter include the electricity to move the air and water through the system, the water consumed, the operating and maintenance labor (and overhead), the chemicals and nutrients, and the annualized cost of media removal, disposal, and replacement (biofilter media usually last about 4–5 years). The indirect or fixed costs are related to overhead and capital recovery. For easily degraded compounds, total operating costs are in the range of 25–50 cents per 1000 m^3 of gas treated, with the fixed costs being slightly more than half of the total. For hard-to-treat compounds, the costs are about double those for easily degraded compounds.

♦ PROBLEMS ♦

14.1 Convert the Henry's constant for ethane at 20 C (given in Appendix B) into units more suitable for use in analyzing a biofilter. Is ethane a good candidate for biofiltration? Why or why not?

14. 2 Consider the following pilot-plant data for controlling ethanol emissions by biofiltration. The gas flow was 15 m^3/min, $T = 28$ C. Bed dimensions = 3 m × 2 m × 1 m (deep).

C_i, ppm	Efficiency, %
200	98
300	98
350	98
400	98
450	89
500	81
550	74
600	68
650	63

Develop an elimination capacity chart from these data.

14.3 A whiskey distillery emits air with 400 ppm ethanol. The air flows out several roof vents at a total flow rate of 2000 m^3/min (at 35 C and 1 atm). The distillery is required to collect all the vented air and route it to a biofilter to control total emissions to less than 50 kg of ethanol per day. Calculate the required control efficiency.

14.4 Management wants you to investigate using a biofilter to control the ethanol emissions from the distillery of Problem 14.3. Someone has suggested a roof-mounted biofilter to save space in the plant yard. Use the pilot-plant data given in Problem 14.2 to design a biofilter (size the bed) to control the distillery's ethanol emissions. What is your initial reaction to the roof-mounted idea? What other data will you need to gather to develop a better opinion?

14.5 The air coming out of the distillery has 60% relative humidity. How much water will you need to saturate this airstream? After saturation, what will be the temperature and the total volumetric flow rate? Should you also provide soaker hoses in your design? Discuss the reasons why you will or will not include soaker hoses.

14.6 It has been reported that a peat biofilter treating styrene had a maximum elimination capacity of 100 g/m^3-hr, and a critical load of between 60–75 g/m^3-hr (Togna and Folsom 1992). Further investigation reveals that a 97% efficiency was achieved at a load of 40 g/m^3-hr. Assume that to meet a local odor ordinance you must design for at least a 90% removal efficiency. Use this information to estimate the dimensions of the biofilter needed to treat 10,000 m^3/hr of air (at 30 C and 1 atm) containing 200 ppm styrene being emitted from a boat-manufacturing plant. Assume the bed must be at least 1 meter deep, and that the surface loading cannot exceed 1.5 m/min.

14.7 Calculate the mass volume loadings of the biofilters in the case study in this chapter.

14.8 Go to the library or the Internet and find a technical article on biofiltration for H$_2$S control (other than one referenced in this chapter). Compare that application with the Orange County case study, especially with regard to *EBRT*, *ML$_V$*, and removal efficiency.

14.9 Based on the simple data given in this chapter, estimate the capital cost of the biofilter described in Problems 14.2 through 14.5.

14.10 Five organic compounds (A through E) were tested for suitability for biofiltration control. Their retardation ratios (in order) were 5, 8, 13, 21, and 34. Which of these is most suitable for a biofilter and why?

14.11 Assuming that Eq. (14.10) holds, what would be the average efficiency of the H$_2$S biofilters 2 and 3 (as described in the case study) if the bed depth had been (a) 3.0 feet or (b) 2.5 feet?

14.12 Consider the data in Table 14.5. Ignoring (for biofilter 1) those months when the inlet concentration was below 1 ppm, compare the efficiencies of biofilter 1 with biofilters 2 and 3. Do these efficiencies make sense in light of Eq. (14.10)? What else might account for the differences?

14.13 Size a spray chamber to humidify the air of Problem 14.5.

REFERENCES

American Society of Civil Engineers. *Odor Control in Wastewater Treatment Plants*, WEF Manual of Practice No. 22, ASCE Manuals and Reports on Engineering Practice No. 82, 1995.

Cooper, C. D., Godlewski, V. J., Hanson, R., Koletzke, M., and Webster, N. "Odor Investigation and Control at a WWTP in Orange County, Florida," *Environmental Progress, 20* (3), 2001.

Deshusses, M. A. Personal communication with author, 2001.

Deshusses, M. A., and Cox, H. H. J. "Biotrickling Filters for Air Pollution Control," in *The Encyclopedia of Environmental Microbiology*, G. Bitton, Ed. New York: John Wiley & Sons, 2002.

Deshusses, M. A., and Johnson, C. T. "Development and Validation of a Simple Protocol to Rapidly Determine the Performance of Biofilters for VOC Treatment," *Environmental Science and Technology, 34* (3), 2000.

Devinny, J. S., Deshusses, M. A., and Webster, T. S. *Biofiltration for Air Pollution Control*. Boca Raton, FL: Lewis Publishers, CRC Press, 1999.

Heinsohn, R. J., and Kabel, R. L. *Sources and Control of Air Pollution*. Upper Saddle River, NJ: Prentice Hall, 1999.

Koe, L. C. C., and Chew, S. H. *Control of Odorous Emissions at Wastewater Treatment Plants—Singapore Experience*, a paper presented at the 91st Annual Meeting of the Air & Waste Management Association, San Diego, CA, June 14–18, 1998.

Kosteltz, A. M., Finkelstein, A., and Sears, G. *Proceedings of the Air & Waste Management Association 89th Annual Conference and Exhibition*. Air & Waste Management Association, Pittsburgh, PA, paper #96-RA87B.02, 1996.

Mukhopadhyay, N., and Moretti, E. C. *Current and Potential Future Industrial Practices for Reducing and Controlling Volatile Organic Compounds*. 1993.

Pope, R. J., et al. *Deep Tunnel Mysteries—Solving Odor/Corrosion Problems*, a paper presented at the 91st Annual Meeting of the Air & Waste Management Association, San Diego, CA, June 14–18, 1998.

Togna, A. P., and Folsom, B. R. *Removal of Styrene from Air Using Bench-Scale Biofilter and Biotrickling Filter Reactors*, a paper presented at the 85th Annual Meeting of the Air & Waste Management Association, 1992.

CONTROL OF SULFUR OXIDES

The effects of acid rain are so far-reaching and so permanent in character that we cannot afford to let more time pass without moving to reduce the acidity even as we continue our investigation into these effects. We need, as soon as possible in both countries, an initial significant reduction in sulfur dioxide emissions from thermal power plants and smelters to slow the rate of environmental deterioration.

John Roberts, Minister of the Environment, Canada, 1981

♦ 15.1 INTRODUCTION ♦

SO_2 and SO_3 are two sulfur oxides that are formed whenever any material that contains sulfur is burned. Sulfur oxides (SO_x) emissions in the United States peaked in 1973 at about 32 million tons per year. By the late 1970s, increased numbers of SO_2 control systems and decreased sulfur contents of fuel oils had reduced emissions significantly. It is estimated that in 1981, SO_x emissions decreased below 25 million tons for the first time in more than a decade and decreased below 20 million in 1995 (U.S. Environmental Protection Agency 2000). About two-thirds of the sulfur oxides in the United States are emitted from coal-fired power plants. Industrial fuel combustion and industrial processes (primarily petroleum refining, H_2SO_4 manufacturing, and smelting of nonferrous metals) account for most of the remainder of the sulfur oxides. In Canada, where much electricity is generated by hydropower, metal smelting is the predominant source of SO_x. Table 15.1 presents some data on U.S. emissions of SO_x for the past thirty years.

As discussed in Chapter 1, there are several serious effects of emissions of SO_2 and SO_3 to the atmosphere. SO_2 is an eye, nose, and throat irritant, and has been correlated with respiratory illnesses. SO_2 also causes loss of chlorophyll in green plants. One of the major effects of SO_x emissions is acidic deposition (acid rain, acid snow). Acidic deposition has destroyed aquatic ecosystems in many lakes in

Table 15.1   Trend in U.S. Emissions of Sulfur Oxides (in teragrams*/year)

Source	1970	1975	1980	1985	1990	1998
Transportation						
Highway vehicles	0.3	0.3	0.4	0.5	0.6	0.3
Non-road sources	0.3	0.3	0.5	0.4	0.3	1.0
Subtotal	0.6	0.6	0.9	0.9	0.9	1.3
Stationary Fuel Combustion						
Electrical utilities	15.8	16.6	15.5	14.2	14.2	12.0
Industries	4.1	2.7	2.4	2.2	2.3	2.6
Commercial/institutional	0.9	0.7	0.7	0.4	0.4	0.4
Residential	0.5	0.3	0.2	0.2	0.2	0.1
Subtotal	21.3	20.3	18.8	17.0	17.1	15.1
Industrial Processes	6.4	5.0	3.8	3.2	3.1	1.4
Total	28.3	25.9	23.5	21.1	21.1	17.8

*1 teragram = 1 million metric tons = 10^9 kg.
Adapted from U.S. Environmental Protection Agency, 1991 and 2000.

Scandinavia, Canada, and the United States. Acidic deposition has damaged thousands of hectares of forests in northern Europe (including the famous Bavarian Forest in West Germany) as well as in North America, and has corroded artifacts and structures in industrialized centers throughout the world.

Although there has been much debate over the exact quantitative relationship between reductions in SO_x emissions and reductions in acidic deposition, there can be little doubt that control of SO_x emissions is important to mitigate the problem of acid rain. The Acid Rain Provisions of the 1990 Clean Air Act include a goal of reducing U.S. SO_x emissions by 10 million tons relative to 1980 levels. As a consequence, it is estimated that flue gas desulfurization installed capacity in the United States (both new and retrofit) about doubled from 1990 to 2000.

◆ 15.2 OVERVIEW OF CONTROL STRATEGIES ◆

The two basic approaches to controlling SO_x emissions are (1) remove sulfur from fuel before it is burned, or (2) remove SO_2 from the exhaust gases. Within those two overall strategies, there are hundreds of methods that have been researched, and dozens of methods that have been demonstrated. Table 15.2 summarizes the options and lists some of the better-known processes. Srivastava and Jozewicz (2001) summarize several of these processes. Brief descriptions of the better-known processes are presented in the following pages. In Section 15.3, we will discuss in detail the most widely used SO_2 scrubbing technique (limestone scrubbing).

Table 15.2 Options for Control of SO_2 Emissions

Main Option	Suboption	Examples of Processes
Do not create SO_2	Switch to a low-sulfur fuel	
	Desulfurize the fuel	Oil desulfurization
		Coal cleaning
SO_2 scrubbing:		
Throwaway	Wet scrubbing	Lime
		Limestone
		Forced oxidation
		Inhibited oxidation
		Dual alkali
		Magnesium Enhanced
		Lime (MEL)
		Seawater
	Dry scrubbing	Lime spray drying
		Lime injection
		Trona
		Nahcolite
		Circulating fluidized bed
Regenerative	Wet processes	Absorption with water
		(smelters)
		Wellman-Lord
		MgO
		Citrate
		Carbonate
		Sulfite
		Forced oxidation (with
		gypsum sales)
	Dry processes	Activated carbon adsorption
		Copper oxide adsorption

FUEL DESULFURIZATION

Many crude oils are "sour;" that is, they contain sulfur (usually in the range from 1 to 3% by weight). In fact, the world pool of crude oil is becoming more sour as oil companies preferentially use up the "sweet" (less than 0.5% sulfur) crude oils. Consumers generally prefer sweet oil products (such as gasoline, kerosene, heating oils, and low-sulfur fuel oils). Hence, many oil refineries operate desulfurization units. In simple terms, a desulfurization unit removes organic sulfur from oil via a catalytic reaction with hydrogen; that is,

$$R\text{–}S + H_2 \rightarrow H_2S + R \tag{15.1}$$

where R represents any organic group.

Up until the mid-1970s, the H_2S produced from desulfurization units often was blended into the refinery fuel gas systems. So, when the H_2S was burned as fuel, it produced SO_2. Thus, there was no reduction in national SO_x emissions—just a redistribution. As the world pool of crude oil became more sour and the demand for sweet products further increased, environmental considerations generated a great incentive to prevent the emission of SO_2 from refineries. Sulfur recovery units were then built that could capture over 98% of the H_2S fed to them.

Most sulfur recovery units use the Claus process, whereby part of the H_2S is burned to SO_2, and then the two compounds are combined over a catalyst to simultaneously oxidize and reduce each other. The net reactions are

$$H_2S + 3/2O_2 \rightarrow H_2O + SO_2 \qquad \text{(15.2)}$$

$$2H_2S + SO_2 \rightarrow 2H_2O + 3S \qquad \text{(15.3)}$$

The elemental sulfur is separated in the molten state and sold as a by-product, and the emissions of either H_2S or SO_2 from a refinery are reduced substantially. The amount of sulfur recovered from refineries more than doubled from 1975 to 1990 and now accounts for a significant portion of the U.S. sulfur supply. Furthermore, the amount of sulfur recovered from refineries and natural gas plants continues to increase. United States government statistics show that in 1989, the U.S. oil and gas industry produced and sold more than 6.5 million metric tons of sulfur (about 56% of the U.S. market). Sulfur recovery from oil refineries throughout the industrialized world will undoubtedly increase significantly from present levels.

Many natural gas fields contain significant concentrations of H_2S (from less than 10% to over 70%) mixed with methane and other light hydrocarbons. Sulfur is recovered from these sources by a Claus process before the gas is admitted to pipelines. Canada is the world's leading producer of sulfur from natural gas (about 5–6 million metric tons per year), with the United States, France, Germany, and several other countries producing significant quantities as well. In a very interesting article, Armstrong and Westmoreland (1985) describe some of the difficulties encountered (and how they were overcome) in designing and building a large sour-gas treatment plant in a remote, mountainous area of Wyoming. In the Wyoming plant, approximately 1000 tons/day of sulfur are being recovered from gas varying in sulfur content from 0.3% to 22% H_2S. Because of stringent air quality requirements, a tail-gas cleanup unit that brings the overall sulfur recovery to 99.7% was added to the normal Claus plant.

Much of the coal burned in the United States has a high sulfur content (2% or more). Unlike oil or gas, a significant fraction (30–70%) of the sulfur in coal is in mineral form, either as pyrites (FeS_2) or as mineral sulfates. Much of that type of sulfur can be removed relatively easily by washing or other physical cleaning processes (at the risk of creating a water pollution or solid waste problem). However, organic sulfur cannot be removed by physical cleaning, and, because coal is a solid, chemical desulfurization processes are extremely costly. Coal gasification and liquefaction processes can remove much of the organic sulfur from fuel, but at a substantial energy penalty. Since sufficient desulfurization of coal to meet air quality requirements cannot be achieved by physical cleaning, for a majority of the SO_x emissions, we must consider some form of flue gas desulfurization.

SO_2 REMOVAL TECHNIQUES

The smelting of sulfide ores of copper, zinc, lead, and nickel involve a step wherein the ores are "roasted." Roasting removes the sulfur from the metal and produces concentrated streams of SO_2. The usual method of SO_2 control at a smelter is to oxidize the SO_2 to SO_3 gas, and then absorb the SO_3 in water to make H_2SO_4, which can be sold as a by-product. Depending on several factors (distance to markets, transportation costs, investment costs, required degree of SO_2 removal, and others), the production of sulfuric acid might or might not be profitable for the smelter. However, note that such operations in the United States and Canada alone produce about six million metric tons of sulfuric acid every year. The environmental protection agencies in both countries are requiring even further reductions in smelter SO_2 emissions, both on new units and on existing plants. Smelting operations also produce significant amounts of sulfuric acid in Europe and Japan.

Flue gas desulfurization (SO_2 scrubbing) and absorption of smelter off-gases to make sulfuric acid both result in reduced SO_2 emissions. The main difference lies in the concentration of SO_2 in the gas. Typically, absorption at a smelter involves a concentrated stream of SO_2 (about 10% or 100,000 ppm SO_2), whereas flue gas desulfurization (FGD) is applied to a stream of gas that is dilute in SO_2 (about 0.2% or 2000 ppm SO_2). The two basic methods of classifying FGD systems are (1) throwaway or regenerative, and (2) wet or dry. A process is of the *throwaway* type if the sulfur removed from the flue gas is discarded. A process is considered *regenerative* if the sulfur is recovered in a usable form. *Wet or dry* refers to the phase in which the main reactions occur, and both wet and dry processes can result in either a throwaway product or a marketable by-product.

THROWAWAY PROCESSES

The wet throwaway processes discussed in this section include conventional limestone slurry (or limestone scrubbing), lime scrubbing, dual alkali, forced oxidation (external), and inhibited oxidation. The two dry throwaway processes that are discussed are lime-spray drying and nahcolite/trona injection. Dalton (1992) reported on the evolution of FGD design, and reviewed the status of FGD systems in the United States and the world. The United States generates about one-fourth of all the electricity in the world, and has about one-half of the world's installed FGD capacity. (Germany and Japan are second and third, respectively, in FGD capacity.) More than half of U.S. electricity generation in 1992 came from coal-fired power plants, and FGD systems are installed on more than 20% of those plants (some 72,000 MW of power-generating capacity). As of 1998, worldwide, there was about a quarter-million MW of power-generating capacity (fossil-fueled electricity generating plants) equipped with flue gas desulfurization systems (Srivastava and Jozewicz 2001). Table 15.3 gives the breakdown of this capacity by type of process (wet throwaway, dry throwaway, or regenerative) and by location (U.S. or rest of world).

During the 1990s, many new power plants were built that fired natural gas. Natural gas was relatively cheap and is much cleaner than coal or oil. For example, SO_2 and PM emissions from uncontrolled natural gas-fired turbines are an order of magnitude lower than emissions from coal-fired boilers with air pollution control equipment from the 1980s. However, in the years 2000–2001, the demand for gas increased rapidly, while world market prices for energy were higher, causing natural gas prices to about triple, making this option less attractive. It now appears that coal again will be the fuel of choice for large electric power plants, thus requiring new flue gas desulfurization systems. Essentially all of these FGD systems are expected to be limestone-scrubbing systems, the oldest and most widely used FGD system in the United States. We will now briefly describe the major throwaway processes (including limestone scrubbing), and will discuss conventional limestone scrubbing in more detail in Section 15.3.

In **limestone scrubbing,** a limestone slurry is contacted with the flue gas in a spray tower. The sulfur dioxide is absorbed, neutralized, and partially oxidized to calcium sulfite and calcium sulfate. The overall stoichiometry can be represented by

$$CACO_3(s) + H_2O + 2SO_2 \rightarrow Ca^{+2} + 2HSO_3^- + CO_2(g) \quad (15.4)$$

$$CACO_3(s) + 2HSO_3^- + Ca^{+2} \rightarrow 2CaSO_3 + CO_2 + H_2O \quad (15.5)$$

$$2CaCO_3(s) + H_2O + 2SO_2 \rightarrow 2CaSO_3 + 2CO_2 + H_2O$$

Table 15.3   Electric Power Capacity (MW) with FGD Technology

Technology	United States	All Other Countries	Total World
Throwaway–Wet	82,092	114,800	196,892
Throwaway–Dry	14,081	10,654	24,735
Regenerative	2,798	2,394	5,192
Total FGD	98,971	127,848	226,819

Adapted from Srivastava and Jozewicz, 2001.

A simplified schematic diagram of a limestone-scrubbing system is presented in Figure 15.1. The major advantage of limestone scrubbing is that the absorbent is abundant and inexpensive, and it has a wide range of commercial acceptance. The disadvantages include scaling inside the tower, equipment plugging, and corrosion.

Lime scrubbing is very similar in equipment and process flow to limestone scrubbing, except that lime is a much more reactive reagent than limestone. The net reactions for lime scrubbing are

$$CaO + H_2O \rightarrow Ca(OH)_2 \tag{15.6}$$

$$SO_2 + H_2O \rightleftharpoons H_2SO_3 \tag{15.7}$$

$$H_2SO_3 + Ca(OH)_2 \rightarrow CaSO_3 \cdot 2H_2O \tag{15.8}$$

$$CaSO_3 \cdot 2H_2O + \tfrac{1}{2}O_2 \rightarrow CaSO_4 \cdot 2H_2O \tag{15.9}$$

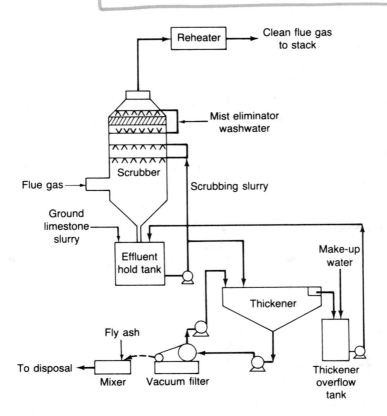

Figure 15.1
Schematic process flow diagram for a limestone-based SO_2 scrubbing system.
(Adapted from Henzel et al., 1981.)

The advantages of lime scrubbing include better utilization of the reagent and more flexibility in operations. The major disadvantage is the high cost of lime relative to limestone. Whereas limestone-scrubbing systems are capable of 90% removal of SO_2, lime-scrubbing systems can routinely achieve 95% removal of SO_2 (Jahnig and Shaw 1981a).

The **dual alkali system** was developed to eliminate the main problems encountered with lime and limestone scrubbing (namely, scaling and plugging inside the scrubbing tower). The dual alkali system uses two reagents and two process loops to remove the SO_2, as shown in Figure 15.2.

A solution of sodium sulfite/sodium hydroxide provides the absorption/neutralization of SO_2 inside the tower. Because both sodium

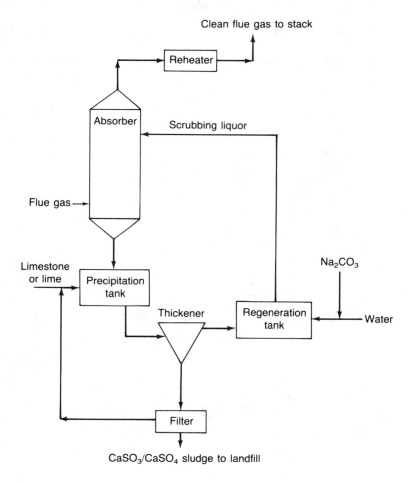

Figure 15.2
Schematic process flow diagram for a dual alkali flue gas desulfurization system.

sulfite and sodium sulfate are soluble in water, no precipitation occurs inside the scrubber. However, disposal of large volumes of sodium sulfite/sodium sulfate solution would pose a water pollution problem and would be very expensive because sodium hydroxide is expensive. Therefore, in a separate vessel, lime or limestone is added to the scrubber effluent along with make-up sodium hydroxide or soda ash (Na_2CO_3). (Make-up sodium is required because the water associated with sludge disposal contains significant amounts of sodium.) The lime or limestone simultaneously precipitates the sulfite/sulfate ions and regenerates the sodium hydroxide. The major advantages of the dual alkali system are reduced scaling and plugging and lower maintenance costs. A major disadvantage is that in some locations, an elaborate treatment or disposal system might be needed for the soluble sodium salts that are discharged with the sludge. Furthermore, a lime prescrubber is needed to remove HCl and any remaining fly ash, both of which add greatly to sodium consumption.

In the **forced oxidation (external) system,** the scrubbing step is accomplished by the same method as with conventional limestone. The $CaSO_3/SO_4$ waste slurry is oxidized (via excess air contact in a separate tank) to a gypsum sludge. The oxidized waste is more easily de-watered and is more stable in sludge disposal ponds. However, the forced oxidation process has high pumping costs and is prone to plugging and scaling.

Inhibited oxidation, a process modification developed recently, involves the use of thiosulfate to inhibit the oxidation of SO_2 to sulfate. The thiosulfate is formed in the scrubber liquid by adding a fine water emulsion of elemental sulfur. Inhibited oxidation works extremely well to prevent scaling; moreover, this process modification requires less operating power (no fans needed to force air into the solution), requires less maintenance, and uses less fresh water than forced oxidation (Dalton 1992).

Lime-spray drying could be termed a wet/dry process. In lime-spray drying, a lime slurry is sprayed into the absorption tower, and SO_2 is absorbed by the slurry, forming $CaSO_3/CaSO_4$. However, the liquid-to-gas ratio is such that the water evaporates before the droplets reach the bottom of the tower. The dry solids are carried out with the gas and collected in a baghouse with the fly ash. The advantages of lime-spray drying include few maintenance problems, low energy usage, and low capital and operating costs. A disadvantage is the potential to blind the fabric if the temperature of the flue gas approaches the dew point.

The direct injection of pulverized lime or limestone into the furnace/boiler or into the downstream duct is a form of **dry scrubbing**. In the dry-scrubbing process, pulverized reagent (lime, trona, or nahcolite) is injected into the flue gas. Dry sorption occurs and the solid particles are collected in a baghouse. Further SO_2 removal occurs as the flue gas flows through the filter cake on the bags. In the past, the main disadvantage of direct injection has been that large quantities of reagent were necessary because only the surfaces of the particles are reactive. However, creating very small particles gives much more surface area per unit weight, resulting in more efficient use of the lime. Dry scrubbing also includes either **trona or nahcolite injection.** Trona is naturally occurring Na_2CO_3. Trona is mined commercially (primarily in the western states) and sold as a bulk commodity. Nahcolite is naturally occurring $NaHCO_3$ (found primarily in Colorado). The advantages of dry scrubbing are low capital costs and low maintenance requirements. The disadvantages are high reagent costs (including transportation) and possible waste disposal problems (leaching of soluble sodium salts).

REGENERATIVE PROCESSES

In general, at the present time, regenerative processes have higher costs than throwaway processes (Radcliffe 1992). However, regenerative processes might be chosen if space is limited, disposal options are limited, and/or markets for the recovered sulfur products are readily available. Regenerative processes produce a reusable sulfur product—conceptually more satisfying than throwing away a resource such as sulfur. In Japan, where FGD is mandated by the government, regenerative processes are used almost exclusively. As a result, Japan produces enough sulfuric acid for its internal industrial consumption and still has some for export. On the other hand, in Germany, where by law all FGD systems must produce a saleable by-product (typically gypsum), the gypsum market has become saturated.

The **Wellman-Lord (W-L) process** is the best-known regenerative FGD process. Therefore, we will discuss the W-L process in more detail than the other regenerative processes. Brief descriptions of several other approaches (namely, magnesium oxide, citrate scrubbing, an activated carbon process and adsorption by a bed of copper oxide) will be presented following the discussion of the Wellman-Lord process.

The four subprocesses of the Wellman-Lord process are (1) flue gas pretreatment, (2) SO_2 absorption by sodium sulfite solution, (3) purge treatment, and (4) sodium sulfite regeneration (U.S. Environmental Protection Agency 1979b). A fifth step that is often included but is not mandatory with the W-L process is the processing of the concentrated stream of SO_2 gas produced via the W-L process to a marketable product such as elemental sulfur or sulfuric acid. The W-L process has been applied successfully at a 115-MW coal-fired power plant in Gary, Indiana, and at four units (totaling 1800 MW) in Waterflow, New Mexico (Electric Power Research Institute 1983). In addition, several refineries in the United States and abroad have installed W-L systems.

A simplified schematic diagram of the W-L process is presented in Figure 15.3. Flue gas from the ESP is blown through a venturi prescrubber. The prescrubber removes most of the remaining particles as well as any existing SO_3 and HCl, which would upset the SO_2 absorption chemistry. The prescrubber also cools and humidifies the flue gas. Typical inlet temperatures and relative humidities are 300 F and 20%, and outlet values are 125 F and 95% (U.S. Environmental Protection Agency 1979b). A liquid purge stream from the prescrubber removes solids and chlorides.

Tray towers are the most common SO_2 absorber units. The flue gas is contacted with aqueous sodium sulfite and the SO_2 is absorbed and reacted in the clear liquid to form sodium bisulfite. The reaction is as follows:

$$Na_2SO_3 + SO_2 + H_2O \rightarrow 2NaHSO_3 \qquad (15.10)$$

Some of the sulfite is oxidized to sulfate by oxygen. Also, any sulfur trioxide that passes through the prescrubber results in aqueous sulfate. The reactions are as follows:

$$Na_2SO_3 + \tfrac{1}{2}O_2 \rightarrow Na_2SO_4 \qquad (15.11)$$

$$2Na_2SO_3 + SO_3 + H_2O \rightarrow Na_2SO_4 + 2NaHSO_3 \qquad (15.12)$$

The sodium sulfate does not contribute to further SO_2 absorption and must be removed. Excessive sulfate accumulation is prevented by a continuous purge from the bottom of the absorber. However, the stream from the bottom of the absorber (the "bottoms") is rich in bisulfite, so most of it is routed for further processing.

Part of the bottoms is sent to the chiller/crystallizer where the less soluble sodium sulfate crystals are formed. From there, the slurry is centrifuged, and the solids are dried and discarded. The centrifugate, which is rich in bisulfite, is returned to the process. The rest of the bottoms is sent to a heated evaporator/crystallizer, where SO_2 is liberated and sodium sulfite crystals are regenerated. The reaction is

$$2NaHSO_3 \xrightarrow[\text{heat}]{} Na_2SO_3 + SO_2 + H_2O \qquad (15.13)$$

The water vapor is condensed and recovered, producing a concentrated stream of SO_2 gas (consisting of about 85% SO_2 and 15% H_2O) (U.S. Environmental Protection Agency 1979b). The SO_2 gas can be reduced to elemental sulfur or oxidized to sulfuric acid either on-site or at a nearby chemical plant. Because some of the sodium is removed from the process via the sodium sulfate purge, soda ash (Na_2CO_3) is added to provide make-up sodium. The soda ash reacts readily in the absorber tower as follows:

$$Na_2CO_3 + SO_2 \rightarrow Na_2SO_3 + CO_2 \qquad (15.14)$$

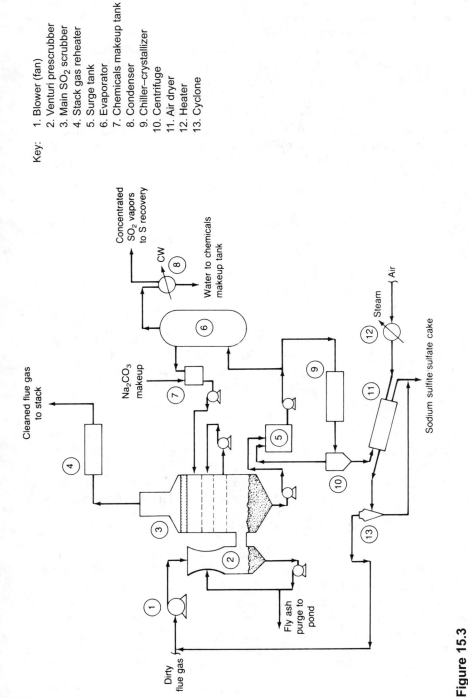

Figure 15.3

Schematic process flow diagram of the Wellman-Lord SO$_2$ scrubbing and recovery system.

(Adapted from U.S. Environmental Protection Agency, 1979b.)

Key: 1. Blower (fan)
        2. Venturi prescrubber
        3. Main SO$_2$ scrubber
        4. Stack gas reheater
        5. Surge tank
        6. Evaporator
        7. Chemicals makeup tank
        8. Condenser
        9. Chiller–crystallizer
        10. Centrifuge
        11. Air dryer
        12. Heater
        13. Cyclone

A typical make-up rate is 1 mole of soda ash per 42 moles of SO_2 removed (Electric Power Research Institute 1983).

Several other regenerative FGD processes have been developed. Although none have been commercialized to the same extent as the Wellman-Lord process, recent economic and technical feasibility studies have shown that other processes compare favorably with the W-L process (Radcliffe 1992).

The **magnesium oxide (MgO) process** has an absorption step similar to that of lime or limestone scrubbing. Wet scrubbing with a slurry of $Mg(OH)_2$ produces $MgSO_3/MgSO_4$ solids. The solids are then calcined (fired in a kiln in the presence of coke or some other reducing agent), generating SO_2 and regenerating MgO. The advantages of the MgO process relative to the W-L process are that there is little (if any) solid waste, and there is better surge capacity owing to the storage of solid sorbent. The primary disadvantages are that a high-temperature calciner is needed, and that the SO_2 product stream is only about 15% SO_2, which virtually mandates that only sulfuric acid (and not elemental sulfur) can be produced.

There are two major **citrate-scrubbing processes,** both of which involve SO_2 absorption with a buffered solution of citric acid and sodium citrate. The citrate ions in solution increase the effective solubility of SO_2 by binding some of the hydronium ions [shown as H^+ in Eqs. (15.16) to (15.18)] created when SO_2 absorbs into water; that is,

$$SO_2\,(g) + H_2O \rightleftharpoons H_2SO_3 \qquad (15.15)$$

$$H_2SO_3 \rightleftharpoons H^+ + HSO_3^- \qquad (15.16)$$

$$Ci^{-3} + H^+ \rightleftharpoons HCi^{-2} \qquad (15.17)$$

$$HCi^{-2} + H^+ \rightleftharpoons H_2Ci^- \qquad (15.18)$$

where Ci^{-3} = citrate ion

As the citrate ions react with and remove hydronium ions in solution, the equilibria of Reactions (15.15) and (15.16) are shifted to the right, promoting further SO_2 absorption.

Once the citrate solution becomes "loaded" with SO_2, there are two methods of regenerating the solution (and recovering the SO_2). In the **U.S. Bureau of Mines process** (*Clean Power from Coal* 1978; see Figure 15.4), the SO_2 is reduced with H_2S to elemental sulfur in a liquid-phase reaction. The H_2S can be generated on-site by reduction of some of the product elemental sulfur using methane and steam, or it can be obtained from an oil refinery if one is nearby. The elemental sulfur is separated from the regenerated citrate solution by air floation.

The other major citrate process is the **Flakt–Boliden process,** which has been applied commercially in Sweden (Electric Power

Research Institute 1983). In the Flakt–Boliden process, SO_2 is recovered rather than elemental sulfur. The recovery is accomplished by heating and steam-stripping the SO_2-loaded citrate solution. By heating the solution, Reactions (15.15) and (15.16) are reversed, and then gaseous SO_2 is carried away from the citrate solution by the steam. The stream of steam and SO_2 is cooled to condense the water. The resulting SO_2-rich stream can be further processed to elemental sulfur via a Claus reaction or to sulfuric acid by oxidation and absorption.

Several **activated carbon systems** have been developed for adsorption of SO_2. The carbon catalyzes the reaction of SO_2 to H_2SO_4, thus preventing desorption of the SO_2. The carbon can be regenerated by water washing, producing a dilute sulfuric acid stream that often is neutralized and discarded. Brown et al. (1972) have reported on a dry system developed by Westvaco that produces elemental sulfur.

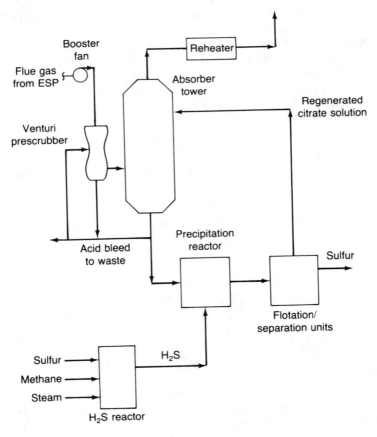

Figure 15.4
Simplified schematic diagram of the U.S. Bureau of Mines citrate FGD process.

The **Westvaco process** consists of four main steps, all of which occur in a continuous, countercurrent, multistage fluidized-bed operation. First, SO_2 is adsorbed onto carbon:

$$SO_2 + H_2O + \tfrac{1}{2}O_2 \xrightarrow[\text{150-300 F}]{\text{Activated Carbon}} H_2SO_4 \qquad \textbf{(15.19)}$$

The carbon then flows to the next vessel, where the sulfuric acid is reacted with hydrogen sulfide, resulting in adsorbed elemental sulfur.

$$H_2SO_4 + 3H_2S \xrightarrow[\text{300 F}]{\text{Activated Carbon}} 4S + 4H_2O \qquad \textbf{(15.20)}$$

In the next bed, the temperature is increased to vaporize and recover one-fourth of the sulfur. The remainder is reacted with purchased hydrogen to generate the H_2S needed by Reaction (15.20). The regenerated carbon is then recycled back to the start.

Adsorption of SO_2 by a bed of copper oxide provides another method of dry SO_2 removal (Jahnig and Shaw 1981a). NO_x can also be removed simultaneously by injecting ammonia into the flue gas. Copper oxide adsorption is one of the few processes that can control both SO_2 and NO_x, and so is of considerable interest for high-sulfur coal-fired power plants. SO_x adsorption and reaction occurs at 750 F to form copper sulfate. The bed is regenerated with a reducing gas (H_2 or $H_2 + CO$) to form a concentrated stream of SO_2. During regeneration, the bed is reduced to copper, but is later oxidized to copper oxide when in the adsorption mode.

♦ 15.3 LIMESTONE SCRUBBING ♦

Limestone scrubbing has become widely accepted by the coal-fired power industry for flue gas desulfurization because limestone scrubbing has an overall lower cost and is simpler to operate than the other processes. Nevertheless, limestone scrubbers are large, complex, expensive units. A photograph of a limestone scrubber on a 364-MW coal-fired power plant is shown in Figure 15.5.

In this section, we will present a more detailed discussion of limestone FGD, along with some example design problems. Most of the material presented in this section is derived from the EPA document *Limestone FGD Scrubbers: User's Handbook,* by Henzel et al. (1981). It should be noted that several major improvements have been made to the limestone FGD process since the handbook was published. As better understanding of the chemistry has been achieved, designs have become simpler and now include less redundancy. Also, equipment has become less costly to purchase and easier to operate, and performance and reliability have improved. To further illustrate, open countercurrent spray towers are now standard, FGD system capital costs have decreased or held constant in the last ten years, and scrubbers can now achieve maximum SO_2 removals of 98–99%, rather than the 88–90% that were typical of the 1980s.

Figure 15.5
A limestone flue gas desulfurization unit. (This combination spray and tray tower unit treats approximately one million cfm to 85% SO_2 removal. Limestone use is about 100,000 tons per year.)

PROCESS CHEMISTRY AND OPERATIONAL FACTORS

Before proceeding to the material and energy balances necessary for process design, it is essential that the process chemistry be understood. The chemistry should be optimized to (1) maximize SO_2 removal, (2) avoid *scaling* (precipitating $CaSO_3$ and $CaSO_4$ inside the scrubber), and (3) maximize the *utilization* of the limestone. The first two objectives are met by providing two separate vessels (the scrubber and the effluent hold tank), and the third objective is met by proper pH control and by using finely ground limestone and a sufficiently high liquid/gas ratio. In this section, our discussion will be brief and limited to a few important operating considerations.

Overall, sulfur dioxide absorption is a two-step process, with the first step occurring in the tower and the second step in the effluent hold tank (see Figure 15.1). In the tower, one mole of $CaCO_3$ reacts with two moles of SO_2 as follows:

$$CaCO_3 + 2SO_2 + H_2O \rightarrow Ca^{+2} + 2HSO_3^- + CO_2 \quad \textbf{(15.21a)}$$

If an excessive amount of $CaCO_3$ is input, the bisulfite ion will not be stable, and $CaSO_3$ will precipitate inside the scrubber (forming scale). A high pH (6.0–6.5) is an indicator of $CaCO_3$ precipitation. However, if the pH is too low (less than about 4.5), SO_2 absorption will be adversely affected.

In the *effluent hold tank* (EHT), more limestone is added to cause the precipitation of $CaSO_3$. The reaction is

$$CaCO_3 + 2HSO_3^- + Ca^{+2} \rightarrow 2CaSO_3 + CO_2 + H_2O \quad \textbf{(15.21b)}$$

Reaction (15.21b) requires an adequate residence time in the EHT (hence, an adequate EHT volume) and a high concentration of solids in the slurry. Reactions (15.21a) and (15.21b), when added together, portray the overall stoichiometry for SO_2 removal from flue gas. Thus, the overall stoichiometric reaction is:

$$CaCO_3 + SO_2 \rightarrow CaSO_3 + CO_2 \quad \textbf{(15.22)}$$

Excess oxygen in the flue gas results in oxidation of some of the $CaSO_3$ to $CaSO_4$ (gypsum). Gypsum precipitation in the scrubber can be prevented by maintaining a high liquid/gas ratio so that the pH of the solution remains fairly constant when SO_2 is absorbed. Gypsum precipitation in the EHT can be promoted by forced oxidation (introduction of additional air).

Inhibited oxidation (discussed in the previous section) is a recently developed alternative to forced oxidation, with the main objective being to prevent scaling in the scrubber. By adding a small amount of EDTA (ethylenediaminetetraacetic acid, $C_{10}H_{16}N_2O_8$) to this system, scaling is reduced even further. Inhibited oxidation is now being favored because it requires less operating power and maintenance, uses less fresh water, and generally costs less than forced oxidation (Dalton 1992).

SO_2 MASS TRANSFER

At the gas–liquid interface, we assume local equilibrium, which can be described by Henry's law as

$$\overline{P}_{SO_2(i)} = H_{SO_2} C_{SO_2(i)} \quad \textbf{(15.23)}$$

where

$\overline{P}_{SO_2(i)}$ = partial pressure of SO_2 at the interface, atm

H_{SO_2} = Henry's law constant, atm/(mol/L)

$C_{SO_2(i)}$ = aqueous SO_2 concentration at the interface, mol/L

The flux of SO_2 in the gas film is

$$\frac{N}{A} = k_g \left(\overline{P}_{SO_2} - \overline{P}_{SO_2(i)} \right) \quad \textbf{(15.24)}$$

where

$\dfrac{N}{A}$ = flux of SO_2, mol/s-cm^2

k_g = local mass transfer coefficient, mol/(s-cm^2-atm)

$\overline{P}_{SO_2}$ = partial pressure of SO_2 in the bulk gas, atm

The flux of SO_2 in the liquid film is

$$\frac{N}{A} = \phi k_l \left(C_{SO_2(i)} - C_{SO_2} \right) \tag{15.25}$$

where

ϕ = an enhancement factor to account for chemical reactions that permit SO_2 to diffuse through the liquid film as bisulfite or sulfite species as well as SO_2

k_l = local mass transfer coefficient, mol/(s-cm^2-mol/L)

C_{SO_2} = concentration of SO_2 in the bulk liquid, mol/L

Combining Eqs. (15.23) through (15.25) to eliminate the interfacial concentrations, we obtain

$$\frac{N}{A} = K_g \left(\overline{P}_{SO_2} - H_{SO_2} C_{SO_2} \right) \tag{15.26}$$

where K_g = overall gas phase transfer coefficient

In Eq. (15.26), K_g is given by

$$\frac{1}{K_g} = \frac{1}{k_g} + \frac{H_{SO_2}}{\phi k_l} \tag{15.27}$$

The two main enhancement reactions are those that convert SO_2 to bisulfite in the liquid film. The reactions are

$$SO_2 + H_2O \rightleftharpoons H^+ + HSO_3^- \tag{15.28}$$

and

$$SO_2 + SO_3^{-2} + H_2O \rightleftharpoons 2HSO_3^- \tag{15.29}$$

The enhancement factor can have a wide range of values, depending on several parameters. For enhancement factors greater than 20, the gas film resistance controls the overall mass transfer rate; for enhancement factors less than 20, both gas and liquid film resistances are important. Table 15.4 lists some chemical parameters and their effects on the enhancement factor.

The use of small amounts of chemical additives can significantly improve SO_2 scrubbing efficiency. These additives work by changing

Table 15.4 SO_2 Scrubber Chemical Parameters and Their Effects on the Enhancement Factor, ϕ

Parameter	Effect on ϕ
Gas composition	ϕ decreases as SO_2 concentration increases; for example, at pH 5.8,
	<table><tr><td>$\overline{P}_{SO_2}$, ppm</td><td>ϕ</td></tr><tr><td>500</td><td>10</td></tr><tr><td>1000</td><td>7.5</td></tr><tr><td>2000</td><td>5.8</td></tr></table>
Bulk liquid sulfite and bisulfite concentrations	ϕ increases as $C_{SO_3^{-2}}$ increases; ϕ decreases as $C_{HSO_3^-}$ increases
pH	ϕ increases as pH increases. (Note: because the solution tends to be in equilibrium with $CaSO_3$ solids, $C_{HSO_3^-}$ decreases as pH increases; thus, the effect of pH is essentially that of $C_{HSO_3^-}$)
Alkali additives	ϕ increases as alkali species increase
Buffer additives (organic acids)	ϕ increases as buffer additives increase (ϕ values of 20–30 are achieved with as little as 10–15 milli-mol/L of adipic acid)

the liquid phase chemistry and thus increasing the rate of SO_2 dissolution. Adipic acid (or adipic acid combined with succinic and glutaric acids) is very effective. For example, according to Dalton (1992), a 90% efficient limestone scrubber can be improved to 95% efficiency with as little as 200 ppm of additives, and to 99% with 500 ppm!

PHYSICAL FACTORS

In addition to the previously discussed chemical factors, there are several important physical factors that affect SO_2 absorption. One of the most important is the liquid/gas ratio (L/G). In general, the greater the L/G ratio, the greater the SO_2 absorption efficiency. However, high L/G ratios result in high operating costs because of higher pumping energy needs and the greater pressure drop in the absorber. Typically, L/G ratios range from 40 to 100 gal/1000 acf. Proper gas and liquid flow distribution is important to prevent alternating wet and dry spots, to prevent scaling, and to promote good gas–liquid contact. The limestone should be ground finely (90% passing a 325-mesh screen) for good utilization. As discussed in Chapter 13, the gas velocity affects both the scrubber diameter and the pressure drop. Loading or flooding can be a problem in limestone scrubbers. Good pH control helps to prevent scaling; scale formation at pHs of 6 or less is only about 5% of that at pHs of 6.2 or greater (Henzel et al. 1981). The power consumption of a limestone FGD unit is large—on the order of 3–6% of the power generated by the plant.

MATERIAL AND ENERGY BALANCES

Henzel et al. (1981) have provided complete examples for performing the material balance calculations and estimating the FGD system energy usage for a typical high-sulfur coal and a typical low-sulfur coal. In this section, we will present example problems to demonstrate some of their calculations for the case with a high-sulfur coal. A simplified process flow diagram is presented in Figure 15.6. The coal contains 3.7% sulfur, and has a heating value of 11,115 Btu/lb_m. The firing rate of coal is 210 tons/hour to generate 500 MW (gross). For our example problems, we will assume that we have already performed the initial coal combustion calculations and have established the rate and composition of the flue gas entering the scrubber (see Table 15.5). Certain other data are needed and/or assumptions must be made to perform the FGD system material balances (see Table 15.6). Together, Tables 15.5 and 15.6 make up the *design basis* for the calculations. The following example problems will illustrate these calculations.

Table 15.5 Flue Gas Characteristics for FGD Design Example Problems

Parameter		Value
Flow rate, acfm		1,615,600
Flow rate, lb_m/hour		5,337,650
Components	**lb_m/hr**	**Mole Percent**
Particulate matter	138	—
Carbon dioxide	942,500	11.76
Hydrogen chloride	432	0.01
Nitrogen	3,746,000	73.45
Oxygen	324,000	5.57
Sulfur dioxide	31,080	0.27
Moisture	293,500	8.94
SO_2 inlet loading		6.6 lb_m /10^6 Btu
Particulate loading		0.03 lb_m /10^6 Btu
Temperature		290 F

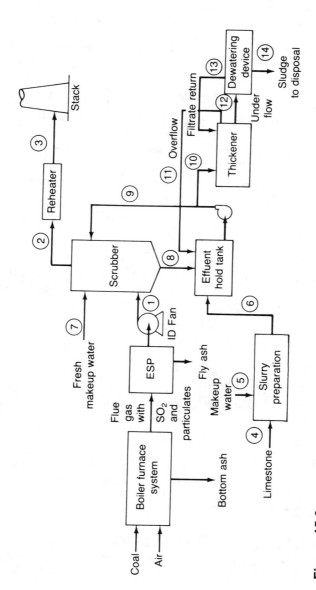

Figure 15.6
Simplified process flow diagram for material balance calculations.
(Adapted from Henzel et al., 1981.)

Table 15.6   Design Basis Assumptions/Data

1. The flue gas flow rate and composition are constant.
2. The limestone is composed of 94% $CaCO_3$, 1.5% $MgCO_3$ and 4.5% inert substances.
3. All inlet sulfur is as SO_2 (no SO_3 gas inlet).
4. The alkalinity supplied for SO_2 and HCl absorption is equal to 1.10 times the stoichiometric ratio.
5. All HCl generated by combustion of coal is completely absorbed.
6. Twenty percent of the absorbed SO_2 is oxidized to sulfate; 80% remains as sulfite.
7. The liquid/gas ratio is set at 75 gal/1000 acf (based on the saturated gas flow rate exiting the scrubber).
8. The gas becomes completely humidified in the scrubber, with no mist carryover out of the scrubber.
9. The use of reheat is to be considered. If used, assume that flue gas temperature must be increased by 40 F.

▶ ▶ ▶ ▶ ▶ ▶ ▶ ▶ ▶

Example 15.1

For the process described in the preceding discussion, calculate the amount of SO_2 that must be removed by the scrubber, and the amount emitted to the atmosphere.

Solution

Under current NSPSs, the FGD scrubber must be 90% efficient (see Chapter 1). Thus,

$$SO_2 \text{ absorbed} = 31{,}080 \text{ lb}_m/\text{hr} \times 0.90 = 27{,}972 \text{ lb}_m \text{ } SO_2/\text{hr}$$
$$SO_2 \text{ emitted} = 31{,}080 \text{ lb}_m/\text{hr} \times 0.10 = 3{,}108 \text{ lb}_m \text{ } SO_2/\text{hr}$$

◀ ◀ ◀ ◀ ◀ ◀ ◀ ◀ ◀

▶ ▶ ▶ ▶ ▶ ▶ ▶ ▶ ▶

Example 15.2

Calculate the limestone feed rate.

Solution

The $CaCO_3$ and $MgCO_3$ in the limestone must neutralize 432 lb_m/hr of HCl and 27,972 lb_m/hr of SO_2. On a molar basis, the number of moles of alkalinity required per hour are

$$\text{for } SO_2 : \ 27{,}972\frac{\text{lb}_m}{\text{hr}} \times \frac{1 \text{ lbmol}}{64.08 \text{ lb}_m SO_2} \times \frac{1 \text{ lbmol alk}}{\text{lbmol } SO_2} = 436.52\frac{\text{lbmol}}{\text{hr}}$$

$$\text{for HCl: } 432\frac{\text{lb}_m}{\text{hr}} \times \frac{1 \text{ lbmol}}{36.5 \text{ lb}_m HCl} \times \frac{0.5 \text{ lbmol alk}}{1 \text{ lbmol HCl}} = 5.92\frac{\text{lbmol}}{\text{hr}}$$

stoichiometric alkalinity required = 436.52 + 5.92 = 442.4 lbmol/hr

from Table 15.6, actual alkalinity = 1.10 × 442.4 = 486.6 lbmol/hr

Since 100 lb_m of purchased limestone contains 94.0 lb_m of $CaCO_3$, 1.5 lb_m of $MgCO_3$, and 4.5 lb_m of inert substances, the "active ingredients" per 100 lb_m are 0.9392 lbmol $CaCO_3$ and 0.0178 lbmol $MgCO_3$. Of the available alkalinity, $CaCO_3$ provides a mole fraction of 0.9392/(0.9392 + 0.0178) = 0.9815. Thus, the rate of $CaCO_3$ provided can be calculated as

$$486.6\frac{\text{lbmol alk}}{\text{hr}} \times 0.9815 \times \frac{100.09\ lb_m\,CaCO_3}{\text{lbmol } CaCO_3} = 47{,}803\frac{lb_m\ CaCO_3}{\text{hr}}$$

Similarly, the rate of $MgCO_3$ is

$$486.6 \times 0.0185 \times \frac{84.33\ lb_m\,MgCO_3}{\text{lbmol } MgCO_3} = 759\frac{lb_m\,MgCO_3}{\text{hr}}$$

The total alkalinity provided is 47,803 + 759 = 48,562 lb_m/hr, representing only 95.5% (by weight) of the purchased limestone. Therefore, the limestone feed rate is

$$48{,}562/0.955 = 50{,}850\ lb_m/\text{hr}$$

◀ ◀ ◀ ◀ ◀ ◀ ◀ ◀

▶ ▶ ▶ ▶ ▶ ▶ ▶ ▶ ▶

Example 15.3

Estimate the amount of water evaporated in the scrubber to humidify the flue gas. Also, calculate the composition, temperature, and volumetric flow rate of the cleaned flue gas leaving the scrubber.

Solution

Use of a psychrometric chart permits rapid estimation of the humidity and temperature of the saturated flue gas leaving the scrubber, but the values must be corrected for the differences in the molecular weights of air and flue gas.

From Table 15.5, the inlet humidity is

$$\mathscr{H}_{\text{inlet}} = 293{,}500/(5{,}337{,}650 - 293{,}500 - 138)$$
$$= 0.0582\ lb_m\,\text{water/}lb_m\ \text{dry gas}$$

The molar average molecular weight of the dry inlet flue gas is 30.5 lb_m/lbmol, so the equivalent humidity on a psychrometric chart for air is

$$\mathscr{H}_{\text{inlet (air basis)}} = 0.0582 \times \frac{30.51}{28.97} = 0.0613$$

Figure 15.7 (an abbreviated psychrometric chart for air–water) depicts the inlet gas (point A) and the adiabatic saturation that occurs in the scrubber. From point B on the chart, the outlet gas temperature is 128 F and the humidity is 0.104 lb_m water/lb_m dry air. The reheating process then occurs at constant absolute humidity to point C.

To correct the outlet humidity from air to flue gas, we need the molecular weight of the outlet gas (on a dry basis). The dry gaseous components exiting the scrubber are virtually the same as those entering, except that the absorbed SO_2 has been replaced mole for mole by CO_2, and all the HCl has been removed. The composition of the "cleaned" outlet flue gas (on a dry basis) is as follows:

Component	Flow Rate, lb_m/hr	Flow Rate, lbmol/hr	Mole Percent
Nitrogen	3,746,000	133,690	80.67
Oxygen	324,000	10,125	6.11
Sulfur dioxide	3,108	48.51	0.0293
Carbon dioxide	961,654	21,851	13.19
Total	5,034,762	165,714.5	

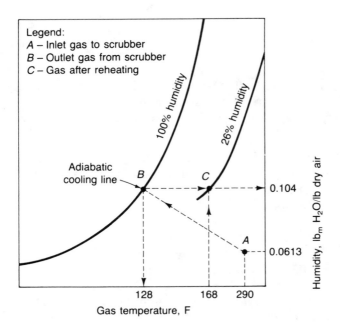

Figure 15.7
Psychrometric chart (not to scale) for Example 15.3.
(Adapted from Henzel et al., 1981.)

From the preceding table, we can calculate the molecular weight as 30.38, and the corrected outlet humidity is

$$\mathscr{H}_{\text{outlet (flue gas basis)}} = 0.104 \times \frac{28.97}{30.38} = 0.0992 \ \text{lb}_m \, \text{water/lb}_m \, \text{dry gas}$$

The outlet moisture flow rate is

$$0.0992 \frac{\text{lb}_m \, \text{water}}{\text{lb}_m \, \text{dry flue gas}} \times 5{,}034{,}762 \frac{\text{lb}_m}{\text{hr}}$$

$$= 499{,}450 \ \text{lb}_m/\text{hr of water vapor}$$

Thus, the rate of evaporation of water in the scrubber is

$$499{,}450 - 293{,}500 = 205{,}950 \ \text{lb}_m/\text{hr} = 412 \ \text{gpm}$$

all of which must be made up by fresh water input into the system. Finally, the volumetric flow rate of gases leaving the scrubber is

$$Q = \frac{\left(165{,}714.5 + \dfrac{499{,}450}{18.02}\right) \dfrac{\text{lbmol}}{\text{hr}} \times 0.7302 \, \dfrac{\text{ft}^3\text{-atm}}{\text{lbmol-}^\circ\text{R}} \times 588 \ \text{R} \times \dfrac{1 \ \text{hr}}{60 \ \text{min}}}{1.005 \ \text{atm}}$$

$$= 1{,}377{,}298 \ \text{acfm} \quad (\text{at } 128 \ \text{F})$$

◀ ◀ ◀ ◀ ◀ ◀ ◀ ◀

▶ ▶ ▶ ▶ ▶ ▶ ▶ ▶ ▶

Example 15.4

Calculate the sludge production rate (100% solids basis) and the free water consumption in the sludge assuming 60% solids is achieved. Calculate the total water make-up rate to the FGD system.

Solution

To perform the sludge calculations, we must know or assume the chemical formula of each compound precipitated. We must also include the unreacted $CaCO_3$ and the inert substances that entered with the limestone. For the sake of brevity, only the results of those calculations made by Henzel et al. (1981) are presented here.

Waste Sludge Solids	
Component	**Mass Flow Rate,** lb_m/hr
$CaCO_3$	4,430
$CaSO_3 \cdot \tfrac{1}{2}H_2O$	35,837
$CaSO_4 \cdot CaSO_3 \cdot \tfrac{1}{2}H_2O$	18,403
$CaSO_4 \cdot 2H_2O$	2,986
Inert substances	2,290
	63,946

The amount of free water contained in a 60%-solids sludge is

$$63,946 \frac{lb_m}{hr} \times \frac{1}{0.60} \times 0.40 = 42,630 \ lb_m \ / \ hr = 85 \ gpm$$

The total flow rate of make-up water is

Humidification	412 gpm
Sludge waste	85 gpm
Total	497 gpm

The mist eliminator wash water is set at 300 gpm, and the balance of the make-up water is used to slurry the limestone.

Energy consumption of an FGD system is primarily due to the fan, the pumps, and the stack gas reheater. The fan power depends on the volumetric flow rate and the pressure drop, as shown in Table 15.7. The pumping power depends primarily on the slurry recirculation rate (L/G ratio). Stack gas reheat can be provided by steam coils, by a fired heater, or by flue gas by-pass. The reheat required is often such that the stack gas is 40–50 F° hotter than the scrubber exhaust. FGD systems in the 1970s often included reheat in an attempt to prevent corrosion in the stack and to give the stack gases more buoyancy. However, reheat is very energy intensive and costly. Recently, U.S. designs have gone exclusively to the use of a "wet stack" (no reheat). It has been found that the wet stack approach can boost the overall thermal efficiency of the power plant by a full percentage point, saving over a million dollars per year in fuel costs. In addition, the capital costs of a reheater are now being saved, and corrosion problems have actually decreased. All other equipment combined (thickener, mill, conveyors, and so forth) uses relatively little energy, consuming about 20% of the power consumed by the fan plus slurry recirculation pumps.

Table 15.7
Typical Limestone FGD System Pressure Drops

Component	Pressure Drop, in. H$_2$O
SO$_2$ scrubber:	
Mobile bed or tray tower	6–8
Spray tower*	2–3
Mist eliminator	0.3–1
Reheater	1
Ductwork	2

* Although gas pressure drop is much lower in a spray tower than in a mobile bed or tray tower, spray towers incur much higher pumping costs owing to higher L/G ratios and high spray nozzle pressure drops.

▶ ▶ ▶ ▶ ▶ ▶ ▶ ▶ ▶

Example 15.5

Estimate the power requirements of the FGD system described in Examples 15.1–15.4. Assume a system pressure drop of 11 in. H_2O. Note that the fan is located between the ESP and the scrubber (Q is 1,616,000 acfm at 290 F and 0.98 atm). Assume that a reheat of 40 F° is needed (T_{stack} = 168 F), and that the heat is added using a 90%-efficient heat exchanger. Assume that the slurry recirculation pumps deliver 90 feet of head, and that the specific gravity of the slurry is 1.09. Assume that all the other pumps and mechanical equipment consume about 20% of the power consumed by the fan and recirculation pumps. Assume that the fan and the pumps have mechanical efficiencies of 70%.

Solution
Fan power

$$P_F = \frac{0.0001575 \text{ hp}}{\text{in. } H_2O\text{-cfm}} \times \frac{1,616,000 \text{ cfm}}{0.7} \times 11 \text{ in. } H_2O$$

$$= 4000 \text{ hp} = 2983 \text{ kW}$$

Slurry recirculation pumps

$$L = \frac{75 \text{ gal}}{1000 \text{ acf}} \times 1,377,300 \text{ acfm} = 103,300 \text{ gpm}$$

$$P_p = \frac{0.0002527 \text{ hp}}{\text{ft } H_2O\text{-gpm}}$$

$$\times \frac{90 \text{ ft slurry} \times 1.09 \text{ ft } H_2O/\text{ft slurry} \times 103,300 \text{ gpm}}{0.7}$$

$$= 3658 \text{ hp} = 2729 \text{ kW}$$

All other mechanical equipment

$$P_o = 0.20 \times (4000 + 3658) \text{ hp}$$
$$= 1532 \text{ hp} = 1141 \text{ kW}$$
$$\text{total electric power} = 2983 + 2729 + 1141$$
$$= 6853 \text{ kW} = 6.85 \text{ MW}$$

The thermal energy required to generate 6.85 MW (at 35% efficiency) is

$$Q_p = 6853 \text{ kW} \times \frac{3412 \text{ Btu/hr}}{\text{kW}} \times \frac{1}{0.35} = 6.682(10)^7 \text{ Btu/hr}$$

Reheat energy

$$Q_H = \frac{(5{,}034{,}762 + 499{,}450)\,\text{lb}_\text{m}/\text{hr} \times 0.25\ \text{Btu/lb}_\text{m}\text{-}°\text{F} \times 40\ \text{F}}{0.9}$$

$$= 6.15(10)^7\ \text{Btu/hr}$$

Total energy consumption of the FGD system

$$Q_T = 6.68(10)^7 + 6.15\ (10)^7\ \text{Btu/hr}$$

$$= 1.28(10)^8\ \text{Btu/hr} \qquad \text{(of which reheat is about half)}$$

For a 35%-efficient power plant, the heat input rate is

$$Q_i = \frac{500\ \text{MW}}{0.35} \times \frac{1000\ \text{kW}}{1\ \text{MW}} \times \frac{3412\ \text{Btu/hr}}{\text{kW}}$$

$$= 4.87(10)^9\ \text{Btu/hr}$$

Thus, this FGD system consumes about

$$\frac{1.28(10)^8}{4.87(10)^9} \times 100\% = 2.6\%$$

of the total energy (coal) purchased. Note that if a reheater is not chosen, then the FGD system only consumes about 1.3% of the total energy purchased.

The preceding examples have illustrated how plant engineers estimate limestone needs, water requirements, and sludge production for large FGD systems. The ability to perform such material balances is an important tool for an engineer, and should be used not only in design but also in monitoring plant operations. A partially completed material balance table for the preceding design is presented in Table 15.8. The stream numbers refer to those labeled in Figure 15.6. Balances around individual pieces of equipment are necessary to solve for the flow rates and compositions of all of the streams numbered in Figure 15.6. After completion of all the material balances, the next step is sizing of the main pieces of equipment.

Table 15.8   Partial Material Balance Table for Limestone FGD Example
Problems (500-MW plant, 3.7%-sulfur coal, 90% SO_2 removal)

Gas Streams

Stream Number*	1	2	3
Flow, 1000 acfm	1615.6	1377.3	1471.0
Flow, 1000 lb_m/hr	5328	5534.4	5534.4
Temperature, °F	290	128	168
SO_2, lb_m/hr	31,080	3108	3108
HCl, lb_m/hr	432	0	0
CO_2, 1000 lb_m/hr	942.5	961.7	961.7
N_2, 1000 lb_m/hr	3746	3746	3746
O_2, 1000 lb_m/hr	324	324	324
H_2O, 1000 lb_m/hr	293.5	499.5	499.5
Particulate matter, lb_m/hr	138	138	138

Liquid and Solid Streams

Stream Number*	4	5	7	14
Flow, gal/min	—	197	300	—
Flow, 1000 lb_m/hr	50.9	98.5	150	106.6
$CaCO_3$, 1000 lb_m/hr	47.8	0	0	4.43
$MgCO_3$, 1000 lb_m/hr	0.8	0	0	—
$CaSO_3$ and $CaSO_4$ compounds, 1000 lb_m/hr	0	0	0	57.2
Inert substances, 1000 lb_m/hr	2.3	0	0	2.3
Free water, 1000 lb_m/hr	0	98.5	150	42.6

*Refer to Figure 15.6 for stream identification.
Adapted from Henzel et al., 1981.

◆ 15.4 COSTS ◆

Cost estimates of FGD systems can be highly variable depending on factors such as type of process, size of plant, type of project (new or retrofit), percentage of sulfur in the coal, location of the plant, costs of raw materials, values of by-products (if any), type of ultimate disposal required for waste products, and so forth. The technology of FGD systems is now more advanced than it was in the 1970s, and costs have decreased (in constant dollars). Nevertheless, FGD systems still represent a huge investment (as much as 20% of the capital cost of a new coal-fired power plant). It had been estimated that the capital expenditures for FGD (new and retrofit) systems in the United States during the 1990s would be 6–8 billion dollars, and by the year 2000, utilities would be spending some 3 billion dollars per year to operate these systems (Dalton 1992). Actual expenditures have not been this high. Differences in cost estimates can arise because of the different assumptions used in different studies for inflation, plant life, cost of capital, prices of raw materials or by-products, and so forth. As an example, in one study (Jahnig and Shaw 1981b), projected delivered prices of $60/ton and $45/ton were used to assign by-product credits to elemental sulfur and sulfuric acid, respectively. However, significant changes occurred in supply and demand for those two products, and by 1984, actual prices for sulfur and sulfuric acid were approximately $135/ton and $55/ton, respectively.

Economic and technical evaluations of several FGD systems were conducted in a study by the Electric Power Research Institute (EPRI) and reported by Radcliffe (1992). In that study, costs were developed for a hypothetical, moderately difficult retrofit of a single 300-MW unit burning a 2.6%-sulfur bituminous coal. Each FGD process removed 90% of the SO_2 except for certain dry injection processes, which removed 50%. A retrofit project was selected as the base case because of the high demand that had been created for these projects by the Clean Air Act Amendments (CAAA) of 1990. Radcliffe (1992) reported that approximately 15,000 MW of FGD retrofits had been ordered as of September 1992, and estimated that another 40,000 MW will be ordered in response to Phase 2 of the CAAA.

As part of the EPRI study, a comprehensive cost-estimating computer program was developed. This menu-driven model computes cost estimates for a number of different FGD processes using user-input site-specific data, and built-in data for the FGD processes. Over 100 EPRI member utilities have used the program, and many have reported that the computer-generated cost estimates were within 5% of their final vendor quotes (Radcliffe 1992). The capital cost estimates are considered to have an absolute accuracy of ± 20% and a relative accuracy of ± 10%.

Costs were developed for each alternative, starting with a detailed process flow sheet, material balance, equipment list, and utility consumption list. Capital and levelized control costs were computed in 1990 dollars. The *levelized control cost* is a present worth cost (in 1990 dollars) assuming a 15-year plant life. A comparison of the projected costs with a previous EPRI study (1983) shows that the 1990 capital costs are not much different from the 1982 costs projected in that study. The fact that the costs have not escalated as might be expected can be explained by two interesting developments. First, the maturation of FGD technology has led to reduced engineering fees and less complex designs. This has produced more efficient FGD systems, which cost less to build and operate. Second, intense competition among FGD system (and component) suppliers has kept prices down.

Hesketh and Cross (1994) reported FGD retrofit costs adjusted to 1992 dollars, using data from the same 1990 EPRI study. A summary comparison of the capital and levelized operating costs of several retrofit FGD systems are presented in Table 15.9. As can be seen for the scenario used in the EPRI study, limestone FGD is not the cheapest throwaway system, considering either capital investment or levelized operating costs. Another conclusion that can be drawn from Table 15.9 is that the regenerative systems have significantly higher capital costs and net operating costs than the throwaway processes. The difference in costs between throwaway and regenerative processes is significant. An additional conclusion is that lime-spray drying had lower capital costs than the wet scrubbing systems, and had lower operating costs per ton of SO_2 removed from the flue gas. Among the regenerative processes, the MgO and Wellman-Lord processes were very similar in cost. Remember that for this hypothetical 300-MW plant retrofit, a difference of only \$33/kW in capital cost is equivalent to 10 million dollars.

Recently, the U.S. Environmental Protection Agency has supported the development of two computer models to provide air pollution control cost analyses for coal-fired power plants. The models are CUECost (Coal Utility Environment Cost–Keeth et al. 1999) and SUSCM (State-of-the-art Utility Scrubber Cost Model–Srivastava 2000). CUECost3 is an Excel 5.0 workbook and is available for free from EPA (http://www.epa.gov/ttn/catc/products.html#software).

CUECost3 provides preliminary cost estimates (± 30% accuracy) and is relatively easy to use. The input parameters are those that have the most effect on costs—plant capacity, heat rate, sulfur content, coal heating value, capacity factor, and disposal mode. For example, using CUECost3, the capital cost estimates for a limestone (forced oxidation) system for a 500-MW power plant range from \$191/kW to \$212/kW. SUSCM is an enhancement of the CUECost algorithms to better reflect recent improvements in technology (such as rubber-lined

Table 15.9   FGD Retrofit System[1] Cost Estimates (in 1992 Dollars)[2]

Process	Investment Cost, $/kW	Levelized Operating Cost, $/ton SO_2 removed
Throwaway		
Limestone (gypsum product)	225	550
Limestone (forced oxidation)	210	540
Dual alkali (limestone)	190	480
Dual alkali (lime)	185	505
Lime-spray drying	150	470
Regenerative		
Wellman-Lord	270	600
MgO	275	615

[1] Data are for a 300-MW plant burning 2.6%-sulfur bituminous coal, retrofitted with FGD.
[2] The levelized cost data were calculated by the present worth method, assuming a 15-year plant life and no inflation (refer to the original article for details).

Adapted from Hesketh and Cross, 1994.

carbon steel for the scrubber instead of an alloy). As predicted by SUSCM, the capital cost estimates are very sensitive to plant capacity, dropping from over \$500/kW for a 100-MW plant to about \$150/kW for a 900-MW plant.

Srivastava and Jozewicz (2001) built on the CUECost algorithms to develop state-of-the-art cost models for FGD systems. Their models performed well in predicting costs that were reasonably close to those actually incurred at large power plants. In general, the reported costs of limestone with forced oxidation applications varied between about \$180/kW to \$350/kW, while the cost of lime-spray drying applications varied between about \$170/kW to \$230/kW.

◆ PROBLEMS ◆

15.1 Assuming that only $CaSO_3$ is produced, calculate the daily production rate (in tons/day) of a 55%-solids sludge from a 90%-efficient limestone FGD system on a 600-MW power plant burning 3.5%-sulfur coal. The plant has a thermal efficiency of 35%, and the coal has a heating value of 12,000 Btu/lb_m. Assume that the limestone is 95% $CaCO_3$ and 5% inert substances, and that the ratio of the actual amount of limestone fed to the theoretical amount required is 1.15.

15.2 Rework Problem 15.1 using lime (CaO) as the reagent. Assume the lime is 98% CaO and 2% inert compounds, and that its feed ratio (actual to theoretical) is 1.03.

15.3 Estimate a typical make-up rate (in lb_m/hr) of sodium carbonate to a Wellman-Lord FGD system for the power plant of Problem 15.1.

15.4 Estimate the water evaporation rate (in gpm) in a scrubber that treats 1,000,000 acfm of flue gas. The flue gas enters at 240 F and 1 atm with 5% relative humidity, and leaves at 1 atm with 100% relative humidity. The molecular weights of the entering and exiting flue gas streams are 30.5 and 30.3, respectively (dry basis).

15.5 Estimate the energy requirements of an FGD system. The scrubber exhaust is 1,000,000 acfm at 130 F and 1 atm. The fan moves 1,200,000 acfm at an actual pressure rise of 10 in. H_2O. The L/G ratio is 65 gal/1000 acf. The pump delivers 75 ft of head, and the slurry specific gravity is 1.10. The required reheat is 50 °F. Make other reasonable assumptions as necessary.

15.6 Estimate annual revenues from sulfur sales for a U.S. Bureau of Mines citrate FGD process that is removing 90% of the SO_2 from a 40%-efficient 400-MW power plant. The coal is 5% sulfur and has a heating value of 20,000 kJ/kg. Sulfur can be sold for \$150/ metric ton, and H_2S can be purchased from a nearby refinery for

$160/metric ton. Also, calculate the net reduction of operating costs for the system (in mills/kWh) due to the net sulfur revenues.

15.7 A power plant produces a flue gas with a mass flow rate of 2 million pounds per hour. Of that, 150,000 lb/hr is water vapor, and 15,000 lb/hr is SO_2. All the other gases that are not SO_2 or water have an average molecular weight of 28. (a) Calculate the concentration of SO_2 in the flue gas in ppm(v) (dry basis). Regulations require that 92% of the SO_2 be captured, and a limestone scrubber will be used to do the job. Assume that scrubber exit gases are saturated with water vapor at 125 F. (b) Estimate the SO_2 concentration of the exiting gases in ppm(v) (dry basis). Assume that there is no HCl in the gases, and that the limestone is 96% $CaCO_3$ (with a MW of 100) and 4% inerts. The alkalinity supplied will be 1.12 times the stoichiometric ratio. (c) Calculate the limestone feed rate, lb/hr. Assume that the flow rate exiting the scrubber is 800,000 acfm and the scrubber pressure drop is 10 in. H_2O. Calculate the fan power used just by the scrubber, in hp.

REFERENCES

Armstrong, W. H., and Westmoreland, J. "The Carter Creek Gas Plant," *Chemical Engineering Progress, 81*(2), February 1985.

Brown, G. N., Torrence, S. L., Repik, A. J., Stryker, J. L., and Ball, F. J. "SO_2 Recovery via Activated Carbon," *Chemical Engineering Progress, 68*(8), August 1972.

Clean Power from Coal: The Bureau of Mines Citrate Process. Washington, DC: U.S. Department of the Interior, 1978.

Dalton, S. M. *Evolution of Flue Gas Desulfurization Design—Recent Changes in the U.S.A.*, a paper presented at the Symposium on Energy and Environment: Transitions in Eastern Europe, Prague, Czechoslavakia, April 1992.

Electric Power Research Institute. *Economic Evaluation of FGD Systems.* Vol. II, *Regenerable FGD Processes, High-Sulfur Coal*, CS-3342, Research Project 1610–1, Final Report, prepared by Stearns-Roger Engineering Corporation, Palo Alto, CA, December 1983.

Henzel D. S., Laseke, B. A., Smith, E. D., and Swenson, D. O. *Limestone FGD Scrubbers: User's Handbook*, EPA-600/8-81-017. Washington, DC: U.S. Environmental Protection Agency, August 1981.

Hesketh, H. E., and Cross, F. L., Jr., Eds. *Sizing and Selecting Air Pollution Control Systems.* Lancaster, PA: Technomic Publishing, 1994, p. 150.

Jahnig, C. E., and Shaw, H. "A Comparative Assessment of Flue Gas Treatment Processes, Part I—Status and Design Basis," *Journal of the Air Pollution Control Association, 31*(4), April 1981(a).

——. "A Comparative Assessment of Flue Gas Treatment Processes, Part II—Environmental and Cost Comparison," *Journal of the Air Pollution Control Association, 31*(5), May 1981(b).

Keeth, R., Blagg, R., Burklin, C., Kosmicki, B., Rhodes, D., and Waddell, T. *Coal Utility Environmental Cost (CUECost) Workbook User's Manual,*

Version 1.0, EPA-600/R-99-056. National Risk Management Research Lab, Research Triangle Park, NC, June 1999.

Laseke, B. A., and Dewitt, T. W. "Status of Flue Gas Desulfurization," *Chemical Engineering Progress, 75*(2), February 1979.

Radcliffe, P. "FGD Economics," *EPRI Journal 17*(6), September 1992.

Srivastava, R. K. *Controlling SO$_2$ Emissions: A Review of Technologies.* EPA-600/R-00-093. National Risk Management Research Lab, Research Triangle Park, NC, October 2000.

Srivastava, R. K., and Jozewicz, W. "Flue Gas Desulfurization: The State of the Art," *Journal of the Air and Waste Management Association, 51*(12), December, 2001.

U.S. Environmental Protection Agency. *Research Summary—Acid Rain,* EPA-600/8-79-028. Washington, DC, 1979(a).

———. *Sulfur Oxides Control Technology Series: Flue Gas Desulfurization—Wellman-Lord Process.* Technology Transfer Summary Report, EPA-625/8-79-001. Research Triangle Park, NC, 1979(b).

———. *National Air Pollutant Emission Estimates, 1940–1990,* EPA-450/4-91-026. Research Triangle Park, NC, 1991.

———. *National Air Quality and Emissions Trends Report, 1998,* EPA-454/R-00-003. Research Triangle Park, NC, 2000.

 CHAPTER 16

CONTROL OF NITROGEN OXIDES

Air pollution in the Los Angeles area is characterized by a decrease in visibility, crop damage, eye irritation, objectionable odor, and rubber deterioration. These effects are attributed to the release of large quantities of hydrocarbons and nitrogen oxides to the atmosphere. The photochemical action of the nitrogen oxides oxidizes the hydrocarbons and thereby forms ozone.

A. J. Haagen-Smit, 1952

◆ 16.1 INTRODUCTION ◆

In the late 1940s, A. J. Haagen-Smit and coworkers discovered that a certain type of smog resulted from atmospheric reactions involving nitrogen oxides and reactive hydrocarbons. After that revelation, most of the initial efforts at controlling photochemical smog were directed toward reducing hydrocarbon emissions, perhaps because that was a much easier problem to tackle. However, research was undertaken toward reducing nitrogen oxides emissions as well. During the past two decades, a better understanding of the formation of nitrogen oxides has been gained, and that knowledge has led to the development of some innovative control technologies for nitrogen oxides. Additionally, interest in NO_x control has increased in recent years owing to the persistence of photochemical smog in many urban areas.

The seven oxides of nitrogen that are known to occur are NO, NO_2, NO_3, N_2O, N_2O_3, N_2O_4 and N_2O_5. Of these seven oxides of nitrogen, NO (nitric oxide) and NO_2 (nitrogen dioxide) are the two most important air pollutants because they are emitted in large quantities. The term "NO_x" can refer to all of the oxides of nitrogen, but in air pollution work NO_x generally refers only to NO and NO_2. Nitrogen oxides are emitted in the

493

United States at a rate of about 22 million metric tons per year, about 50% of which is emitted from mobile sources. (Mobile sources include automobiles, trucks, buses, and so forth, and will be discussed in detail in Chapter 18.) Of the 9–10 million metric tons of nitrogen oxides that originate from stationary sources, about 30% is the result of fuel combustion in large industrial furnaces and 60% is from electric utility furnaces (U.S. Environmental Protection Agency 1991, 2000). Table 16.1 presents some data on U.S. emissions of NO_x for the past thirty years.

About 95% of all NO_x from stationary combustion sources is emitted as NO. Nitric oxide is formed by either or both of two mechanisms—*thermal NO_x* or *fuel NO_x* (U.S. Environmental Protection Agency 1983). **Thermal NO_x** is the NO_x formed by reactions between nitrogen and oxygen in the air used for combustion. The rate of formation of thermal NO_x is extremely temperature sensitive, and becomes rapid only at "flame" temperatures (3000–3600 F). **Fuel NO_x** results from the combustion of fuels that contain organic nitrogen in the fuel (primarily coal or heavy oil). Fuel NO_x formation is dependent on local combustion conditions (oxygen concentration and mixing patterns) and on the nitrogen content of the fuel.

Because electric utility boilers emit large quantities of NO_x, we will briefly review some typical emission factors. There are several types of furnace/boilers as distinguished by the firing mode. The major firing modes are single or opposed wall, cyclonic, and tangentially

Table 16.1   Trend in U.S. Emissions of Nitrogen Oxides (in teragrams*/year)

Source	1970	1975	1980	1985	1990	1998
Transportation						
Highway vehicles	6.3	7.6	7.8	7.0	6.4	7.0
All other vehicles						
(including non-road)	1.6	1.8	2.0	2.9	4.4	4.8
Subtotal	7.9	9.4	9.8	9.9	10.8	11.8
Stationary Fuel Combustion						
Electric utilities	4.4	5.2	6.4	6.7	6.0	5.5
Industries	3.9	3.4	3.1	2.9	2.7	2.7
Commercial/institutional	0.3	0.3	0.3	0.2	0.3	0.3
Residential	0.4	0.4	0.4	0.4	0.7	0.7
Subtotal	9.0	9.3	10.2	10.2	9.7	9.2
Industrial Processes	0.7	0.7	0.7	0.6	0.6	0.8
All Other Sources						
(forest fires, open burning,						
and solid waste incineration)	0.8	0.3	0.3	0.2	0.3	0.3
Total	18.4	19.7	21.0	20.9	21.4	22.1

*1 teragram = 1 million metric tons = 10^9 kg.
Adapted from U.S. Environmental Protection Agency, 1991, 2000.

fired units. Also, depending on the ash characteristics, the furnace can be a dry-bottom furnace (solid ash removal) or a wet-bottom furnace (molten ash removal). Cyclonic-fired boilers are used primarily for burning coal; the other types can burn coal, oil, or gas. Owing to the declining availability and higher costs of oil and gas, most new utility boilers are being designed to burn pulverized coal.

Emissions of NO_x vary depending on the type of fuel and the type of firing. (The factors influencing NO_x emissions will be explored in more detail in Section 16.2.) Some examples of emission factors (without controls) are presented in Table 16.2. Figure 16.1 presents typical ranges of NO concentrations in various types of coal-fired utility boilers as functions of power output.

Successful control of NO_x depends on an understanding of the fundamental principles of NO_x formation. In this chapter, we will present a brief introduction to the chemistry of NO_x formation. However, note that NO_x control is still a field of very active research, and the detailed mechanisms of NO_x formation are not precisely known. In the past, NO_x has been the most difficult and expensive pollutant to control.

Table 16.2   Selected NO_x Emission Factors (without controls)

Type of Boiler	Type of Fuel	Emissions Factor[a] kg NO_x/Mg of Coal (or as indicated)
Pulverized coal	Anthracite	9
Pulverized coal		
Dry bottom	Bituminous and subbituminous	$10.5(7.5)^b$
Wet bottom	Bituminous and subbituminous	17
Cyclone (crushed coal)	Bituminous and subbituminous	18.5
Pulverized coal	Lignite	$6\text{–}7(4)^b$
Utility boilers		
Tangentially fired	Residual oil	5^c
Vertically fired	Residual oil	12.6^c
All (in general)	Natural gas	8.80^d
Industrial boilers	Residual oil	6.6^e
	Distillate oil	2.4^c
	Natural gas	2.24^d

[a] Total nitrogen oxides are expressed as NO_2.
[b] The values in parentheses are for tangentially fired units.
[c] Emissions are expressed as kg NO_2/1000 liters of oil burned.
[d] Emissions are expressed as kg NO_2/1000 cubic meters of gas burned.
[e] NO_x emissions in industrial boilers depend strongly on the nitrogen content of the residual oil.

Adapted from U.S. Environmental Protection Agency, 1991.

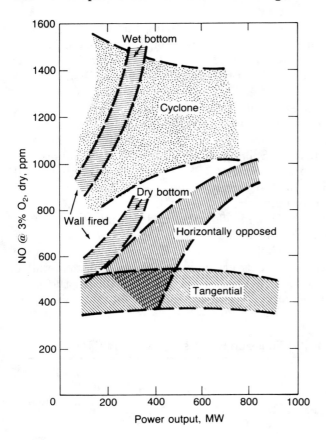

Figure 16.1
Typical (uncontrolled) NO flue gas concentrations for coal-fired boilers.
(U.S. Environmental Protection Agency, 1983.)

Thus, the techniques and technology for NO_x control are relatively new. The material here will serve as an overview of NO_x control. We will not attempt to provide definitive design information in this chapter.

♦ 16.2 CHEMISTRY OF NO$_x$ FORMATION ♦

THERMAL NO$_x$

Both thermodynamic (equilibrium) and kinetic (rate) information are important in the understanding of NO_x formation. The main chemical reactions responsible for NO_x formation were first proposed by Zeldovich (1946). The reactions in Zeldovich's model are as follows:

$$N_2 + O \rightleftharpoons NO + N \qquad (16.1)$$

$$N + O_2 \rightleftharpoons NO + O \qquad (16.2)$$

Reactions (16.1) and (16.2) are supplemented by the reaction

$$N + OH \rightleftharpoons NO + H \qquad (16.3)$$

Reactions (16.1) and (16.2) are the most important reactions in Zeldovich's model. In presenting Reactions (16.1)–(16.3), it is assumed that the fuel combustion reactions (between C, H, and O) have reached equilibrium and that the concentrations of O, H, and OH can be described by equilibrium equations.

If we first consider only the thermodynamics of NO$_x$ formation, we can begin our discussion with the stoichiometric relationships, which are

$$N_2 + O_2 \rightleftharpoons 2NO \qquad (16.4)$$

$$NO + \tfrac{1}{2}O_2 \rightleftharpoons NO_2 \qquad (16.5)$$

The equilibrium constants for Reactions (16.4) and (16.5) (respectively) are

$$K_{P_1} = \frac{\left(\overline{P}_{NO}\right)^2}{\overline{P}_{N_2}\,\overline{P}_{O_2}} = \frac{\left(y_{NO}\right)^2}{y_{N_2}\,y_{O_2}} \qquad (16.6)$$

and

$$K_{P_2} = \frac{\overline{P}_{NO_2}}{\left(\overline{P}_{NO}\right)\left(\overline{P}_{O_2}\right)^{1/2}} = \left(P_T\right)^{-1/2}\frac{y_{NO_2}}{\left(y_{NO}\right)\left(y_{O_2}\right)^{1/2}} \qquad (16.7)$$

where

K_P = equilibrium constant
$\overline{P}_i$ = partial pressure of component i, atm
y_i = mole fraction of component i
P_T = total pressure, atm

Table 16.3 presents data for the equilibrium constants K_{P_1} and K_{P_2} for various temperatures at atmospheric pressure.

Table 16.3   Equilibrium Constants for the Formation of NO and NO$_2$

Temperature		K_P	
°K	°F	$N_2 + O_2 \rightleftharpoons 2NO$	$NO + \tfrac{1}{2}O_2 \rightleftharpoons NO_2$
300	80	10^{-30}	$1.4(10)^6$
500	440	$2.7(10)^{-18}$	$1.3(10)^2$
1000	1340	$7.5(10)^{-9}$	$1.2(10)^{-1}$
1500	2240	$1.1(10)^{-5}$	$1.1(10)^{-2}$
2000	3140	$4.0(10)^{-4}$	$3.5(10)^{-3}$
2200	3500	$3.5(10)^{-3}$	$2.6(10)^{-3}$

Adapted from *Joint Army, Navy, Air Force Thermochemical Tables*, 1986.

▶ ▶ ▶ ▶ ▶ ▶ ▶ ▶ ▶

Example 16.1

Considering only Reaction (16.4), calculate the equilibrium concentration of NO for a flue gas at 1500 K and 1 atm total pressure with a composition of 76% N_2, 4% O_2, 8% CO_2, and 12% H_2O.

Solution

From Table 16.3, the value of K_P for the reaction $N_2 + O_2 \rightleftharpoons 2NO$ at 1500 K is $1.1(10)^{-5}$. Therefore,

$$\frac{\left(\bar{P}_{NO}\right)^2}{\bar{P}_{N_2}\bar{P}_{O_2}} = 1.1(10)^{-5}$$

Thus,

$$\left(\bar{P}_{NO}\right)^2 = 1.1(10)^{-5}(0.76)(0.04)$$

$$\bar{P}_{NO} = 5.78(10)^{-4} \text{ atm}$$

The mole fraction of NO ($\bar{P}_{NO}/P_T$) is $5.78(10)^{-4}$, and the equilibrium concentration of NO is 578 ppm. Because this concentration of NO is such a small fraction of the total, it is not necessary to correct the stated composition of the gases. However, if a large amount of NO had been produced, the concentrations of the reactants (N_2 and O_2) would have had to be adjusted downward.

◀ ◀ ◀ ◀ ◀ ◀ ◀ ◀ ◀

Table 16.4 presents calculated equilibrium concentrations of NO and NO_2 at various temperatures in heated air and in flue gas (with assumed concentrations of 76% N_2 and 3.3% O_2).

Table 16.4   Calculated Equilibrium Concentrations (in ppm) of NO and NO_2 in Air and Flue Gas

Temperature		Air		Flue Gas	
°K	°F	NO	NO_2	NO	NO_2
300	80	$3.4(10)^{-10}$	$2.1(10)^{-4}$	$1.1(10)^{-10}$	$3.3(10)^{-5}$
800	980	2.3	0.7	0.8	0.1
1400	2060	800	5.6	250	0.9
1873	2912	6100	12	2000	1.8

Note: The reactions considered are $N_2 + O_2 \rightleftharpoons 2NO$ and $NO + \frac{1}{2}O_2 \rightleftharpoons NO_2$. The flue gas is defined to contain 76% N_2 and 3.3% O_2.

Adapted from U.S. Environmental Protection Agency, 1970.

A review of Tables 16.3 and 16.4 leads to the following "expectations":

1. At flame-zone temperatures (3000–3600 F), we expect to observe very high NO_x concentrations (6,000–10,000 ppm), and ratios of NO/NO_2 ranging from 500:1 to 1000:1.

2. At flue gas exit temperatures (300–600 F), we expect to observe very low NO_x concentrations (<1 ppm), and ratios of NO/NO_2 ranging from 1:10,000 to 1:10.

In actual furnaces, we observe neither of the preceding two cases. NO_x concentrations exiting from large coal-fired power plants at typical flue gas temperatures range from 300–1200 ppm, and the ratios of NO/NO_2 range from 10:1 to 20:1. We must consider factors other than equilibrium to explain such behavior.

The rate of NO formation is the other major factor influencing the actual NO_x concentrations. Values of the kinetic constants for Reactions (16.1)–(16.3) are given in Table 16.5. Consideration of these reactions, with the assumption that O and H atoms and OH radicals in the post-flame zone are at equilibrium values, leads to the development of a theoretical rate expression for NO formation. The rate is a strong function of the temperature, as well as of the nitrogen and oxygen concentrations.

We start by writing expressions for the net rates of formation of NO and N based on Reactions (16.1)–(16.3):

$$r_{NO} = k_{+1}[N_2][O] - k_{-1}[NO][N] + k_{+2}[N][O_2] - k_{-2}[NO][O] \\ + k_{+3}[N][OH] - k_{-3}[NO][H] \tag{16.8}$$

$$r_N = k_{+1}[N_2][O] - k_{-1}[NO][N] - k_{+2}[N][O_2] + k_{-2}[NO][O] \\ - k_{+3}[N][OH] + k_{-3}[NO][H] \tag{16.9}$$

Reaction	Rate constant k, m^3/mol-s
$(1)\ N_2 + O \overset{k_{+1}}{\underset{k_{-1}}{\rightleftharpoons}} NO + N$	$k_{+1} = 1.8(10)^8\ e^{-38,370/T}$ $k_{-1} = 3.8(10)^7\ e^{-425/T}$
$(2)\ N + O_2 \overset{k_{+2}}{\underset{k_{-2}}{\rightleftharpoons}} NO + O$	$k_{+2} = 1.8(10)^4\ T\, e^{-4680/T}$ $k_{-2} = 3.8(10)^3\ T\, e^{-20,820/T}$
$(3)\ N + OH \overset{k_{+3}}{\underset{k_{-3}}{\rightleftharpoons}} NO + H$	$k_{+3} = 7.1(10)^7\ e^{-450/T}$ $k_{-3} = 1.7(10)^8\ e^{-24,560/T}$

Table 16.5   Rate Constants for the Zeldovich Mechanism (for NO_x formation)

Note: T is in degrees Kelvin.
Adapted from Flagan and Seinfeld, 1988.

Because Reaction (16.1) has a high activation energy, most of that reaction occurs only after the fuel combustion reactions are complete (but before the gases flow out of the hot flame zone). Thus, we are justified in assuming that the concentrations of O, H, and OH are at equilibrium levels. In addition, since free nitrogen atoms are much more reactive than NO, the N atoms are consumed about as fast as they are generated. This is the quasi-steady-state assumption—that the concentration of N atoms is about constant (at $[N]_{ss}$) throughout most of the reaction time.

Setting the net rate of production of N atoms (r_N) equal to zero, Eq. (16.9) can be solved for $[N]_{ss}$.

$$[N]_{ss} = \frac{k_{+1}[N_2][O] + k_{-2}[NO][O] + k_{-3}[NO][H]}{k_{-1}[NO] + k_{+2}[O_2] + k_{+3}[OH]} \qquad \textbf{(16.10)}$$

This result can then be substituted into Eq. (16.8) to get an expression for r_{NO} that depends only on the concentrations of N_2, NO, and O_2, and the equilibrium concentrations of O, H, and OH.

$$r_{NO} = k_{+1}[N_2][O] - k_{-2}[NO][O] - k_{-3}[NO][H] + \qquad \textbf{(16.11)}$$
$$\left(-k_{-1}[NO] + k_{+2}[O_2] + k_{+3}[OH]\right) \times \left([N]_{ss}\right)$$

The interesting thing about Eq. (16.11) is that the initial rate of formation of NO (when concentrations of NO are small) is just equal to twice that of Reaction (16.1). That is

$$r_{NO_{initial}} = 2\,k_{+1}[N_2][O] \qquad \textbf{(16.12)}$$

▶ ▶ ▶ ▶ ▶ ▶ ▶ ▶ ▶

Example 16.2

Given a hydrocarbon flame at 1870 C where the mole fractions of N_2 gas and O atoms are 0.75 and $9.5(10)^{-4}$ respectively, (a) calculate the initial rate of NO formation (in mol/m^3-s); and (b) if this rate holds constant for 0.03 seconds, calculate the concentration (in ppm) of NO in the gases leaving the flame zone.

Solution

(a) At T = 1870 C (2143 K),

$$k_{+1} = 1.8(10)^8\,e^{-38,370/2143} = 3.015 \text{ m}^3/\text{mol-s}$$

Assuming P = 1 atm, the molar density of the gases is

$$\rho_M = P/RT = \frac{1 \text{ atm}}{8.206(10)^{-5}\dfrac{\text{m}^3\text{-atm}}{\text{mol-K}}\,2143 \text{ K}}$$

$$\rho_M = 5.686 \text{ mol/m}^3$$

Therefore,

$$[N_2] = 0.75 \times 5.686 = 4.26 \text{ mol/m}^3$$

$$[O] = 9.5(10)^{-4} \times 5.686 = 0.0054 \text{ mol/m}^3$$

The initial rate of NO formation is

$$r_{NO} = 2(3.015)(4.26)(0.0054)$$

$$= 0.1388 \text{ mol/m}^3\text{-s}$$

(b) If this initial rate holds constant for 0.03 seconds, then

$$[NO] = 0.1388(0.03)$$

$$= 4.16(10)^{-3} \text{ mol/m}^3$$

Converting to ppm,

$$[NO] = \frac{4.16(10)^{-3}}{5.686}(10)^6 = 732 \text{ ppm}$$

◀ ◀ ◀ ◀ ◀ ◀ ◀ ◀

Several investigators (Bartok et al. 1972; Malte and Pratt 1975; Sarofim and Pohl 1973; and others) have observed experimentally that NO concentrations in the flame zone are significantly higher than could have been formed by the Zeldovich mechanism. Some investigators believe that such "prompt" NO formation is due to super-equilibrium radical concentrations that are likely to exist in hydrocarbon flames (Bowman 1975; Fenimore 1971).

Fenimore (1971) has observed concentrations of "prompt" NO (NO formed in the first five milliseconds) in the range from 40 ppm to over 100 ppm. Duterque, Avegard, and Borghi (1981) reported on the difficulties of trying to sample from very fast gas-phase reactions at reactor residence times less than four milliseconds. Prompt NO is more important in fuel-rich combustion zones. It is likely that the intermediate, hydrogen cyanide (HCN), is formed when N_2 is attacked by hydrocarbon radicals that are formed during "flame combustion." HCN combines with OH to form CN, and then CN is oxidized to NO. Also, N atoms and NH radicals are formed and can be oxidized to NO in the flame (Flagan and Seinfeld 1988).

MacKinnon (1974) experimentally studied heated mixtures of N_2, O_2, and Ar, as well as air. MacKinnon showed that below 1600 C, concentrations of less than 200 ppm NO were formed. Above 1800 C, concentrations of several thousand ppm NO were formed, and at about 1950 C, concentrations as high as 12,000–13,000 ppm could be formed. Furthermore, MacKinnon observed a peak NO concentration at 3615 F ± 46 F° (1990 C ± 26 C°). Above 3700 F, the net formation of NO decreases (because of decomposition reactions). NO concentrations increased rapidly with time up to about 4–5 seconds, after which no further increases were observed. MacKinnon developed a mathematical rate expression for NO formation from his experimental results. His global model predicts the concentration of NO as a function of temperature, nitrogen and oxygen concentrations, and time. At a pressure of 1 atm, the model is

$$C_{NO} = 5.2(10)^{17} [\exp - 72,300/T] y_{N_2} y_{O_2}^{1/2} t \qquad (16.13)$$

where

C_{NO} = NO concentration, ppm

y_i = mole fraction of component i

T = absolute temperature, °K

t = time, s

In actual flame systems, the presence of H, C, OH, S, and other atoms and radicals can significantly affect the NO formation rate, and a global model is no substitute for a detailed mechanistic model. However, consideration of Eq. (16.13) is sufficient to explain (at least qualitatively) the differences between observed and expected NO_x concentrations. The gases are only in the flame zone for a short time (for about 0.5 seconds). Although actual concentrations do not reach equilibrium concentrations, an appreciable amount of NO_x is formed owing to the very rapid rates of the reactions. Once the gases have moved away from the hot flame zone, they cool rapidly, reducing the rates of reactions by orders of magnitude, and effectively "freezing" the concentrations of NO and NO_2 at the levels that were formed initially.

Published information (U.S. Environmental Protection Agency 1970) can be used to illustrate how rapidly NO can form by comparing the time required to obtain a concentration of 500 ppm NO for various temperatures. At 3600 F, a concentration of 500 ppm NO would be formed in 0.12 seconds; at 3200 F, it would take 1.1 seconds; and at 2800 F, it would take 16 seconds. The oxidation of NO to NO_2 is less temperature sensitive than the formation of NO. Even so, as the temperature drops quickly (as it does when the gases move away from the hottest part of the combustion zone), both reaction rates slow down so much that the amount of NO emitted is essentially that which is originally formed.

Breen et al. (1971) have addressed the problem of NO formation in power-plant boilers. They used a computer to simulate about 50 simultaneous chemical reactions in the boiler. They investigated the effects of time, temperature, flue gas recirculation, combustion air preheat, and the *fuel equivalence ratio*. (The **fuel equivalence ratio** ϕ is a multiple of the theoretical fuel/air ratio, and is the inverse of the *stoichiometric ratio*. The **stoichiometric ratio** is unity when the actual air/fuel ratio equals the theoretical air/fuel ratio needed for complete combustion with no excess oxygen.) Some of the results of Breen et al. are presented in Figure 16.2. Figure 16.2(a) depicts the relationship between NO concentration, time, and temperature in a natural-gas–fired furnace. Figure 16.2(b) shows the effect of the fuel equivalence ratio on nitric oxide concentrations.

Consideration of Eq. (16.13) can provide insight into some of the strategies for control of thermal NO$_x$. The peak temperature should be reduced, the gas residence time at the peak temperature should be reduced, and the oxygen concentration in the zone of highest temperature should be reduced. Engineered methods of achieving these strategies (and their effectiveness) will be discussed in Section 16.3.

Fuel NO$_x$

When a fuel contains organically bound nitrogen (as do most coals and residual fuel oils), the contribution of the fuel-bound nitrogen to the total NO$_x$ production is significant. The N–C bond is considerably weaker than the N–N bond in molecular nitrogen, so it is not surprising that fuel nitrogen can be oxidized to NO. Both laboratory (Pershing et al. 1975) and full-scale experiments (Thompson and McElroy 1976) have shown that fuel-bound nitrogen can account for over 50% of the total NO$_x$. The kinetic mechanisms of fuel nitrogen oxidation are currently an area of active research.

Not all organic nitrogen is converted to NO$_x$. The nitrogen content of most U.S. coals ranges from 0.5% to 2%, whereas that of residual fuels ranges from 0.1% to 0.5%. Conversion efficiencies of fuel nitrogen to NO$_x$ for coals and residual fuel oils have been observed between 10% and 60% (U.S. Environmental Protection Agency 1983). Possible fates of fuel nitrogen are summarized in Figure 16.3.

The oxidation of fuel nitrogen to NO is highly dependent on the air/fuel ratio. In Figure 16.4 (on page 506), data are plotted showing the percent conversion of fuel nitrogen to NO$_x$ as a function of the fuel equivalence ratio ϕ. The fuel equivalence ratio primarily affects the oxidation of the volatile R–N fraction (where R represents an organic fragment) rather than the nitrogen remaining in the char. The degree of fuel-air mixing also strongly affects the percent conversion of fuel nitrogen to NO, with greater mixing resulting in greater percent conversion.

Small temperature changes do not seem to affect production of NO$_x$ from fuel nitrogen (Pershing 1976). This behavior is in direct contrast to thermal NO$_x$ production, which is highly sensitive to temperature.

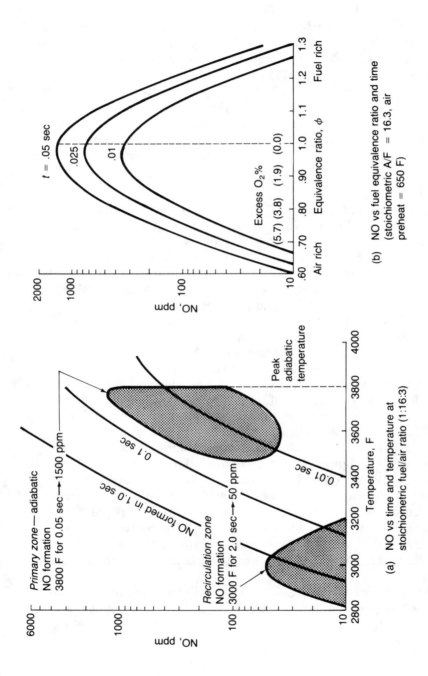

Figure 16.2
Simulated effects of time, temperature, and fuel/air ratio on nitric oxide formation in a natural-gas-fired boiler.
(Adapted from Breen et al., 1971.)

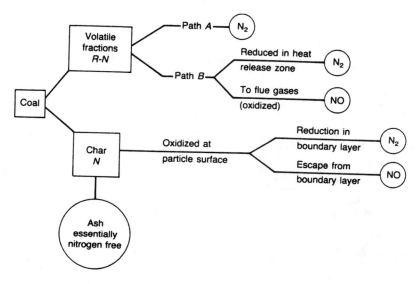

Figure 16.3
Possible fates of nitrogen contained in coal.
(Adapted from Heap et al., 1976.)

▶ ▶ ▶ ▶ ▶ ▶ ▶ ▶ ▶

Example 16.3

A fluidized-bed combustor burns 1000 kg/hr of coal with 2% (by weight) nitrogen at 1600 K, using an air/fuel ratio that is twice the stoichiometric ratio. Use Figure 16.4 (ignoring any thermal NO$_x$ formation) to estimate the mass emission rate of nitrogen oxides.
(a) Express the emissions as NO. (b) Also express the emissions as NO$_2$.

Solution

(a) The fuel equivalence ratio is the inverse of the stoichiometric ratio, and thus equals ½. From Figure 16.4, we can estimate that 35% of the fuel nitrogen is converted to NO. Thus, expressed as NO, the mass emissions rate is

$$\dot{M}_{NO_x} = 0.35 \times 0.02 \times 1000 \frac{kg}{hr} \times \frac{30 \text{ kg NO}}{14 \text{ kg N}} = 15 \frac{\text{kg NO}_x \, (\text{as NO})}{hr}$$

(b) Expressed as NO$_2$, the mass emissions rate is

$$\dot{M}_{NO_x} = \frac{15 \text{ kg NO}_x \,(\text{as NO})}{hr} \times \frac{46 \text{ kg NO}_2}{30 \text{ kg NO}} = 23 \frac{\text{kg NO}_x \,(\text{as NO}_2)}{hr}$$

◀ ◀ ◀ ◀ ◀ ◀ ◀ ◀

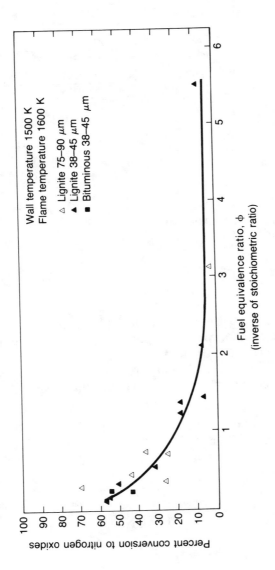

Figure 16.4
Conversion of fuel nitrogen to NO_x (for certain pulverized coals).
(Adapted from Pohl and Sarofim, 1976.)

16.3 NO$_x$ CONTROL: ♦ STATIONARY SOURCES

♦

The two broad categories of NO$_x$ controls are combustion modifications and flue gas treatment techniques. Combustion modifications are used to limit the formation of NO$_x$ during the actual combustion. Flue gas treatment techniques are used to remove NO$_x$ from flue gases after the NO$_x$ has been formed. Although flue gas treatment systems are in commercial use in Japan (Parkinson 1981), some observers in the United States believe that combustion controls alone might be sufficient to reduce NO$_x$ emissions. Current studies indicate that combustion modifications are able to reduce NO$_x$ emissions from full-scale power plants by 30–50% (Kokkinos et al., 1992).

▶ ▶ ▶ ▶ ▶ ▶ ▶ ▶ ▶

Example 16.4

A coal-fired power plant has a flue gas concentration of 1000 ppm NO$_x$. The anthracite coal has a heating value of 12,000 Btu/lb$_m$. Flue gas is produced in the ratio of 20 lb$_m$ flue gas/lb$_m$ coal, and the flue gas has a molecular weight of 28.0. Calculate the percentage reduction in NO$_x$ required to meet the NSPS for NO$_x$ emissions.

Solution

From Chapter 1, the NSPS for this case is 0.6 lb$_m$ NO$_x$/10^6 Btu of heat input (expressed as NO$_2$). Expressing all the NO$_x$ as NO$_2$, the allowable concentration of NO$_2$ in the exhaust gases is

$$\frac{0.6 \text{ lb}_m \text{NO}_2}{10^6 \text{Btu}} \times \frac{12,000 \text{ Btu}}{\text{lb}_m \text{coal}} \times \frac{1 \text{ lb}_m \text{ coal}}{20 \text{ lb}_m \text{ gas}} \times \frac{28 \text{ lb}_m \text{ gas}}{\text{lbmol gas}}$$

$$\times \frac{1 \text{ lbmol NO}_2}{46 \text{ lb}_m \text{NO}_2} \times 1(10)^6 = 219 \text{ ppm}$$

$$\text{Percent reduction} = \frac{1000 \text{ ppm} - 219 \text{ ppm}}{1000 \text{ ppm}} \times 100\% = 78.1\%$$

◀ ◀ ◀ ◀ ◀ ◀ ◀ ◀ ◀

COMBUSTION MODIFICATIONS

As discussed in Section 16.2, there are several factors that contribute to high NO$_x$ formation. Combustion controls reduce NO$_x$ formation by one or more of the following strategies:

1. Reduce peak temperatures of the flame zone
2. Reduce gas residence time in the flame zone
3. Reduce oxygen concentrations in the flame zone

The preceding changes to the combustion process can be achieved by either (1) modification of operating conditions on existing furnaces, or (2) purchase and installation of newly designed (low-NO_x) burners and/or furnaces.

Both process modifications and new burner/furnace designs rely on the following concepts to implement the three main strategies for reducing NO_x emissions (U.S. Environmental Protection Agency 1983):

1. Reduce peak temperatures by
   - using a fuel-rich primary flame zone
   - increasing the rate of flame cooling
   - decreasing the adiabatic flame temperature by dilution
2. Reduce the gas residence time in the hottest part of the flame zone by
   - changing the shape of the flame zone
   - using the steps listed in Strategy 1
3. Reduce the O_2 content in the primary flame zone by
   - decreasing the overall excess air rates
   - controlled mixing of fuel and air
   - using a fuel-rich primary flame zone

Modification of Operating Conditions.   During the 1980s, intensive research and development efforts led to a number of successful tactics that can be used to reduce NO_x formation without buying new burners or furnaces. These tactics include

1. Low-excess-air firing (LEA)
2. Off-stoichiometric combustion (OSC) (includes overfire air)
3. Flue gas recirculation (FGR)
4. Gas reburning
5. Reduced air preheat and/or reduced firing rates
6. Water injection

Low-excess-air (LEA) firing is a very simple yet effective technique. Forty years ago it was not uncommon to see furnaces operating with 50–100% excess air. (**Excess air** is the amount of air in excess of that stoichiometrically required for 100% combustion; 50% excess air corresponds to a stoichiometric ratio of 1.5 or an equivalence ratio of 0.667.) As fuel prices escalated, the percent of excess air (EA) was decreased to about 15–30% to save money. This level of EA was considered a practical limit because it gave good combustion but did not require extensive furnace monitoring. Owing to less-than-perfect mixing of air and fuel, there must be some EA present at all times to ensure good fuel use and to prevent smoke formation. In recent years, development of advanced instrumentation has allowed continuous automatic furnace monitoring and control of EA, and the percent EA can now be reduced below the 15–30% limit. The EPA conducted tests on various boilers with various fuels. The results indicate that a 19% average reduction of NO_x (from 495 ppm to 408 ppm, both corrected to dry flue gas at 3% O_2) can be achieved by reducing the percent of excess air from an average of 20% to an average of 14% (Lim et al. 1980).

Off-stoichiometric combustion (OSC) (often called *staged combustion*) combusts the fuel in two or more steps. The initial or primary flame zone is fuel-rich, and the secondary (and following) zones are fuel-lean. Without retrofitting with specially designed burners, OSC can be accomplished (1) by firing some of the burners (usually the lower row) fuel-rich and the rest fuel-lean, or (2) by taking some of the burners out of service and allowing them only to admit air to the furnace, or (3) by firing all of the burners fuel-rich and admitting the remaining air over the top of the flame zone (*overfire air*). In a test of 31 boilers burning coal, oil, or gas, NO_x emissions were reduced an average of 34% using Method (2) (Lim et al. 1980). Careful monitoring of the flue gas is necessary with this method to protect against CO, smoke, and possible coal slagging.

Flue gas recirculation (FGR) is simply the rerouting of some of the flue gas back to the furnace. Usually, flue gas from the economizer outlet is used, and so the furnace air temperature and the furnace oxygen concentration are reduced simultaneously. In retrofit applications FGR can be very expensive. In addition to requiring new large ducts, major modifications to the fans, dampers, and controls might be required. Furthermore, the additional gas flow through the firebox and flues might cause operating and maintenance problems. FGR is usually better applied to new designs than to existing furnaces.

Gas reburning involves the injection of natural gas or some other fuel into the boiler above the main burners at a rate equal to about 10–25% of the total heat input (Kokkinos 1992). This creates a fuel-rich "reburn" zone in the middle of the boiler above the primary combustion zone. NO_x-laden combustion gases from the primary zone flow into the reburn zone and mix in the fuel-rich environment with the reburning gases. The reburn fuel does not form a well-defined flame, but rather burns uniformly throughout the zone. Hydrocarbon radicals (formed as intermediates of the rich combustion process) react with the NO_x from below and chemically reduce the NO_x to molecular nitrogen. The combustion process is completed in a burnout zone above the reburn zone where fresh air is injected to complete the oxidation of the remaining hydrocarbons and CO.

Pilot- and full-scale tests cosponsored by the Electric Power Research Institute indicate that reburning can reduce NO_x emissions by 40–60% without increasing CO emissions (Kokkinos 1992). This technique can only be applied to boilers tall enough to provide sufficient residence time (1.5 seconds) for both the reburn and the burnout processes to be completed. Reburning is the primary NO_x control technique for "cyclone"-type coal-fired boilers.

Reduced air preheat lowers peak temperatures in the flame zone, thus reducing thermal NO_x. However, unless an alternative means of recovering heat is available, a substantial overall energy penalty results. A similar statement can be made for **reduced firing rates**. By reducing the firing rate (derating the boiler), we reduce the heat release per unit volume (combustion intensity). Reducing the firing rate generally reduces thermal NO_x formation, but creates several problems. Besides the obvious penalty of reducing unit capacity, low load operation usually requires increased excess air to control smoke and CO emissions. Also, operating flexibility is reduced.

Water injection (or steam injection) can be an effective means of reducing flame temperatures, thus reducing thermal NO_x. Water injection has been shown to be very effective for gas turbines, with NO_x reductions of about 80% for a water injection rate of 2% of the combustion air (Crawford et al. 1977). The energy penalty for a gas turbine is about 1% of its rated output, but for a utility boiler, it can be as high as 10%.

Low-NO_x Equipment. New low-NO_x burners represent the most common equipment design change for reducing NO_x formation. Low-NO_x burners are not only effective on new power plants, but also can be readily applied to older facilities as retrofit projects. Basically, low-NO_x burners inhibit NO_x formation by controlling the mixing of fuel and air. Different burner manufacturers use different hardware to control the fuel-air mixing, but all designs essentially automate two effective tactics described in the previous section—LEA and OSC. Tests indicate that low-NO_x burners reduce NO_x emissions by 40–60% compared with older, conventional burners. A schematic diagram showing some of the general features found in several low-NO_x burner designs is presented in Figure 16.5.

Acceptance of low-NO_x burners by the power industry has been good. In fact, low-NO_x burners are now a standard part of new designs. However, a large percentage of utility coal-fired NO_x emissions are from plants built prior to 1971. These older plants are not now regulated, but many of them could be retrofitted with low-NO_x burners for a significant overall reduction in U.S. NO_x emissions.

Burner spacing can also affect NO_x formation. In a large power boiler, there may be fifteen to twenty burners on the front wall. With all the flames merging because of tight spacing, higher temperatures (hence, greater NO_x formation) result, especially in the central portion of the burner matrix. In newer designs, greater distances between burners are being provided. This strategy produces less interaction among flames, and more radiant cooling of individual flames.

Furnaces themselves are being designed in larger sizes. Larger furnaces inhibit NO_x formation in similar ways as increased burner

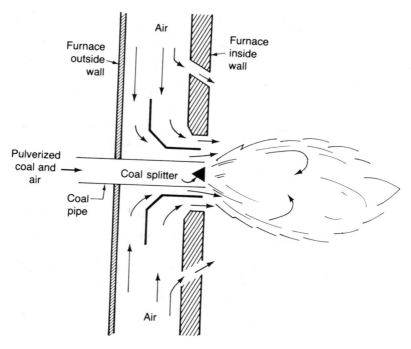

Figure 16.5
Schematic drawing of a low-NO$_x$ pulverized coal burner.

spacing. Larger enclosures provide more time for complete combustion using off-stoichiometric combustion burners. Also, larger enclosures provide more waterwall tubes to achieve the same amount of heat transfer from lower-temperature flames. Finally, larger enclosures reduce turbulent mixing of the fuel and air, which also inhibits NO$_x$ formation (particularly fuel NO$_x$)

FLUE GAS TREATMENT TECHNIQUES

Flue gas treatment (FGT) to remove NO$_x$ is useful in cases where higher removal efficiencies are required than can be achieved with combustion controls. Flue gas treatment is also used where combustion controls are not applicable (such as in controlling NO$_x$ emissions from HNO$_3$ plants). FGT for NO$_x$ control has been emphasized much more in Japan than in the United States, and most of the current processes were developed in Japan. FGT techniques are broadly classified as dry or wet techniques; the dry techniques include catalytic reduction, non-catalytic reduction, and adsorption, and the wet process is absorption.

Catalytic Reduction.   The most advanced FGT method is **selective catalytic reduction** (SCR). There are more than 70 full-scale

SCR units operating in Japan (U.S. Environmental Protection Agency 1983). In selective catalytic reduction, only the NO_x species are reduced (ultimately to N_2 gas). Nonselective catalytic reduction can be used, but is less desirable than selective catalytic reduction because free oxygen as well as NO_x is consumed by the reductant.

With a suitable catalyst, NH_3, H_2, CO, or even H_2S could be used as the reducing gas, but the most commonly used material is NH_3. The catalyst is a mixture of titanium and vanadium oxides and is formulated in pellets (for gas-fired units) or honeycomb shapes (for coal- or oil-fired units, which might have particulates in the flue gas). The stoichiometries of the reactions (U.S. Environmental Protection Agency 1983) are

$$4NO + 4NH_3 + O_2 \rightarrow 4N_2 + 6H_2O \qquad \textbf{(16.14)}$$

$$2NO_2 + 4NH_3 + O_2 \rightarrow 3N_2 + 6H_2O \qquad \textbf{(16.15)}$$

The best temperature range for SCR catalyst activity and selectivity is from 300 to 400 C (600 to 800 F). Ammonia is vaporized and injected downstream from the economizer (boiler feedwater preheater) as shown in Figure 16.6. SCR units typically achieve about 80% NO_x reduction. Fouling of the catalyst is a significant concern with coal-fired power plants.

▶ ▶ ▶ ▶ ▶ ▶ ▶ ▶ ▶

Example 16.5

A power plant operates with 800 ppm NO_x in the flue gas. The flue gas flow rate is $2.0(10)^6$ acfm at 300 C and 1 atm. An SCR system is being designed for 75% removal of NO_x. Calculate the stoichiometric amount of ammonia required in kg/day.

Solution

Assuming 100% NO, one mole of NH_3 is required per mole of NO. Therefore,

$$\dot{M}_{NO} = \frac{(1 \text{ atm})\left[2.0(10)^6 \text{ ft}^3/\text{min}\right]}{(1.314 \text{ atm-ft}^3/\text{lbmol-}°K)(573 \text{ K})} \times 800(10)^{-6} = 2.12 \frac{\text{lbmol}}{\text{min}}$$

$$\dot{M}_{NH_3} = 0.75 \times 2.12 \frac{\text{lbmol}}{\text{min}} \times \frac{17 \text{ lb}_m}{\text{lbmol}} \times \frac{1 \text{ kg}}{2.205 \text{ lb}_m} \times \frac{1440 \text{ min}}{\text{day}}$$

$$= 17,700 \frac{\text{kg NH}_3}{\text{day}}$$

◀ ◀ ◀ ◀ ◀ ◀ ◀ ◀ ◀

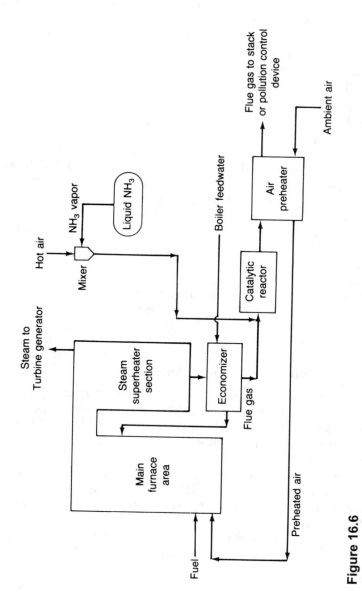

Figure 16.6

Schematic flow diagram for the selective catalytic reduction (SCR) method of NO$_x$ control.

Selective Noncatalytic Reduction (SNR).   At   temperatures of 900–1000 C, NH_3 will reduce NO_x to N_2 without a catalyst. At NH_3: NO_x molar ratios of 1:1 to 2:1, about 40–60% NO_x reduction can be achieved (U.S. Environmental Protection Agency 1983). If only 40–60% NO_x reduction is needed, SNR might be preferred over SCR because of the operating simplicity and lower cost of SNR. Potential problems with SNR include incomplete mixing of NH_3 with the hot flue gas, and improper temperature control. If the temperature is too low, unreacted ammonia will be emitted; if the temperature is too high, NH_3 will be oxidized to NO.

Adsorption.   Several **dry sorption** techniques have been proposed and demonstrated for simultaneous control of NO_x and SO_x. One type of system uses activated carbon with NH_3 injection to simultaneously reduce the NO_x to N_2 and oxidize the SO_2 to H_2SO_4. The carbon must be operated in the temperature range of 220–230 C, and must be regenerated to remove the H_2SO_4. Thermal regeneration (to produce a concentrated stream of SO_2) appears to be economically preferred to removing the acid by washing (U.S. Environmental Protection Agency 1983). If no sulfur is present in the fuel, the carbon acts as a catalyst for NO_x reduction only.

Another adsorption system uses a copper oxide catalyst. The copper oxide adsorbs SO_2 to form copper sulfate. Both copper oxide and copper sulfate are reasonably good catalysts for the selective reduction of NO_x with NH_3. The catalyst beds are regenerated with hydrogen to yield an SO_2-rich stream that can be further processed to elemental sulfur or H_2SO_4. This process has been installed on a 40-MW oil-fired boiler in Japan, and has shown 90% SO_x removal along with 70% NO_x reduction (U.S. Environmental Protection Agency 1983).

Wet Absorption.   Wet absorption or wet-scrubbing processes usually remove SO_x as well as NO_x. The main disadvantage of wet absorption of NO_x is the low solubility of NO. Indeed, the NO must often be oxidized to NO_2 in the flue gas before a reasonable degree of absorption can occur in water.

Counce and Perona (1983) published a detailed theoretical model along with experimental data to explain the NO_x–HNO_x–H_2O system. They combined chemical reaction rates, mass transfer rates, and equilibrium data to develop their model, which matched experimental results over the range of feed NO_x concentrations studied (0.01 to 0.10 atm partial pressure). The reactions in their model include the following:

Gas phase

$$NO + \tfrac{1}{2}O_2 \rightarrow NO_2 \qquad \textbf{(16.16)}$$

$$2NO_2 \rightleftharpoons N_2O_4 \qquad \textbf{(16.17)}$$

$$NO + NO_2 \rightleftharpoons N_2O_3 \qquad \textbf{(16.18)}$$

$$NO + NO_2 + H_2O \rightleftharpoons 2HNO_2 \qquad \textbf{(16.19)}$$

Liquid phase

$$N_2O_3 + H_2O \rightarrow 2HNO_2 \tag{16.20}$$

$$N_2O_4 + H_2O \rightarrow HNO_2 + HNO_3 \tag{16.21}$$

$$2NO_2 + H_2O \rightarrow HNO_2 + HNO_3 \tag{16.22}$$

$$3HNO_2 \rightleftharpoons HNO_3 + H_2O + 2NO \tag{16.23}$$

Counce and Perona conducted their experimental work at gas-phase concentrations of NO$_x$ that are much higher than are found in flue gases from combustion sources. To effectively absorb NO at normal flue gas concentrations, chemical enhancement is required. One technique for chemical enhancement involves the gas-phase oxidation of NO to NO$_2$ followed by absorption of NO$_2$ in a caustic scrubbing solution. Another technique is absorption of NO in a caustic scrubbing solution followed by oxidation in the liquid phase.

Uchida, Kobayashi, and Kageyama (1983) conducted an experimental study in which they investigated NO absorption into both aqueous KMnO$_4$/NaOH and Na$_2$SO$_3$/FeSO$_4$ solutions. Their basic chemical equations for aqueous KMnO$_4$/NaOH absorption of NO are as follows:

At high pH

$$NO + MnO_4^- + 2OH^- \rightarrow NO_2^- + MnO_4^{-2} + H_2O \tag{16.24}$$

At low or neutral pH

$$NO + MnO_4^- \rightarrow NO_3^- + MnO_2 \tag{16.25}$$

As reported by Uchida, Kobayashi, and Kageyama (1983), Reaction (16.25) results in solid MnO$_2$, which floats on the surface of the solution and reduces gas transfer rates, and is therefore not practical.

Absorption of NO in a solution of Na$_2$SO$_3$/FeSO$_4$ proceeds according to the following reactions:

$$FeSO_4 + NO \rightleftharpoons Fe(NO)SO_4 \tag{16.26}$$

$$Fe(NO)SO_4 + 2Na_2SO_3 + 2H_2O \rightarrow Fe(OH)_3 + Na_2SO_4 + NH(SO_3Na)_2 \tag{16.27}$$

Uchida et al. (1983) conducted trials with NO concentrations of 399 ppm, 900 ppm, and 1790 ppm, which are typical flue gas NO concentrations. In general, they found that the rate of NO absorption in the KMnO$_4$/NaOH solution was in the range of 1–10 mol/s-cm^2, and the rate in the Na$_2$SO$_3$/FeSO$_4$ solution was in the range of 1–4 mol/s-cm^2. In the future, wet scrubbing techniques might or might not gain widespread acceptance. However, note that if wet SO_x scrubbing is practiced at a particular location, it might be practical to consider simultaneous NO_x scrubbing as well.

It was mentioned earlier that NO can be oxidized in the flue gas to make the NO_x more soluble and amenable to scrubbing. There has been considerable research into ways to accomplish this. During the past five to ten years, several investigators have studied the injection of hydrogen peroxide as an oxidizer for NO and other pollutants in the flue gas (Kasper et al. 1996; Zamansky et al. 1996; Collins et al. 2001). In this process, liquid hydrogen peroxide is sprayed into hot flue gases, and the peroxide splits into OH radicals that begin a radical chain reaction oxidizing the NO to NO_2 and HNO_3. Some of the key reactions are as follows (Lee, Pennline, and Markussen 1990; Chao 1994; Zamansky et al. 1996):

$$H_2O_2 \rightleftharpoons 2OH \tag{16.28}$$
$$OH + NO \rightleftharpoons HNO_2 \tag{16.29}$$
$$OH + NO_2 \rightleftharpoons HNO_3 \tag{16.30}$$
$$OH + SO_2 \rightleftharpoons HSO_3 \tag{16.31}$$
$$OH + H_2O_2 \rightleftharpoons HO_2 + H_2O \tag{16.32}$$
$$HO_2 + NO \rightleftharpoons NO_2 + OH \tag{16.33}$$
$$HO_2 + SO_2 \rightleftharpoons SO_3 + OH \tag{16.34}$$
$$OH + HO_2 \rightleftharpoons H_2O + O_2 \tag{16.35}$$

The optimum temperature for this sequence of reactions is about 500 °C. Collins et al. (2001) showed in a pilot-scale reactor that the NO can be converted to products with about 90% efficiency using a mole ratio of H_2O_2 to NO_x of about 1.0, a level that might make the process cost-competitive with SCR (Haywood and Cooper 1998).

♦ 16.4 COSTS ♦

With the recent and ongoing development (and rapid changes) in NO_x control technology, information on costs must be considered tentative at best. Low-NO_x burners appear to be very cost effective, yielding 50% NO_x reductions on coal-fired boilers at capital costs of about $8.50–$13/kW (Kokkinos et al. 1992). Gas reburn retrofits have been estimated to cost between $15 and $45/kW. Remember that for a 1000-MW plant, a capital cost of $20/kW is equal to $20 million.

Selective catalytic reduction and selective noncatalytic reduction systems generally cost more (per ton of NO_x removed) than low-NO_x burners. In the 1980s, SCR system capital costs were greater than $100/kW (Moore 1984). Maxwell, Burnett, and Faucett (1980) reported preliminary capital investments for various NO_x and NO_x/SO_x removal systems. The average of the estimated costs of three SCR systems in 1980 was $140/kW (excluding the cost of the ESP). The range of estimated costs of three wet scrubbing systems for removing both NO_x and SO_x was from $205/kW to $482/kW ($205 million to $482 million for a 1000-MW plant). However, costs are significantly lower today than they were 20 years ago.

Because of improved designs and more familiarity with the technology, costs for NO_x controls have been coming down in the past decade. The U.S. EPA (1999) reports that for a coal-fired power plant, SCR capital costs dropped to the range of $70/kW in the late 1990s. "Cost effectiveness" is a term used by EPA to measure the cost to control NO_x; it is usually reported as $/ton of NO_x destroyed. Table 16.6 summarizes some cost estimates for three popular forms of NO_x control. However, EPA (1999) points out that it is likely that costs will decline further since there is now increased competition among vendors who supply these systems.

Table 16.6   Estimated Costs of NO_x Controls in 1997

Process	Capital Cost $/kW	Percent Control %	Cost Effectiveness $/ton of NO_x destroyed
Low-NO_x Burners	15–30	40–65	15–600
SNR	10–20	35–40	700–1300
SCR	69–72	70–80	500–2800

U.S. Environmental Protection Agency, 1999.

◆ Problems ◆

16.1   Considering only Reaction (16.4), calculate the equilibrium concentration of NO (in ppm) for a flue gas at 1500 K and 1 atm pressure with a composition of 77% N_2, 8% O_2, 6% CO_2, and 9% H_2O.

16.2   Rework Problem 16.1 if the flue gas is at 2000 K and 1 atm.

16.3   Rework Problem 16.1 for a flue gas at 2000 K with a composition of 75% N_2; 1% O_2, 8% CO, and 16% H_2O.

16.4   Estimate the total uncontrolled emission rate of NO_x (in kg/min) from a pulverized-coal boiler that is 40% efficient and produces 400 MW of electricity. The anthracite coal has a heating value of 26,000 kJ/kg. Also estimate the percent reduction in NO_x required to meet the NSPS.

16.5   Estimate the concentration (in ppm) and the mass emission rate of NO_x (in kg/hr, expressed as NO) if a coal containing 1.5% nitrogen is combusted at 1400 K using an air/fuel ratio that is 1.5 times the stoichiometric ratio. Consider only fuel NO_x and assume that only NO is formed. The combustion rate of coal is 600 kg/hr, and the flue gas (with an average molecular weight of 27.5) is produced at the rate of 13,000 kg/hr.

16.6 Express the results for Problem 16.5 as NO_2.

16.7 Calculate the stoichiometric amount of ammonia needed to reduce 900 ppm NO to 250 ppm NO in a flue gas flowing at 10,000 actual m^3/min at 300 C and 1 atm. Give your result in kg NH_3/day.

16.8 Rework Problem 16.7 for an NO reduction from 1100 ppm to 250 ppm.

16.9 Assuming a fuel oil composition can be represented by the formula $C_{11}H_{20}N$, calculate the stoichiometric amount of air required for complete combustion to CO_2, H_2O, and NO (assuming no thermal NO is formed). Give your result in kg air/kg fuel.

16.10 For Problem 16.9, calculate the concentration of NO in the flue gas assuming that the actual amount of air used is 1.15 times the stoichiometric amount. Give your result in $\mu g/m^3$.

16.11 In the design of a boiler, it has been calculated that a stable flame will exist at 1927 C, with *initial* mole fractions of N_2 and O_2 (meaning *after* complete fuel combustion but *before* any NO_x formation) in the flame gases of 0.73 and 0.05 respectively.

   a. Calculate the *equilibrium* concentration of NO (ppm) and the *final* mole fractions of N_2 and O_2 in the flame gases.

   b. If the mole fraction of O atoms in the flame is $1.0 (10)^{-3}$, calculate the initial rate of NO formation (in mol/m^3-s), considering only the Zeldovich reactions.

   c. Assuming that the initial rate calculated in Part b holds constant for 0.02 seconds, after which time the gases exit the flame zone, calculate the NO concentration (in ppm) in those exiting gases.

16.12 Suppose the boiler of Problem 16.11 has been built. If the actual concentration of NO leaving the boiler is 1400 ppm, explain qualitatively why this value is different from the values calculated in either Part a or Part c of Problem 16.11.

16.13 Recalculate the answer to Part c of Problem 16.11 using Eq. (16.13) from the text and using the initial mole fractions of N_2 and O_2. Why do you think there is such a large difference in the answers?

REFERENCES

Bartok, W., Engleman, V. S., Goldstein R., and Del Valle, E. G. "Basic Kinetic Studies and Modeling of NO formation in Combustion Processes," *American Institute of Chemical Engineers Symposium Series*, 126(68), 1972.

Bowman, C. T. "Non-Equilibrium Radical Concentrations in Shock-Initiated Methane Oxidation," *15th Symposium (International) on Combustion*. Pittsburgh, PA: The Combustion Institute, 1975.

Breen, B. P., Bell, A. W., De Volo, N. B., Bagwell, F. A., and Rosenthal, K. "Combustion Control for Elimination of Nitric Oxide Emissions from Fossil-Fuel Power Plants," *13th Symposium (International) on Combustion.* Pittsburgh, PA: The Combustion Institute, 1971, p. 391.

Chao, Der-Chen. *Kinetic Modeling of the H₂O₂ or O₃ Enhanced Incineration of NO and/or CO*, M.S. Thesis, University of Central Florida, 1994.

Collins, M. M., Cooper, C. D., Dietz, J. D., Clausen, C. A., and Tazi, L. "Pilot-Scale Evaluation of H_2O_2 Injection to Control NO_x Emissions," *ASCE Journal of Environmental Engineering, 127*(4), April 2001, pp. 329–336.

Counce, R. M., and Perona, J. J. "Scrubbing of Gaseous Nitrogen Oxides in Packed Towers," *Journal of the American Institute of Chemical Engineers, 29*(1), January 1983, pp. 26–32.

Crawford, A. R., Manny, E. H., and Bartok, W. "Field Testing: Application of Combustion Modifications to Power Generating Combustion Sources," in *Proceedings of the Second Stationary Source Combustion Symposium.* Vol. II, *Utility and Large Industrial Boilers*, EPA–600/7–77–073b. Washington, DC: U.S. Environmental Protection Agency, 1977.

Duterque, J., Avegard, N., and Borghi, R. "Further Results on NO_x Production in Combustion Zones," *Combustion Science and Technology, 25,* 1981, p. 85.

Fenimore, C. P. "Formation of Nitric Oxide in Premixed Hydrocarbon Flames," *13th Symposium (International) on Combustion.* Pittsburgh, PA: The Combustion Institute, 1971, p. 373.

Flagan, R. C., and Seinfeld, J. H. *Fundamentals of Air Pollution Engineering.* Englewood Cliffs, NJ: Prentice Hall, 1988.

Haagen-Smit, A. J. "Chemistry and Physiology of Los Angeles Smog," *Industrial and Engineering Chemistry, 44*(6), June 1952.

Haywood, J., and Cooper, C. D. "The Economic Feasibility of Using Hydrogen Peroxide for the Enhanced Oxidation and Removal of Nitrogen Oxides from Coal-Fired Power Plants," *Journal of the Air and Waste Management Association, 48*(3), 1998, pp. 238–246.

Heap, M. P., Tyson, T. J., Carver, G. P., Martin, G. B., and Lowes, T. M. "The Optimization of Burner Design Parameters to Control NO_x Formation in Pulverized Coal and Heavy Oil Flames," *Proceedings of the Stationary Source Combustion Symposium.* Vol. II, *Fuels and Process Research and Development*, EPA–600/2–76–152b. Washington, DC: U.S. EPA, 1976.

Joint Army, Navy, Air Force Thermochemical Tables (3rd ed.). Washington, DC: American Chemical Society, 1986.

Kasper, J. M., Clausen, C. A., and Cooper, C. D. "Control of Nitrogen Oxide Emissions by Hydrogen Peroxide-Enhanced Gas-Phase Oxidation of Nitric Oxide," *Journal of the Air and Waste Management Association, 46,* 1996, pp. 127–133.

Kokkinos, A. "Reburning for Cyclone Boiler Retrofit NO_x Control," *EPRI Journal,* December 1992.

Kokkinos, A., Cichanowicz, J. E., Eskinazi, D., Stallings, J., and Offen, G. "NO_x Controls for Utility Boilers: Highlights of the EPRI July 1992 Workshop," *Journal of the Air Pollution Control Association, 42*(11), November 1992, p. 1498.

Lee, Y. J., Pennline, H. W., and Markussen, J. M. *Flue Gas Cleanup with Hydroxyl Radical Reactions*, DOE/PETC/TR-90/6, February 1990.

Lim, K. J., Waterland, L. R., Castaidini, C., Chiba, Z., and Higginbotham, E. B. *Environmental Assessment of Utility Boiler Combustion Modification*

NO$_x$ Controls. Vol. 1, *Technical Results,* EPA–600/7–80–075a. Washington, DC: U.S. Environmental Protection Agency, 1980.

MacKinnon, D. J. "Nitric Oxide Formation at High Temperatures," *Journal of the Air Pollution Control Association, 24*(3), March 1974.

Malte, P. C., and Pratt, D. T. "Measurement of Atomic Oxygen and Nitrogen Oxides in Jet-Stirred Combustion," *15th Symposium (International) on Combustion.* Pittsburgh, PA: The Combustion Institute, 1975.

Maxwell, J. D., Burnett, T. A., and Faucett, H. L. *Preliminary Economic Analysis of NO$_x$ Flue Gas Treatment Processes,* EPA–600/7–80–021. Washington, DC: U.S. Environmental Protection Agency, February 1980.

Moore, T. "The Retrofit Challenge in NO$_x$ Control," *Electric Power Research Institute Journal,* November 1984, pp. 26–33.

Parkinson, G. "NO$_x$ Controls: Many New Systems Undergo Trials," *Chemical Engineering,* March 9, 1981, pp. 39–43.

Pershing, D. W. *Nitrogen Oxide Formation in Pulverized Coal Flames,* Ph.D. Dissertation, University of Arizona, 1976.

Pershing, D. W., Martin, G. B., and Berkau, E. E. "Influence of Design Variables on the Production of Thermal and Fuel NO from Residual Oil and Coal Combustion," *American Institute of Chemical Engineers Symposium Series, 148*(71), 1975.

Pohl, J. H., and Sarofim, A. F. "Fate of Coal Nitrogen during Pyrolysis and Oxidation," *Proceedings of the Stationary Source Combustion Symposium.* Vol. I, *Fundamental Research,* EPA-600/2-76-152a. Washington, DC: U.S. Environmental Protection Agency, 1976.

Sarofim, A. F., and Pohl, J. H. "Kinetics of Nitric Oxide Formation in Premixed Laminar Flames," *14th Symposium (International) on Combustion.* Pittsburgh, PA: The Combustion Institute, 1973.

Thompson, R. E., and McElroy, M. W. *Effectiveness of Gas Recirculation and Staged Combustion in Reducing NO$_x$ in a 560 MW Coal-Fired Boiler,* EPRI FP–257, Nat. Tech. Information Service PB 260582, 1976.

Uchida, S., Kobayashi, T., and Kageyama, S. "Absorption of Nitrogen Monoxide into Aqueous KMnO$_4$ /NaOH and Na$_2$SO$_3$/FeSO$_4$ Solutions," *Industrial and Engineering Chemistry, 22*(2), 1983, pp. 323–329.

U.S. Environmental Protection Agency. *Control Techniques for Nitrogen Oxide Emissions from Stationary Sources,* Pub. AP–67. Washington, DC: National Air Pollution Control Administration, 1970.

———. *Control Techniques for Nitrogen Oxide Emissions from Stationary Sources* (2nd ed.), EPA-450/3-83-002. Research Triangle Park, NC, 1983.

———. *National Air Pollution Emission Estimates, 1940–1990,* EPA-450/4-91-026. Research Triangle Park, NC, 1991.

———. *Nitrogen Oxides (NO$_x$), Why and How They are Controlled,* EPA 456/F-99-006R. Research Triangle Park, NC, 1999, pp. 32–33.

———. *National Air Quality and Emissions Trends Report, 1998,* EPA-454/R-00-003. Research Triangle Park, NC, 2000.

Zamansky, V. M., Ho, L., Maly, P. M., and Seeker, W. R. "Gas Phase Reactions of Hydrogen Peroxide and Hydrogen Peroxide/Methanol Mixtures with Air Pollutants," *The 26th International Symposium on Combustion.* Pittsburgh, PA: The Combustion Institute, 1996, pp. 2125–2132.

Zeldovich, J. "The Oxidation of Nitrogen in Combustions and Explosions," *Acta Physicochimica URSS, 21*(4), Moscow, 1946.

CHAPTER 17

A VAPOR CONTROL
PROBLEM

Press on: Nothing in the world can take the place of perseverance. Talent will not; nothing is more common than unsuccessful men with talent. Genius will not; unrewarded genius is almost a proverb. Education will not; the world is full of educated derelicts. Persistence and determination alone are omnipotent.

Calvin Coolidge

◆ 17.1 INTRODUCTION ◆

In Chapters 11–16, we have considered the technology available for application in the control of gaseous emissions. This chapter brings together some of these ideas through a stepwise description of the design of four alternative systems for controlling solvent emissions from a plastic-film printing operation. Because this problem is more involved and open-ended than the particulate problem we presented in Chapter 9, it lends itself to a team approach. As a semester project, the instructor may find it appropriate to assign the problem to design groups of three to four students for a period of 8 weeks.

◆ 17.2 PROBLEM STATEMENT ◆

The problem description is presented in the form of five interoffice memos, which illustrate the flow of information that might occur during such a design project. The first memo assigns four alternative solutions to design groups 1–4 and introduces the "players" for the memos that follow.

Clemson Printing Company
Interoffice Memorandum
May 10, 2002

TO: Design Groups 1, 2, 3, and 4
FROM: F. C. Smith, Manager, Process Engineering
SUBJECT: Solvent emission control for Newry Plant

Recent policy statements and legal actions by our state regulatory agency require that solvent emissions from the Newry, SC, plant be reduced by 95% by December 31, 2003. To meet these requirements, we must initiate a preliminary design evaluation of a control system for this plant immediately.

Our Corporate Environmental Group (CEG) has suggested that we look at four processes, which will be designated Process A, Process B, and so on. Processes A through C use established control technology for hydrocarbon vapors and all can be designed to meet the 95% requirement. Process D, on the other hand, is based on technology proven for odorous and other VOCs, but has seen limited use for treating the types of hydrocarbons emitted from our plant. Nevertheless, it has potential due to its low operating costs. Our Engineering Department has been assigned the task of evaluating and comparing these technologies, and providing a sound basis for selecting one of these processes for the Newry Plant. Group 1 is assigned Process A, and Groups 2, 3, and 4 are assigned Processes B, C, and D, respectively.

Process A is a fixed-bed carbon adsorption system with necessary storage capacity (tankage) to hold recovered solvents for monthly shipment to a commercial reclaiming company. Process B uses thermal vapor incineration with heat recovery. Process C uses catalytic vapor oxidation with heat recovery. "Heat recovery" may include preheating the incoming air or heating water and/or generating steam. Process D is based on degradation of the hydrocarbons in a biofilter. CEG recognizes that a biofilter may not be feasible for these particular solvents due to the expected high concentrations of solvent vapors and the large space required to treat these particular compounds. However, they desire to have a thorough review of the biofiltration option.

I expect your preliminary design reports by December 1, 2002, in order to make a final recommendation to management by January 31, 2003. I am requesting (by copy of this memo) that David Jones, Utilities Manager at Newry, send you the results of a recent emissions study of the printing operation and a layout drawing showing vent locations and the site selected for the control system. Please note that if more space is needed (e.g., for a large biofilter), CEG has informed us that additional land can be purchased adjacent to the Control System Site for $50,000 per acre. Feel free to contact our corporate Research Group and Economics Group for more information as needed.

I have attached a recently updated report format guide for your use. Good luck with your projects.

F. C. Smith
Process Engineering Manager

c: David Jones, Utilities Manager, Newry Plant
Attachment: Memo dated 3/25/02, Format Guide for Technical Reports

Clemson Printing Company
Interoffice Memorandum
March 25, 2002

TO: All Design Groups
FROM: Manager, Process Engineering
SUBJECT: Standard Design Report Format

The following format is suggested for all preliminary cost-estimate reports.

1. *Executive summary:* A brief description of the project and results of the study, including conclusions and recommendations. The summary should not exceed two pages and should be written so that it can be transmitted as is.

2. *Introduction:* A concise statement of the problem, covering background and objectives.

3. *Technical information:* A description of the proposed process, including a flow sheet showing details of flow rates, concentrations, and equipment sizes. Summaries of equipment specifications and costs, capital cost estimate (± 25%), and projected operating expenses. Brief discussion of economic analysis, including return-on-investment evaluation, where applicable.

4. *Conclusions and recommendations*

5. *Appendix:* Calculations, graphs, and an explanation of all assumptions made.

F. C. Smith

Clemson Printing Company
Newry Plant
Newry, SC 29633
May 12, 2002

TO: Design Groups 1, 2, 3, and 4, Central Engineering
FROM: Dave E. Jones, Utilities Manager
SUBJECT: Printing operation emissions survey and press vent layout

The attached table summarizes the results of a recent VOC emission survey. In addition to this study, we have reviewed operating data on each press and conclude that on the average (based on 24-hr operation seven days per week) each press operates 60% of the time, but there are times when all presses are operating simultaneously. The entire plant is down three weeks per year.

Vent locations are shown in the attached figure. Each vent extends 6 ft above the roof line, which is 25 ft above ground level. Each vent is equipped with a damper that automatically closes on press shutdown. The site available for the emission control system is also shown. Keep in mind the statement by CEG about additional land being available for purchase. This site is the nearest space available and will require a 200 ft duct from the closest press vent to the control system.

Please contact me if you need additional information.

Dave E. Jones

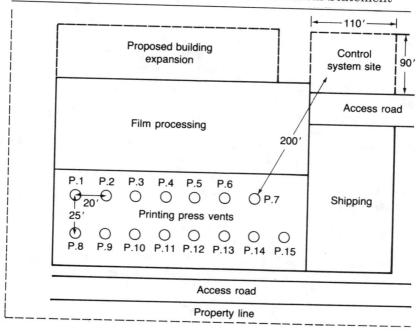

Figure 17.1
Layout of printing press vents and control system site (not to scale).

Table 17.1 Volatile Organic Emissions from Printing Area

Press No.	Exhaust Temperature, °F	Flow, acfm	Hexane lb$_m$/hr	Heptane lb$_m$/hr	Total lb$_m$/hr
1	100	3,520	4.0	6.9	10.9
2	125	6,020	10.1	14.4	24.5
3	115	3,070	3.1	6.1	9.2
4	100	5,815	14.0	14.1	28.1
5	90	4,820	8.0	9.5	17.5
6	102	3,060	2.0	4.9	6.9
7	95	3,390	7.5	5.9	13.4
8	115	4,310	3.5	6.1	9.6
9	115	3,010	5.0	5.3	10.3
10	105	4,830	11.7	10.0	21.7
11	80	7,540	10.3	15.2	25.5
12	100	1,720	4.1	5.9	10.0
13	99	5,810	8.9	8.2	17.1
14	105	5,340	10.1	8.0	18.1
15	90	8,870	13.2	13.9	27.1
Total		71,125	115.5	134.4	249.9

Clemson Printing Company
Interoffice Memorandum
May 15, 2002

TO: Design Groups 1, 2, 3, and 4
FROM: Manager, Economics Group
SUBJECT: Economic basis for solvent emission control project for Newry plant

The following data are supplied for your use on subject project. Notice that they have been adjusted to the project year.

Basis year: 2002
Utilities costs:
  Fuel—$8.00/$10^6$ Btu
  Steam—$8.50/1000 lb_m
  Electricity—9.0¢/kWh
  Cooling water—$0.25/1000 gal (available at 85 F, returned at 115 F)
  Waste water treatment: 30¢/lb_m COD (chemical oxygen demand)
Operating labor: $17.50/hr
Value of recovered mixed solvent: $1.35/gal
Capital related costs:
  Depreciation (for pollution control projects)—8 yr, straight line
  Taxes and insurance—2% of total capital investment
  Maintenance (labor + materials)—7% of total capital investment
  Pre-tax ROI (if applicable)—20%
  or interest on capital (if applicable)—8%

_____

John E. Cash

Clemson Printing Company
Interoffice Memorandum
May 14, 2002

TO: Process Engineering, Design Groups 1, 2, 3, and 4
FROM: Manager, Engineering Research
SUBJECT: Control of printing ink solvent vapors

This memo is in response to your recent requests for information regarding the subject. Our group has been very interested in carbon adsorption and incineration of printing-ink solvent vapors as a means of controlling these emissions. The following is the information you requested on the adsorption of hexane and heptane onto XYZ activated carbon:

Solvent removal from air at 40 C: 99.99% with a very steep breakthrough curve.

Capacity (at 40 C) for either solvent: 0.08 g/g carbon on a regenerated bed

Regeneration steam: 0.3 kg of steam/kg of carbon

Carbon cost: $5.00/kg

Carbon service life: 4 yrs

Regeneration can be with steam or hot nitrogen. The downstream separation processes would be different in each case. If you plan to condense the steam and solvent vapors after regenerating the carbon bed, you should keep in mind that at 40 C the solubilities of hexane and heptane in water are 0.004 and 0.005 mole percent, respectively.

Our experience with incineration of normal hexane and heptane indicates you should design for the following conditions to achieve 95% removal:

	Thermal	**Catalytic**
Operating temperature	1300 F	750 F
Residence time	0.60 sec	0.60 sec
Heat loss	10%	5%
Pressure drop (including heat recovery)	8.0 in. H_2O	10.0 in. H_2O
Heat transfer coefficients		
air-to-air	8 Btu/hr-ft^2-°F	8 Btu/hr-ft^2-°F
air-to-water	10 Btu/hr-ft^2-°F	

For your preliminary cost estimate, assume air-to-air heat exchangers cost $21/ft^2 installed and air-to-water exchangers cost $30/ft^2 installed.

Tom A. Edison

♦ 17.3 SELECTING A DESIGN FLOW RATE ♦

In this problem it is necessary to design a system for an exhaust stream that varies in flow rate and concentration with time. The probability that all presses will be in operation at any time is exceedingly small, so the system should be designed for the maximum flow that can reasonably be expected (not that with all presses running at once). An estimate of the maximum flow can be obtained by summing the flows from all the presses operating at any given time, based on the probability that each press is operating at that time. The status of each press, whether running or not, is determined by generating a random number between 0 and 1.0. If the number is equal to or less than the fraction of the time the press operates (0.6 in this case), the press is operating and its flow rate is included in the sum. The summation is repeated 100 times or more, and the maximum flow rate and solvent emission rate calculated represents the probable maximum value of each that can be expected to occur. As an example, the computer program shown in Table 17.2 calculates the maximum flow and solvent emission rate for an operating frequency of 0.6. The calculated values are more reliable if an individual operating frequency for each press is known.

General Requirement For All Groups

1. Estimate the total exhaust flow rate and solvent emission rate to be treated for the specified operating frequency.

2. Prepare an engineering sketch of the ductwork required to transport the exhaust gas to the control system; show the location of the blower.

3. Based on information contained in the interoffice memos dated May 12–15, 2002, estimate the blower horsepower requirement for each control option.

Questions

1. What assumptions did you make, if any, regarding the average temperature of the exhaust?

2. What static pressure rise is required of the blower?

3. Would you suggest installing a spare blower?

4. Should round or rectangular ducts be specified?

At this point, each design group is working on a separate problem leading to a final estimate of capital and operating costs. The following discussion will outline by "requirements" the work to be accomplished in designing the activated carbon system, the two incineration systems, and the biofilter system.

Table 17.2   FORTRAN Program to Estimate the 1 in 100 Chance Maximum Combined Flow Rate from a Number of Independent Sources

```
        DIMENSION PRESS (25), FLOW (25), SOLV (25)
        READ (5,*) NP, PCTON
        DO 5 I = 1, NP
        READ (5,*) FLOW (I), SOLV (I)
5       CONTINUE
C
C
C       NP = NUMBER OF PRESSES, PCTON = % ON, FLOW = ACFM,
C          SOLV = SOLV RATE
C
        TEST1 = -1.
        TEST2 = -1.
        N = 34589
        DO 15 J = 1, 100
        TOTFLO = 0.0
        TOTSOL = 0.0
        DO 10 I = 1, NP
        CALL RAND (N, L, FLAG)
        N = L
        PRESS(I) = 0.0
        IF (FLAG .LE. PCTON) PRESS(I) = 1.0
        TOTFLO = TOTFLO + PRESS(I)*FLOW(I)
        TOTSOL = TOTSOL + PRESS(I)*SOLV(I)
10      CONTINUE
        IF (TOTFLO .GE. TEST1) TEST1 = TOTFLO
        IF (TOTSOL .GE. TEST2) TEST2 = TOTSOL
15      CONTINUE
        WRITE (6,20) TEST1, TEST2
20      FORMAT (1H1, ////, 5X, 'MAX. FLOW RATE (ACFM) =
1         ', F10.1, 5X,
2         ' MAX. SOLVENT RATE (POUNDS/HR) = ', F10. 1)
        STOP
        END
        SUBROUTINE RAND (N, L, FLAG)
        L = N*1220703125
        IF (L) 1, 2, 2
1       L = 2147483647 + 1 + L
2       FLAG = L
C
C       2147483647 = 2**31
C
        FLAG = FLAG*0.4656613E-9
        N = L
        RETURN
        END
```

◆ 17.4 CARBON ADSORPTION SYSTEM ◆

First Requirement

1. Prepare a preliminary flow diagram for a fixed-bed carbon adsorption system that includes facilities for storing and shipping recovered solvent.

2. Estimate the total mass of carbon required based on 1-, 2-, and 3-hour regeneration cycles.

3. Select the most economic bed size and regeneration cycle.

4. Calculate steam requirements.

5. Prepare a final material balance for the system.

6. Size the vapor condenser and the decanter. Base the decanter size on a 30-minute residence time.

7. Size the blower needed to supply cooling air.

8. Size the recovered solvent storage tank and transfer pump.

Questions

1. If more than two carbon beds are used, will the beds be operated in series or parallel?

2. What fire protection facilities are required?

3. Should the system be housed in a building?

4. Are any special materials of construction required?

Second Requirement

1. Estimate the installed cost of the carbon system, ductwork and blowers, and recovered solvent storage and shipping facilities.

2. Estimate the total cost of the system.

Third Requirement

1. Estimate the variable operating costs for the carbon system.

2. Estimate the fixed charges.

3. Estimate the annual credit for recovered solvent.

4. Calculate the total annual operating cost for the system.

5. What is the sensitivity of the total AOC to the value of the recovered solvent (e.g., what if the value of the recovered solvent drops to 50 cents/gallon or to zero)?

◆ 17.5 THERMAL INCINERATION SYSTEM ◆

First Requirement

1. Draw a preliminary sketch of the thermal incineration system, including heat recovery.

2. Estimate operating conditions for an inlet air preheater that will recover 30, 50, and 70% of the incinerator exhaust heat.

3. Based on the rule of thumb that states pressure drops of 6, 8, and 10 in. H_2O can be expected for heat recoveries of 30, 50, and 70%, select a preheater size, remembering that the preheater must meet return-on-investment requirements.

4. Calculate fuel gas requirements for the incinerator.

5. Prepare a material balance for the incinerator.

6. Prepare incinerator specifications.

7. Recheck preheater specifications.

8. Select an incinerator stack diameter and height.

9. Prepare a final flow sheet showing incinerator and all duct-work. Identify all streams giving composition and temperature.

Questions

1. What materials of construction do you suggest for the air pre-heater?

2. Is secondary heat recovery—say production of hot water—after the preheater justified?

Second and Third Requirements

The second and third requirements essentially will be the same as those for the carbon adsorption system. There will be no credit for recovered solvent, but there may be a credit for hot water produced.

◆ 17.6 CATALYTIC INCINERATION SYSTEM ◆

The requirements for the catalytic system will follow closely those outlined for the thermal system. Catalyst life and replacement costs also must be estimated. In this application, normal catalyst life (4 years) should be expected, and catalyst amount and cost can be estimated as discussed in Chapter 11.

◆ 17.7 BIOFILTER SYSTEM ◆

First Requirement

1. Draw a preliminary sketch of the biofilter system showing all inputs and outputs.

2. Calculate the exhaust flow rate and contained solvent flow rates to the biofilter.

3. Estimate liquid rates to and from the spray tower and biofilter.

4. Size the spray tower and biofilter.

5. Size required blowers and pumps.

6. Prepare a final material balance for the system.

Questions

1. Are special materials of construction required?

2. Are seasonal ambient temperature variations going to significantly affect the operations of the biofilter (particularly, its removal efficiency)?

Second and Third Requirements

These requirements will be essentially the same as for the incineration systems. There should be a charge added for disposal of the blowdown water from the spray tower and the leachate from the biofilter. If additional land is needed, assume that you must purchase 25% more than the amount needed just for the biofilter (to provide access to all sides of the biofilter, and a buffer zone to the property line).

◆ 17.8 SUMMARY ◆

Each design group will prepare a preliminary design report using the format suggested in the memo from F. C. Smith dated March 25, 2002. Keep in mind that your report "sells" your work to upper management. Write clearly and concisely, and refer frequently to your process flow diagram.

CHAPTER 18

MOBILE SOURCES

AN OVERVIEW

I will build a motor car for the great multitude. It will be constructed of the best materials, by the best men, after the simplest designs that modern engineering can devise. But it will be so low in price that no man making a good salary will be unable to own one—and enjoy with his family the blessing of hours of pleasure in God's great open spaces.

Henry Ford, 1908

♦ 18.1 INTRODUCTION ♦

Cars, trucks, buses, motorcycles, airplanes, boats, trains, bulldozers, tractors—in short, all things that move and emit air pollutants are *mobile sources*. *On-road* vehicles (as opposed to *non-road* vehicles) emit the lion's share of pollution, but emissions from aircraft, vessels, construction equipment, and the like are significant. Although the United States has more cars per capita than any other nation, mobile source pollution is not limited to this country. All around the world, major urban areas have become congested with motor vehicles and many urban airsheds have become severely polluted. In 1999, about 77% of carbon monoxide (CO) in the United States was emitted from mobile sources, with about 66% of the mobile source emissions coming from on-road vehicles (U.S. EPA 2001). However, in certain cities, estimates of urban-area CO emissions attributable to mobile sources have ranged as high as 95% (U.S. EPA 2000). Scenes like the one shown in Figure 18.1 have become all too common around the world, and now the vision of Henry Ford sometimes seems like a cruel hoax.

The proliferation of mobile sources (particularly on-road vehicles) has fundamentally altered the field of air pollution control around the

Figure 18.1
Rush-hour traffic. This scene could be from any of a number of cities around the world, but happens to be from Boston, Massachusetts.
(Courtesy of the International Road Federation, Washington, D.C.)

world. There are several key reasons for this change. First, huge numbers of small, diverse, decentralized sources that move around are far more difficult to regulate and control than a much smaller number of the larger stationary sources. Second, the *sources* of the emissions (the vehicles on roadways) are usually very close to the *receptors* of interest (people living and working in urban areas). Third, cars have become economic status symbols, and driving is now an integral part of the lifestyles of hundreds of millions of people.

Control of mobile source pollution was initially the legislatively mandated responsibility of the automobile manufacturers. In the past twenty years or so, it has become recognized that control of new cars alone is not sufficient to ensure clean air. Control strategies now include state and local regulation of in-use vehicle maintenance practices, of fuel quality and composition, and even of individual driving habits. During the past two decades, both pollution control technology and regulatory strategies have developed mainly in the United States. Happily, the better approaches are being adopted by other developed (and upper-income developing) nations. Unhappily, it seems that almost as fast as reductions in per vehicle emissions are achieved, those gains are offset by the continuing growth in the number and usage of motor vehicles.

▶ ▶ ▶ ▶ ▶ ▶ ▶ ▶ ▶

Example 18.1

Between 1980 and 1990, the average CO emission factor (EF) of the vehicle fleet in Orange County, Florida, dropped by almost half, from about 65 to 34 grams/veh-mile. However, the total miles driven in the county increased by 60% during this same time period. Did county-wide emissions of CO go up or down, and by how much?

Solution

$$\text{Emissions}_{1990} = \text{Emissions}_{1980} \times (\text{EF}_{90} / \text{EF}_{80}) \times \text{Vehicle growth}$$
$$= \text{Emissions}_{1980} \times (34 / 65) \times 1.60$$
$$= \text{Emissions}_{1980} \times 0.837$$

So emissions decreased from 1980, but only by about 16%.

◀ ◀ ◀ ◀ ◀ ◀ ◀ ◀ ◀

◆ 18.2 MAGNITUDE OF THE PROBLEM ◆

As was seen in Chapter 1, motor vehicles emit all of the primary pollutants, and contribute mightily to the formation of ozone in urban areas. In the United States, the vehicular pollutants of more concern are carbon monoxide (CO), nitrogen oxides (NO_x), and volatile organic compounds (VOCs). Emissions of sulfur dioxide (SO_2) and particulate matter (PM) from vehicles are of less concern in the United States, owing to the high degree of fuel desulfurization, the relatively low percentage of diesel vehicles, and the relatively good condition of the vehicle fleet. In other countries (and in certain U.S. cities) mobile source emissions of SO_2 and PM are significant. Emissions of lead (Pb) in the United States during the past 25 years have declined dramatically owing to the EPA-mandated phase-out of leaded gasolines. In other countries, lead emissions from motor vehicles still pose a significant problem. Summaries of mobile source emissions in the United States for both 1980 and 1999 are presented in Table 18.1. Data on emissions of air pollutants from various countries (with percentage contributions from mobile sources) are presented in Table 18.2.

As Faiz et al. (1992) discuss, there has been a strong worldwide movement toward motorization since the end of World War II. Since 1970, the fastest growth rates have been observed in the upper-income developing countries of the world. Total world production of motor vehicles (cars, trucks, and buses) grew from about 10 million vehicles per year in 1950 to about 50 million per year in 1990. (Nearly half of that production was accounted for by just five companies—General Motors, Ford, Toyota, Nissan, and Volkswagen.) In addition, in 1988, nearly 11 million motorcycles, motorscooters, and three-wheeled vehicles were produced, mostly in the Asian countries. The total number of

Table 18.1 Mobile Source Emissions of Air Pollutants in the U.S. in 1980 and 1999

Source Category	Pollutant Emissions, millions of metric tons/year											
	CO		VOC		NO$_x$		PM-10		SO$_x$		Lead[1]	
	1980	1999	1980	1999	1980	1999	1980	1999	1980	1999	1980	1999
Mobile Sources												
On-Road Vehicles	78.0	50.0	9.0	5.3	8.6	8.6	0.4	0.3	0.5	0.3	60.5	0.0
Gasol. Cars & Motorcycles	53.6	27.4	5.9	2.9	4.4	2.9	0.1	0.1	0.1	0.1	47.2	0.0
Gasol. Pick-ups & SUVs	16.2	16.1	2.1	1.7	1.4	1.6	0.1	0.0	0.1	0.1	11.7	0.4
Heavy Trucks & Diesels	8.3	6.5	1.0	0.7	2.8	4.1	0.2	0.2	0.3	0.1	1.6	0.0
Aircraft	0.7	1.0	0.1	0.2	0.1	0.2	0.0	0.0	0.0	0.0	0.9	0.5
Railroads	0.1	0.1	0.0	0.0	0.7	1.2	0.0	0.1	0.1	0.1	na	na
Marine Vessels	0.1	0.1	0.0	0.0	0.5	1.0	0.0	0.0	0.1	0.2	na	na
Other Non-Road[2]	13.6	24.0	2.2	3.0	2.2	3.1	0.4	0.4	0.1	0.0	3.3	0.0
Total Mobile Sources	92.5	75.2	11.3	8.5	12.1	14.1	0.8	0.8	0.8	0.6	64.7	0.5
Total All Sources[3]	117.4	97.4	26.3	18.1	23.4	23.6	6.3	3.0	25.9	21.1	74.2	4.2
Percent Mobile Sources	79	77	43	47	52	60	13	27	3	3	87	12

[1] For lead, units are thousands of tons per year.
[2] Other non-road category includes: construction equipment, lawn and garden tools, recreational boats, industrial and commercial vehicles, etc.
[3] For PM-10, totals do not include fugitive dust from paved and unpaved roads, open fields, construction sites, etc.
na = not available

All data extracted from U.S. Environmental Protection Agency, EPA-454/R-01-004, March 2001.

Table 18.2   Estimated Annual* Emissions of Air Pollutants in Selected Countries in the Late 1980s—Total Anthropogenic and Percent from Mobile Sources

Total Emissions, millions of metric tons/yr and (Percent from Vehicles)						
Country	CO	VOC	NO_x	PM	SO_x	Pb**
Developed Countries						
Canada	10.1 (66)	2.3 (40)	1.9 (64)	1.6 (13)	3.7 (3)	0.2
USA	61.2 (67)	18.6 (33)	19.8 (41)	6.9 (20)	20.7 (4)	6.9
France	6.4 (70)	2.1 (38)	1.7 (60)	0.4 (20)	1.6 (5)	8.0
Germany	8.9 (74)	2.5 (52)	3.0 (61)	0.6 (13)	2.3 (5)	na
Italy	5.5 (82)	0.8 (56)	1.5 (43)	0.4 (23)	2.0 (7)	7.0
United Kingdom	5.5 (85)	1.8 (30)	2.5 (45)	0.5 (34)	3.7 (1)	3.1
Eastern Europe						
Hungary	1.7 (60)	0.3 (28)	0.4 (30)	0.5 (10)	1.5 (1)	0.5
Poland	0.9 (40)	0.2 (37)	1.6 (33)	3.4 na	4.2 (3)	1.0
Russia	43.4 (66)	14.0 (40)	6.3 (29)	14.0 (30)	17.6 na	na
Asia						
South Korea	3.3 (25)	0.2 (57)	0.4 (85)	0.5 (8)	0.5 (8)	na
Taiwan	3.2 (46)	0.9 (53)	0.6 (50)	5.9 (1)	1.4 (14)	na
Malaysia	0.5 (50)	na	0.1 (36)	na	0.2 (1)	na
Thailand	0.7 (60)	0.04 (46)	0.2 (23)	0.3 (3)	0.3 (15)	1.5

Percent Contribution to World Emissions						
	CO	VOC	NO_x	PM	SO_x	Pb
USA						
all sources	35	40	29	12	21	14
motor vehicles	24	30	28	20	20	10
Other Developed Countries						
all sources	36	27	25	11	19	20
motor vehicles	49	43	47	30	20	15
Rest of World						
all sources	29	33	46	77	60	66
motor vehicles	27	27	25	50	60	75

* Year of estimate varies among entries but generally is 1986, 1987 or 1988.
** Thousands of metric tons per year (percents from vehicles not available)
na = not available

Adapted from Faiz et al., 1992.

motor vehicles in use in the world in 1988 was estimated at 630 million (including some 90 million motorcycles and three-wheelers). During the 1990s, economic growth continued in many regions of the world, resulting in substantial further increases in vehicle ownership. Although the less-developed countries generally cannot afford sophisticated pollution control technology, the fact that only a few multinational companies account for so much of the world vehicle production offers hope for global cooperation to reduce motor vehicle pollution.

During the past 50 years, overall growth in the number of motor vehicles in the world (which is up by a factor of 10) has far exceeded the growth in the world population (which has doubled). Projections made by the United Nations show this trend continuing for some time into the future. There are several reasons for this phenomenal growth in vehicles, especially in developing nations: increasing urbanization, increasing real incomes, decreasing real costs of vehicles, the desire for improving social status, the desire for independence, and others (Faiz et al. 1992). Although the total numbers of vehicles are up all over the world, the growth in motor vehicles *by type* is quite different in different regions of the world. Table 18.3 shows the regional distribution of motor vehicles by type in 1988.

Several important points can be gleaned from Table 18.3. *First,* the OECD nations have far more cars, trucks and buses than the non-OECD countries, but the numbers of two- and three-wheelers are about equal. [OECD stands for Organization for Economic Cooperation and Development; the OECD countries include most of Western Europe, Canada, the United States, Japan, Australia, and New Zealand. The non-OECD countries are sometimes called the developing countries.] Based on the numbers in Table 18.3, the developing countries have 20% of all the cars, 30% of the trucks and buses, but 47% of the two- and three-wheelers. *Second,* several Asian countries have a far greater proportion of trucks and buses in their vehicle fleet than does the United States or Europe. *Third,* although a region like Asia has far fewer cars than the United States (which in fact has the most cars of any country), this part of the world has far more two- and three-wheelers than the United States. In fact, three countries in Asia—Thailand, Malaysia, and Indonesia—had over 10 million two- and three-wheeled vehicles in use in 1988!

These trends, along with certain features of vehicles found in the developing countries, combine to create significant air pollution problems in urban centers in those countries. Some of the important characteristics of the non-OECD vehicle fleet are:

1. Most of the two- and three-wheeled vehicles are equipped with highly polluting two-stroke engines.

2. Many of the fleets in developing countries are much older than those in the OECD countries, and are poorly maintained.

3. There are higher proportions of trucks and buses (most without pollution control equipment) in the non-OECD countries.

4. Poorer quality fuels are sold in the developing countries resulting in more pollution emissions.

For these and other reasons, vehicular emissions produce high concentrations of several pollutants in crowded urban centers. This is now being viewed as a very serious problem for many developing nations.

▶ ▶ ▶ ▶ ▶ ▶ ▶ ▶ ▶

Example 18.2

(a) According to Table 18.2, which country emitted more CO from mobile sources—France or Italy? (b) Using data for the United States from this same table, estimate the total world emissions of CO from all sources in the late 1980s.

Solution

(a) From the top half of the table (abbreviating millions of metric tons as mmt),

Table 18.3   Types and Numbers of Vehicles in Use, 1988

	Millions of Vehicles			
	Cars	Trucks & Buses	2- & 3-Wheelers	TOTAL
OECD Countries*				
USA	141.3	43.1	7.1	191.5
Canada	11.9	4.0	0.4	16.3
Europe	138.5	18.0	22.0	178.5
Japan	30.8	21.7	18.2	70.7
Australia	8.8	2.3	0.4	11.5
Sub-total OECD	331.3	89.1	48.1	468.5
Non-OECD Countries*				
Eastern Europe	30.0	11.7	3.9	45.6
Latin America	27.3	7.9	4.3	39.5
Africa	8.0	4.4	1.2	13.6
Middle East	6.6	3.7	0.6	10.9
Asia	10.2	10.0	33.3	53.5
Sub-total non-OECD	82.1	37.7	43.3	163.1
TOTAL	413.4	126.8	91.4	631.6

* OECD stands for Organization for Economic Cooperation and Development; the OECD countries include most of Western Europe, Canada, the United States, Japan, Australia, and New Zealand. The non-OECD countries are sometimes called the developing countries.

Adapted from Faiz et al., 1992

France: 6.4 mmt CO × 0.70 = 4.48 mmt CO
(from mobile sources)

Italy: 5.5 mmt CO × 0.82 = 4.51 mmt

France and Italy emitted essentially equal amounts of CO from mobile sources.

(b)

United States: 61.2 mmt CO total

61.2 mmt × 0.67 = 41.0 mmt
(mobile sources)

From the bottom half of Table 18.2, the 61.2 mmt was 35% of the world's CO emissions considering all sources, so

61.2 mmt/0.35 = 174.9 mmt
(world total CO)

Note that if we had used just the U.S. mobile source emissions, a similar answer would have resulted:

41.0 mmt/0.24 =170.8 mmt
(world total CO)

◀ ◀ ◀ ◀ ◀ ◀ ◀ ◀ ◀

◆ 18.3 CHARACTERISTICS OF ENGINES ◆
IN MOBILE SOURCES

The prime mover of mobile sources is the **internal combustion engine**. Most cars are powered by a conventional four-stroke, gasoline-burning internal combustion engine (see Figure 18.2). In this engine, a mixture of air and gasoline is drawn into any given cylinder during the *intake stroke* (a stroke is the travel of the piston up or down the length of the cylinder), then is compressed to a pressure of several atmospheres during the *compression stroke* before being ignited by a spark from the spark plug. The fuel burns rapidly (one might say explosively!), releasing a great amount of heat in a very short time. This heat rapidly expands the gases, driving the piston downwards during the *power stroke*. After this second downstroke, the piston is driven upward again by the crankshaft, and the exhaust gases are then pushed out of the cylinder through the exhaust valves during this fourth *exhaust stroke*. This process is repeated several hundred times per minute for each cylinder during driving.

The gasoline combustion reactions happen very quickly and go nearly to completion. However, to the extent that the combustion is not complete, CO and VOCs (unburned hydrocarbons or partially oxidized fuel fragments) are emitted. Several factors prevent the combustion reactions from going to 100% completion, including quenching by the cylinder walls, incomplete mixing of all of the fuel and air, vehicle operating parameters (such as acceleration or load), and poor engine maintenance.

Without getting into the details of the reaction mechanisms whereby gasoline (a mixture of dozens of hydrocarbon compounds) is converted (by hundreds of individual reactions) into end products, let us consider only the overall stoichiometry of the reaction. For this purpose, we will assume that gasoline can be represented by the compound octene. Assuming complete combustion with the stoichiometric amount of air, and assuming for the moment that nitrogen is completely inert, the reaction is

$$C_8H_{16} + 12\,O_2 + 45.1\,N_2 \rightarrow 8\,CO_2 + 8\,H_2O + 45.1\,N_2 \quad \text{(18.1)}$$

Eq. (18.1) gives us the molar ratio of reactants directly. However, it is much more common to speak of the mass **air-to-fuel ratio (AFR)**. For the reaction given by Eq. (18.1), the stoichiometric AFR is 14.7, which is a typical value for many distillate hydrocarbon fuels. The **stoichiometric ratio (SR)** can be defined as the actual AFR divided by the stoichiometric AFR, as illustrated by the following equation:

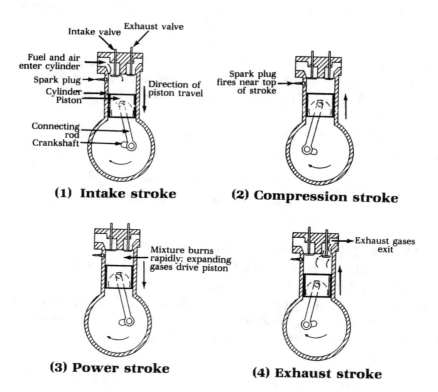

(1) Intake stroke

(2) Compression stroke

(3) Power stroke

(4) Exhaust stroke

Figure 18.2
Schematic of a four-stroke internal combustion engine.

$$SR = \frac{(A/F)_{\text{actual}}}{(A/F)_{\text{stoich}}} \qquad\qquad (18.2)$$

where

SR = stoichiometric ratio

A = air intake rate, g/s

F = fuel usage rate, g/s

Mixtures for which ER < 1 are called fuel rich (or simply rich); when ER > 1, the mixture is termed fuel lean (or lean).

Example 18.3

(a) Calculate the stoichiometric air-to-fuel ratio for the combustion of benzene, one of the aromatic compounds found in gasoline. (b) For a stoichiometric ratio of 1.10, calculate the volume of exhaust gases produced (at STP) per kg of benzene burned.

Solution

(a) First, write the reaction identifying reactants and products only

$$C_6H_6 + O_2 + N_2 \rightarrow CO_2 + H_2O + N_2$$

Next, balance it stoichiometrically, starting with the carbon atoms, followed by the hydrogen, oxygen, and nitrogen, in that order. Keep in mind that 3.76 moles of nitrogen are present in air for every mole of oxygen. The stoichiometric equation is

$$C_6H_6 + 7.5\,O_2 + 28.2\,N_2 \rightarrow 6\,CO_2 + 3\,H_2O + 28.2\,N_2$$

The stoichiometric AFR is

$$AFR = \frac{(7.5 \text{ moles } O_2 \times 32 \text{ g/mole}) + (28.2 \text{ moles } N_2 \times 28 \text{ g/mole})}{1 \text{ mole benzene} \times 78 \text{ g/mole}}$$

$$= 13.2 \text{ g air per g benzene}$$

(b) For an SR of 1.10, the re-balanced equation is

$$C_6H_6 + 8.25\,O_2 + 31.0\,N_2 \rightarrow 6\,CO_2 + 3\,H_2O + 31.0\,N_2 + 0.75\,O_2$$

78 g   264 g   868 g   264 g   54 g   868 g   24 g

In this case, for every 78 g of benzene, there are 1132 g of air, and 1210 g of exhaust gases. Because 40.75 moles of gases are produced in this reaction, the volume of gases produced is

$$V = \frac{40.75 \text{ moles gas}}{0.078 \text{ kg fuel}} \times \frac{24.45 \text{ L}}{\text{mole}} \times \frac{1 \text{ m}^3}{1000 \text{ L}} = \frac{12.77 \text{ m}^3}{\text{kg fuel}}$$

where the constant 24.45 liters/mole is the molar volume of gas at STP (25 C and 1 atm).

◀ ◀ ◀ ◀ ◀ ◀ ◀ ◀ ◀

It turns out that the AFR is a critical parameter to the efficient operation of an internal combustion engine. For operation at less than the stoichiometric ratio (SR < 1), there is not enough oxygen for the complete combustion of the fuel, and CO and VOC emissions are high. As SR increases towards unity, CO and VOC emissions decrease rapidly. However, as the mixture gets leaner (SR > 1) the engine operates "rougher." At even greater values of SR, the combustion becomes unstable and CO and VOC emissions increase again.

Based on the above discussion, it would seem obvious that we ought to operate an engine at SR = 1. However, the temperature of the combustion gases is maximized at stoichiometric burning. As we learned in Chapter 16, nitric oxide (NO) formation is a strong function of temperature. In Eq. (18.1) we have assumed that the nitrogen present in the air did not participate in the reactions occurring in the cylinder. In fact, some N_2 and some O_2 do react to form nitric oxide, and NO emissions from uncontrolled gasoline engines are significant.

Recall that NO production is maximized at high temperatures in the presence of extra oxygen. These conditions are met at air-to-fuel ratios of 15 to 16. Figure 18.3 shows how the exhaust gas concentration of each of three major pollutants varies with the AFR in a four-stroke, gasoline-fueled internal combustion engine.

Rather than having separate strokes for intake and exhaust, a **two-stroke, gasoline-driven internal combustion engine** combines these functions with the compression and power strokes. In the two-stroke engine, the spark plug fires every time the piston approaches the top of the cylinder rather than every other time as with the four-stroke engine. Another important difference is that lubrication of the internal working parts of the smaller, simpler two-stroke engine are met by mixing the motor oil with the fuel in the gas tank. The oil, of course, is more difficult to burn completely than the gasoline.

Consideration of the two-stroke engine cycle begins with ignition of the air-fuel mixture, the combustion of which causes the gases to expand driving the piston downward and conveying power to the crankshaft. During part of its downward travel, exhaust ports are uncovered and spent gases leave the cylinder. Next, the air ports are uncovered and a new charge of fresh air and fuel is drawn into the cylinder by the piston as it continues downward. Next, on the upward stroke, all the ports are closed and the piston compresses the fresh air-fuel mixture as it travels up to top dead center and readies the charge for the next ignition and power stroke.

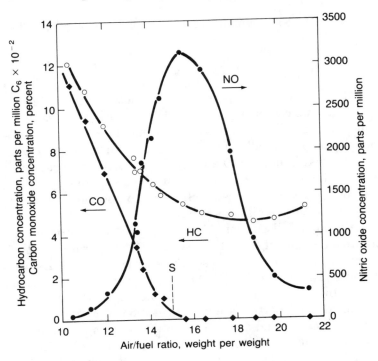

Figure 18.3
The effects of the air/fuel ratio on hydrocarbon, carbon monoxide, and nitric oxide exhaust emissions.
(Adapted from Agnew, 1968.)

The two-stroke engine has several apparent economic advantages over a four-stroke engine. It is smaller, lighter, less expensive, simpler to operate, and easier to maintain than its four-stroke counterpart. Hence, it is very popular for small motorcycles, motorscooters, and three-wheelers, especially in the less affluent countries. However, these apparent advantages do not consider the external costs associated with air pollution.

The emissions of air pollutants from two-stroke engines can be large per unit amount of fuel burned, and average from 75% to 750% greater than emissions from comparable four-stroke engines for CO and VOCs respectively. NO_x emissions from two-stroke engines tend to be a bit lower, but PM emissions—soot and unburned oil droplets—can be as much as 1000% greater. Two-stroke engines are still used in the United States on outboard motor boats, and on some lawn and garden equipment. During warm weather, in areas with many lakes, air pollution emissions from these sources can be significant.

It was shown in Table 18.3 that small two- and three-wheeled vehicles (most of which are powered by two-stroke engines) are very popular in Asia. They are also popular in some cities in eastern Europe. One reason is the low initial cost of these small vehicles. Another reason is that in urban centers in many non-OECD countries, the streets are narrow and highly congested, leading to a very real advantage for people who drive small vehicles. Even in industrialized Asian countries like Japan and Taiwan, two-stroke engines power a huge number of small vehicles.

Malaysia is typical of the urbanizing Asian countries, where motorcycles play a dominant role in transport. Over 1 million motorcycles were registered in Peninsular Malaysia in 1980 (Barwell 1985) representing about 10 motorcycles per 100 people, compared with about 5 cars per 100 people. The most popular version is a 70 cubic centimeter (cc) displacement, two-stroke motorbike of Japanese manufacture. These motorbikes also are prevalent in rural areas where they are used for such things as transporting entire families on their weekly visits to the city or for carrying large loads of rubber to sell at the local collection point. Of course, such severe overloading greatly increases the emissions from these vehicles.

Two other types are engines are widely used in transport, and will be discussed briefly. **Diesel engines** are mostly found on larger load-carrying vehicles like trucks and buses (and non-road vehicles like trains and boats). They are sometimes used on passenger cars. Unlike gasoline engines (either four- or two-stroke), diesel engines do not require spark ignition. Rather, in the diesel cycle, the air-fuel mixture is compressed to a significantly higher pressure than in a gasoline engine, causing the temperature of the mixture to increase until the autoignition temperature of the fuel is reached. The fuel burns and drives the piston downward in the power stroke.

Diesel fuel is a distillate fuel. Like gasoline, it is a mixture of hydrocarbons but is slightly heavier and less volatile than gasoline. However, in a diesel engine, the operating AFR is much leaner than for gasoline engines, ranging from 15 to 100 (Wark and Warner 1981). Hence, the emissions of CO and VOCs from diesel engines are characteristically lower than from gasoline engines. On the other hand, because of the diesel engine's higher pressures and temperatures, its NO_x emissions tend to be higher. In addition, odors and PM (smoke) are problems for diesel engines. Black smoke results when the engine "lugs down" (gets overloaded), which occurs typically during acceleration from a stop. The black smoke is primarily unburned carbon, or soot, and indicates engine overload. There are also high emissions of CO and VOCs during those times. No visible emissions or a small amount of white or gray smoke represents normal cruising operation for diesel engines.

Another type of engine that is widely used in transportation is not found on any highway. We refer to the **turbojet engine** that powers most modern aircraft. In a jet engine, the fuel (a kerosene-like mixture of hydrocarbons, heavier than gasoline but lighter than diesel fuel) is sprayed into the front of the combustion chamber mixing with enough primary air to burn most of the fuel. Excess air is admitted in separate zones to complete the burnout of the fuel in an extremely lean environment. There are no pistons, but rather the hot, expanding gases flow through the combustion chamber (mixing with more excess air), then into the turbine section where the energetic gases turn the fan, and finally exhaust at high velocity from the rear of the engine, driving the plane forward.

On an annual mass emission basis, pollution from jet engines is small relative to that from highway vehicles. Yet, because of the visibility of the smoke, and because of the concentration of aircraft activity at airports, aircraft emissions get much attention. Airports also attract large numbers of on-road vehicles—primarily private cars, buses, taxis, and delivery trucks. Understanding and controlling pollution emissions from all sources associated with airports can be a key part of a local air quality management plan.

According to Wayson and Bowlby (1988), airports typically contain six types of sources of air pollution—motor vehicles, aircraft, ground service vehicles, fuel storage and handling facilities, small combustion sources (like furnaces and boilers), and miscellaneous aircraft maintenance operations (such as parts plating, solvent degreasing, or painting). One large airplane can emit large amounts of several air pollutants, as shown in Table 18.4. However, it is important to keep in mind that aircraft often emit less than half of the total emissions coming from within the property lines of an airport. Ground service vehicles, public transportation vehicles, and personal cars may contribute more than the aircraft to localized high ground level concentrations of CO near an airport.

Table 18.4   Pollutant Emissions from Jet-Engine-Powered Aircraft

Aircraft	Engine	Mode	Fuel Flow (lb/hr)	Rate of Emissions Per Aircraft, lb/hr				
				CO	NO_x	VOC	SO_x	Particulate
B-727-200	JT8D-17	Idle	3450	117	11.7	30.3	3.6	1.2
	(3 engines	Takeoff	29,900	21	608	1.5	30.0	11.1
	per aircraft)	Climbout	23,700	23.7	370	1.2	23.7	7.8
		Approach	8430	60.6	58.2	4.2	8.4	4.5
B-747-200	JT9D-70	Idle	7200	245	23.2	49.6	7.2	8.8
	(4 engines	Takeoff	77,500	15.6	2400	11.6	77.6	15.2
	per aircraft)	Climbout	63,900	19.2	1550	9.6	64	16.0
		Approach	23,400	30.4	190	10.4	23.6	9.2
DC-10	CF6-50C	Idle	3620	264	9.0	109	3.6	0.12
	(3 engines	Takeoff	56,700	1.2	2010	0.6	56.7	1.62
	per aircraft)	Climbout	46,900	14.1	1386	0.6	46.8	1.62
		Approach	15,800	68.1	158	0.3	15.9	1.32

◆ 18.4 VEHICLE EMISSIONS ◆ AND EMISSION CONTROLS

There are a number of factors that affect how much or how little air pollution will be produced from motor vehicles. These can be arbitrarily separated into the following groups: (1) engine design and operating features, (2) driver operating and maintenance practices, (3) fuel composition, (4) add-on pollution control technology, and (5) environmental conditions. Each group of factors is discussed briefly in the following paragraphs.

The **engine design feature** that influences emissions most strongly is the *air-to-fuel ratio*. Current practice for modern vehicles equipped with catalytic converters is to burn at a precisely controlled, slightly rich mixture (ER = 0.98–0.99). This prevents excessive NO_x formation, without producing excessive amounts of CO and VOCs. With this overall target in mind, the methods of controlling the flow rate of fuel to provide the right macromixture (overall AFR) and the methods of mixing the air and fuel to provide a homogeneous micromixture in every part of the cylinder are very important. For years, the macromixture was controlled (and not very well) by carburetors. Today, achieving the desired AFR is accomplished by fuel injectors with computer-controlled fuel and air injection rates, which are monitored and adjusted by the computer to optimize the combustion mixture for different driving conditions. Chamber designs that induce a swirl to the incoming fuel and air have been shown to provide even better micromixing of the reactants, thus minimizing the formation of CO and VOCs.

An older modification of the conventional internal combustion engine is the *stratified charge engine*. It employs the concept of two-stage combustion (two-stage combustion in power plants was discussed in Chapter 16). The stratified charge engine is built and operated similarly to a conventional four-stroke engine except that there are two combustion chambers for each cylinder. In the first (smaller) chamber, a fuel-rich mixture is ignited by the spark. The flame front advances and ignites a very lean mixture in the second (larger) chamber. The gases then expand and drive the piston. As expected, two-stage combustion avoids the very high temperatures of stoichiometric combustion (thus preventing excessive NO_x formation), and yet results in very complete combustion (thus producing low amounts of CO and VOCs).

A more recent innovation is the *extra-lean-burn engine*. In this engine, combustion in a standard cylinder is sustained by forming an air/fuel mixture that is richer at the plug and much leaner elsewhere in the cylinder. This is achieved with chamber design, intake timing, and swirl patterns (Grable 1993). This engine, which mimics the stratified charge concept without the mechanical problems associated with two combustion chambers, is available on some cars now. It uses air-to-fuel ratios as high as 25 and so emits very little CO and VOCs. Coupled with a higher drive ratio, the engine is forced to operate in the low-rpm/high mechanical efficiency portion of its power band, giving excellent gas mileage. Lean-burn engines demand even more precise control of fuel injection than standard engines, so a direct air/fuel sensor is used instead of the customary oxygen sensor. As with all lean-burn engines, increased NO formation is a drawback.

Another important engine design feature is the *compression ratio*. A higher compression ratio yields more power but produces higher temperatures and thus more NO_x. The surface-to-volume ratio of the combustion chamber also affects emissions. A larger cylinder wall surface area relative to the cylinder volume means more premature cooling of the gases and quenching of the combustion reactions on the walls, and thus higher emissions of CO and VOCs.

The timing of the spark relative to the stroke of the piston is a significant engine operating parameter. Normally, the spark occurs at about 20° before top dead center to provide the best performance. However, both VOCs and NO can be reduced by more than 50% by retarding the spark advance to top dead center (at the expense of good performance).

One other engine modification is *exhaust gas recirculation* (EGR). In this modification, part of the exhaust gas is recirculated to the air intake manifold. The exhaust gases mix with air and then flow into the cylinders where they dilute and cool the combusting gases in the cylinders. EGR is a very effective way to reduce NO_x formation.

Typically, in two-stroke engines the fuel and air mixing characteristics are poorer than in larger, better-controlled engines. This results in numerous fuel-rich pockets within the cylinders and consequent high CO and VOC emissions. These engines often are too small to support much in the way of pollution control equipment. In any case, unburned oil mists in the exhaust would foul most pollution control equipment.

The mechanical design of diesel engines affects emissions—particularly soot—in complicated ways. *Soot* is small particles of carbon onto which heavy hydrocarbons are adsorbed. Soot production can increase significantly when poor micromixing results in fuel-rich pockets even though the overall AFR may be correct.

Obviously, no matter how well the engine has been designed, the vehicle must be **operated and maintained** properly, or emissions will be higher than they should be. Good operation includes driving sensibly, not overloading the vehicle, using the proper fuel, and using the proper oil/fuel ratio (for two-stroke engines). Proper maintenance includes keeping the engine in tune, changing the oil regularly, keeping the spark plugs clean, making sure the EGR valve is functioning properly, and even keeping proper air pressure in the tires.

Fuel composition plays an important role in air pollution emissions. An obvious variable is the level of impurities in the fuel. Sulfur is an undesirable impurity for at least two reasons. The first is that sulfur in the fuel leads directly to formation of the criteria pollutant SO_2. A second important reason is that sulfur temporarily poisons the catalyst in automobile catalytic converters, and thus contributes to increased emissions of CO and VOCs.

In the United States, motor vehicle fuels (both gasoline and diesel) tend to be more stringently *refined* (processed in an oil refinery) than in other countries. One step in the refining process is *hydrodesulfurization,* in which organic sulfur is removed from the various hydrocarbon feedstocks and products by reaction with hydrogen. Because of this refining process, fuels in the United States have less sulfur than fuels produced in other parts of the world, and motor vehicles here produce much lower SO_x emissions than elsewhere.

Lead could also be considered an "impurity" in gasoline, although in truth it has been added during the final blending of gasoline at the refinery as one way to boost octane. (As little as 2–3 g of tetraethyl lead per gallon of gasoline can boost the octane rating by several points.) As mentioned before, the enforced steady reduction in the overall lead content of gasoline from 1975 through 1995 (when it effectively went to zero) in the United States has dramatically reduced lead emissions into urban air. However, this decrease in lead has forced oil companies to add more aromatics and olefins (higher octane stocks) to the blend to boost the octane of the final product. This process has led to higher motor vehicle emissions of "air toxics" and a higher degree of photochemical activity of the exhausted VOCs. Additionally, aromatics have a greater tendency to form soot than alkanes.

Another important characteristic of gasoline is its *volatility,* as measured by the *Reid Vapor Pressure* (RVP). The RVP is increased to its final specification limit when butane is added to the gasoline blend—more being added in the winter to help provide quick starts in cold weather. The higher the RVP, the greater is the tendency for the fuel to evaporate. Higher RVPs produce higher VOC emissions from cars during driving, from refueling operations, and even from fuel storage (at gasoline stations and bulk terminals). In recent years, EPA mandates have resulted in gasolines with lower RVPs than in the past, especially during the summer months (the peak ozone season).

In the past twenty years or so, much attention has been focused on *oxygenated fuels*—primarily methanol and ethanol—as substitutes or extenders for gasoline. These fuels produce fewer CO and VOC emissions, and will be discussed in more detail later in this chapter.

Add-on pollution control technology substantially reduces the final level of emissions from a vehicle. The most obvious technology on cars of the last three decades is the *catalytic converter,* which is used to treat the engine exhaust emissions. Such treatment includes the oxidation of CO and VOC to end products (CO_2 and H_2O), and the chemical reduction of NO_x to N_2 and O_2. Originally, this was accomplished by operating the engine fuel-rich and routing the exhaust gases over a reducing catalyst to control NO_x. Then additional air was injected into the exhaust system, and the gases were passed over an oxidizing catalyst. However, such a design wasted fuel and was not well accepted by the public. A three-way catalytic converter using platinum and rhodium metals supported on alumina now does both jobs simultaneously in one unit. This three-way catalyst requires precise computer control of the engine AFR and real-time oxygen sensors in the exhaust. Figure 18.4 shows that tight control of the AFR is needed for the three-way catalyst to function properly. In fact, modern vehicles (including large trucks as well as passenger cars) are equipped with on-board computers that now control AFR, spark timing, EGR rate, idle speed, and other operating parameters.

Two items of equipment that have not been mentioned yet are the *carbon canister* and the *positive crankcase ventilation* (PCV) *valve*. The carbon canister is a small carbon bed adsorber that collects evaporative emissions of VOCs from the hot engine after the vehicle has been shut down. (Formerly, gasoline fumes would simply escape into the air.) Later—the next time the car is operated—the carbon bed is desorbed by passing some of the intake air through the carbon canister, and the VOCs are routed back into the cylinders. The PCV valve's function is to route the air from the engine crankcase (which contains oil vapors and mists) into the cylinders for burning.

A device that is starting to be used on diesel vehicles is the *trap-oxidizer*. This device is a combination catalytic converter and particulate filter. Soot and other diesel emissions are caught on the trap and later oxidized.

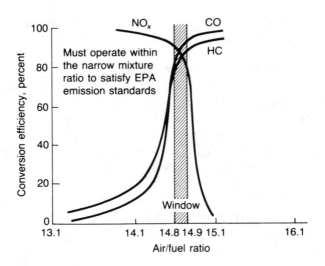

Figure 18.4
Effect of the air-to-fuel ratio on conversion efficiencies of a three-way catalyst.
(Adapted from Niepoth et al., 1978.)

It is important to note at this point that operator *tampering* with the on-board vehicle pollution control equipment can defeat the best of designs. Tampering includes disconnecting the carbon canister, jamming the EGR valve, disconnecting the oxygen sensor, removing the catalytic converter, removing the fuel inlet restrictor, and so forth. Tampering can increase emissions as much as 700% relative to a similar untampered vehicle.

According to several EPA studies, tampering rates are about 20–25% nationwide. In one study conducted by the Florida Department of Environmental Protection (Nichols 1992), a total of 5414 vehicles were inspected during unannounced visits to 544 used car lots in six central Florida counties between 1990 and 1992. The overall tampering rate for these vehicles was reported to be 17.2%. The tampering rate for the 1671 vehicles manufactured between 1975 and 1980 was 29.4%, while for the 1981 and later model cars it was 11.8%. However, as Nichols (1992) points out, these tampering rates were probably low because not all the pollution control devices were inspected.

Misfueling—which means putting leaded gasoline into cars designed for unleaded—is a form of tampering that was particularly troubling during the 1970s and 1980s. Just to save a few pennies per gallon on gasoline, a significant percentage of drivers punched out the fuel inlet restrictor and put leaded gasoline into cars equipped with catalytic converters. In addition to being responsible for much of the particulate lead in urban air, this practice permanently deactivated hundreds of thousands of catalytic converters. Filling a tank with leaded gasoline as few as ten times can permanently destroy the catalyst's ability to control pollutants. Fortunately, leaded gasoline has been phased out in the United States, and many other countries are following suit.

Environmental conditions such as ambient temperature and pressure can affect vehicle emission rates significantly. As mentioned, engines operate best at a precise air-to-fuel ratio (recall that the AFR is a *mass* ratio). The engine consumes liquid fuel and ambient air, each on a volumetric basis. The density of each changes with ambient temperature (the air more so than the liquid fuel) so the mass of air relative to the mass of fuel changes even when the volumetric flow rate of each stays the same. In addition, cold temperatures extend engine warm-up time, during which the cold cylinder surfaces more quickly quench the combustion reactions, thus increasing emissions. Furthermore, because new vehicles are tested in a 75 F chamber to see if they meet EPA emission limits, there is a tendency for manufacturers to optimize the engine and pollution control system to minimize emissions at 75 F. Actual vehicle emissions tend to increase at temperatures both above and below 75 F. In addition to ambient temperature itself, such factors as humidity, cloud cover, and solar loading can greatly affect the use of air conditioning. Air conditioning places a significant added load on the engine, which results in increased emissions of pollution.

Another temperature effect that must be mentioned is the engine operating temperature. It has been well documented that emission rates are significantly higher during the first 3–8 minutes after start up (when the engine is cold and before the catalytic converter reaches its operating temperature).

Ambient pressure affects air density directly but has almost no effect on fuel density. The effect on air density is very noticeable when there is a large change in altitude. Cars are designed for "standard" air with a density appropriate to sea level; however, air is considerably "thinner" in cities like Denver and Mexico City, where the same volumetric flow of air may contain 20% less oxygen by mass than at sea level. Another very important effect of reduced ambient pressure is the incorrect action of various diaphragms, dashpots, and sensors that help monitor and control engine performance.

▶ ▶ ▶ ▶ ▶ ▶ ▶ ▶ ▶

Example 18.4

The design AFR for a particular vehicle burning a certain grade of gasoline is 14.7, based on an air density of 1.20 kg/m^3. If this same car is driven in Mexico City with no adjustments (that is, it draws in the same volume of air per unit of liquid fuel), calculate its actual AFR and the stoichiometric ratio. Mexico City has an elevation of 2240 m above sea level, and its air has a density of 1.00 kg/m^3.

Solution

Choose a basis of 1 kg of fuel burned. At sea level, with an AFR of 14.7, the volume of air taken in is

$$1 \text{ kg fuel} \times \frac{14.7 \text{ kg air}}{1 \text{ kg fuel}} \times \frac{1 \text{ m}^3 \text{ air}}{1.20 \text{ kg air}} = 12.24 \text{ m}^3$$

In Mexico City, the mass of air in this volume of air is

$$12.24 \text{ m}^3 \times \frac{1.00 \text{ kg air}}{1 \text{ m}^3 \text{ air}} = 12.24 \text{ kg air}$$

The actual AFR and SR are

    AFR = 12.24 kg air/kg fuel; and

    SR = $\dfrac{12.24}{14.7}$ = 0.83 (approximately 17% below stoichiometric)

The jets in the carburetor (or the fuel injectors) of this vehicle should be replaced.

◀ ◀ ◀ ◀ ◀ ◀ ◀ ◀ ◀

The fleet of vehicles on the road in any given year is a mixture of many vehicle types (cars, trucks, buses, and so forth), which use different fuels (gasoline, diesel, or others). Additionally, the vehicles have a wide age distribution. Some cars in the fleet will be the most recent model year, a significant proportion of the others will represent each model year for the last ten years, and other vehicles will be twenty years old or older! In the United States, during any given year many people trade in their vehicles. The net result is that each year some of the oldest vehicles are removed from the fleet and replaced by the newest models via a free-market chain reaction that effectively maintains certain percentages of each model year in the fleet year after year. Of course, the newest models are built to meet the most recent (most stringent) emission standards and are the least polluting if maintained and operated properly.

The history of tailpipe emission standards in the United States for new passenger cars or light duty vehicles (LDVs) is presented in Table 18.5, and U.S. standards for heavy duty vehicles (HDVs) are presented in Table 18.6. Selected examples of LDV emission standards of various countries are shown in Table 18.7.

Table 18.5   History of New Light Duty Vehicle Exhaust Emission Standards in the U.S.[a]

	New Vehicle Emission Standards, grams/mile (gpm) or as stated			
Year	Carbon Monoxide	Hydrocarbons	Oxides of Nitrogen	Particulate Matter
Prior to controls	3.4%[b]	850 ppm	1000 ppm	–
1968–69	2.0%	350 ppm	–	–
1970–71	23	2.2	–	–
1972[c]	39	3.4	–	–
1973–74	39	3.4	3.0	–
1975–76	15	1.5	3.1	–
1977–79	15	1.5	2.0	–
1980	7.0	0.41	2.0	–
1981	3.4	0.41	1.0	–
1982–86	3.4	0.41	1.0	–
1987–93	3.4	0.41	1.0	0.60[d]
1994+	3.4	0.41	1.0	0.20[d]
	10.0[e]	0.41	0.4	0.08[f]

[a] The federal test procedure varied during the years prior to 1975. Since 1975, the procedure has been CVS-75, a constant volume sample test that includes hot and cold starts.
[b] % = percent by volume; ppm = parts per million.
[c] Change of test procedure this year; the apparent relaxation of standards is not real.
[d] Applies to diesels only.
[e] This additional standard is for 20 F; 3.4 grams per mile still applies at 75 F.
[f] All LDVs.

Adapted from U.S. Environmental Protection Agency, 1992. Numerous other notes and comments are included in the original table.

The standards in Table 18.5 are presented as **emission factors**. In general, an emission factor (EF) is a measure of an *average* rate of emission of a pollutant for a defined activity rate. For an average vehicle, an EF is the *average* rate of emission of a particular pollutant when the vehicle is driven in a specified manner. EFs usually are given in units of grams of pollutant per mile driven for LDVs, in grams per brake horsepower-hour for HDVs, and in grams per minute for idling vehicles of both types. For *used* vehicles, EFs are convenient *unitized* measures of current emission rates. For *new* vehicles, EFs are measured on a subset of newly made cars and are used to gauge whether the car manufacturer is meeting the EPA standards for new vehicle performance (the emission standards are set in the same units as the EFs).

These new vehicle performance standards are based on a standardized test cycle—the Federal Test Procedure (FTP)—that approximates average urban driving. The FTP is a demanding test which involves numerous acceleration and deceleration cycles on a chassis dynamome-

Table 18.6   History of New Heavy Duty Vehicle Exhaust Emission Standards in the U.S.

	Exhaust Emission Standards, grams/brake horsepower-hour (g/bhp-hr)						
	HDGV[a]			HDDV[b]			
Year	CO	HC	NO$_x$	CO	HC	NO$_x$	PM
Prior to controls	155	10.9	6.71	–	–	–	–
1974–78	40	–	–	40	–	–	–
1979	25	1.5	–	25	1.5	–	–
1980–83	25	1.5	–	25	1.5	–	–
1984	25	1.5	–	25	1.5	10.7	–
1985–86	37.1	1.9	10.6	15.5	1.3	10.7	–
1987[c]	37.1	1.9	10.6	15.5	1.3	10.7	–
	14.4	1.1	10.6	–	–	–	–
1988–90[c]	37.1	1.9	6.0	15.5	1.3	6.0	0.6
	14.4	1.1	6.0	–	–	–	–
1991–93[c]	37.1	1.9	5.0	15.5	1.3	5.0	0.10[d]
	14.4	1.1	5.0	–	–	–	0.25[e]
1994 +[c]	37.1	1.9	5.0	–	–	–	–
	14.4	1.1	5.0	15.5	1.3	5.0	0.10[f]

[a] HDGV = heavy duty gasoline vehicles.
[b] HDDV = heavy duty diesel vehicles.
[c] Higher standards for HDGV are for vehicles > 14,000 pounds gross vehicle weight rating.
[d] Urban buses.
[e] All HDDV other than urban buses.
[f] All HDDV.

Adapted from U.S. Environmental Protection Agency, 1992. Numerous other notes and comments are included in the original table.

ter meant to simulate the stop-and-go driving patterns observed in urban areas (the test is based on Los Angeles in the early 1970s). The U.S. FTP has been adopted by Canada, Mexico, Korea, Taiwan, and a number of other countries. Japan and the European Economic Community each have their own standardized tests, patterned after the FTP, but more representative of cities in those nations.

If one simply averages the emissions from a large, representative sample of vehicles on the road, a composite or **fleet-averaged** emission factor results. This fleet-averaged EF, given in units of grams per mile (g/mi) per average vehicle, is very useful in predicting total emissions from vehicles located at such facilities as roadways, intersections, or parking lots (and subsequently predicting concentrations near these facilities). With a fleet-averaged EF, one simply uses the number of vehicles driving on the road without worrying about which individual vehicles are passing by.

Table 18.7   Selected Examples of Motor Vehicle Emission Standards from Around the World

	CO	HC	NO_x	PM
European Economic Community				
1990, LDV, g/test	25	3.0	3.5	1. 1 (diesels)
1993, HDV, g/kWh	4.5	1.1	8.0	0.36
Australia				
1986, LDV, g/km	6.3	0.93	1.93	
Brazil				
1988, LDV, g/km	24.0	2.1	2.0	
1992, LDV, g/km	12.0	1.2	1.4	
Canada				
1990, Cars, g/km	2.11	0.25	0.62	0.12 (diesels)
1988, HDV, g/km	37.1	1.9	6.0	
India				
1989, 2- and				
3-wheelers g/km[a]	12–30	8–12	–	
1989, cars, g/km[a]	14–27	2.0–2.9		
Japan				
1991, LDV, g/km	2.7	0.39	0.48	
Mexico				
1989, cars, g/mi	35.2	3.20	3.68	
1991, cars, g/mi	11.2	1.12	2.24	
1993, cars, g/mi	3.4	0.40	1.00	

[a] Sliding scale depending on weight of vehicle.
Adapted from Faiz et al., 1992.

Many variables influence the numerical value of an EF—two of the more important being vehicle type and vehicle age. Type should be obvious, as heavy vehicles ought to emit at rates different from lighter vehicles. Age is a factor even for a vehicle with a well-designed engine, equipped with modern pollution control equipment, that has been properly operated and maintained, and that has used clean fuels. As the vehicle ages, its mechanical equipment wears down, and its emission rate increases. This natural deterioration increases the average emissions per vehicle each year.

▶ ▶ ▶ ▶ ▶ ▶ ▶ ▶ ▶

Example 18.5

Assume that the vehicle mix in a particular city is well represented as 90% passenger cars and 10% trucks. Given the following table of new vehicle EFs (applicable only at one speed and one temperature), deterioration rate factors (EF multipliers), and fraction of VMT (vehicle miles traveled) distributed by age, calculate the fleet-averaged EF for this city in 1992 (for this one speed and one temperature).

Solution

For cars, multiply the first three columns for each year (for example, for 1992, $12.0 \times 1.07 \times 0.112 = 1.438$), then sum the results for all

Data for Example 18.5

	Cars			Trucks		
Model Year	New EF g/mi	Deterioration Factor	Fraction of VMT	New EF g/mi	Deterioration Factor	Fraction of VMT
1992	12.0	1.07	0.112	25.0	1.09	0.109
1991	12.0	1.12	0.142	25.0	1.13	0.105
1990	12.0	1.16	0.130	35.0	1.17	0.103
1989	12.0	1.20	0.121	35.0	1.21	0.099
1988	15.0	1.23	0.108	50.0	1.25	0.095
1987	15.0	1.26	0.094	50.0	1.29	0.087
1986	15.0	1.28	0.079	50.0	1.32	0.082
1985	24.0	1.30	0.063	50.0	1.35	0.072
1984	24.0	1.32	0.047	80.0	1.38	0.062
1983	24.0	1.34	0.032	80.0	1.41	0.051
1982	30.0	1.35	0.025	80.0	1.44	0.040
1981	30.0	1.36	0.020	80.0	1.46	0.031
1980	35.0	1.37	0.015	95.0	1.48	0.024
1979	35.0	1.38	0.006	95.0	1.50	0.017
1978	50.0	1.39	0.003	95.0	1.52	0.011
1977	50.0	1.40	0.002	95.0	1.53	0.007
1976	50.0	1.41	0.001	95.0	1.54	0.005

the years. The fleet-averaged EF for cars thus obtained is 19.9 g/mi. Repeat this process for the trucks. The averaged truck EF is 66.1 g/mi. The overall fleet-averaged EF thus is

$$(0.9 \times 19.9) + (0.1 \times 66.1) = 24.5 \text{ g/mi}$$

◀ ◀ ◀ ◀ ◀ ◀ ◀ ◀ ◀

If one can predict correct values for all the variables that influence the EF, one can project EFs into any given year in the future. Because the EF is reported in g/mi, it is a *strong* function of the vehicle's average speed. (Lower average speeds tend to include longer idling times and more starts and stops which increases the total emissions [and thus the EF] over a given distance.) Other major factors include the percentages of all vehicle types in the local fleet, the calendar year (determines vehicle age by model year), the local ambient temperature, the local altitude, the percent cold starts, and the percent tampering.

▶ ▶ ▶ ▶ ▶ ▶ ▶ ▶ ▶

Example 18.6

A vehicle emits CO at an almost constant rate of 6.0 g/min while cruising at speeds between 20 mph and 40 mph, and emits 4.0 g/min while idling. Ignoring acceleration and deceleration, calculate the average EF (g/mi) for travel on a three-mile stretch of road with traffic signals for the following two cases:

Case a.

light traffic—speed while traveling = 30 mph; total stopped time = 2 minutes

Case b.

heavy traffic—speed while traveling = 20 mph; total stopped time = 5 minutes.

Also, calculate the average travel speed for each case.

Solution
Case a.

$$\text{Cruising time} = \frac{3 \text{ miles}}{30 \text{ mph}} \times \frac{60 \text{ min}}{1 \text{ hr}} = 6 \text{ minutes}$$

$$\text{EF} = \left[(6 \text{ min} \times 6.0 \text{ g/min}) + (2 \text{ min} \times 4.0 \text{ g/min})\right]/3 \text{ miles}$$
$$= 14.7 \text{ g/mi}$$

Average travel speed is

$$\frac{3 \text{ miles}}{8 \text{ minutes}} \times \frac{60 \text{ min}}{1 \text{ hour}} = 22.5 \text{ mph}$$

Case b.

$$\text{Cruising time} = \frac{3 \text{ miles}}{20 \text{ mph}} \times \frac{60 \text{ min}}{1 \text{ hr}} = 9 \text{ minutes}$$

$$\text{EF} = \left[(9 \text{ min} \times 6.0 \text{ g/min}) + (5 \text{ min} \times 4.0 \text{ g/min}) \right] / 3 \text{ miles}$$
$$= 24.7 \text{ g/mi}$$

Average travel speed is

$$\frac{3 \text{ miles}}{14 \text{ minutes}} \times \frac{60 \text{ min}}{1 \text{ hour}} = 12.9 \text{ mph}$$

◀ ◀ ◀ ◀ ◀ ◀ ◀ ◀ ◀

As stated earlier, in order to correctly model vehicle emissions, all the factors that affect emissions must be taken into account. For years, the EPA has supported a massive testing and modeling effort to predict fleet-averaged emission factors from U.S. vehicles. They have developed a comprehensive computer program that allows the user to change those variables for which local data is available and to use national average (default) values for other variables. As of December 2001, the official EPA version of this program was **MOBILE5a** (which was released in 1993). In January 2002, just as this textbook was in final editing, EPA released **MOBILE6**. Although MOBILE6 represents a significant change in format and input requirements from MOBILE5a, and although it gives somewhat different answers for the EFs (Keely and Cooper 2002), the principles of this kind of emission factor model are well illustrated by a sample output from MOBILE5a, along with several graphs from MOBILE5a representing the behavior of the EFs.

An output from MOBILE5a is presented in Figure 18.5. Curves showing the behavior of EFs as a function of selected variables were developed from repeated runs of MOBILE5a and are shown in Figure 18.6. As can be seen, the EFs are strong functions of calendar year, speed, and temperature (as well as many other variables not shown). As mentioned above, MOBILE6 has now replaced MOBILE5a, and will be used in the same manner as was MOBILE5a.

In addition to predicting vehicle EFs for past years, MOBILE6 has built-in forecasts of future vehicle-type distributions and can predict fleet-averaged EFs for fifty years into the future. Thus, it can be used in the very important role of comparing the effects of different control strategies on air quality in future years. Another important use for MOBILE6 is to help planners and regulators develop emission inventories and budgets for state implementation plans (SIPs). Mobile source emissions can account for 40% or more of all the statewide emissions of VOCs, NO_x, and CO. In addition, when doing regional

MOBILE5A OUTPUT (EDITED)

Minimum Temp: 50. (F)   Maximum Temp: 75. (F)
Period 1 RVP: 9.0   Period 2 RVP: 9.0   Period 2 Yr: 1993
VOC HC emission factors include evaporative HC emission factors. Emission factors are as of Jan. 1st of the indicated calendar year.
Region: Low   Altitude: 500. Ft.   I/M Program: No
Ambient Temp: 70.0 / 70.0 / 70.0 F   Anti-tam. Program: No   Operating Mode: 21.0 / 27.0 / 21.0
Reformulated Gas: Yes   ASTM Class: C

Vehicle Speed: 20 miles/hr

Calendar Year: 1975

Veh. Type:	LDGV	LDGT1	LDGT2	LDGT	HDGV	LDDV	LDDT	HDDV	MC	All Veh
Veh. Spd.:	20.0	20.0	20.0		20.0	20.0	20.0	20.0	20.0	
VMT Mix:	0.715	0.140	0.074		0.027	0.001	0.000	0.033	0.010	
Composite Emission Factors (Gm/Mile)										
VOC HC:	9.01	9.32	15.99	11.64	27.34	1.50	0.00	5.36	12.06	9.97
Exhst CO:	72.21	72.71	100.50	82.37	276.94	2.95	0.00	14.77	37.64	77.58
Exhst NOX:	3.83	3.78	5.85	4.50	9.56	1.56	0.00	28.96	0.30	4.92

Calendar Year: 1990

Veh. Type:	LDGV	LDGT1	LDGT2	LDGT	HDGV	LDDV	LDDT	HDDV	MC	All Veh
Veh. Spd.:	20.0	20.0	20.0		20.0	20.0	20.0	20.0	20.0	
VMT Mix:	0.655	0.161	0.082		0.031	0.009	0.002	0.052	0.008	
Composite Emission Factors (Gm/Mile)										
VOC HC:	3.16	4.15	6.21	4.84	11.19	0.70	1.01	3.36	5.02	3.81
Exhst CO:	27.18	34.48	49.40	39.50	136.41	1.63	1.93	13.59	21.69	32.50
Exhst NOX:	1.90	2.23	2.98	2.48	6.70	1.62	1.91	21.57	0.87	3.19

Calendar Year: 2005

Veh. Type:	LDGV	LDGT1	LDGT2	LDGT	HDGV	LDDV	LDDT	HDDV	MC	All Veh
Veh. Spd.:	20.0	20.0	20.0		20.0	20.0	20.0	20.0	20.0	
VMT Mix:	0.601	0.196	0.087		0.031	0.002	0.002	0.075	0.006	
Composite Emission Factors (Gm/Mile)										
VOC HC:	1.96	2.32	3.10	2.56	4.02	0.51	0.69	2.09	4.72	2.22
Exhst CO:	17.01	19.54	26.30	21.61	26.52	1.41	1.55	11.05	18.02	18.11
Exhst NOX:	1.36	1.61	2.21	1.79	4.35	1.12	1.24	8.23	0.88	2.09

Vehicle Speed: 25 miles/hr

Calendar Year: 1975

Veh. Type:	LDGV	LDGT1	LDGT2	LDGT	HDGV	LDDV	LDDT	HDDV	MC	All Veh
Veh. Spd.:	25.0	25.0	25.0		25.0	25.0	25.0	25.0	25.0	
VMT Mix:	0.715	0.140	0.074		0.027	0.001	0.000	0.033	0.010	
Composite Emission Factors (Gm/Mile)										
VOC HC:	7.91	8.22	14.24	10.31	21.89	1.26	0.00	4.50	10.67	8.71
Exhst CO:	59.72	60.13	83.97	68.41	217.30	2.33	0.00	11.67	32.12	63.91
Exhst NOX:	4.07	4.02	6.15	4.76	9.98	1.44	0.00	26.73	0.33	5.09

Calendar Year: 1990

Veh. Type:	LDGV	LDGT1	LDGT2	LDGT	HDGV	LDDV	LDDT	HDDV	MC	All Veh
Veh. Spd.:	25.0	25.0	25.0		25.0	25.0	25.0	25.0	25.0	
VMT Mix:	0.655	0.161	0.082		0.031	0.009	0.002	0.052	0.008	
Composite Emission Factors (Gm/Mile)										
VOC HC:	2.69	3.59	5.33	4.18	9.00	0.59	0.85	2.82	4.66	3.25
Exhst CO:	22.63	29.23	40.90	33.15	107.03	1.29	1.52	10.74	17.29	26.89
Exhst NOX:	1.93	2.36	3.12	2.61	6.99	1.50	1.77	19.90	0.96	3.17

Calendar Year: 2005

Veh. Type:	LDGV	LDGT1	LDGT2	LDGT	HDGV	LDDV	LDDT	HDDV	MC	All Veh
Veh. Spd.:	25.0	25.0	25.0		25.0	25.0	25.0	25.0	25.0	
VMT Mix:	0.601	0.196	0.087		0.031	0.002	0.002	0.075	0.006	
Composite Emission Factors (Gm/Mile)										
VOC HC:	1.66	2.00	2.65	2.20	3.37	0.43	0.58	1.75	4.45	1.88
Exhst CO:	13.09	15.41	20.84	17.08	20.81	1.11	1.23	8.73	14.37	14.10
Exhst NOX:	1.40	1.62	2.23	1.81	4.55	1.03	1.15	7.60	0.97	2.07

Figure 18.5
A MOBILE5a output.

ozone modeling (see Chapter 19), detailed estimates of mobile source emissions (by hour of the day and by location within an urban area) are some of the inputs required for a numerical photochemical model to predict hourly ozone concentrations throughout a metropolitan (or much larger) region.

In addition to the regional modeling described above, MOBILE6 can be used for microscale modeling even though this usage was not the original intent (MOBILE6 gives average trip EFs, and includes start emissions). Carbon monoxide emission factors can be input into Gaussian dispersion models (see Chapter 20) to predict CO concentra-

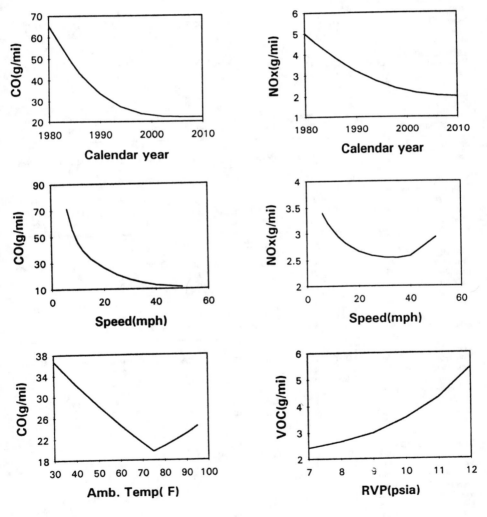

Figure 18.6
Typical behavior of U.S. light duty vehicles fleet-averaged emission factors as functions of several parameters.

tions near roadways, and thus help assess the potential air quality impacts of new roadway projects. Recently, a microscale emission factor model has been described (Singh and Huber 2000, 2001) that shows good promise for predicting real-time CO emission factors for use in microscale dispersion models. However, at the present time, the only accepted mobile source emissions model is EPA's MOBILE model (currently MOBILE6).

◆ 18.5 OTHER MEASURES FOR CONTROL ◆ OF AIR POLLUTION FROM VEHICLES

Once vehicles are sold, individuals assume the responsibility for their proper operation and maintenance, including the pollution control equipment. However, regulatory control of mobile sources as a group becomes the responsibility of local and state governmental agencies. Over the past thirty years, it has become well recognized that organized efforts by state and local governments are necessary to ensure that some of our urban areas achieve their air quality goals. States, under the authority of the Clean Air Amendments, and with EPA approval, establish State implementation Plans (SIPs) that detail the steps to be taken either statewide or in individual counties to help achieve and/or maintain ambient air quality standards. As regards mobile sources, these steps fall into three broad categories—inspection and maintenance programs, transportation control measures, and changes in automotive fuels.

Inspection and maintenance (I/M) programs involve regular vehicle inspections to ensure that vehicles meet minimum in-use air pollution standards. These minimum standards are purposely set high; only highly polluting vehicles do not pass. The inspection stations are usually operated by private contractors under state supervision. Citizens are required to get annual inspection stickers, much like annual tag renewals. The inspection covers the key pollution control equipment and checks tailpipe emissions. I/M programs help identify those vehicles in which engines are not operating properly or which have other problems that are causing high exhaust emissions. Cars that do not meet the limits must be repaired to correct the problems and be reinspected in order to receive their "operating permit" (sticker). I/M programs can be effective at reducing in-use fleet emissions by identifying and removing from the road "super-emitters" (vehicles that emit up to 100 times the average) and by reducing tampering rates.

A good I/M program includes each of the following:

1. A suitable standardized tailpipe test procedure
2. Effective enforcement for full compliance
3. Qualified mechanics and continuing training programs for diagnostics and repairs

4. Different but effective in-use standards for vehicles of different model years

5. Minimization of waivers and exemptions

6. Periodic evaluation and review by the state regulatory agency

Transportation control measures (TCMs) and traffic management techniques (TMTs) can be very effective at reducing air pollution within urban areas. These plans are usually included in a document called a Transportation Improvement Plan that coordinates such efforts area-wide. Transportation control measures and traffic management techniques result in fewer vehicles on the roads and/or in smoother traffic operations. In either case, emissions (and concentrations) are reduced. Section 108(f) of the Clean Air Act Amendments of 1990 requires the EPA (in concert with the U.S. Department of Transportation) to prepare information "regarding the formulation and emission reduction potential of transportation control measures related to criteria pollutants and their precursors." The sixteen categories of TCMs listed in Section 108(f) are presented in Table 18.8. While a comprehensive list of all measures that have been used throughout the world would be exceedingly long, it would generally include steps from one or more of the following categories: *transit options* (more buses, exclusive bus lanes, park-and-ride lots), *economic incentives/penalties* (high-priced parking, car pool subsidies, higher gasoline taxes, road-use taxes), and *regulatory steps* (parking bans, auto-free zones, gasoline rationing, even restricted driving days—rationed by odd or even numbers on the license plates). The economic and regulatory measures are much more effective in the United States than are the transit options. In other countries, transit is more widely

Table 18.8   Transportation Control Measures

1. Trip reduction ordinances
2. Vehicle-use limitations/restrictions
3. Employer-based transportation management
4. Improved public transit
5. Parking management
6. Park and ride/fringe parking
7. Flexible work schedules
8. Traffic flow improvements
9. Area-wide rideshare incentives
10. High-occupancy vehicle facilities
11. Major activity centers
12. Special events
13. Bicycling and pedestrian programs
14. Extended vehicle idling
15. Extreme cold starts
16. Voluntary removal of pre-1980 vehicles

accepted, but other steps are necessary also. For example, city leaders in Amsterdam voted to stop all private traffic in the urban core in the early 1990s, and in February 2000, the Italian government banned the use of private cars on Sundays in Rome and 150 other cities in Italy. In Rome, on Sundays, public transportation is free.

Traffic management refers to the control of the *flow of vehicles on the streets*, as well as to the *management of travel demand*. Techniques to control vehicle flow include such steps as timing and sequencing traffic signals, improving intersections, widening streets, adding protected left-turn lanes and free-flowing right-turn curves, creating one-way street pairs, designating high-occupancy-vehicle (HOV) lanes, planning truck routes and excluding trucks from certain streets. All these steps tend to reduce congestion and allow for smoother, more free-flowing traffic. As stated earlier, as the average traffic speed increases, vehicle emission factors go down.

Travel demand can be altered by encouraging staggered work hours (which reduces peak hour demand and improves level of roadway service). Encouraging businesses to adopt four-day work weeks (with different days for different sets of workers) spreads out the travel demand. Finally, by developing sectors in an urban area with housing and working areas located in close proximity, the total miles traveled can be reduced.

Changes in motor vehicle fuels include modifications to gasolines, replacements for gasolines, or nonhydrocarbon fuel options. The biggest success story of modern times as regards gasoline modifications has been the dramatic reduction in the lead content of all gasoline sold in the United States (a 99% decrease from 1970 to 1994). Recently, reductions in the maximum allowable RVP of gasoline have proven very effective at reducing total emissions of VOCs from gasoline-powered vehicles as well as evaporative losses from gasoline stations and storage areas, especially in the hotter, ozone-prone summer months.

Currently, **oxygenated fuels** are of high interest. The main benefit of fuels with one or more oxygen atoms embedded in the fuel molecules is a significant reduction in CO when compared with straight gasoline. It has been shown that a 30% reduction in CO emissions can be achieved by switching to a gasoline blend with 3% oxygen by weight. VOC emissions are also reduced somewhat with oxygenated fuels, but NO_x emissions tend to increase slightly, which could exacerbate ozone problems. Gasolines blended with methanol, ethanol, tertiary butyl alcohol, or methyl tertiary butyl ether (MTBE) have been tested and used with success in some cities in the United States. However, MTBE has been cited for groundwater contamination in several cases. Several U.S. cities (Denver, Colorado is one example) require that oxygenates be used during the winter months.

Replacements for gasoline, or *alternative fuels*, include compressed natural gas, liquefied petroleum gas, and pure methanol or ethanol. Some properties of these fuels and conventional fuels are shown in Table 18.9. As can be seen from the table, the alcohols have significantly less energy per unit mass than the pure hydrocarbon fuels. Methane and propane both have high energy content per unit mass, but require compression and/or refrigeration to maintain them as liquids. Nevertheless, at the present time, these fuels are being explored as reasonable alternatives or supplements to gasoline.

Each fuel has advantages and disadvantages compared with gasoline. **Compressed natural gas** and **liquefied petroleum gas** produce about 50% less CO and VOCs, and, more importantly, they produce little or no photochemically active VOCs and no air toxics (such as benzene). The main disadvantages of these fuels are related to on-board fuel storage and handling, and refueling. The **alcohols** produce less CO and very few photochemically reactive compounds, but formaldehyde emissions are much higher. Other disadvantages of alcohols include fuel handling (methanol dissolves several types of rubber and other gasket materials, and can be absorbed through the skin) and the reduced range of the vehicle (lower energy content per volume of fuel). All alternative fuels share the disadvantage (compared with gasoline) of lack of infrastructure for refueling stations. Although these so-called "clean" fuels do not appear ready to displace gasoline completely, there may be an excellent niche for them in terms of fueling in-city fleet vehicles such as buses, delivery trucks, or city maintenance vehicles.

Nonhydrocarbon fuel options include hydrogen, electric vehicles, and solar-powered cars. **Hydrogen** has many appeals—it produces no carbon monoxide, no carbon dioxide, no hydrocarbons, and has the highest energy content per mass of any combustible fuel. However, there are still many technological and safety obstacles to overcome before hydrogen becomes a practical commercial alternative for highway vehicles. **Solar-powered cars** are still in the experimental stage. Only **electric-powered vehicles** are of practical interest at the present time.

Table 18.9   Selected Properties of Motor Vehicle Fuels

Property	Fuel					
	Methane	**Propane**	**Methanol**	**Ethanol**	**Gasoline**	**Diesel**
Boiling Point, °C	−162	−42	65	79	35–200	120–350
Energy Content (LHV)*, kJ/g	50	46	20	27	44	42
Liquid Density, kg/liter	0.42	0.51	0.79	0.78	0.72–0.78	0.84–0.88
Octane Number, (RON+ MON)*/2	120	105	99	97	87–93	~25

*LHV = lower heating value; RON = research octane number; MON = motor octane number.

Electric vehicles requiring tracks or overhead electric lines have been used for decades (trolleys, for example). Today, however, there is greater interest in electric vehicles that do not require these devices. Battery-operated cars and light duty vans have been developed and are being sold to the public. These battery-powered vehicles give acceptable driving performance. They operate essentially pollutant-free on the street, and with very low noise levels. Of course, it is recognized that there are air pollution emissions (at a fossil-fueled power plant) associated with the generation of the electricity required to recharge the batteries; however, these air pollutants generally are well controlled, and certainly are far removed from urban streets. At present, the limitations of a battery powered vehicle include the limited range (80–150 miles), the long time required to recharge the battery (6–8 hours), and the high cost.

A recent development with a more reasonable cost is the **hybrid car** (a combination of gasoline and electric power). These vehicles, which are available today, get very good gas mileage and reduce idling emissions essentially to zero. They function by automatically turning off the gasoline engine when the car comes to a stop (such as at a traffic signal), using the electric motor for routine driving, and automatically restarting the gasoline engine only when the driver presses the accelerator and needs more power than the electric motor can deliver. When braking, much of the braking energy is captured and used to recharge the batteries.

◆ PROBLEMS ◆

18.1 From 1975 to 1990, the population (and number of vehicles) in a certain metropolitan area doubled. Additionally, the average miles driven per person increased by 20%. Using the EFs in Figure 18.5, estimate the level of annual emissions of ozone precursors (VOCs and NO_x) from mobile sources in 1990 as a percentage of the 1975 level (assume average speed remained constant at 25 mph).

18.2 If the same growth rates as in Problem 18.1 were observed over the same time period in an urban area in a *developing* country that did not have a concurrent decline in the EFs, estimate that country's 1990 level of emissions of ozone precursors (as a percentage of the 1975 level).

18.3 Rework Problem 18.1 assuming that the growth in traffic in the city caused a citywide reduction in average travel speed from 25 mph to 20 mph.

18.4 Calculate the percent decrease or increase in emissions from U.S. on-road vehicles from 1980 to 1999 for each of the pollutants in Table 18.1. Comment on the NO_x trend.

18.5 Consider the data in Table 18.2. What were the *mobile source* annual emissions of CO from France, Thailand, and the United States, respectively? Calculate the approximate ratios of CO emissions (France/U. S., Thailand/U. S., and Thailand/France).

18.6 Assuming that three-fourths of the *lead* emissions shown in Table 18.2 for each country originated from motor vehicles, calculate the same ratios for lead emissions as requested for CO emissions in Problem 18.5. How would you account for any differences in the CO ratios versus the Pb ratios?

18.7 Assume the following gross estimates for annual vehicular CO_2 emissions: cars—1400 kg/car; trucks and buses—7000 kg/vehicle; two- and three-wheelers—400 kg/vehicle. Using the data in Table 18.3, estimate the world motor vehicle emissions of CO_2, and the percent contributions from vehicles in the United States, in the rest of the OECD countries, and in all the other countries in the world.

18.8 Assume that diesel fuel can be represented as $C_{15}H_{30}$, and that air is 79% N_2 and 21% O_2. Calculate the stoichiometric AFR.

18.9 Assume that gasoline can be represented by the formula C_7H_{13}. Calculate the stoichiometric AFR using the composition of air given in Problem 18.8.

18.10 Calculate the exhaust gas volume (in m^3, at STP) per kg of gasoline for the conditions of Problem 18.9. For a liquid gasoline density of 0.75 kg/liter, express your answer as a ratio—m^3 exhaust gases (at STP)/liter of gasoline burned.

18.11 A vehicle traveling along a highway at 55 mph emits CO and NO in the ratios of 80 and 10 g/liter of fuel burned, respectively. Assume that the vehicle travels 20 miles on the highway per gallon of gasoline burned. Calculate the CO and NO emission factors in g/mi.

18.12 Assume the exhaust gas ratio requested in Problem 18.10 is 9.3 std m^3/liter at an AFR of 14.0, and the ratio is 10.2 std m^3/liter at an AFR of 14.9, which is the point marked with an S (for stoichiometric) in Figure 18.3. Use Figure 18.3 to estimate the concentrations of CO and NO in the exhaust gases, then estimate the mass emissions of CO and NO (in grams/liter of gasoline burned) for these two air-to-fuel ratios.

18.13 Calculate the stoichiometric AFR for (a) methanol, (b) ethanol, (c) methane, and (d) hydrogen.

18.14 Calculate the stoichiometric AFR for a blend of 60% (by weight) gasoline (C_7H_{13}) and 40% methanol (CH_3OH).

18.15 Calculate the actual AFR for a fuel (C_6H_{14}) when the equivalence ratio is (a) 0.90, (b) 1.10, and (c) 1.30.

18.16 Assume that an alternate-fuel vehicle burns rubbing alcohol (C_3H_7OH).

  a. Write a balanced chemical equation for the stoichiometric combustion of this fuel, and calculate the stoichiometric air-to-fuel ratio required for this fuel.

  b. Assuming that one-fourth of one percent (0.25%) of the carbon atoms in the fuel produce CO instead of CO_2, but that the air input stays the same as in the stoichiometric equation developed in part (a), calculate the concentration of CO in the vehicle exhaust, in ppm.

18.17 Assume that a vehicle burns pure ethyl alcohol (C_2H_5OH) as the fuel.

  a. Calculate the stoichiometric air-to-fuel ratio (mass ratio).

  b. Assume the vehicle is traveling at 30 miles/hr, and gets 15 miles/gallon "gas mileage." Also, assume that the density of ethyl alcohol is 6.5 lbs/gal. Calculate the volumetric flow rate of air required to provide 5% *excess air* for combustion. Give your answer in scfm (std ft^3/min) of air.

18.18 A mobile source emissions estimate is needed for a revised SIP. Assume that in the year 2005, there will be 100,000 vehicles in a certain county, and each one will travel an average of 9,000 miles per year at an average speed of 20 mph. Calculate the annual emissions of CO, VOC, and NO_x (in tons per year). Use Figure 18.5 to get the emission factors.

REFERENCES

Agnew, W. G. Research Publication GMR-743. Warren, MI: General Motors Corporation, 1968.

Barwell, G. A. *Rural Transport in Developing Countries.* Boulder, CO: Westview Press, 1985.

Faiz, A., Weaver, C., Sinha, K., Walsh, M., and Carbajo, J. *Air Pollution from Motor Vehicles.* Washington, DC: The World Bank, 1992.

Grable, R. "Honda Civic VX," *Motor Trend*, January 1993.

Keely, D. K., and Cooper, C. D. *Using MOBILE6 to Get Emissions Factors for a Microscale CO Analysis*, a paper submitted for presentation at the 95th Annual Conference of the Air and Waste Management Association, Baltimore, MD, June 2002.

Nichols, L. A. "Used Car Lots Tampering Inspections—Central Florida District," an informal report to the Florida Department of Environmental Regulation, June 1992.

Niepoth, G. W., Gumbleton, J. J., and Haefner, D. R. *Closed Loop Carburetor Emission Control Systems.* Warren, MI: General Motors Corporation, June 1978.

Singh, R. B., and Huber, A. H. "Development of a Microscale Emission Factor Model for CO for Predicting Real-Time Motor Vehicle Emissions," *Journal of the Air and Waste Management Association*, 50(11), November 2000.

Singh, R. B., and Huber, A. H. "Sensitivity Analysis and Evaluation of Micro-FacCO: A Microscale Motor Vehicle Emission Factor Model for CO Emissions," *Journal of the Air and Waste Management Association, 51*(7), July 2001.

U.S. Environmental Protection Agency. *Compilation of Air Pollutant Emission Factors,* Volume II—*Mobile Sources* (4th ed.), AP-42, NTIS No. PB 87-205266. Ann Arbor, MI, 1985.

———. *Mobile Source Emissions Standards Summary.* Washington, DC: Office of Mobile Sources, July 16, 1992.

———. *National Air Quality and Emissions Trends Report, 1998*, EPA-454/R-00-003, EPA-450/4-91-023. Research Triangle Park, NC, March 2000.

———. *National Air Quality and Emission Trends Report, 1999*, EPA-454/R-01-004. Research Triangle Park, NC, March 2001.

Wark, K., and Warner, C. F. *Air Pollution, Its Origin and Control* (2nd ed.). New York: Harper & Row, 1981.

Wayson, R. L., and Bowlby, W. "Inventorying Airport Air Pollutant Emissions," *American Society of Civil Engineers Journal of Transportation Engineering, 114*(1), January 1988.

CHAPTER 19

AIR POLLUTION AND
METEOROLOGY

The smoke from chimneys straight ascends,
Then spreading back to earth it bends.

When the ditch and pond offend the nose,
Then look for rain and stormy blows.

Old weather proverbs

♦ 19.1 INTRODUCTION ♦

Meteorology is the study and forecasting of weather changes resulting from large-scale atmospheric circulation. Meteorological principles and a knowledge of both macro- and micro-scale circulation patterns are major factors in effective air pollution control. Faith (1959) listed the following requisites for every air pollution problem:

1. There must be a pollutant emission into the atmosphere.

2. The emitted pollutant must be confined to a restricted volume of air.

3. The polluted air must interfere with the well-being of people.

Requisite 2 occurs during periods of adverse weather conditions that restrict the mixing and dispersion of pollutants. Since we have no control over the weather, emission rates must be controlled so that pollution problems are less inclined to develop during adverse weather periods. Figure 19.1 is a photograph that demonstrates the result of pollutant emissions accumulating in a volume of air restricted by an atmospheric inversion. This chapter will review atmospheric circulation and its impact on air pollution control requirements.

569

Figure 19.1
Atmospheric inversion over a city. This photo is of Hartford, Connecticut, viewed from the Talcott Mountain Science Center, about 6 miles away. An energetic steam plume was able to pierce the inversion but dissipated within 3–4 minutes. The photograph was taken about 8:00 A.M. on a cold day; by 12:00 the pollution layer had completely obscured any view of the city.

◆ 19.2 GENERAL ATMOSPHERIC ◆ CIRCULATION PATTERNS

If the earth's surface were completely smooth and of a uniform composition, average surface winds would be approximately those shown in Figure 19.2. This figure also shows average circulation patterns with altitude. In the equatorial regions, air is heated and rises well above the friction layer (3000 ft); as it flows north the earth's rotation deflects it in an easterly direction. An accumulation of air in the regions of 30° N and S latitudes produces a high-pressure belt. Surface winds from these high-pressure areas provide the prevailing southwesterly flow over the region from 30° N to 50° N. Northerly winds from the polar high-pressure regions converge with the south-westerlies producing a storm belt with variable winds from 50° N to 70° N.

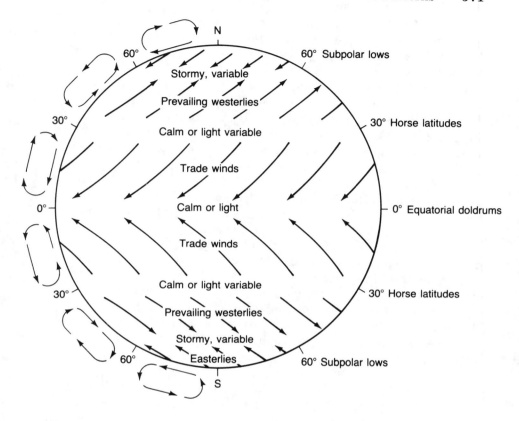

Figure 19.2
General air circulation if the earth were smooth and of homogeneous composition. (Adapted from Petterssen, 1958.)

While actual surface winds in many areas may be similar to those predicted for a smooth earth, many factors work to produce very erratic surface wind patterns. Some of the more important factors affecting surface circulation are:

1. topography

2. diurnal variation and seasonal variation in surface heating

3. variation in surface heating owing to the presence of ground cover and proximity to large bodies of water

Mountain ranges have a significant effect on circulation patterns and result in rather dramatic climatic changes in regions only a few hundred miles wide. With land masses near the ocean, their faster heating and cooling (compared with the relatively constant temperature of the ocean) produces seasonal and diurnal changes in surface winds. Temperatures may vary as the ground cover varies from snow to vegetation to desert.

Above 6000 ft, circulation patterns are much more uniform. In the northern hemisphere, air generally flows in a northerly direction from high altitudes and high pressures over the subtropical regions toward low pressures over the polar region, but it is then deflected to a prevailing west-to-east direction by the earth's rotation. The general westerly circulation aloft can be changed considerably by the intrusion of cold polar air masses across the United States. Figure 19.3, for instance, shows a constant-pressure chart with lines of constant elevation representing the topography of the 500 millibar (mb) surface (1 atmosphere of pressure equals 1013 mb). Note the deep wedge of cold air over the central United States. Winds aloft flow essentially parallel to the contours of the constant pressure surface. This type of plot is a valuable tool for the weather forecaster because the flow aloft provides a relatively accurate projection of the movement of surface storms.

With a description of global circulation patterns in mind, we will now find it useful to discuss the interaction of the various forces that determine the wind velocity and direction. The important forces to be considered are the pressure gradient force, F_P; the deflective force or Coriolis force, F_D; the friction force, F_F; and the centrifugal force, F_C. The driving force for air movements is the existence of pressure gradients resulting in airflow from high to low pressure. The other forces act to modify the direction and velocity of flow produced by the gradient force, as will be discussed in the following paragraphs.

The Coriolis force that results from the earth's rotation deflects air movement to the right in the northern hemisphere and to the left in the southern hemisphere. This force can be visualized by considering a cone rotating in a counterclockwise direction, the direction the earth rotates when viewed from the North Pole. If a thin stream of fluid were introduced at the top of the rotating cone, it would flow down the cone and would be deflected to the right of its path with respect to a point on the surface of the cone. Thus, in the northern hemisphere, a north wind shifts until it is coming from a northeast direction and a south wind becomes a southwest wind.

Friction acts as a drag force opposed to the direction of the wind. This force increases as the wind velocity and the roughness of the terrain over which the air is moving increase. The effect of friction decreases rapidly with elevation, and above 3000 to 4000 ft it is essentially negligible.

The centrifugal force exerted on an air parcel is proportional to the curvature of the path of the parcel and the velocity of the parcel. Thus, this force is effective in the case of circulation around well-developed high- and low-pressure cells. Centrifugal force is directed away from the center of curvature.

The simplest interaction between the pressure gradient force and the Coriolis force can be illustrated by neglecting friction and assuming the airflow path is straight (no centrifugal force). In this case, F_P is approximately balanced by F_D, as shown in Figure 19.4. The resultant

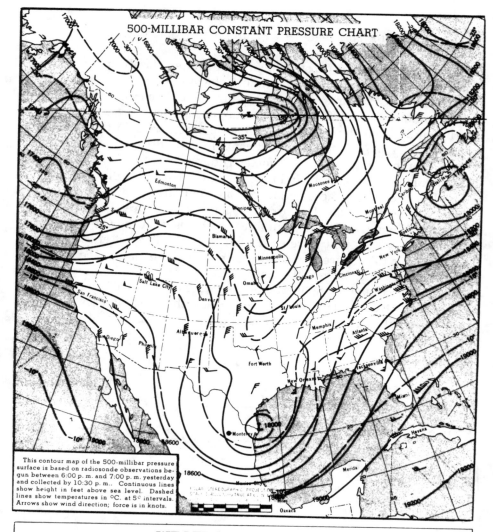

500-MILLIBAR CONSTANT PRESSURE CHART

This contour map of the 500-millibar pressure surface is based on radiosonde observations begun between 6:00 p. m. and 7:00 p. m. yesterday and collected by 10:30 p. m.. Continuous lines show height in feet above sea level. Dashed lines show temperatures in °C. at 5° intervals. Arrows show wind direction; force is in knots.

WEATHER FORECASTS
U. S. WEATHER BUREAU, WASHINGTON, D. C.
FRIDAY, JANUARY 24, 1958

DISTRICT OF COLUMBIA AND VICINITY
Today... cloudy this morning, rain or snow this afternoon, highest near 40°.
Tonight... rain or snow with some chance of heavy snow, lowest near 30°.
Saturday... windy and cold, some snow likely.

MARYLAND, becoming cloudy this morning, followed by rain or snow this afternoon, high temperatures 35° to 42°; windy tonight with snow in the west and central portions and rain or snow elsewhere, chance of heavy snow in the west and central portions, low temperatures 25° to 30° in the west and 28° to 35° elsewhere; Saturday windy and cold with snow and rain ending; gale warnings displayed.

VIRGINIA, gale warnings displayed on the Chesapeake Bay and Atlantic Coast; cloudy and colder with rain or snow in the southwest portion, spreading over the east and north portions today and continuing tonight, chance of heavy snow in the west and north portions and most likely rain in the southeast portion, highest 40° to 45° in the southeast and in the 30's in the west and north portions today, lowest tonight in the 30's, except 26° to 32° in the west portion and rather strong winds, windy and cold with some snow likely Saturday, mostly in the west and north portions.

Figure 19.3

A 500 mb contour map showing a deep cold-air trough over the central United States.
(Courtesy of the U.S. Weather Bureau.)

wind, u_g, is called the geostrophic wind and often closely approximates the actual wind, particularly at higher elevations.

If the path of a moving parcel of air is highly curved, such as the flow around a well-defined low-pressure or high-pressure cell, the centrifugal force on the parcel, F_C, must be considered. In this case, as Figure 19.5 shows, the combined centrifugal and Coriolis forces just balance the pressure gradient force. The resultant gradient wind, u_g, approximates the actual wind flow where the path is highly curved.

The effect of surface friction, F_F, on wind direction and velocity is shown in Figure 19.6. The frictional force, F_F, is directly opposed to the actual wind, u. The friction force combines with the Coriolis force to produce a resultant force which balances F_P. The net effect of friction then is to produce a wind direction that slants across the lines of constant pressure toward the lower pressure.

The frictional force increases as the roughness of the terrain increases. Over a relatively smooth surface, the deflection toward low pressure will average 20° to 25° and u will be 65 to 75% of u_g. Very

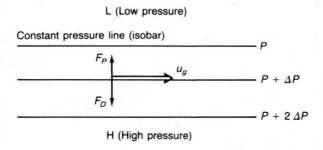

L (Low pressure)

Constant pressure line (isobar) ────────────────── P

F_P

u_g ────── P + ΔP

F_D

──────────── P + 2 ΔP

H (High pressure)

Figure 19.4
Geostrophic wind resulting from a balance of the pressure gradient and Coriolis forces.

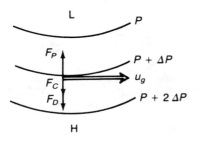

L   P

F_P P + ΔP

F_C u_g
F_D P + 2 ΔP

H

Figure 19.5
Gradient wind resulting from a balance of the pressure gradient force and combined Coriolis and centrifugal forces.

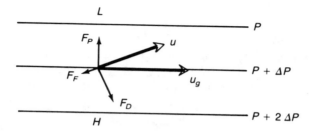

Figure 19.6
The effect of friction on wind direction and velocity.

uneven surface features may produce a deflection of 40° and reduce u to 35% of u_g.

The surface map corresponding to the upper air chart in Figure 19.3 is shown in Figure 19.7. Plotted on the surface chart are lines of constant surface pressure referred to as **isobars.** The isobars are plotted at 4 mb intervals. Wind speed and direction are indicated by the flag at each station. Each bar on the flag indicates 10 knots and no flag indicates calm conditions. Surface friction produces sufficient deflection so that the wind direction is not parallel to the isobars but is 25° to 40° toward low pressure.

Wind velocity generally increases with height, and in the lower elevations this increase can be approximated by a simple power law as shown below:

$$\frac{u_2}{u_1} = \left(\frac{z_2}{z_1}\right)^p \tag{19.1}$$

where

u_2, u_1 = wind velocity at higher and lower elevation, m/s

z_2, z_1 = higher and lower elevation, m

p = function of stability (defined later in chapter), dimensionless

$p \approx 0.5$ for very stable conditions

$p \approx 0.15$ for very unstable conditions

The wind direction normally changes in a clockwise manner as elevation increases. A clockwise change is referred to as **veering** and counterclockwise change is **backing.** Veering wind direction with elevation can be seen by comparing surface winds in Figure 19.7(top) with the winds near the 18,000 ft level in Figure 19.3. Wind direction is designated by the direction from which the wind is blowing.

We can see from Figure 19.7(top) that a low-pressure area (referred to as a storm or cyclone) has developed in the Gulf along the shear line between the cold northwesterly flow and the warm southeasterly flow. A cross-section through the cyclone is shown in Figure

19.7(bottom). Warm moist air is lifted over colder dry air, and the resulting cooling produces cloud formation and precipitation. The leading edge of the southerly cold air movement is shown as a cold front, $\bigtriangledown\bigtriangledown\bigtriangledown$, and the corresponding surface boundary of the northerly moving warm air is shown as a warm front, $\frown\frown\frown$. The precipitation ahead of a warm front may extend several hundred miles and

SOURCE: Courtesy of the U.S. Weather Bureau.

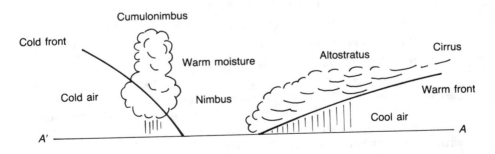

Figure 19.7

(Top) surface map showing low-pressure system associated with trough aloft shown in Figure 19.3; (bottom) cross-section through the cyclone shown on top, with cloud types and precipitation patterns.

(Courtesy of the U.S. Weather Bureau.)

will be fairly continuous. The cold frontal zone is much narrower, seldom exceeding 50 miles. Weather associated with the cold front may include thunderstorms and severe upper air turbulence. It is apparent that weather conditions associated with fronts and storm centers provide excellent mixing conditions for rapid air pollution dispersion.

◆ 19.3 LOCAL CIRCULATION EFFECTS ◆

Two local circulation patterns result from unequal surface heating and cooling. These are the land-sea breeze and the mountain-valley wind. Local winds must be taken into account in plant siting and emission control requirements.

THE LAND-SEA BREEZE

The mechanics of the land-sea breeze are illustrated in Figure 19.8. In the daytime, the land area is heated rapidly, which in turn heats the air just above it. The water temperature remains relatively constant. The air over the heated land surface rises, producing a low pressure relative to the pressure over the water. The resulting pressure gradient produces a surface flow off the water toward the land (a sea breeze). Initially the flow will be onto the land, but as the breeze develops, the Coriolis force will gradually shift the direction so that the flow is more parallel with the land mass. After sunset and several hours of cooling by radiation, the land

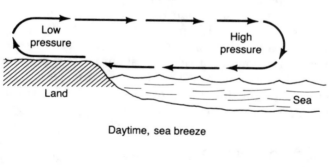

Daytime, sea breeze

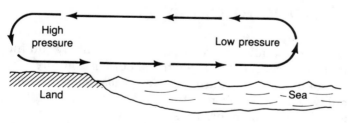

Nighttime, land breeze

Figure 19.8
Schematic of land-sea breeze development.

mass will be cold compared to the water temperature. Then the reverse flow pattern will develop, resulting in a wind off the land (a land breeze).

MOUNTAIN-VALLEY WINDS

Figure 19.9 is a sketch of the flow patterns resulting from the uneven heating and cooling of mountain slopes and adjacent valleys. During daylight hours, the air adjacent to the mountain slope heats rapidly and rises. This air then settles over the cooler valley, producing an up-slope wind during the day. At night, the cooler air on the mountain slope flows down into the valley.

WIND ROSES AND LOW-LEVEL
POLLUTANT DISPERSION PATTERNS

An effective way to present graphically the average wind data for a specific location is with a wind rose, as shown in Figure 19.10. In this plot, the average wind direction is shown as one of sixteen compass points, each point separated by 22.5° measured from true north. The length of the bar plotted for a given direction indicates the percentage of time the wind came from that direction. Since wind direction is constantly changing, the time percentage for a specific compass point actually includes those times for wind directions 11.25° on either side of the point. The percentage of time for a given velocity range is shown by the thickness of the direction bar. Referring to Figure 19.10, we see that the average wind direction was from the southwest 19% of the time; 7% of the time the southwesterly wind velocity was 16–30 mph.

A wind rose provides a good estimate of local particulate dispersion patterns. Figure 19.11 shows a particulate fallout pattern around an emission source and a wind rose based on the same time period. In this case, the wind rose was modified to show the direction in which the wind was blowing.

◆ 19.4 ATMOSPHERIC STABILITY ◆
AND VERTICAL MIXING

Everyone at times has observed that smoke from burning leaves or a barbecue grill will rise a few feet vertically then seemingly run into an invisible barrier that causes the smoke to fan out horizontally. This phenomenon is usually apparent during late afternoon on a clear and relatively calm day. The plume from a tall stack may occasionally be seen to behave in a similar manner. These observations indicate that under certain conditions the lower layer of the atmosphere can resist vertical mixing. Resistance to vertical mixing is referred to as stability. The invisible barrier that at times impedes vertical air movement and produces adverse vertical mixing is an abrupt change in the verti-

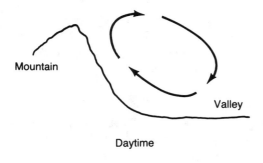

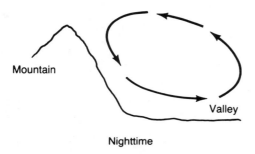

Figure 19.9
Schematic of mountain-valley wind development.

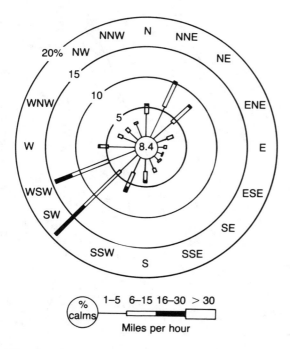

Figure 19.10
Wind rose, showing direction and velocity frequencies.

cal temperature profile. Some basic thermodynamic principles can be used to relate vertical stability and temperature.

THE HYDROSTATIC EQUATION

Consider the column of air shown in Figure 19.12. The parcel of air in the volume Adz will remain stationary if the sum of all vertical forces on the parcel is zero. Assigning forces in the downward direction a negative sign and remembering that force is the product of pressure times area, we can write:

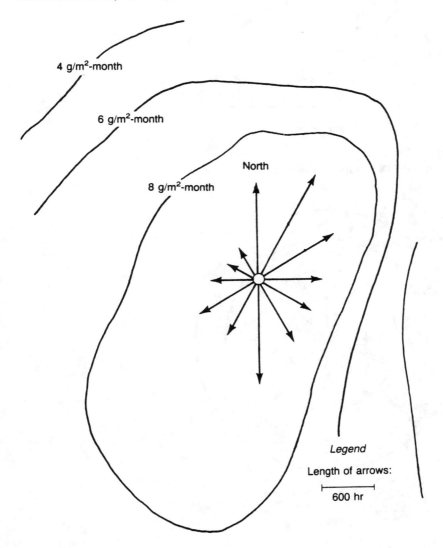

Figure 19. 11
Wind rose and corresponding particulate fallout pattern.

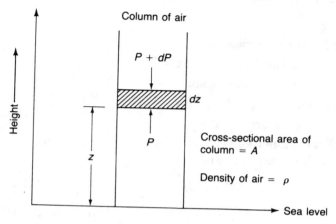

Figure 19.12
Conditions at hydrostatic equilibrium.

$$PA - (P + dP)A - g\rho A \; dz = 0 \qquad \textbf{(19.2)}$$

where

P = pressure, Pa

A = area, m^2

g = acceleration of gravity, m/s^2

ρ = density, kg/m^3

z = elevation, m

The first two terms represent the difference in pressure on the lower and upper surfaces of the parcel; the third term is the gravitational force on the mass of air in the parcel.

We see that the area A in each term cancels, which results in

$$dP = -g\rho \; dz \qquad \textbf{(19.3)}$$

Equation (19.3) demonstrates that atmospheric pressure decreases with height. Since air is a compressible fluid, the variation of ρ with height must be known to integrate Eq. (19.3).

ADIABATIC HEATING AND COOLING DURING VERTICAL AIR MOVEMENT

An **adiabatic process** is one in which no heat is supplied or withdrawn. A parcel of air that is not immediately next to the earth's surface is sufficiently well insulated by its surroundings that either compression or expansion of the parcel may be assumed to be adiabatic.

The first law of thermodynamics for a nonflow system is written as

$$dH = C_p \; dT - V \; dP \qquad \textbf{(19.4)}$$

where

dH = heat added per unit mass, J/kg

C_p = specific heat at constant pressure, J/kg-°K

dT = incremental temperature change, °K

V = specific volume, m³/kg

dP = incremental pressure change, Pa

If a parcel of dry air moves vertically under adiabatic conditions (no heat is added or lost), Eq. (19.4) becomes

$$C_p \, dT = V \, dP \tag{19.5}$$

Since $V\rho = 1$, Eqs. (19.3) and (19.5) can be combined to eliminate dP and give

$$-\frac{dT}{dz} = \frac{g}{C_p} = \gamma_d \tag{19.6}$$

where γ_d = dry adiabatic lapse rate, °K/m

Equation (19.6) shows that dry air cools adiabatically with height at a constant rate equal to γ_d, the dry adiabatic lapse rate. The **dry** adiabatic lapse rate calculated from Eq. (19.6) is 0.98 C per 100 m or roughly 5.4 F per 1000 ft. This means that a parcel of dry air lifted 100 m will cool 0.98 C owing to adiabatic expansion. If the air is saturated with water vapor and the saturated water content in the air is $\mathscr{H}_s$ in kg H₂O per kg of dry air, then Eq. (19.4) can be written

$$-\Delta H_v d\mathscr{H}_s = C_p \, dT - V \, dP \tag{19.7}$$

where

ΔH_v = latent heat of vaporization, J/kg

$d\mathscr{H}_s$ = change in water content owing to condensation, kg H₂O/kg dry air ($d\mathscr{H}_s < 0$ for condensation)

Combining Eq. (19.7) with Eq. (19.3) to eliminate dP results in the following relationship for the wet adiabatic lapse rate, γ_s:

$$-\frac{dT}{dz} = \gamma_s = \gamma_d + \frac{\Delta H_v \, d\mathscr{H}_s}{C_p \, dz} \tag{19.8}$$

Equation (19.8) shows that, as air rises and condensation occurs, $\mathscr{H}_s$ decreases, producing a slower cooling rate and consequently a lower lapse rate than for dry air. In very cold air, $\mathscr{H}_s$ is low, and γ_s approaches γ_d. The **wet** adiabatic lapse rate is approximately 3 F per 1000 ft and the **actual** atmospheric lapse rate, γ, will average around 3.5 F per 1000 ft.

▶ ▶ ▶ ▶ ▶ ▶ ▶ ▶ ▶

Example 19.1

During a set of upper-air measurements the lapse rate was constant at 1.0 C per 100 m. If the atmosphere is assumed to behave as a perfect gas and the sea-level temperature and pressure were 15 C and 1 atm, at what altitude was the pressure one-half the sea-level pressure?

Solution

Use Eq. (19.3), the ideal gas law, and Eq. (19.6). For an ideal gas, $\rho = PMW/RT$, then Eq. (19.3) becomes

$$\frac{dP}{P} = -\frac{gMW}{RT}dz$$

Since T changes with height at a constant rate

$$T = (15 + 273) - \frac{1.0 \cdot z}{100} = 288 - 0.01z$$

Substituting into the equation above gives

$$\frac{dP}{P} = -\frac{gMW}{R}\frac{dz}{288 - 0.01z}$$

Integrating between $P = 1.0$ and $P = 0.5$ and $z = 0$ and z,

$$\int_{0.5}^{1.0}\frac{dP}{P} = +\frac{gMW}{R}\int_{0}^{z}\frac{dz}{288 - 0.01z}$$
$$R = 8314 \text{ J/kgmol-}^\circ\text{K}$$
$$MW = 29 \text{ kg/kgmol}$$
$$\ln\left(\frac{1.0}{0.5}\right) = \frac{9.81 \times 29}{8314 \times 0.01}\ln\left(\frac{288}{288 - 0.01z}\right)$$
$$0.693 = 3.42\ln\left(\frac{288}{288 - 0.01z}\right)$$
$$\ln\left(\frac{288}{288 - 0.01z}\right) = 0.2026$$
$$288 = 352.7 - 0.0122z$$
$$z = 5303 \text{ m}$$

◀ ◀ ◀ ◀ ◀ ◀ ◀ ◀ ◀

THE EFFECT OF LAPSE RATE ON VERTICAL STABILITY

Vertical stability is directly related to the actual lapse rate in the following manner:

Lapse Rate	Stability Condition
$\gamma > \gamma_d$	Unstable
$\gamma = \gamma_d$	Neutral
$\gamma < \gamma_d$	Stable

These stability conditions are illustrated in Figure 19.13. During unstable conditions, a parcel of air moved upward cools at a slower rate than the surrounding air and is accelerated upward by buoyant forces. Moved downward, the parcel warms at a slower rate and is accelerated downward. When $\gamma = \gamma_d$, upward and downward motion result in the parcel temperature changing at the same rate as the surrounding air, with no resultant buoyant forces. During stable conditions, upward movement produces a parcel cooler than the surroundings, hence the parcel will settle back to its original elevation. Downward motion produces a warmer parcel, which will rise to its original elevation. Under stable conditions, vertical air movement is actually dampened out by adiabatic warming or cooling; just the opposite is true during unstable conditions.

TEMPERATURE INVERSIONS

Since we have shown that $\gamma < \gamma_d$ results in atmospheric stability, maximum stability occurs when γ becomes negative (that is, when temperature increases with elevation). This condition is referred to as an **inversion** and produces the most adverse mixing conditions, which can effectively trap emissions near the ground. Three types of inversions can occur, and each is associated with a particular set of weather conditions.

Frontal Inversions. These inversions are usually at relatively high altitudes and result when a warm air mass overruns a cold air mass below. The inversion occurs in the mixing zone between the two air masses. Since the inversion is associated with a front and normally moderate to high winds, it cannot significantly affect the excellent mixing conditions present in a frontal zone. It is therefore not important from a pollution control standpoint.

Subsidence Inversions. This type of inversion is of major importance in pollution control because it can affect large areas for a period of several days. The subsidence inversion, from the word *subsiding,* is associated with either a stagnant high-pressure cell or a flow aloft of cold dry air from an ocean onto a land mass surrounded by mountains. The former situation occurs frequently over the eastern and southeastern United States in the fall and in parts of Europe and Great Britain in early winter, and the latter occurs often in the Los Angeles basin area. The subsidence inversion mechanism is illus-

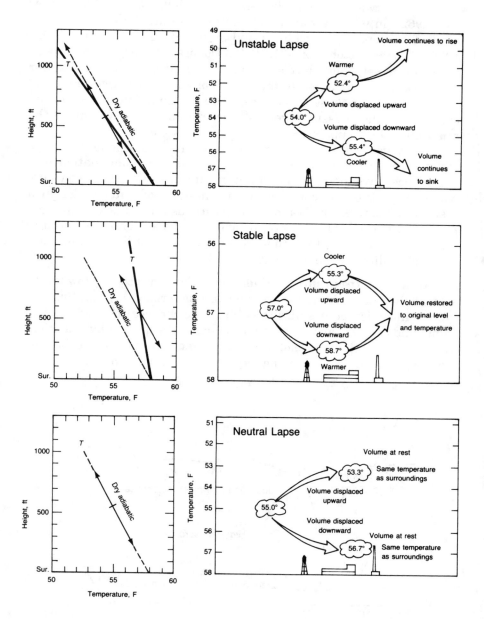

Figure 19.13

The effect of the actual lapse rate on vertical stability.

(Adapted from American Industrial Hygiene Association, 1960.)

trated in Figure 19.14. The frequency of periods of stagnating high pressure cells over the eastern United States is shown in Figure 19.15. The subsidence inversion in conjunction with light or calm surface winds can lead to excessive pollutant buildup over wide areas. London has suffered two acute episodes—December 5–9, 1952, and December 3–7, 1962—in which a number of deaths were attributed to the high level of suspended particulates and sulfur oxides in the air (Stern 1968). During October 27–31, 1948, Donora, Pennsylvania, experienced extremely high levels of suspended particulates and sulfur oxides, which contributed to the deaths of 20 persons (Schrenk et al. 1949). In each of these incidences, pollution emission levels did not change, but the emissions were restricted to a limited volume of the atmosphere by an upper level subsidence inversion.

Los Angeles has received notoriety for its photochemical smogs, which became noticeable in the early 1950s. The combination of a subsidence inversion, surrounding mountains, and light onshore winds traps the exhaust from a vast number of automobiles in a restrictive volume. The trapped hydrocarbons and oxides of nitrogen react to produce a bluish haze, which contains a number of eye irritants and compounds toxic to many species of plant life. Photochemical smog is discussed in more detail in Section 19.5.

Radiation Inversions.   Radiation inversions occur at low levels, seldom above a few hundred feet, and dissipate relatively quickly. This type of inversion occurs during periods of clear weather and light to calm winds and is caused by rapid cooling of the ground by radiation. The inversion begins developing at dusk and continues until the surface warms again the following day. Initially only the air close to the

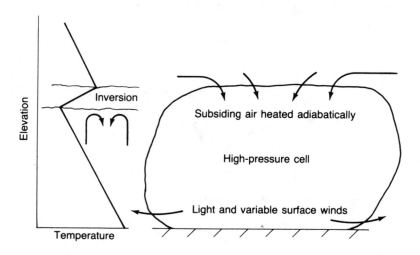

Figure 19.14
Formation of a subsidence inversion.

surface is cooled, but after several hours the cool layer and hence the top of the inversion can extend to 500 ft. Emissions from stacks below the inversion will be trapped in a very shallow layer. A schematic of the development of a radiation inversion is shown in Figure 19.16.

A condition referred to as **fumigation** may occur during early to mid-morning when a strong radiation inversion begins to break up owing to surface heating. A plume from a stack may be carried to the ground within a few hundred yards of the stack by strong convective currents. This condition can result in very high ground-level pollutant concentrations for short time intervals (see Figure 19.17). Radiation inversions occur frequently during the year and are particularly prevalent during periods of stagnating high pressure. Hence, radiation inversions cause further deterioration of poor pollutant dispersion

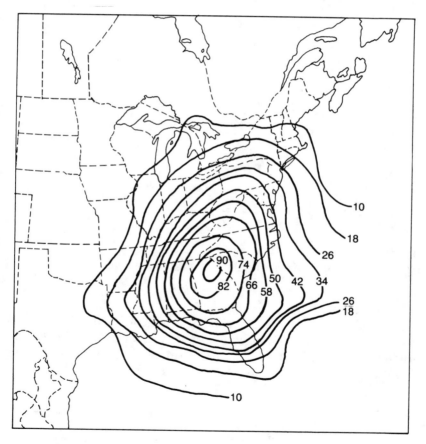

Figure 19.15
Frequency of stagnating high-pressure cells over the eastern United States. The contours give the number of periods of four or more successive days between 1936–1965.
(Adapted from Kershover, 1967.)

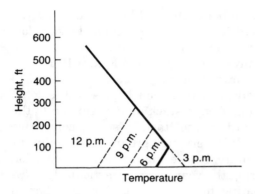

Figure 19.16
Development of a radiation inversion.

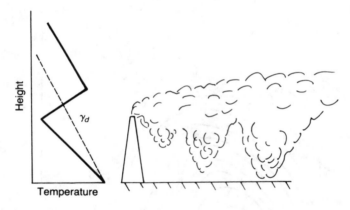

Figure 19.17
Fumigation resulting from the breakup of a radiation inversion.

during periods of upper-level subsidence inversion. Ground cover has a significant effect on radiation rates because more rapid cooling occurs in open areas. Evaporation after a rain will increase the cooling rate, and fog formation will have the opposite effect.

◆ 19.5 PHOTOCHEMISTRY AND SMOG ◆

A few years after the end of World War II, when automobiles and gasoline had become plentiful again, people living in the Los Angeles basin began experiencing a particularly irritating type of atmospheric pollution. From the beginning, the term *smog* was used to describe the glaring bluish-brownish, ozone-laden haze that enveloped the basin from late morning to mid-afternoon when a certain combination of weather conditions existed. The word *smog* had been coined earlier in Great Britain to describe the apparent combination of "smoke and fog" that prevailed during periods of excessively high levels of suspended particulates and oxides of sulfur. Although we now know there is little similarity between the British- and Los Angeles-type smogs, we still refer to the Los Angeles phenomenon as smog (or more precisely— photochemical smog). While Los Angeles has received most of the publicity, virtually all metropolitan areas in the world that have high automobile densities have experienced smog during warm and stagnant atmospheric conditions.

CAUSES OF PHOTOCHEMICAL SMOG

The increasing incidence of photochemical smog and its irritating and damaging effects spurred a flurry of research activities in the early 1950s that continued at a record pace through the 1960s. Haagen-Smit (1952) is credited with first reporting that most of the injurious effects of smog could be attributed to oxidants (mainly ozone) produced in the lower atmosphere by a complex, photochemically initiated process involving oxides of nitrogen and various hydrocarbons. He and his coworkers were able to closely simulate the smog process by irradiating automobile exhaust with visible light in a large irradiation chamber. During the ten years following Haagen-Smit's work, researchers, including Schuck (1957), Leighton (1961), and Stephens (1966), developed a good working knowledge of the overall smog process and identified many of the chemical reactions occurring as well as the atmospheric conditions and pollution levels that were requisites for smog formation.

A qualitative relationship between the major chemical and atmospheric variables active in the smog formation process was summarized by Stern et al. (1984) in the following manner:

$$PPL = \frac{(ROG)(NO_x)(\text{Light Intensity})(\text{Temperature})}{(\text{Wind Velocity})(\text{Inversion Height})} \qquad (19.9)$$

where

PPL = photochemical pollution level

ROG = concentration of reactive organic gases

NO_x = concentration of oxides of nitrogen

More recently, the factors affecting photochemical air quality were reviewed by Brown (1992); a list of these important factors is presented in Table 19.1.

Keep in mind that Eq. (19.9) is qualitative. Quantifying the emissions of reactive precursors, their transport, and the photochemical reactions occurring in a large urban airshed, and trying to resolve

Table 19.1   The Major Factors Affecting Photochemical Air Quality

The spatial and temporal distribution of emissions of NO_x and VOCs (both anthropogenic and biogenic).

The composition of the emitted VOCs.

The spatial and temporal variations in the wind fields.

The dynamics of the boundary layer, including stability and the level of mixing.

The chemical reactions involving VOCs, NO_x, and other important species.

The diurnal variation of solar insolation and temperature.

The loss of ozone and ozone precursors by dry deposition.

The ambient background of VOCs, NO_x, and other species in, immediately upwind, and above the regions of study.

these processes spatially and temporally has been the subject of intense modeling efforts for more than three decades.

The model currently in favor with the U.S. EPA is a massive computer program—the Urban Airshed Model (UAM) (Morris et al. 1990). The UAM is a three-dimensional photochemical grid model designed to calculate the concentrations of ozone and other chemically reactive pollutants by simulating the physical and chemical processes in the atmosphere that affect pollutant concentrations. The basis for the UAM is the species continuity equation. This equation represents a mass balance on a volume element of air in which all of the relevant emission, transport, diffusion, chemical reaction, and removal processes are expressed in mathematical terms as follows:

$$\frac{\partial c_i}{\partial t} + \frac{\partial (uc_i)}{\partial x} + \frac{\partial (vc_i)}{\partial y} + \frac{\partial (wc_i)}{\partial z}$$

$$= \frac{\partial}{\partial x}\left(K_H \frac{\partial c_i}{\partial x}\right) + \frac{\partial}{\partial y}\left(K_H \frac{\partial c_i}{\partial y}\right) + \frac{\partial}{\partial z}\left(K_v \frac{\partial c_i}{\partial z}\right) + R_i + S_i - L_i$$

(19.10)

where

c_i = concentration of species i, $\mu g/m^3$

u, v, w = horizontal and vertical wind speed components, m/s

K_H, K_v = horizontal and vertical turbulent diffusion coefficients, m^2/s

R_i = net production rate of pollutant i by chemical reactions within the volume element, $\mu g/s\text{-}m^3$

S_i = net emission rate of pollutant i into the volume element, $\mu g/s\text{-}m^3$

L_i = net rate of removal of pollutant i by surface uptake processes per unit volume, $\mu g/s\text{-}m^3$

The UAM employs finite differencing numerical techniques for the solution of Eq. (19.10). The model is complicated and difficult to learn, and requires significant computer resources, but does produce a comprehensive analysis of ozone pollution within an urban area. Given enough time to gather accurate and voluminous input data (especially on emissions), UAM can produce excellent simulation results. As an example, Figure 19.18 presents ozone concentrations at three locations in central Florida as predicted from a UAM simulation of the Orlando metropolitan area compared with hourly monitoring data (Cooper et al. 1993).

A detailed discussion of UAM is not appropriate in this text; however, we will look at the photochemical process in general by assessing the importance of the parameters in Eq. (19.9) and Table 19.1. Then we will present and discuss the sequence of atmospheric chemical reactions that are believed to produce the damaging secondary pollutants in smog. Although several oxidants are produced, the pollutant of main concern is ozone.

Equation (19.9) indicates that the severity of smog formation is proportional to the atmospheric concentrations of both reactive organic gases and oxides of nitrogen plus the intensity of the sunlight irradiating this mixture. The denominator of Eq. (19.9) is simply a restatement of the second requisite for any pollution problem— namely, that the emitted pollutants must be confined to a restricted volume. In this case, Requisite 2 refers to a stagnant air condition that must confine and maintain reactant levels for a sufficient time for solar irradiation to take place and to initiate the smog process. Temperature must be included as a parameter because of its effect on the rates of the reactions. It was shown long ago that the photochemical process essentially stops at ambient temperatures below about 15 C (Alley and Ripperton 1962).

Even with only a rudimentary understanding of the complex chemical reaction sequence leading to photochemical smog, it soon became clear to early researchers why Los Angeles was plagued with frequent smogs. The highly populated Los Angeles basin is surrounded on three sides by mountains up to 8000 feet in elevation. Millions of automobiles and various industrial sources release large quantities of various reactive organic gases and oxides of nitrogen into the atmosphere. A relatively stationary high pressure area to the west provides an easterly flow of cool upper air which subsides over the basin causing an inversion as was shown previously in Figure 19.14. The resulting inversion, combined with a light sea breeze and the surrounding mountains, effectively traps the emitted pollutants and provides a massive photochemical reaction vessel. An additional effect of the Pacific high pressure cell is to provide the area with clear skies some 180 to 220 days per year, thus ensuring the high levels of solar radiation necessary for smog production.

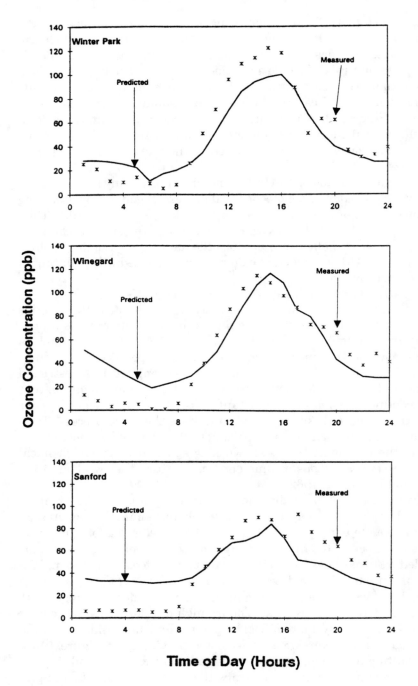

Figure 19.18
Urban Airshed Model results—modeled versus measured ozone concentrations in central Florida on August 7, 1990.
(Adapted from Cooper et al., 1993.)

ANATOMY OF A SMOG EPISODE

A typical smoggy day often begins bright and clear with an early morning temperature in the range of 60 F to 75 F. Figure 19.19 presents typical levels of NO_x and ozone as the smog process progresses. During the hours of darkness, from roughly 1900 hours to 0500 hours, the concentration of oxides of nitrogen and hydrocarbons (hydrocarbons are not shown in Figure 19.19) remain at normal urban ambient levels. As commuter traffic builds toward a maximum at around 0800 hours, there is an attendant rapid increase in the concentration of NO and hydrocarbons. The increase in NO and hydrocarbons occurs first, as these are emitted from sources directly. The increase in NO_2 lags by a short period because it is formed by atmospheric reactions. Shortly after sunrise and the onset of the photochemical reaction sequence, the NO concentration begins a rapid decrease and the oxidant (predominantly ozone) concentration increases at a similar rate. By 0900 hours the photochemical process is in full swing. There is the noticeable pungent odor of ozone in the air, one's eyes begin to water and burn, and the panoramic view of the city begins to fade in a brownish aerosol haze that will increase until about 1100 to 1200 hours (noon). For most people the peak period of discomfort will last from about noon until 1500 hours.

By 1600 hours the intensity of solar radiation has decreased to a low level and the photochemical process has virtually shut down. Oxidant levels drop rapidly as their production rates go to zero. By 2400 hours (midnight) the concentration of oxides of nitrogen and hydrocar-

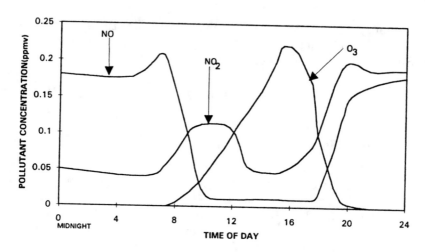

Figure 19.19

Concentrations of NO_x and O_3 during one day of a photochemical smog episode in St. Louis, Missouri, in 1962.

(Adapted from U.S. Environmental Protection Agency, 1962.)

bons have again reached normal urban ambient levels (for these stagnant air conditions) and are ready to participate in more smog formation reactions the next day.

THE CHEMISTRY OF PHOTOCHEMICAL SMOG

The task of identifying the individual chemical reactions that constitute the smog process has been extremely difficult and after some fifty years of work, the job is still far from complete. The following discussion is intended to provide only a brief overview of the current understanding of the process. The reader is referred to Guderian (1985) for a more detailed review of photochemistry and the effects of smog.

Once the photochemical nature of the process was recognized, it was quickly shown that, of all the reactants involved in the chemistry of photochemical smog, only NO_2 could absorb light over a wavelength band that would provide sufficient energy to bring about its photochemical dissociation. A photon of light can be represented as $h\nu$, where h is Planck's constant, and ν is the frequency of the light. The energy of a photon is directly proportional to the frequency. Further, since

$$\nu = c / \lambda \qquad (19.11)$$

where

c = velocity of light

λ = wavelength of light

then

$$h\nu = hc / \lambda \qquad (19.12)$$

The energy required to dissociate a gmol of NO_2 is the amount carried by $6.0225(10)^{23}$ photons and this quantity is referred to as an Einstein. Substituting the values of c and h, we find that:

$$E = 2.859(10)^5 / \lambda \qquad (19.13)$$

where

E = energy of dissociation, kcal/gmol

λ = wavelength of light, Å

The energy required to dissociate NO_2 is approximately 72 kcal/gmol (Perkins 1974). It can be seen from Eq. (19.13) that this level of energy requires that the wavelength must be approximately 4000 Å or lower. Nitrogen dioxide does in fact absorb at wavelengths of roughly 4500 Å and below. Furthermore, very little ultraviolet radiation ($\lambda < 3000$ Å) reaches the earth's surface, because of absorption by stratospheric ozone. This means that energy absorbed in the 3000 Å to 4000 Å range must bring about the dissociation of NO_2. In fact, 90% of the NO_2 molecules absorbing in the 3000 Å to 3700 Å band do dissociate (Seinfeld 1975).

THE NITROGEN DIOXIDE PHOTOLYTIC CYCLE

As the previous section indicates, the onset of photochemical smog depends on a starting concentration of NO_2 and solar insolation. The atomic oxygen produced by the photolysis of NO_2 is very reactive and rapidly combines with O_2 in the air to produce O_3. In the presence of NO, however, the O_3 will immediately decompose, regenerating the nitrogen dioxide. This nitrogen dioxide photolytic cycle is summarized in the following three reactions.

$$NO_2 + h\nu \rightarrow NO + O \qquad \textbf{(19.14)}$$

$$O + O_2 \rightarrow O_3 \qquad \textbf{(19.15)}$$

$$O_3 + NO \rightarrow NO_2 + O_2 \qquad \textbf{(19.16)}$$

Hence, while the presence of NO_2 is required to form O_3, the nitrogen dioxide photolytic cycle by itself does not generate net ozone, and cannot explain ozone accumulation.

Equations (19.14)–(19.16) indicate that in order for O_3 to accumulate, an additional pathway for conversion of NO to NO_2 is needed that will not destroy O_3. This alternate NO-to-NO_2 conversion pathway is provided by the atmospheric photochemical oxidation of reactive organic gases (ROGs). ROGs—primarily olefinic hydrocarbons, aromatic compounds, and aldehydes—are emitted from many sources into urban air. The reactions of these materials with atomic oxygen and ozone produce a myriad of reactive free radicals.

$$ROG + O + O_3 \rightarrow R\cdot\ + R\dot{C}O + RO\cdot + OH\cdot \qquad \textbf{(19.17)}$$

where R represents any organic group, and can simply be H

A major reaction of ROGs (sometimes referred to as hydrocarbons—although not all hydrocarbons are reactive) is to react with hydroxyl radicals ($OH\cdot$) to produce peroxy radicals ($RO_2\cdot$). The peroxy radicals then react rapidly with NO as shown in the alternate NO-to-NO_2 conversion pathway below.

$$RO_2\cdot + NO \rightarrow RO\cdot + NO_2 \qquad \textbf{(19.18)}$$

This alternate NO-to-NO_2 conversion pathway is a key feature, because it enables the net accumulation of ozone, as shown schematically in Figure 19.20.

The free radicals may further react with more ROGs and acid gases to produce the chain propagating and chain terminating reaction steps that make up the smog process. Some of these reactions are illustrated by the following equations.

$$ROG + OH\cdot \rightarrow R\cdot + R\dot{C}O + H_2O \tag{19.19}$$

$$R\cdot + O_2 \rightarrow RO_2\cdot \tag{19.20}$$

$$R\dot{C}O + O_2 \rightarrow R\overset{\overset{O}{\|}}{C}OO\cdot \tag{19.21}$$

The peroxy radicals formed by several reactions can also remove NO_2 from the atmosphere, without consuming ozone, by the following reactions:

$$RO_2\cdot + NO_2 \rightarrow ROONO_2 \tag{19.22}$$

$$R\overset{\overset{O}{\|}}{C}OO\cdot + NO_2 \rightarrow R\overset{\overset{O}{\|}}{C}OONO_2 \tag{19.23}$$

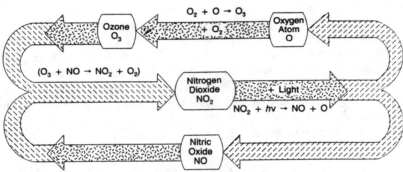

(a) Steady state reactions involving NO_2, NO and O_3 (no organics)

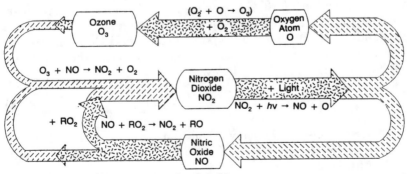

(b) Reactions (including organics) that allow O_3 buildup

Figure 19.20
Schematic of nitrogen dioxide photolytic cycle—without and with the alternative NO-to-NO_2 pathway enabling the net accumulation of ozone.
(Adapted from Morris et al., 1990.)

Hydrocarbon Oxidation Cycle. Hydrocarbons are ultimately oxidized in the atmosphere to form carbon dioxide (CO_2) and water (H_2O). Intermediate steps in this overall oxidation process typically involve cyclical reactions driven by hydroxyl radical attack on either the parent hydrocarbon or on partially oxidized intermediate compounds. Two essential characteristics of reactive hydrocarbons (vis-à-vis smog formation) are: (1) reaction with the hydroxyl radical to yield a peroxy radical, and (2) photolysis to yield hydrogen-containing radicals. The first reaction is common to virtually all organic molecules, but the second usually involves oxygenated intermediates containing the carbonyl (C = O) bond.

The production of the hydroxyl radical shown by Eq. (19.17) is an important step in the overall process because this intermediate reacts readily with hydrocarbons in a chain reaction that oxidizes two molecules of NO and removes one molecule of hydrocarbon as indicated below:

$$ROG + OH\cdot \rightarrow R\cdot + H_2O \tag{19.24}$$

$$R\cdot + O \rightarrow RO_2\cdot \tag{19.25}$$

$$RO_2\cdot + NO \rightarrow NO_2 + RO\cdot \tag{19.26}$$

$$RO\cdot + O_2 \rightarrow RCHO + HO_2\cdot \tag{19.27}$$

$$HO_2\cdot + NO \rightarrow NO_2 + OH\cdot \tag{19.28}$$

Formaldehyde (HCHO) is a stable intermediate of the above reactions and is also the simplest organic molecule exhibiting the two essential characteristics of hydrocarbon reactivity (smog formation potential). Thus, the chemistry of formaldehyde is common to virtually all mechanisms of atmospheric chemistry. Formaldehyde reacts in the atmosphere as follows. It can react with an OH· radical

$$HCHO + OH\cdot \rightarrow H\dot{C}O + H_2O \tag{19.29}$$

Also, photolysis can occur:

$$HCHO + h\nu \rightarrow H + H\dot{C}O \tag{19.30}$$

or

$$HCHO + h\nu \rightarrow H_2 + CO \tag{19.31}$$

where $h\nu$ denotes the absorption of light (ultraviolet in this case). The hydrogen atom produced in Reaction (19.30) combines immediately with O_2 to yield $HO_2\cdot$.

$$H + O_2 \rightarrow HO_2\cdot \tag{19.32}$$

The formyl radical ($\overset{\bullet}{H}CO$) produced by Reactions (19.29) and (19.30) also reacts very rapidly with O_2:

$$\overset{\bullet}{H}CO + O_2 \rightarrow HO_2 \cdot + CO \qquad (19.33)$$

Note that the essential peroxy radical ($HO_2 \cdot$) generated by Reactions (19.32) and (19.33) readily oxidizes NO to NO_2, regenerating an OH·, via the NO-to-NO_2 conversion pathway essential to ozone formation as previously discussed. Also note that the carbon monoxide (CO) generated by Reaction (19.33) can participate in the smog process, acting like an organic molecule to yield a peroxy radical, which then oxidizes NO to NO_2:

$$OH\cdot + CO \rightarrow H + CO_2 \qquad (19.34)$$

$$H + O_2 \rightarrow HO_2\cdot \qquad (19.35)$$

$$HO_2\cdot + NO \rightarrow NO_2 + OH\cdot \qquad (19.36)$$

Literally hundreds of organic gases have been measured in urban ambient air. Various organic gases have been found to differ widely in ozone-forming potential (reactivity). Glasson and Tuesday (1970) reported that disubstituted internal olefins were the most photochemically reactive compounds, while parafinic compounds such as methane and ethane were essentially nonreactive. Compounds showing intermediate levels of reactivity included unsubstituted olefins, cyclopentene, and cyclohexene, and some substituted benzenes. In most, but not all, cases the reactivity of a VOC can be characterized through experimental measurement of its rate of reaction with the OH radical. Specifically, it can be related to the rate at which the reactive material disappeared during radiation chamber experiments. For some VOCs, however (that is, high-molecular-weight aliphatics), the OH· reaction rate constant is not a valid index of ozone-forming potential (Morris et al. 1990), and use of more complex experimental/modeling techniques is required.

NOₓ and Radical Sink Reaction. A major chain terminating reaction is that between NO_2 and hydroxyl radicals to form nitric acid:

$$OH\cdot + NO_2 \rightarrow HONO_2 \qquad (19.37)$$

This reaction removes both a hydroxyl and an NO_2 molecule.

The equations presented in the past few pages are adequate to explain most of the chemical aspects of the formation of photochemical smog. A route is provided to the production of observed entities such as ozone, aldehydes, and peroxyacetyl nitrate (PAN), and the necessary photochemical initiating sequence is modeled.

The route whereby the aerosol haze is produced during the smog process is more speculative. In the early 1960s, research at the Taft

Engineering Center in Cincinnati showed that the amount of aerosol produced during smog chamber experiments greatly increased with the addition of SO_2 to the reaction mixture. Other studies further indicated that the peak ozone and NO_2 concentrations were lowered by the addition of SO_2. Analysis of filter samples collected during smog episodes showed significant sulfate concentrations. SO_2 may be converted to a sulfate by oxidation to SO_3. The oxidation can be achieved by several intermediate and end products of the smog reaction including O_3 and peroxy radicals. SO_2 may also be converted to SO_3 by photolysis and reaction with atomic oxygen. The SO_3 is then absorbed by liquid droplets and converted to H_2SO_4. SO_2 absorbed directly into liquid droplets can be catalyzed to SO_3 by several metallic ions (often present in urban air). The complexity of aerosol fluid mechanics in the atmosphere coupled with the numerous reaction possibilities involved in the generation of aerosols during the smog process provides a challenging research area.

THE EFFECTS OF SMOG

The temporary physiological effects of smog such as eye and respiratory irritation were mentioned earlier. Assessment of the seriousness of long-term effects of any type of pollutant on humans, plants and animals, and materials and structures provides a basis for suggesting cost-effective control measures. In the case of smog, just the short-term eye and respiratory irritation along with the demoralizing effect of reduced visibility are sufficient for most of the affected population to want action to be taken to solve the problem.

It is difficult to estimate the cost of smog damage to human health because neither the short-term nor the long-term effects are clearly understood. There is no doubt that there are long-term effects of exposure to even low concentrations of ozone, such as premature aging of the lungs. Lippmann (1989) suggests that more research is needed to determine if existing ambient standards are adequate.

Some efforts have been made, however, to assign an approximate monetary value to the damage to forests, agricultural products, and personal property. The EPA has estimated that crop damage by ozone alone may exceed \$2 to \$3 billion annually (Spensley 1992). With the development of tires with high-mileage tread wear, the service life of these tires in many applications (recreational vehicles, for example) is now determined by sidewall deterioration due to oxidants in the air rather than tread wear. Other evidence of smog damage includes fading and cracking of paints and accelerated corrosion of metals.

THE CONTROL OF PHOTOCHEMICAL SMOG

The approach to controlling smog has been to reduce emissions of its precursors (ROGs and NO_x). Smog chamber experiments have been used to investigate the effects on ozone concentrations of reducing precursor NO_x and hydrocarbon concentrations. ROG reductions (with constant NO_x) always lead to a slowing of the ozone production process and lower peak ozone concentrations. NO_x reductions (with constant ROG) can lead to a speeding up of the ozone production process, and can *increase* or *decrease* peak ozone values depending on the ROG-to-NO_x ratio.

Thus, whereas ROG control is never detrimental, NO_x control can be detrimental, particularly in the central cores of urban areas (Dimitriades 1989). A combination of ROG and NO_x reductions can lead to little change in the timing of the chemistry of photochemical smog (the slowing-down effect of ROG reduction balancing the speeding-up effect of NO_x reduction) but will usually result in a lowering of the peak ozone concentration (Seinfeld 1988). Seinfeld demonstrated this fact by examining the competition between ROGs and NO_x for hydroxyl radicals. The approximate rate constants are:

$$ROG + OH\cdot \rightarrow RO_2 \qquad 3110 \text{ ppm}^{-1} \text{ min}^{-1}$$
$$NO_2 + OH\cdot \rightarrow HNO_3 \qquad 17{,}000 \text{ ppm}^{-1} \text{ min}^{-1}$$

Thus, when the ratio of ROG to NO_x is approximately between 5 and 6, the two species have equal rates of reacting with the $OH\cdot$. If this ratio is much larger than 6, there is a shortage of NO that can be oxidized to NO_2, and ozone production is controlled by the amount of NO_x available. In this range, decreasing NO_x leads to a decrease in the peak ozone. On the other hand, when ROG/NO_x is on the order of 5 or less, the ready availability of NO_x makes the formation rate of ozone dependent on ROG. NO will scavenge O_3 faster than it reacts with $RO_2\cdot$, and also NO_2 will react with $OH\cdot$ to give nitric acid. Decreasing NO_x can lead to an increase in peak ozone as the efficiency of ozone formation increases.

Ozone is an extremely reactive pollutant and it can be scavenged by the very same pollutants that produce it. The reactions that produce it are numerous and complex. It is for these reasons that the O_3 concentrations do not respond linearly to precursor controls. The key requisite to solving urban O_3 problems is a good, quantitative understanding of the precursor-to-O_3 relationships (spatially and temporally), which in turn requires a detailed understanding of the atmospheric ozone-forming processes (Dimitriades 1989).

There are some 100 urban areas in the U.S. with a combined population of approximately 100 million people that do not meet the existing ambient air quality standards for ozone (Spensley 1992). The control of VOCs, NO_x and other photochemical smog precursors are a major thrust of the Clean Air Act Amendments of 1990. Implementation of these much stricter federal regulations combined with lowered automotive emissions still may not be sufficient to control smog in congested urban areas. It is very likely that many of the affected areas will ultimately be forced to resort to one or more severe control options including:

1. Bans on automobile usage during periods of high smog potential
2. Limiting traffic in certain cities to electric vehicles or vehicles equipped to burn low-emission fuels
3. Expensive overhauls of existing mass transit systems

In any case, control of photochemical smog during the next two decades will require considerable effort by many people and will undoubtedly be very expensive.

♦ 19.6 METEOROLOGY AND AIR POLLUTION ♦

Some applications of meteorology in air pollution control can now be generalized. The various applications all relate directly or indirectly to Requisite 2 for the development of an air pollution problem.

AIR POLLUTION SURVEYS

Depending on the objective of an air pollution survey, mobile or fixed samplers may be used. In either case, sample location is determined to a large extent by average wind direction. Other meteorological data necessary for sample correlation are temperature, cloud cover, and lapse rate where possible. Local temperatures are used to estimate the contribution of home heating to total pollutant emission rates.

SELECTION OF PLANT SITES

The air pollution climatology of an area should be a major consideration in selecting a plant site. Of particular interest is average wind speed and direction data. Seasonal wind roses should be prepared in order to provide a rough estimate of pollutant dispersion patterns. Wind roses based on average winds excluding frontal weather systems are especially helpful. Frequency of stagnant weather periods must be considered. The effect of topography and local wind systems, such as land-sea breezes and mountain-valley winds, should be investigated with respect to dispersion patterns and nearby residential and industrial areas.

Weather records for a potential plant site may be obtained from the National Climatological Data Center located in Asheville, North Carolina. The center will also contract to prepare specific weather summaries and frequency studies.

SPECIFICATION OF EMISSION RATES

Allowable emission rates are dependent on many factors, including whether or not the source is in an attainment or nonattainment area. Emission rates must be controlled to ensure that problems will not arise even during poor dispersion conditions, which means that knowledge of the frequency of poor dispersion weather is necessary. In addition, weather conditions should be considered when plant start-ups are being scheduled or when major repairs will be undertaken that may produce more emissions.

STACK DESIGN

Meteorological factors are a major consideration in tall stack design. We discuss design procedures in detail in Chapter 20, but here we can list the weather parameters required. These are average wind speed at stack elevation, average temperature, average mixing conditions (stability), and average lapse rate. Stack height design must take into account the average height and frequency of inversions. For major emission sources such as generating stations, ideal stack height should exceed the most frequent inversion height. In addition, we should consider not only the averages of temperature, wind speed, stability, and so forth but also the frequency with which "worst-case" combinations of these parameters may occur. Since air pollution concentrations will be highest during times when the meteorological conditions are the worst (that is, when they favor accumulation of pollutants), we must design for the worst case that can reasonably be expected to occur.

Typically, this worst-case scenario is identified during the permit application process by the technique of using an EPA-approved computerized air pollution dispersion model (such as ISCST—see Appendix D) to predict downwind concentrations at hundreds of locations simultaneously. An initial estimate of the stack height is made as described in the preceding paragraph from an analysis of average weather data. This stack height and the design emission rates are input into the dispersion model along with historical (actual) *hour-by-hour meteorological data for the plant location for a period of 1 to 5 years!* Only if the modeled downwind ground-level concentrations do not violate the NAAQSs can the plant be permitted to be built. Dispersion modeling is the subject of the next chapter.

♦ PROBLEMS ♦

19.1 Explain the basis for the two proverbs at the beginning of the chapter.

19.2 Prepare an annual wind rose for the following data, which are representative of Orlando, Florida, in 1970.

	Number of Hourly Observations				
	Wind Speed, Knots				
Wind Direction	1–3	4–6	7–10	>10	Total
N	295	393	412	157	1257
NE	323	306	212	65	906
E	296	388	465	102	1251
SE	160	237	167	56	620
S	218	325	255	122	920
SW	112	196	162	69	539
W	174	266	221	141	802
NW	105	159	193	94	551
Total	1683	2270	2087	806	6846
Number of calms					1914
					8760

19.3 Rework Example 19.1, using U.S. customary engineering units.

19.4 Refer to Figure 19.13 for the case of the stable lapse rate. What distance vertically did the parcel of air at 57.0 F have to be raised for it to cool to 55.3 F?

19.5 The measured lapse rate at a given location was 3.8 F per 1000 ft, and the surface temperature was 60 F. If an air parcel at the surface was heated to 65 F by solar radiation and accelerated vertically by buoyant forces, at what height would the parcel achieve neutral stability?

19.6 If the wet adiabatic lapse rate is 3.1 F per 1000 ft., calculate the change in water content in the air per 1000 ft.

19.7 Once it was determined that the Los Angeles smog problem was due in part to low on-shore wind velocities and frequent atmospheric inversions, many "solutions" have been suggested to alleviate these conditions. Two of the more unusual solutions include installing huge fans mounted on barges that are anchored offshore to increase wind velocity, and drilling large tunnels through the mountains to permit better air circulation. Another such suggestion is to dissipate the inversion layer by spraying water into it, evaporating the water and thus cooling the warm inversion layer and promoting better vertical circulation. Estimate the amount of water (in millions of gallons) that would be required to reduce the temperature by 5.0 degrees F of a 500-ft-thick layer of air covering the LA basin (the basin area is about 2400 square miles).

19.8 A parcel of air at elevation z_0 is suddenly given a downward momentum. The local atmospheric lapse rate is 6.5 F per 1000 feet. Draw a simple diagram of what will happen to this parcel, and provide a *brief* word-description of the events. Is the atmosphere in this region stable, neutral, or unstable?

19.9 Figure 19.8 depicts the direction of an idealized sea breeze as perpendicular to the land mass. Consider the East Coast beaches of the United States. Assume that the pressure gradient is directly east to west at a beach that runs directly north to south. Considering friction and the Coriolis force, from which direction will the actual sea breeze be blowing?

19.10 Assume that in a certain area, the atmospheric lapse rate (γ_{actual}) is equal to zero from the ground up to 2000 meters. Starting from the hydrostatic equation (given below), derive an equation relating atmospheric pressure to altitude.

$$\frac{dP}{dz} = -g\rho$$

Use your equation to calculate the pressure (in mb) at 1200 m above the ground (assume surface temperature = 15 C and surface pressure = 1000 mb).

REFERENCES

Alley, F. C., and Ripperton, L. A. "The Effect of Temperature on Photochemical Oxidant Production," *Journal of the Air Pollution Control Association,* 12, 1962, p. 464.

American Industrial Hygiene Association. *Air Pollution Manual.* Detroit, MI: American Industrial Hygiene Association, 1960.

Brown, K. A. "Application and Initial Assessment of the Urban Airshed Model (Base Case Scenario) for the East Central Florida Region," master's thesis, Civil and Environmental Engineering Department, University of Central Florida, Orlando, 1992.

Cooper, C. D., Wayson, R. L., Brown, K., Moramganti, R., and Muthineni, V. "Central Florida Ozone Study," a final report to the Florida Department of Environmental Regulation. Orlando, FL: Civil and Environmental Engineering Department, University of Central Florida, 1993.

Dimitriades, B. "Photochemical Oxidant Formation: Overview of Current Knowledge and Emerging Issues," *Atmospheric Ozone Research and Its Policy Implications.* Amsterdam, Netherlands: Elsevier, 1989.

Faith, W. L. *Air Pollution Control.* New York: Wiley, 1959.

Glasson, W. A., and Tuesday, C. S. "Hydrocarbon Reactivities in the Atmospheric Photooxidation of Nitric Oxide," *Environmental Science and Technology,* 4, 1970, p. 916.

Guderian, R. *Air Pollution by Photochemical Oxidants.* New York: Springer-Verlag, 1985.

Haagen-Smit, A. J. "Chemistry and Physiology of Los Angeles Smog," *Industrial and Engineering Chemistry, 44,* 1952, pp. 1342–1346.

Kershover, J. *Climatology of Stagnating Anticyclones East of the Rocky Mountains, 1936–1965,* USPHS 999-AP-34, National Center for Air Pollution Control, 1967.

Leighton, P. A. *Photochemistry of Air Pollution.* New York: Academic Press, 1961.

Lippmann, M. "Health Effects of Ozone—A Critical Review," *Journal of the Air Pollution Control Association, 39,* 1989, pp. 672–695.

Morris, R. E., Myers, T. C., Carr, E. L., Causley, M. C., Douglas, S. G., Fieber, J. L., Gardner, L., Jimenez, M., and Haney, J. L. *User's Guide for the Urban Airshed Model,* Vol. I–V. Research Triangle Park, NC: U.S. Environmental Protection Agency, 1990.

Perkins, H. C. *Air Pollution.* New York: McGraw-Hill, 1974.

Petterssen, S. *Introduction to Meteorology.* New York: McGraw-Hill, 1958.

Schrenk, H. H., Heimann, H., Clayton, G. D., Gafefer, W. M., and Wexler, H. *United States Public Health Service Bulletin No. 306,* 1949.

Schuck, E. A. "Eye Irritation from Irradiated Auto Exhaust," a report to the Air Pollution Foundation, Report No. 18, Los Angeles, CA, 1957.

Seinfeld, J. H. *Air Pollution—Physical and Chemical Fundamentals.* New York: McGraw-Hill, 1975.

———. "Ozone Air Quality Models: A Critical Review," *Journal of the Air Pollution Control Association, 38,* 1988, pp. 616–641.

Spensley, J. W. "An Overview of the Clean Air Act Amendments of 1990," *Shepard's Clean Air Act Reporter, 1*(3), May 1992.

Stephens, E. R. "Reactions of Oxygen Atoms and Ozone in Air Pollution," *International Journal of Air and Water Pollution, 10,* 1966, p. 649.

Stern, A. C., Ed. *Air Pollution,* Vol. I (2nd ed.). New York: Academic Press, 1968.

Stern, A. C., Boubel, R. W., Turner, D. B., and Fox, D. L. *Fundamentals of Air Pollution* (2nd ed.). Orlando, FL: Academic Press, 1984.

U.S. Environmental Protection Agency. *A Regional Air Pollution Study for St. Louis, Missouri,* Research Triangle Park, NC, 1962.

Chapter 20

Atmospheric
Dispersion Modeling

Transport and dilution of pollutants by air motions goes on constantly; and all of humanity, particularly that large segment inhabiting cities and industrial areas, depends strongly on this capability of the air to carry away and dilute the pollutants it receives. What we call "air pollution" occurs when too much waste material is emitted into an air volume for the air's capacity to carry it away and dilute it. Thus we must . . . understand the atmospheric mechanisms that result in transport and dilution. . . .

F. A. Gifford, 1975

◆ 20.1 Introduction ◆

The release of pollutants into the atmosphere is a time-honored technique for "disposing" of them. When people first discovered fire and brought it home, they surely must have discovered chimneys also, or else they quickly experienced the effects of not venting the smoke from their caves. That local air did not become completely unusable long ago is due to several self-cleaning characteristics of the atmosphere. One of these important properties is the atmosphere's ability to disperse highly concentrated streams of gaseous pollutants. Of course, the atmosphere's ability to disperse such streams is not infinite and varies from quite good to quite poor, depending on local meteorological and geographical conditions. Thus, with the onset of the Industrial Revolution and the subsequent exponential growth of energy consumption, people began to recognize the need for more and better pollution control equipment to prevent indiscriminate abuse of the atmosphere. However, we still rely heavily on atmospheric dispersion for final disposal of many pollutants.

As was mentioned in Chapter 1, history has demonstrated that harmful effects occur when pollutants build up to high concentrations in a local area. The accumulation of pollutants in any localized region is a function of emission rates, dispersion rates, and generation or destruction rates (that is, by chemical reaction). The dispersion of pollutants is almost entirely dependent upon local meteorological conditions, such as wind speed and direction, and atmospheric stability (the tendency of air to not mix in the vertical direction). If wind is strong and/or if good vertical mixing in the atmosphere exists, pollutants will be dispersed quickly into a large volume of air, resulting in low concentrations. If the wind is relatively low and an inversion is present, then pollutant concentrations can increase. The ability to model atmospheric dispersion and to predict pollutant concentrations from a proposed new source are important parts of air pollution engineering.

Much progress has been made and sophistication achieved in our models in recent years, but the accuracy of dispersion models is still somewhat limited. Even so, as long as we recognize the limitations, we can use the models successfully. One might ask, "Why do we need to conduct modeling at all—why not simply measure ambient air quality?" Certainly we do have very sophisticated instrumentation nowadays, and we can deploy instruments and personnel to measure the concentrations that result from the totality of emissions from vehicles and large industrial sources. Furthermore, we know that models are imperfect representations, and that modeling is expensive, time consuming, and subject to many possible inaccuracies. Why not avoid those uncertainties, and put all our efforts into measuring "the real world"?

Despite its many shortcomings, modeling is *essential*. There are several reasons why modeling must be conducted, and why there is no substitute for it. First, it is impossible to measure the impact from a facility that will be built in the future. Yet, we need to have a reasonable estimate of that impact before we allow the facility to be constructed. Modeling allows us to do that. Second, comprehensive measurement programs could be 1000 times more expensive than modeling, and would also be subject to errors. Third, modeling is the only practical approach when there are many sources, and when we wish to isolate the potential effects of just one source. Finally, modeling may not be 100% accurate, but it is precise (reproducible). Thus, modeling provides an impartial and reproducible tool for assessing and comparing various alternatives. However, one should always keep in mind the limitations of modeling and use good judgement in interpreting the results of any modeling study. This is especially true when making policy or decisions that could result in extremely high costs for industry and/or the public.

◆ 20.2 A PHYSICAL EXPLANATION ◆
OF DISPERSION

How does dispersion take place? A continuous stream of pollutants released into a steady wind in the open atmosphere first will rise, then bend over and travel with the mean wind, which will dilute the pollutants and carry them away from the source. This plume of pollutants will also spread out or disperse both in the horizontal and vertical directions from its centerline. A schematic of a bent-over plume, depicting the physical stack height (h), the plume rise (Δh), and the effective stack height (H), is presented in Figure 20.1. This dispersion is not surprising; it is intuitively obvious that matter behaves "properly" by moving from a region of high concentration to one of lower concentration. But the spreading of a bent-over plume is due to factors other than simple molecular diffusion.

Recall that any fluid in turbulent flow contains **eddies** (or swirls), which are macroscopic random fluctuations from the "average" flow. An eddy may intercept part of a narrow plume and quickly interchange a "batch" of concentrated pollutants with a batch of clean air some distance away from the plume. The combined effect of many eddies of various sizes is to broaden and dilute the plume. Of course, eddies can operate in both the lateral and vertical directions.

Eddies in the atmosphere result from both thermal and mechanical influences. Energy from the sun is absorbed by the ground and converted into heat. This thermal energy is transferred into the lowest levels of the air by conduction/convection, creating thermal eddies. The stronger the solar insolation, the more eddies are formed. Mechanical eddies result from the shear forces produced when air blows across a rough surface. Surfaces with greater roughness (such as trees or buildings) create more eddies than smooth surfaces (ice or snow).

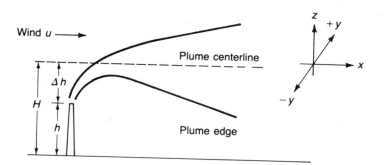

Figure 20.1
The spreading of a bent-over plume.

Another reason for plume spread is the random shifting of the wind. Concentrations of pollutants at a particular spot are measured over some period of time, referred to as the averaging time. But during this time the wind may change direction and blow more or less pollutant toward the detector. The longer the averaging time, the more likely many such shifts will occur. These random fluctuations help spread the plume over a larger downwind area.

Because of eddies and wind fluctuations, the plume must be considered on a time-averaged rather than an instantaneous basis. Consider a period of time during which the wind is reasonably steady at an average speed, u, in the x direction (keep in mind, though, that it really fluctuates in direction and speed). As described by Williamson (1973), the time-averaged pollutant concentration at a given distance, x_0, downwind from the source is normally distributed in the $\pm y$ direction. However, the instantaneous concentration profile (in the y direction) at x_0 is vastly different, as shown in Figure 20.2. It should be pointed out that as the distance x_0 increases, the pollutant spreads further in the y and z directions and the maximum concentration decreases (see Figure 20.3). It should be noted that Figure 20.3 depicts elevated (plume height) concentration profiles in the y direction. The maximums continuously decline and the lateral spread increases with increasing x. The ground-level centerline concentrations behave differently, as we describe later.

A similar spreading of the plume occurs in the vertical direction, resulting in another normal distribution of pollutant concentration. Thus, the distribution of pollutant is termed **binormal**. One method

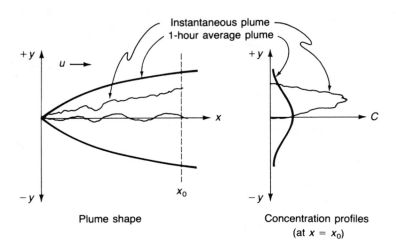

Plume shape

Concentration profiles
(at $x = x_0$)

Figure 20.2
Top view of an instantaneous plume and a 1-hr average plume and their corresponding concentration profiles.

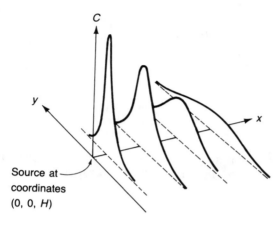

Figure 20.3
Behavior of the downwind, elevated transverse concentration profiles as a function of distance downward.

of developing an equation to model this behavior is to model the wind as being absolutely constant and accounting for the plume spread by eddy diffusivities alone. Based on this approach, a second-order partial differential equation can be derived from material balance considerations. One particular solution to this equation is known as the *Fickian diffusion equation,* which predicts the pollutant concentration to be binormally distributed.

The derivation of the Fickian diffusion equation applied to the atmospheric dispersion problem has been presented by others (Wark and Warner 1981; Williamson 1973). However, this model, which requires the use of average eddy diffusivities and an absolutely steady wind, is only one way to approximate physical reality. Another valid approach is based on the statistical nature of the dispersion process. This model is usually referred to as the **Gaussian dispersion equation** and is discussed in the next section.

◆ 20.3 THE GAUSSIAN MODEL ◆

A Gaussian or normal distribution often results from random processes. A brief review of some of the properties of a Gaussian distribution is presented in Appendix C. In the previous section it was pointed out that the time-averaged concentration profiles about a plume centerline are binormal. Such behavior has been shown by Pasquill (1961) to be well modeled by a double Gaussian equation. This equation (which models the dispersion of a nonreactive gaseous pollutant from an elevated source) is given here in a form that predicts the steady-state concentration at a point (x, y, z) located downwind from the source:

$$C = \frac{Q}{2\pi u \sigma_y \sigma_z} \exp\left(-\frac{1}{2}\frac{y^2}{\sigma_y^2}\right)\left\{\exp\left(-\frac{1}{2}\frac{(z-H)^2}{\sigma_z^2}\right) + \exp\left(-\frac{1}{2}\frac{(z+H)^2}{\sigma_z^2}\right)\right\}$$

$$(20.1)$$

where

C = steady-state concentration at a point (x, y, z), $\mu g/m^3$

Q = emissions rate, $\mu g/s$

σ_y, σ_z = horizontal and vertical spread parameters, m (these are functions of distance, x, and atmospheric stability)

u = average wind speed at stack height, m/s

y = horizontal distance from plume centerline, m

z = vertical distance from ground level, m

H = effective stack height ($H = h + \Delta h$, where h = physical stack height and Δh = plume rise, m)

A view of the double Gaussian distribution in the plume, portrayed by Eq. (20.1) is presented in Figure 20.4.

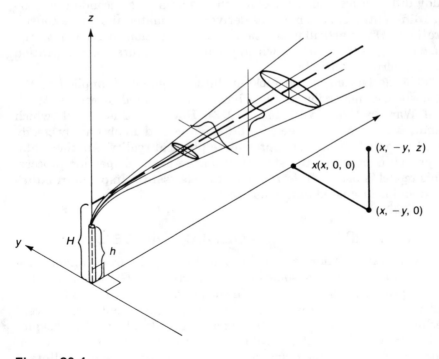

Figure 20.4
Coordinate system showing Gaussian distributions in the horizontal and vertical.
(Adapted from Turner, 1970.)

Note that there are two terms in Eq. (20.1) with an exponential in z, one with $(z - H)$ and one with $(z + H)$. The reason for the two terms is to account for the fact that pollutants cannot disperse underground. A further explanation for this apparently trivial statement is in order. Without the second exponential term in z, Eq. (20.1) would be

$$C = \frac{Q}{2\pi u \sigma_y \sigma_z} \exp\left[-\frac{1}{2} \frac{y^2}{\sigma_y^2} \right] \exp\left[-\frac{1}{2} \frac{(z - H)^2}{\sigma_z^2} \right] \quad \textbf{(20.2)}$$

which is a "true" double Gaussian equation, with a source at $y = 0$, and $z = H$. The situation represented by Eq. (20.2) is depicted in Figure 20.5(a), which shows pollutant dispersing infinitely in the $\pm z$ direction *including underground*. Since this is impossible, the model is corrected by adding a fictitious image source underground (emitting at $-H$), which adds to the above-ground concentration an amount exactly equal to that "lost" underground by the real source. The equation for the image source is

$$C = \frac{Q}{2\pi u \sigma_y \sigma_z} \exp\left[-\frac{1}{2} \frac{y^2}{\sigma_y^2} \right] \exp\left[-\frac{1}{2} \frac{(z + H)^2}{\sigma_z^2} \right] \quad \textbf{(20.3)}$$

When Eqs. (20.2) and (20.3) are added, Eq. (20.1) results. This behavior is depicted in Figure 20.5(b).

As a convenience, we usually just say that the plume reflects at the ground level or exhibits reflection. If the plume were dispersing over water and contained a highly soluble gas, the assumption of reflection might not be justified. However, under usual conditions, it is assumed that pollutants such as CO, VOCs, SO_x, or NO_x do reflect at the ground.

It is important to keep in mind some general relationships indicated by Eq. (20.1):

1. The downwind concentration at any location is directly proportional to the source strength, Q.

2. The downwind ground-level $(z = 0)$ concentration is generally inversely proportional to wind speed. (H also depends on wind speed in a complicated fashion that prevents a strict inverse proportionality.)

3. Because σ_y and σ_z increase as the downwind distance (x) increases, the *elevated* plume **centerline** concentration continuously declines with increasing x. However, *ground-level* centerline concentrations increase, go through a maximum, and then decrease as one moves away from the stack. The reason for this behavior, depicted in Figure 20.6, is because the pollutants initially require some time and distance before they can diffuse to ground level. As they begin to reach the ground, reflection oc-

curs, causing a rapid increase in ground-level concentrations. Finally, as more of the pollutants disperse upwards and outwards, ground-level concentrations begin to decline.

4. The dispersion parameters, σ_y and σ_z, increase with increasing atmospheric turbulence (instability). Thus, unstable conditions decrease average downwind concentrations.

5. The maximum ground-level concentration calculated from Eq. (20.1) decreases as effective stack height increases. The dis-

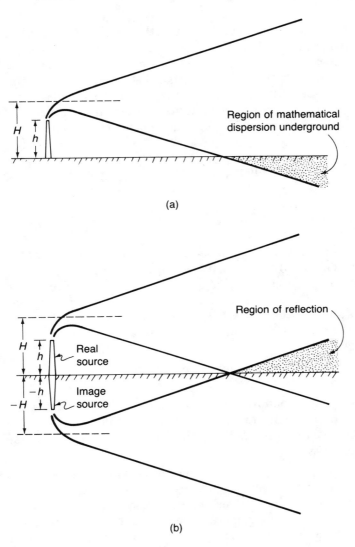

(a)

(b)

Figure 20.5
Schematic diagram depicting (a) mathematical dispersion of pollutants underground and (b) reflection due to an "image source."

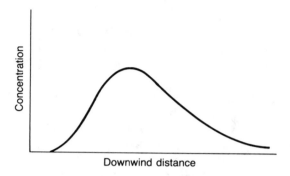

Figure 20.6
Variation of ground-level centerline concentrations with distance downwind from an elevated source.

tance from the stack at which the maximum concentration occurs increases with H.

The Gaussian dispersion equation is extremely important in air pollution work. It is the basis for almost all of the computer programs developed by the U.S. Environmental Protection Agency for atmospheric dispersion modeling. For this reason, a brief discussion of atmospheric stability estimation and some illustrations of the use of the Gaussian equation are presented in the following few paragraphs.

ATMOSPHERIC STABILITY CLASSES

As was explained in Chapter 19, air is termed unstable when there is good vertical mixing. This occurs with strong insolation (solar radiation striking the earth's surface) combined with light winds. As a result of absorbing solar energy, the earth's surface heats and then warms the layers of air near the ground. The warm air rises, promoting vertical mixing. Stable air results when the surface of the earth is cooler than the air above it (such as on a clear, cool night). Then the layers of air next to the earth are cooled, and no vertical mixing can occur. Because the dispersion parameters, σ_y and σ_z, are strong functions of atmospheric stability, as well as downwind distance, it is important to discuss the estimation of these parameters.

A quantitative method for estimating the dispersion parameters was introduced by Pasquill (1961) and made more convenient by Gifford (1961). The Pasquill-Gifford system for dispersion estimates was adopted by the U.S. Public Health Service (Turner 1970) and has been widely used ever since. The correlations, which are presented graphically in Figures 20.7 and 20.8, were developed by application of theoretical principles to the analysis of actual dispersion data. Tests were conducted in level, open terrain, in which short-time average (10 min) concentrations were measured, along with emission rates and wind speed. Extrapolations of the data were made to extend the graphs for values of x greater than 1000 m. Because of the restrictive circumstances under which these curves were developed, predictions of concentrations using the Gaussian model should not be expected to be closer than about $\pm 50\%$ of actual values.

For convenience, atmospheric stability has been broken into six categories, arbitrarily labeled A through F, with A being the most unstable. Turner (1970) has suggested a way to estimate the stability class of the atmosphere based on the angle of the sun, the extent of cloud cover, and the surface wind speed. This system is summarized in Table 20.1. We suggest that the reader carefully review the footnotes to the table.

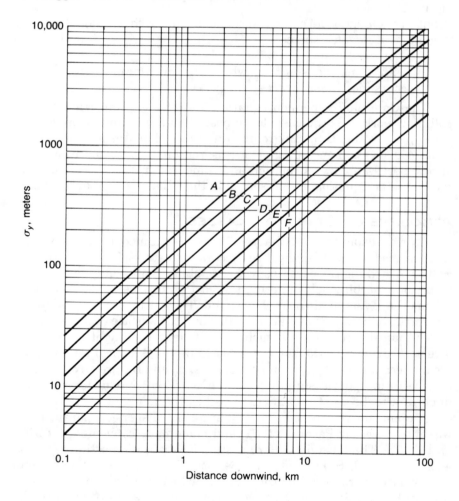

Figure 20.7
Horizontal dispersion coefficient as a function of downwind distance from the source.
(Adapted from Turner, 1970.)

Sometimes it is difficult for different people to obtain consistent readings from Figures 20.7 and 20.8. Furthermore, such graphical representations are inconvenient for use in computer programs. Martin (1976) published equations that give reasonable fits to these curves. The general equations are

$$\sigma_y = ax^b \tag{20.4}$$

and

$$\sigma_z = cx^d + f \tag{20.5}$$

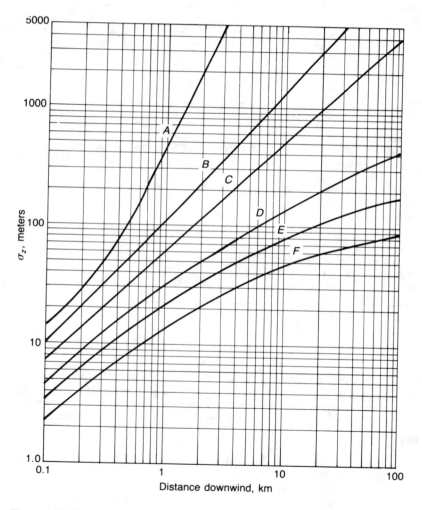

Figure 20.8

Vertical dispersion coefficient as a function of downwind distance from the source.

(Adapted from Turner, 1970.)

where a, b, c, d, and f are constants that are dependent on the stability class and on the distance x (x must be expressed in km).

The numerical values for the constants in Eqs. (20.4) and (20.5) are given in Table 20.2. The following example illustrates typical uses of the information presented in the past few pages.

Table 20.1   Stability Classifications*

Surface Wind Speed[a] m/s	Day Incoming Solar Radiation Strong[b]	Day Incoming Solar Radiation Moderate[c]	Day Incoming Solar Radiation Slight[d]	Night Cloudiness[e] Cloudy ($\geq 4/8$)	Night Cloudiness[e] Clear ($\leq 3/8$)
<2	A	A–B[f]	B	E	F
2–3	A–B	B	C	E	F
3–5	B	B–C	C	D	E
5–6	C	C–D	D	D	D
>6	C	D	D	D	D

[a] Surface wind speed is measured at 10 m above the ground.
[b] Corresponds to clear summer day with sun higher than 60° above the horizon.
[c] Corresponds to a summer day with a few broken clouds, or a clear day with sun 35-60° above the horizon.
[d] Corresponds to a fall afternoon, or a cloudy summer day, or clear summer day with the sun 15–35°.
[e] Cloudiness is defined as the fraction of sky covered by clouds.
[f] For A–B, B–C, or C–D conditions, average the values obtained for each.

* A = Very unstable   D = Neutral
  B = Moderately unstable   E = Slightly stable
  C = Slightly unstable   F = Stable
  Regardless of wind speed, Class D should be assumed for overcast conditions, day or night.

Adapted from Turner, 1970.

Table 20.2   Values of Curve-Fit Constants for Calculating Dispersion Coefficients as a Function of Downwind Distance and Atmospheric Stability

Stability	a	b	$x < 1$ km c	$x < 1$ km d	$x < 1$ km f	$x > 1$ km c	$x > 1$ km d	$x > 1$ km f
A	213	0.894	440.8	1.941	9.27	459.7	2.094	−9.6
B	156	0.894	106.6	1.149	3.3	108.2	1.098	2.0
C	104	0.894	61.0	0.911	0	61.0	0.911	0
D	68	0.894	33.2	0.725	−1.7	44.5	0.516	−13.0
E	50.5	0.894	22.8	0.678	−1.3	55.4	0.305	−34.0
F	34	0.894	14.35	0.740	−0.35	62.6	0.180	−48.6

Adapted from Martin, 1976.

▶ ▶ ▶ ▶ ▶ ▶ ▶ ▶ ▶

Example 20.1

Nitric oxide (NO) is emitted at 110 g/s from a stack with physical height of 80 m. The wind speed at 80 m is 5 m/s on an overcast morning. Plume rise is 20 m. (a) Calculate the ground-level centerline concentration 2.0 km downwind from the stack. (b) Calculate the concentration at 100 m off the centerline at the same x distance.

Solution

As noted in Table 20.1, Class D stability should be used for overcast conditions. From Eqs. (20.4) and (20.5), $\sigma_y = 126$ m and $\sigma_z = 51$ m. Substituting all the proper values into Eq. (20.1) yields the following:

(a)

$$C = \frac{110(10)^6}{2\pi(5)(126)(51)}\exp\left[-\frac{1}{2}\frac{0^2}{126^2}\right]$$

$$\times\left\{\exp\left[-\frac{1}{2}\frac{(-H)^2}{51^2}\right]+\exp\left[-\frac{1}{2}\frac{H^2}{51^2}\right]\right\}$$

$$= 545(1)2\exp\left[-\frac{1}{2}\frac{100^2}{51^2}\right]$$

$$= (545)(0.293)$$

$$= 159 \ \mu g/m^3$$

(b) At $y = 100$, everything else being the same,

$$C = 159\exp\left[-\frac{1}{2}\frac{100^2}{126^2}\right]$$

$$= 159(0.73)$$

$$= 116 \ \mu g/m^3$$

◀ ◀ ◀ ◀ ◀ ◀ ◀ ◀ ◀

In Chapter 19, we introduced the power law to predict the variation of wind speed with height. The equation is repeated below for convenience:

$$\frac{u_2}{u_1} = \left(\frac{z_2}{z_1}\right)^p \tag{19.1}$$

where

z_1, z_2 = elevations 1 and 2

u_1, u_2 = wind speeds at z_1 and z_2

p = exponent

The exponent p varies with atmospheric stability class and with surface roughness. For "rough" surfaces (typical of urban and suburban areas), the U.S. Environmental Protection Agency (1995) recommends the exponent values shown in Table 20.3. For flat, open country and lakes and seas, there is less variation between the surface wind and the geostrophic wind. The wind speed should be adjusted to H, but to be conservative, EPA recommends that u be adjusted to h. Example 20.2 illustrates the use of Eq. (19.1).

Table 20.3   Exponents for Wind Profile (Power Law) Model

Stability Class	Exponent (p)	
	Rough Surface (urban)	Smooth Surface (rural)
A	0.15	0.07
B	0.15	0.07
C	0.20	0.10
D	0.25	0.15
E	0.30	0.35
F	0.30	0.35

Adapted from U.S. Environmental Protection Agency, 1995.

Example 20.2

Consider the same data as in Example 20.1, except that the meteorology is such that the wind speed (at 10 m) is 4 m/s and it is mid-afternoon on a hot summer day. Calculate the ground-level centerline concentration at x = 2.0 km, assuming rough terrain.

Solution

From Table 20.1, Class B stability is estimated, owing to the wind speed and strong insolation. To calculate the wind speed at stack height, we use Eq. (19.1). An estimate of p for Class B stability is 0.15. From Eq. (19.1), we see that the wind speed at 80 m is

$$u = 4\left(\frac{80}{10}\right)^{0.15}$$

$$= 5.5 \text{ m/s}$$

From Figure 20.7 and 20.8, estimates of σ_y and σ_z are 300 m and 230 m, respectively. Substituting into Eq. (20.1), we have

$$C = \frac{110(10)^6}{\pi(5.5)(300)(230)}(1)\exp\left[-\frac{1}{2}\frac{100^2}{230^2}\right]$$

$$= 92.3(.91)$$

$$C = 84 \ \mu g/m^3$$

◀ ◀ ◀ ◀ ◀ ◀ ◀ ◀ ◀

THE DEPENDENCE OF CONCENTRATION ON AVERAGING TIME

The concentration predicted by Eq. (20.1), using the σ_y and σ_z values from Figures 20.7 and 20.8, is a 10-minute-average concentration. As mentioned previously, a longer time-averaged concentration would be expected to be less than a short time-average, owing to wind shifts and turbulent diffusion. For averaging times between 10 min and 5 hr, data reported by Hino (1968) suggest that the concentrations at the two averaging times are related as follows:

$$C_t = C_{10}\left(\frac{10}{t}\right)^{0.5} \tag{20.6}$$

where

t = averaging time, min

C_t = concentration for averaging time t

For averaging times less than 10 minutes, the exponent in Eq. (20.6) is more likely on the order of 0.2 instead of 0.5 (Nonhebel 1960). Others have suggested that the exponent varies with atmospheric stability. For instance, the Texas Air Control Board has used 0.05 to 0.675 for stability classes F–A.

ESTIMATING THE MAXIMUM DOWNWIND GROUND-LEVEL CONCENTRATION

The maximum downwind ground-level concentration always occurs on the centerline. Turner (1970) developed a graphical means of estimating the maximum concentration (C_{max}), and the distance at which it occurs (x_{max}). Both C_{max} and x_{max} depend on stability class, effective stack height, wind speed, and emission rate.

Figure 20.9 presents Turner's results as a plot of x_{max} versus $(Cu/Q)_{max}$. The solid lines represent the various stability classes, and the numbers along the lines represent various values of the effective stack height. Typically, one knows the stability class, the effective height, the wind speed, and the emission rate. From these data and the plots, one may calculate C_{max} and x_{max}.

In order to better understand the qualitative variation of C_{max} and x_{max} with stability class and effective stack height, look at Figure

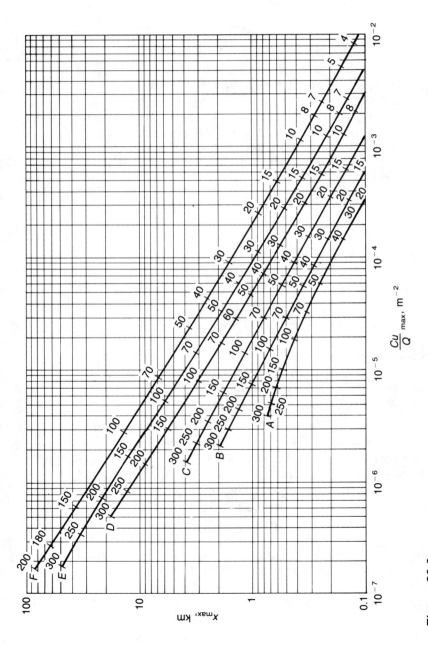

Figure 20.9
Maximum (Cu/Q) and distance to C_{max} as a function of stability (solid lines) and effective stack height (numbers [in meters]).(Adapted from Turner, 1970.)

20.10. Notice that as stability changes from unstable to stable, C_{max} decreases slightly while x_{max} increases. Also, the region of relatively high concentrations is much larger under stable conditions. An increase in H results in a dramatic reduction in C_{max}, while increasing x_{max} slightly. However, note that the concentrations a "long way" away from the source approach the same value regardless of stack height.

The plots in Figure 20.9 were fit to a polynomial equation by Ranchoux (1976), which can be written as

$$(Cu/Q)_{max} = \exp[a + b \ln H + c(\ln H)^2 + d(\ln H)^3] \qquad (20.7)$$

where a, b, c and d are constants that depend on stability class as shown in Table 20.4. In Eq. (20.7), H must be in m and $(Cu/Q)_{max}$ is in m^{-2}.

One of the most straightforward methods to find C_{max} and x_{max} is by the repeated use of Eq. (20.1). Of course, this method requires the

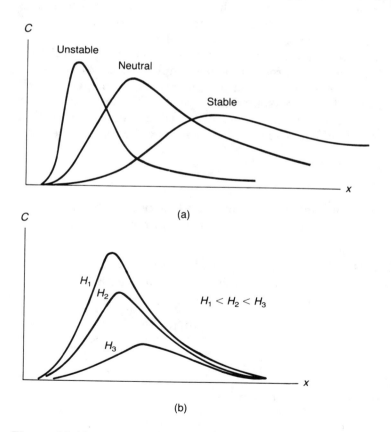

Figure 20.10

Behavior of ground-level centerline concentration from an elevated source as a function of downwind distance and (a) atmospheric stability (at constant H), and (b) effective stack height (at constant stability).

Table 20.4 Values of Curve-Fit Constants for Estimating $(Cu/Q)_{max}$ from H as a Function of Atmospheric Stability

Stability	Constants			
	a	b	c	d
A	−1.0563	−2.7153	0.1261	0
B	−1.8060	−2.1912	0.0389	0
C	−1.9748	−1.9980	0	0
D	−2.5302	−1.5610	−0.0934	0
E	−1.4496	−2.5910	0.2181	−0.0343
F	−1.0488	−3.2252	0.4977	−0.0765

Adapted from Ranchoux, 1976.

use of a computer or programmable calculator. Typically, a very large x interval is chosen, inside of which the maximum is sure to occur. The concentration is calculated at either end of the interval, using Eq. (20.1). Then the interval is progressively narrowed, until the maximum concentration is located within some suitably small x. A listing of a FORTRAN computer program, which employs Eqs. (20.1), (20.4), and (20.5) to calculate ground-level concentrations given x, Q, u, y, H, and the stability class is presented in Appendix D. The program also has a routine to calculate C_{max} and x_{max}, using the Golden Section search method, an efficient scheme for narrowing the search interval.

▶ ▶ ▶ ▶ ▶ ▶ ▶ ▶ ▶

Example 20.3

Given the same data as Example 20.1, estimate the maximum ground-level concentration and the distance at which it occurs.

Solution

For Class D stability and $H = 100$ m, from Figure 20.9, x_{max} is estimated to be 3 km, and $(Cu/Q)_{max}$ is estimated to be $8.1(10)^{-6}$ m^{-2}. Thus

$$C_{max} = 8.1(10)^{-6} \frac{110(10)^6}{5}$$

$$= 178 \ \mu g/m^3$$

Similar results should result from Eq. (20.7)—that is,

$$C_{max} = \frac{110(10)^6}{5} \exp\left[-2.5302 - 1.561\ln 100 - 0.0934(\ln 100)^2\right]$$

$$= 183 \ \mu g/m^3$$

Using the computer program listed in Appendix D, we obtained a value of 184 μg/m^3 for $C_{\max}$, which occurred at a distance of 2.9 km.

◀ ◀ ◀ ◀ ◀ ◀ ◀ ◀ ◀

ESTIMATING THE DOWNWIND CONCENTRATION UNDER AN ELEVATED INVERSION

An elevated inversion can act as a lid to prevent the upward dispersion of pollutants as shown in Figure 20.11. An assumption is usually made that the inversion layer reflects pollutants similar to the way the ground does. With two reflecting surfaces, an infinite number of images is required. The equation that models a trapped plume is

$$C = \frac{Q}{2\pi u \sigma_y \sigma_z} \left[\exp\left(\frac{-1}{2}\frac{y^2}{\sigma_y^2}\right) \right] \sum_{-\infty}^{+\infty} \left\{ \exp\left(\frac{-(z - H + 2jL)^2}{2\sigma_z^2}\right) \right.$$
$$\left. + \exp\left(\frac{-(z + H + 2jL)^2}{2\sigma_z^2}\right) \right\}$$

(20.8)

where L = height from the ground to the bottom of the inversion layer, m. In practice, it is only necessary to vary the summation index, j, from -2 to $+2$ to obtain reasonable convergence of Eq. (20.8).

During the break-up of a ground-based inversion, **fumigation** of a plume can occur. Fumigation describes the situation in which an unstable boundary layer grows up to a fanning plume and quickly mixes it vertically throughout the distance between the ground and the plume. Under these conditions, the equation that best predicts the downwind concentration is

$$C = \frac{Q}{(2\pi)^{1/2} u \sigma_y H} \exp\left(-\frac{1}{2}\frac{y^2}{\sigma_y^2}\right)$$

(20.9)

Turner (1970) suggests that Eq. (20.9) is also valid for the plume trapped under an elevated inversion if L is substituted for H and if x is

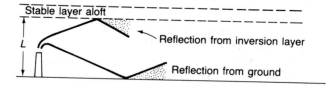

Figure 20.11
Plume dispersion under an elevated inversion.

sufficiently far away from the source. This "sufficient" distance is set equal to twice the distance at which the edge of the plume begins to interact with the inversion layer. If the distance of first interaction is x_L, then according to Turner (1970) σ_z at this distance is given by

$$\sigma_z = 0.47(L - H) \tag{20.10}$$

from which x_L can be determined from Figure 20.8.

At the point of first interaction, no effect of multiple reflections has occurred. Clearly, Eq. (20.1) is valid up to this distance. For $x > 2x_L$, Eq. (20.9) holds (with L substituted for H). Thus, we have an alternative to Eq. (20.8). For $x < x_L$, use Eq. (20.1); for $x > 2x_L$, use Eq. (20.9); and for $x_L < x < 2x_L$, simply interpolate between the concentrations calculated at x_L and $2x_L$.

◆ 20.4 TALL STACKS AND PLUME RISE ◆

Consider Eq. (20.1), written for the ground-level centerline concentration:

$$C = \frac{Q}{\pi u \sigma_y \sigma_z} \exp\left(-\frac{H^2}{2\sigma_z^2}\right) \tag{20.11}$$

Equation (20.11) demonstrates that, with everything else constant, the concentration at any point x should go down very quickly as H increases. According to Wark and Warner (1981), C_{max} decreases approximately as H^2 for many conditions.

This relationship is the reason for designing tall stacks. The objective of stack design is to be able to specify a stack that is tall enough to release a given quantity of pollutants under all meteorological conditions such that the maximum ground-level concentrations do not exceed a certain criterion. Of course, building an excessively tall stack means spending money unnecessarily, hence the need for quantitative design. It should be noted that there is still some debate on the overall worth of a tall stack, since it does nothing to reduce the mass emission rate. It should also be noted that if the mass emission rate increases as the stack gets taller, the main purpose of tall stacks—reducing local concentrations—may be negated.

Before proceeding further, we should emphasize that what we call tall stacks differ from what we call (for lack of a better term) short stacks. Tall stacks are those that are tall (>2.5 times as tall as the tallest of the nearest buildings), have stack gases with high buoyancy and exit velocity (>1.5 maximum average wind speed expected), exhibit significant plume rise, and are on furnaces with large heat emission rates (10 MW or more). Short stacks, on the other hand, are short (<50 m), have stack gases with low buoyancy but perhaps high velocity, exhibit small plume rise, and are on such "small" sources as process furnaces and factories. Emissions from tall stacks have been studied thoroughly and are fairly well modeled by the Gaussian model and its variations. Emissions from short stacks are not as well studied and often exhibit large deviations from the Gaussian model owing to interactions with local terrain and buildings. Tables 20.5 and 20.6 (Schnelle 1976) give some data on tall stacks and other tall things.

DESIGN PROCEDURES

Modern-day design of tall stacks makes considerable use of computers. Computers are invaluable to the stack designer both for their ability to process large quantities of meteorological data and for their ability to quickly calculate expected concentrations for a given set of design conditions. There are many different computer programs available; in Appendix D we present a list and short descriptions of those used by the U.S. Environmental Protection Agency. However, it is important to be aware of the basic procedure for stack design, which is as follows:

1. Analyze the meteorology.

2. Make a preliminary hazard assessment based on meteorology.

3. Test various cases of physical stack parameters (height and diameter), plume rise models, and plant locations in conjunction with the "worst-case" meteorology.

4. Consider effects of local terrain.

5. Review your results—do they make sense?

Table 20.5 Heights of Tall Things

Stacks	ft	Buildings	ft
Inco Smelter (Subdury, Ontario)	1250	Sears Tower	1450
Am. Electric Power Mitchell Plant	1206	Empire State Building	1250
TVA-Cumberland Plant	1000	John Hancock Building	1127
TVA-Paradise Plant	800	Eiffel Tower	984
TVA-Gallatin Plant	500	Washington Monument	555

Adapted from Schnelle, 1976.

Table 20.6 The Inco Copper Smelter Stack, Sudbury, Ontario

Height	1250 ft
Diameter at top and at base	52 ft (top), 116 ft (base)
Inside diameter at top	50 ft
Stack gas velocity	65–70 ft/sec

Adapted from Schnelle, 1976.

Recall that the effective stack height, H, is made up of two components: a physical stack height, h, and a plume rise, Δh. The physical stack height, of course, is a completely free design parameter, but the plume rise is very important and can be larger than the physical stack height in some cases. Failure to account for the beneficial effects of plume rise might result in excessive overdesign of the stack and a considerable unnecessary expense.

PLUME RISE

A plume of hot gases emitted vertically has both a momentum and a buoyancy. As the plume moves away from the stack, it quickly loses its vertical momentum (owing to drag by and entrainment of the surrounding air). As the vertical momentum declines, the plume bends over in the direction of the mean wind. However, quite often the effect of buoyancy is still significant, and the plume continues to rise for a long time after bending over.

The buoyancy term is due to less-than-atmospheric density of the stack gases and may be temperature or composition induced. In either case, as the plume spreads out in the air (all the time mixing with the surrounding air), it becomes diluted by the air.

Modeling the rise of a plume of gases emitted from a stack into a horizontal wind is a complex mathematical problem. Plume rise depends not only on such stack gas parameters as temperature, molecular weight, and exit velocity, but also on such atmospheric parameters as wind speed, ambient temperature, and stability conditions.

Before we present formulas for calculating plume rise, we must stress that many empirical and theoretical methods exist for calculating Δh as a function of source and atmospheric parameters. As Briggs (1975) points out, many of the models do not agree either with each other or with observations other than those used to fit the equations originally. Figure 20.12 compares some of the models for which Δh is inversely proportional to the wind speed to the first power, and directly proportional to the heat emission rate to the first power.

As Briggs (1975) so aptly puts it, "Undoubtedly, many an engineer has felt frustrated after discovering that different formulas give Δh ranging over a factor of 10 or more!" However, it is necessary to account for plume rise, and therefore we discuss some of the better-known methods in the next few paragraphs.

Several published accounts compare plume rise models with each other and with actual observations (Carson and Moses 1969; Thomas, Carpenter, and Colbaugh 1970). We refer the interested reader to the literature for further details.

One well-known plume rise equation is the Holland (1953) formula, which at one time was recommended for use by the U.S. Environmental Protection Agency. The equation is:

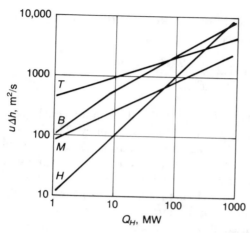

KEY: H = Holland   M = Moses and Strom
        B = Briggs   T = TVA

Figure 20.12
Comparisons of various plume rise models.
(Adapted from Briggs, 1975.)

$$\Delta h = \frac{v_s d_s}{u}\left[1.5 + 2.68(10)^{-3} P_a \left(\frac{T_s - T_a}{T_s}\right) d_s\right] \qquad (20.12)$$

where

v_s = stack gas velocity, m/s

u = mean wind speed at stack height, m/s

d_s = stack inner diameter, m

P_a = atmospheric pressure, mb

T_s = stack gas temperature, °K

T_a = atmospheric temperature, °K

The last term of the Holland formula is directly related to heat emission rate (Wark and Warner 1981), so Eq. (20.12) could be written as

$$\Delta h = 1.5\frac{v_s d_s}{u} + \frac{9.6 Q_H}{u} \qquad (20.13)$$

where Q_H = heat emissions rate, MW

The Holland formula as depicted by Eq. (20.12) or (20.13) is valid for neutral conditions, and the plume rise obtained by it should be corrected for other than neutral stability (Holland 1953). Suggested multiplicative correction factors are 1.1 or 1.2 for Class B or A stability and 0.9 or 0.8 for Class E or F stability.

Another model with reasonable performance is the modified Concawe formula (Thomas, Carpenter, and Colbaugh 1970):

$$\Delta h = \frac{101.2 (Q_H)^{0.444}}{u^{0.694}} \tag{20.14}$$

The Briggs plume rise model (1969, 1970, 1972) is currently recommended by the U.S. Environmental Protection Agency. The Briggs model is used in all EPA computer programs and appears to be better for large thermally dominated plumes. Briggs recognized that even after a plume was bent over by the wind it continued to rise, owing to its thermal buoyancy. Thus, his equations predict Δh as a function of a buoyancy flux term, F_B (which is usually dominated by thermal buoyancy), wind speed, and distance downwind. After a "long enough" travel time (or distance downwind), the plume reaches its final rise.

Different equations are used in the Briggs model, depending on atmospheric stability. First, for neutral or unstable conditions, the downwind distance to the point of final plume rise is x_f, where

$$x_f = 119 (F_B)^{2/5} \quad \text{for } F_B \geq 55 \text{ m}^4/\text{sec}^3 \tag{20.15}$$

or

$$x_f = 49 (F_B)^{5/8} \quad \text{for } F_B < 55 \text{ m}^4/\text{sec}^3 \tag{20.16}$$

The plume rise (as a function of downwind distance, x) is calculated by

$$\Delta h = \frac{1.6 (F_B)^{1/3}}{u} (x_f)^{2/3} \quad \text{for } x \geq x_f \tag{20.17}$$

or

$$\Delta h = \frac{1.6 (F_B)^{1/3}}{u} (x)^{2/3} \quad \text{for } x < x_f \tag{20.18}$$

The buoyancy flux term, F_B, is given by Briggs (1975) as

$$F_B = g \left(1 - \frac{MW_s}{28.9} \right) \left(\frac{T_a}{T_s} \right) \frac{v_s d_s^2}{4} + 8.9 \left(\frac{P_0}{P_a} \right) Q_H \tag{20.19}$$

where

F_B = buoyancy flux, m^4/s^3

g = the gravitational constant, 9.8 m/s^2

MW_s = molecular weight of stack gas

P_0 = standard sea level pressure, mb

If, as in the case for most combustion sources, MW_s is approximately equal to 28.9, then the first term of Eq. (20.19) becomes negligible, and F_B is easily calculated from 8.9 $(P_0/P_a)Q_H$ (Briggs 1975).

For stable conditions, first a stability parameter S is calculated as follows:

$$S = \frac{g}{T_a}\left(\frac{\Delta\theta}{\Delta z}\right) \qquad (20.20)$$

where

$\dfrac{\Delta\theta}{\Delta z}$ = potential temperature gradient, °K/m

S = stability parameter, s^{-2}

If data are not available for the potential temperature gradient, use 0.02 K/m for Class E and 0.035 K/m for Class F. Next, the plume rise is calculated using

$$\Delta h = 2.6\left(\frac{F_B}{uS}\right)^{1/3} \text{ for } u \geq 1.5 \text{ m/s} \qquad (20.21)$$

or

$$\Delta h = 5F_B^{0.25}S^{-0.375} \text{ for } u < 1.5 \text{ m/s} \qquad (20.22)$$

For the final plume rise under stable conditions, use the smaller Δh predicted from either Eq. (20.21) or Eq. (20.22) regardless of wind speed.

▶ ▶ ▶ ▶ ▶ ▶ ▶ ▶ ▶

Example 20.4

For Class D stability, calculate the final plume rise from a power plant stack, given the following information:

v_s = 20 m/s	P_a = 1000 mb
d_s = 5 m	T_a = 280 K
u = 6 m/s	T_s = 400 K
Q_H = 40,000 kJ/s	MW_s = 28.9 g/gmol

Use the Holland formula, the modified Concawe formula, and the Briggs method.

Solution

Using Eq. (20.13) for the Holland formula, and recognizing that 40,000 kJ/s = 40 MW,

$$\Delta h = \frac{1.5(20)5}{6} + \frac{9.6(40)}{6}$$

$$\Delta h = 89 \text{ m (Holland)}$$

From Eq. (20.14),

$$\Delta h = \frac{101.2(40)^{0.444}}{6^{0.694}}$$

$$\Delta h = 150 \text{ m (Concawe)}$$

To use the Briggs method, first calculate the buoyancy flux. Using Eq. (20.19) with the first term equal to zero,

$$F_B = 8.9\left(\frac{1013}{1000}\right)40 = 360.6$$

Next, calculate x_f from Eq. (20.15), since $F_B > 55$:

$$x_f = 119(360.6)^{0.4}$$
$$= 1254 \text{ m}$$

Finally, since the final plume rise is desired, use Eq. (20.17):

$$\Delta h = \frac{1.6(360.6)^{0.333}}{6}(1254)^{0.667}$$

$$\Delta h = 221 \text{ m (Briggs)}$$

Notice the large differences among predicted plume rises using the three methods. For a stack with a physical height of 100 m, the effective height obtained via the Briggs method would be 70% greater than that obtained via the Holland formula.

CRITICAL WIND SPEED

Consider Eq. (20.11) solved specifically at x_{max} for C_{max}. It is obvious that the pre-exponential term is inversely proportional to u. But in the previous section, it was shown that Δh (and thus H) is inversely proportional to u also. Thus, the exponential term increases with increasing u, while the pre-exponential term decreases. The result is that a maximum C_{max} exists at some wind speed, u_c, called the *critical wind speed;* that is, if we solve for C_{max} for various wind speeds and plot the values of C_{max} versus wind speed, the plot would trace a smooth curve passing through a maximum. The critical wind speed depends on the stability and the specific plume rise model used. It is best determined through use of a computer.

Example 20.5

Consider the following data: $Q = 110$ g/s, class $= D$, $h = 100$ m, MW_s $= 28.9$, $P_a = 1000$ mb, and $Q_H = 10$ MW. Using the Briggs plume rise equation, calculate C_{max} as a function of wind speed and find the critical wind speed, the maximum C_{max}, and the distance at which it occurs.

Solution

A computer program that corrected wind speed to stack height and used the Briggs plume rise equation was used to solve this problem. Various surface wind speeds were chosen. The results are presented in the following table.

u_{10}, m/s	u_h, m/s	C_{max}, μg/m³	x_{max}, km	H, m
1.0	1.4	10.3	48	512
2.0	2.8	20.3	18.5	306
3.0	4.2	26.5	11.7	237
4.0	5.6	29.9	8.9	203
5.0	6.9	31.5	7.4	182
6.0	8.3	32.1	6.5	169
7.0	9.7	32.1	5.9	159
8.0	11.1	31.7	5.4	152
10.0	13.9	30.3	4.8	141

From the tabulated results given above, the critical wind speed at stack height for Class D stability lies between 8.3 and 9.7 m/s and results in a C_{max} of 32 μg/m³ at distances between 6 and 6.5 km.

◄ ◄ ◄ ◄ ◄ ◄ ◄ ◄ ◄

OTHER STACK DESIGN CONSIDERATIONS

As air blows past a structure, a low-pressure region forms behind the structure, resulting in recirculating eddies and wake formation. In addition, if the structure is large, disturbances in the flow system are experienced above and on either side of the structure. Typical disturbances from a building and from the stack itself are pictured in Figure 20.13. These effects may result in plume downwash unless the designer takes steps to compensate for them.

The region of turbulence owing to buildings may extend to nearly twice the building height h_B and downwind 5 to 10 times h_B (Turner 1970). A well-known rule of thumb for stack designers is to make the stack height at least 2.5 times the height of the highest building near the stack.

Plume downwash behind a stack can usually be avoided by specifying a stack diameter small enough to ensure that the stack exit velocity is greater than 1.5 times the maximum expected sustained

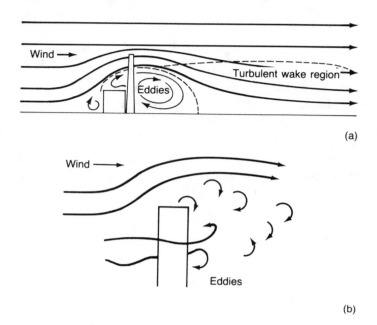

(a)

(b)

Figure 20.13
Typical flow disturbances owing to (a) a building and (b) a stack.

Figure 20.14
A view of a large power plant showing the wind turning vanes on the six steel stacks. Note the electrostatic precipitator on the far left of the photograph.

wind speed at stack height. Also, the capital cost of the stack will be decreased by decreasing the stack diameter. But keep in mind that pressure drop in the stack will increase as the square of the gas velocity, thus increasing fan power costs.

The effects of stack downwash may also be negated by mechanical adaptations on the stack. One such device is a flat disk located circumferentially at the stack top. It has been suggested that the disk diameter be equal to three times the stack top diameter (Schnelle 1976). Another possibility is to include wind turning vanes on the outside of the stack to reduce the low-pressure region behind the stack. Such vanes are shown in Figure 20.14. The next example illustrates the procedures for stack design.

▸ ▸ ▸ ▸ ▸ ▸ ▸ ▸ ▸

Example 20.6

A source emits 20.0 g/s of pollutant contained in 6000 m^3/min of stack gases (at $P = 1$ atm and $T = 390$ K). The maximum expected wind speed is 15 m/s. The heat emission rate is 7 MW. For a wind speed of 8 m/s (at all heights) and Class D stability, design a stack (specify the height and diameter) that will prevent ground-level concentrations from exceeding 50 μg/m^3 anywhere downwind. Use the Holland formula for plume rise.

Solution

The inner stack diameter is calculated first. The stack gas velocity is set by

$$v_s = 1.5 \times 15 \text{ m/s} = 22.5 \text{ m/s}$$

From v_s, the stack diameter is obtained by the continuity equation

$$Q = v_s A_s = v_s \frac{\pi d_s^2}{4}$$

$$d_s = \left(\frac{4Q}{\pi v_s}\right)^{1/2} = \left(\frac{(4)6000/60}{\pi 22.5}\right)^{1/2} = 2.4 \text{ m}$$

For the conditions of this problem,

$$\left(\frac{Cu}{Q}\right)_{\text{max}} = \frac{(50)(8)}{20.0(10)^6} = 2.0(10)^{-5} \text{ m}^{-2}$$

From Figure 20.9 at the intersection of this value of $(Cu/Q)_{\text{max}}$ and the Class D stability line, values for H and x_{max} are found:

$$H = 68 \text{ m}$$

$$x_{\text{max}} = 1.7 \text{ km}$$

From the Holland formula, we have

$$\Delta h = \frac{1.5(22.5)(2.4)}{8} + \frac{9.6}{8}(7) = 18.5 \text{ m}$$

Thus,

$$h = H - \Delta h = 49.5 \text{ m}$$

Specify a stack 50 m tall and with a top inner diameter of 2.4 m.

STACK COSTS

Stacks are arbitrarily classified as either short (less than about 100 feet) or tall (can exceed 1000 feet). The costs of short stacks depend primarily on three factors: stack height, stack diameter, and material of construction. The U.S. EPA has developed a simple cost estimating equation as follows:

$$C = a\, D^b \qquad\qquad\qquad (20.23)$$

where

C = total installed cost in 1993 dollars, \$/ft of stack height

D = stack diameter, in.

a, b = parameters given in Table 20.7

Tall stacks may require additional components, such as ladders, platforms, guy wires, strong foundations, and aircraft warning lights. While short stacks may cost several thousand dollars, tall stacks may run several million dollars. Figure 20.15 can be used to estimate the cost of a tall stack. Whether Eq. (20.23) or Figure 20.15 is used to estimate stack cost, the result must be updated to current year dollars using the methods described in Chapter 2.

Table 20.7 Cost Estimating Parameters for Short Stacks

Material	Equation Parameter[*]		Applicable Range	
	a	b	D, in.	H, feet
PVC[1]	0.393	1.61	12–36	≤ 10
Plate – CS, coated[2]	3.74	1.16	6–84	20–100
Plate – 304 SS[3]	12.0	1.20	6–84	20–100
Sheet – Galv CS[4]	2.41	1.15	8–36	< 75
Sheet – 304 SS[5]	4.90	1.18	8–36	< 75

[*] For use with Equation 20.23
[1] polyvinyl chloride
[2] carbon steel plate, coated with one coat of shop paint
[3] stainless steel plate, type 304
[4] carbon steel sheet, galvanized
[5] stainless steel sheet, type 304

Adapted from Vatavuk, 1996.

20.5 COMPUTER PROGRAMS FOR DISPERSION MODELING (POINT SOURCES)

Atmospheric dispersion modeling is an attempt to represent by mathematical or statistical means the observed phenomena of pollutants being diluted and dispersed in the open atmosphere. Although many different models have been tried, with varying degrees of success, the most widely used today are the Gaussian-based models. Various modifications of the basic Gaussian equation have been made to account for such things as elevated inversions, line or area sources, and

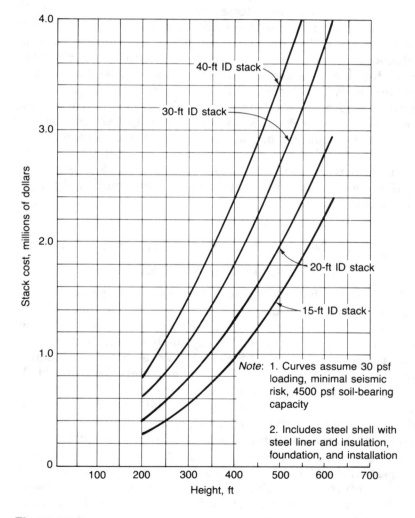

Figure 20.15

Installed costs (in 1977 dollars) for tall steel stacks, with liners and insulation.
(Adapted from Neveril, 1978)

topography. The dispersion parameters were determined empirically over only a limited range of conditions and the Gaussian approach is far from perfect. Nonetheless, it is used because of two main advantages: (1) it is easy to understand and apply, and (2) the mathematics are relatively simple and easily adaptable to computer programs.

One word of caution: As long as you input data in the proper format, a computer program will give you answers. But just because these answers are printed neatly on green-stripe (for those of you old enough to remember green-stripe) paper, you are not necessarily justified in believing them! Be aware of the assumptions and limitations built into the program you are using, and do not use it to try to solve problems for which it was never designed.

Over the past 30 years, the U.S. Environmental Protection Agency has developed many Gaussian-based computer programs, which can be purchased from NTIS (National Technical Information Service) by any interested party. In addition, many models can be downloaded directly from the EPA's electronic bulletin board. These programs are listed by name in Appendix D (several are of historical interest only at this time). All of the programs are written in FORTRAN, all use the Briggs plume rise equation, and all require various sorts of meteorological data.

Note that there are many non-EPA models available. It is always wise to check with state and local agencies before embarking on a large simulation project.

♦ 20.6 MOBILE SOURCES AND ♦ LINE SOURCE MODELS

As discussed in Chapter 18, mobile sources emit large amounts of CO, NO_x, and VOCs. Light-duty vehicles (LDVs) such as automobiles, SUVs, vans, pick-ups, and motorcycles are the largest subgroup of the mobile sources and emit the lion's share of these pollutants. Large trucks and buses (heavy-duty vehicles—HDVs) emit at much higher rates than do LDVs; even though HDVs comprise only about 10% of all vehicles, as a group they emit almost as much as LDVs.

Since emissions of pollutants vary with the type of vehicle, the type of driving, the age of the vehicle, the pollution control equipment, the degree of maintenance, and other factors, one of the problems in the modeling of vehicle emissions, then, is to determine a fleet-averaged emission factor. A fleet-averaged emission factor is calculated for a current or future year, taking into account the age distribution of vehicles, the deterioration in engine and control device efficiency, the driving cycle, and other variables. The EPA computer program MOBILE6 can be used to calculate the emission factor for any particular set of variables.

Once a fleet-averaged emission factor has been obtained, it can be used in a line source dispersion model, assuming that the cars are spread uniformly over the length of roadway being modeled. One model that has been shown to be reasonably accurate is CALINE 3 (Benson 1979). CALINE 3 models a roadway as a series of finite length line sources and allows for winds that are nearly parallel to the road as well as those that are perpendicular. It can handle curves, intersections, and other complex situations. Because the finite length line source (FLLS) approach is much more powerful than the older infinite length line source model, a short discussion of the FLLS method is presented here.

Consider the FLLS shown in Figure 20.16. A uniform average emission rate, q, is defined for the FLLS in units of μg/m-s. The x axis is parallel with the wind and passes through the receptor. The y axis is co-linear with the FLLS, which is perpendicular to the wind. (If the road is not perpendicular, a series of small equivalent FLLS can be drawn that are perpendicular to the wind. This will be demonstrated in the example following this section.)

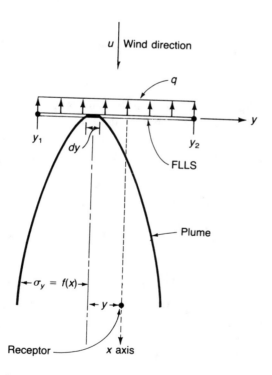

Figure 20.16
The finite length line source (FLLS) model.

The differential concentration at the receptor owing to the differential element of the line (which can be treated as a point source) is given by

$$dC = \frac{q\,dy}{2\pi u \sigma_y \sigma_z} \exp\left(-\frac{y^2}{2\sigma_y^2}\right)\left\{\exp\left[-\frac{(z-H)^2}{2\sigma_z^2}\right]\right.$$
$$\left. + \exp\left[\frac{-(z+H)^2}{2\sigma_z^2}\right]\right\}$$

(20.24)

Note: Because of the increased turbulence caused by cars, the σ_y and σ_z functions must be modified when modeling dispersion from roadways. It is beyond the scope of this text to discuss that modification in detail. For further reading, see Benson (1979).

Since σ_y and σ_z are constant with respect to y, Eq. (20.24) can be integrated over the FLLS and written as follows:

$$C = \frac{K}{2\pi\sigma_y}\int_{y_1}^{y_2} \exp\left(\frac{-y^2}{2\sigma_y^2}\right)dy$$

(20.25)

where

$$K = \frac{q}{u\sigma_z}\left\{\exp\left[\frac{-(z-H)^2}{2\sigma_z^2}\right] + \exp\left[\frac{-(z+H)^2}{2\sigma_z^2}\right]\right\}$$

(20.26)

The reason for leaving the quantity $2\pi\sigma_y$ out of the constant K will become obvious shortly. Define a variable B such that

$$B = \frac{y}{\sigma_y}$$

(20.27)

and substitute into Eq. (20.25) for y. The result is

$$C = \frac{K}{2\pi\sigma_y}\int_{B_1}^{B_2} \exp\left(-\frac{B^2}{2}\right)dB\,\sigma_y$$

(20.28)

Eq. (20.28) can be written as

$$C = \frac{K}{\sqrt{2\pi}}\frac{1}{\sqrt{2\pi}}\int_{B_1}^{B_2} \exp\left(\frac{-B^2}{2}\right)dB$$

(20.29)

or

$$C = \frac{K}{\sqrt{2\pi}}(G_2 - G_1)$$

(20.30)

where G_2, G_1 = the Gaussian distribution function evaluated at B_2 and B_1, respectively.

The Gaussian distribution function is discussed briefly and tabulated in Appendix C. Thus, Eq. (20.30) can be applied to solve the FLLS problem. If a roadway is modeled by several FLLS, then the contributions of all FLLS to the receptor are summed to obtain a total concentration. This method is illustrated in the next example.

▶ ▶ ▶ ▶ ▶ ▶ ▶ ▶ ▶

Example 20.7

Given the geometry of a road segment and receptor shown in Figure 20.17, and the data below, calculate the expected ground-level CO concentration at the receptor. Use the two subsegments shown to get two equivalent finite length line sources, and assume a constant emission rate along each FLLS. CO emission factor = 15 g/km, Class is D, u = 1.0 m/s, $H = 0$ m, and traffic count is 6000 vehicles per hour. Assume that $\sigma_y = 20$ m, and $\sigma_z = 12$ m at $x = 50$ m, and that $\sigma_y = 22$ m, and $\sigma_z = 14$ m at $x = 67.5$ m.

Solution

First, a uniform emission rate on the road is calculated:

$$q = \frac{15 \text{ g}}{\text{km vehicle}} \times \frac{6000 \text{ vehicles}}{\text{hr}} \times \frac{1 \text{ hr}}{3600 \text{ s}}$$
$$\times \frac{1 \text{ km}}{1000 \text{ m}} \times \frac{10^6 \mu\text{g}}{1 \text{ g}}$$
$$q = 25,000 \ \mu\text{g/m-s}$$

Next, equivalent FLLS segments are drawn through the midpoints of the two segments that are perpendicular to the wind direction. From the given geometry, Figure 20.18 is prepared. The length of $FLLS_1$ is 50 m and $FLLS_2$ is $70 \cos 30° = 60.6$ m. The distance d is $70/2 \sin 30° = 17.5$ m. The equivalent emission rate on $FLLS_2$ is

$$q' = 25,000 \times 70 \text{ m}/60.6 \text{ m} = 28,867 \ \mu\text{g/m-s}$$

The coordinates y_1, y_2, and y_3 (relative to the x axis defined by the receptor) are –55.0, –5.0, and +55.6, respectively. The x distances from $FLLS_1$ and $FLLS_2$ to the receptor are 50 m and 67.5 m, respectively. At these distances, σ_{y_1} and σ_{y_2} are 20 and 22 m, respectively, and σ_{z_1} and σ_{z_2} are 12 and 14 m, respectively.

Let us consider $FLLS_1$ now:

$$B_1 = \frac{y_1}{\sigma_{y_1}} = \frac{-55}{20} = -2.75 \qquad B_2 = \frac{-5.0}{20} = -0.25$$

From Eq. (20.26) for $z = 0$ and $H = 0$,

$$K_1 = \frac{25,000}{(1.0)12}(1.0 + 1.0) = 4167$$

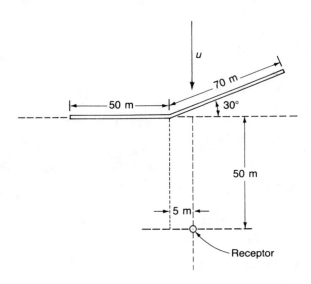

Figure 20.17
Geometry for Example 20.7.

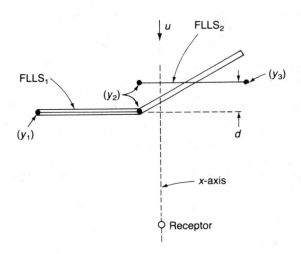

Figure 20.18
Sketch to accompany Example 20.7.

From Eq. (20.30) and the table in Appendix C the concentration at the receptor owing to $FLLS_1$ is

$$C = \frac{4167}{\sqrt{2\pi}}\left[G(-0.25)-G(-2.75)\right] = 1662[0.4013-0.0030] = 662 \ \mu g/m^3$$

Next consider $FLLS_2$:

$$B_1 = \frac{y_2}{\sigma_{y_2}} = \frac{-5.0}{22} = -0.227 \qquad B_2 = \frac{55.6}{22} = 2.53$$

From Eq. (20.26),

$$K_2 = \frac{28,867}{(1)(14)}(1.0+1.0) = 4124$$

From Eq. (20.30),

$$C = \frac{4124}{\sqrt{2\pi}}(0.9943-0.4102) = 961 \ \mu g/m^3$$

The total concentration at the receptor is $662 + 961 = 1623 \ \mu g/m^3$ or 1.42 ppm.

◆ PROBLEMS ◆

20.1 For an emission rate of 200 g/s, an effective stack height of 80 m, Class C stability, and a wind speed at stack height of 8 m/s, calculate the ground-level concentration of a nonreactive pollutant: (a) 1000 m directly downwind; and (b) 5000 m directly downwind.

20.2 Name and briefly discuss the two major factors or influences that result in the formation of atmospheric eddies. Which one predominates in Class A type atmospheric stability?

20.3 On a clear night with a surface wind speed of 2 m/s, 50 g/s of a nonreactive pollutant is released at ground level ($H = 0$). Calculate the ground-level concentration: (a) 500 m directly downwind, and (b) 1000 m directly downwind.

20.4 At noon on a sunny summer day with a surface wind speed of 4 m/s, SO_2 is released over rough terrain from a 90 m stack at a rate of 400 g/s. Assume that the plume rise is 60 m. Calculate the ground-level concentration: (a) 3000 m downwind; (b) 3000 m downwind and 100 m crosswind; and (c) 3000 m downwind and 500 m crosswind.

20.5 The plan and profile views of a pollution source and surroundings are given below. For an emissions rate of 165.0 g/s, Class D stability, rough terrain, and a wind speed of 7 m/s (at 10 m), calculate the ground-level concentration at a receptor (A) that is 2002 m away from the source but not directly downwind (see diagrams). Assume plume rise is 50 m.

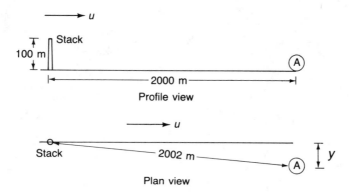

20.6 Rework Problem 20.5 for smooth terrain, a surface wind speed of 3 m/s, and (a) Class B stability; and (b) Class F stability.

20.7 If you measured a 10-min average concentration of 75 µg/m^3 at a certain point downwind from an emitting source, what concentration would you expect for a 20-min averaging time?

20.8 For the data of Problem 20.1, use Figure 20.9 to estimate the maximum downwind ground-level concentration and the distance at which it occurs.

20.9 Check your answer to Problem 20.8 using Eq. (20.7).

20.10 For an effective height of 150 m, an emission rate of 800 g/s, a wind speed of 8 m/s at stack height, and Class D stability, calculate the downwind centerline ground-level concentrations at the following distances: (a) 500 m; (b) 1000 m; (c) 2000 m; (d) 4000 m; (e) 7000 m; (f) 10,000 m; and (g) 15,000 m.

20.11 Use Figure 20.9 to solve Problem 20.10 for C_{max} and x_{max}.

20.12 Develop a spreadsheet to solve Problem 20.10 on a PC.

20.13 For a wind speed of 6 m/s, a stack gas exit velocity of 20 m/s, a stack diameter of 4 m, a heat emission rate of 15 MW, and an ambient pressure of 950 mb, calculate the final plume rise in a neutral atmosphere using (a) the Holland; (b) the modified Concawe; and (c) the Briggs method. Assume the stack gas has a molecular weight of 28.9.

20.14 Rework Problem 20.13 for Class E stability and a wind speed of 3 m/s. Assume the ambient temperature is 280 K.

20.15 For Class B or C stability, the exponents in Eqs. (20.14) and (20.15) are nearly unity. Assume that

$$\sigma_y = ax \text{ and } \sigma_z = cx$$

where a and c are the constants from Table 20.2

Derive expressions for $(Cu/Q)_{max}$ and x_{max} from Eq. (20.1). Hint: Set $y = 0$, substitute the above expressions for σ_y and σ_z into Eq. (20.1), then differentiate with respect to x and set $dC/dx = 0$ to find x_{max} in terms of H. Back substitute to find $(Cu/Q)_{max}$ in terms of a, c, and H.

20.16 The answers to the above problem are $x_{max} = 0.707 \ H/c$ and $(Cu/Q)_{max} = 0.234 \ c/(aH^2)$. Use these relationships to solve Problem 20.8.

20.17 For a steel stack that exhausts 1200 m³/min of gases at 1 atm and 400 K, calculate the inner diameter if you are designing for a maximum expected wind speed at stack height of 12 m/s.

20.18 Estimate the installed cost in 2002 dollars of a 304-stainless steel plate stack that is 20 m tall and 1.5 m in diameter.

20.19 Estimate the installed cost in 2002 dollars of a tall steel stack with liner and insulation. The stack is 500 ft tall and has an inner diameter of 15 ft.

20.20 Using the modified Concawe formula for plume rise, calculate the physical stack height required for a stack that must keep ground-level concentrations below 200 µg/m³ for a source emitting at 1000 g/s with a heat loss rate of 20 MW. Consider only neutral conditions and wind speeds to 20 m/s.

20.21 Design a stack (specify the physical height and diameter) to meet the following conditions: Q_H = 20 MW, total gaseous exhaust rate = 20,000 m³/min at P = 1000 mb and T = 410 K, SO_2 concentration in the stack gases = 2000 ppm, maximum downwind ground-level concentration to be allowed = 500 µg/m³, critical wind speeds (constant at all heights) = 9 m/s for Class D or 5 m/s for Class C, and the maximum wind speed expected = 15 m/s. For simplicity, use the modified Concawe plume rise formula. Assume that Class C or Class D will be the limiting stability.

20.22 Given the following roadway segment and receptor location configuration and other data, calculate the expected ground-level CO concentration at the receptor: CO emission rate = 10,000 µg/m-s; Class is E; u = 2 m/s; and H = 0 m. At x = 30 m, for Class E, assume σ_y = 12 m and σ_z = 8 m. Highway and receptor configuration are as shown on the next page.

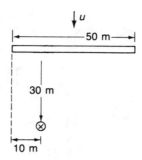

20.23 A proposed source will emit 50 g/s of SO_2 from a 30-m high stack that has an inner diameter of 1.5 m. The stack gas has a velocity of 15 m/s and $Q_H = 0.5$ MW. Determine the critical wind speed for Class D stability (a) using the Holland plume rise equation, and (b) using the modified Concawe method for plume rise.

20.24 It is 12:00 noon on an overcast day in July. The surface wind speed is 3 m/s. What is the stability class?

20.25 Estimate the downwind, centerline ground-level concentration (at a downwind distance of 5000 meters from a stack) of a non-reactive pollutant that is emitted from a 90-meter-tall stack at the rate of 40 g/s. The stack is located in a rural area, surrounded for many miles by open pasture land. The atmospheric stability is Class D, and the surface wind speed is 4.0 m/s. The plume rise is 40 m. Give your answer in $\mu g/m^3$.

20.26 Name three factors that influence the predicted rise of an exhaust plume from a chimney.

REFERENCES

Benson, P. E. *Caline 3—A Versatile Dispersion Model for Predicting Air Pollutant Levels Near Highways and Arterial Streets,* California Department of Transportation FHWA–CA–79–23, Sacramento, CA, November 1979.

Briggs, G. A. *Plume Rise.* AEC Critical Review Series, TID–25075, National Technical Information Service, Washington, D.C., 1969.

————. *Some Recent Analyses of Plume Rise Observations,* a paper presented at the 1970 International Air Pollution Control Conference, Washington, D.C., 1970.

————. "Discussion of Chimney Plumes in Neutral and Stable Surroundings," *Atmospheric Environment,* 6(1), 1972.

————. "Plume Rise Predictions," *Lectures on Air Pollution and Environmental Impact Analyses,* American Meteorological Society, 1975.

Carson, J. E., and Moses, H. "The Validity of Several Plume Rise Formulas," *Journal of the Air Pollution Control Association,* 19(11), 1969.

Gifford, F. A. "Uses of Routine Meteorological Observations for Estimating Atmospheric Dispersion," *Nuclear Safety,* 2(4), 1961.

Gifford, F. A. "Atmospheric Dispersion Models for Environmental Pollution Applications," Chapter 2 in *Lectures on Air Pollution and Environmental Impact Analysis*. Boston: American Meteorological Society, 1975.

Hino, M. "Maximum Ground-Level Concentration and Sampling Time," *Atmospheric Environment, 2*(3), 1968.

Holland, J. Z. *A Meteorological Survey of the Oak Ridge Area*, Report ORO–99. Washington, DC: Atomic Energy Commission, 1953.

Martin, D. O. "The Change of Concentration Standard Deviation with Distance," *Journal of the Air Pollution Control Association, 26*(2), 1976.

Neveril, R. B. *Capital and Operating Costs of Selected Air Pollution Control Systems*, EPA–450/5–80–002. Research Triangle Park, NC: U.S. Environmental Protection Agency, December 1978.

Neveril, R. B., Price, J. U., and Engdahl, K. L. "Capital and Operating Costs of Selected Air Pollution Control Systems—II," *Journal of the Air Pollution Control Association, 28*(9), 1978.

Nonhebel, G. *Journal of the Institute of Fuel, 33*(4), 1960.

Pasquill, F. "The Estimation of the Dispersion of Windborne Material," *Meteorological Magazine, 90*(1063), 1961.

Ranchoux, R. J. P. "Determination of Maximum Ground Level Concentration," *Journal of the Air Pollution Control Association, 26*(11), 1976.

Schnelle, K. B., Jr. *The Engineer's Guide to Air Pollution Meteorology*, Lecture notes for an AICHE Advanced Seminar, American Institute of Chemical Engineers, New York, 1976.

Thomas, F. W., Carpenter, S. G., and Colbaugh, W. C. "Plume Rise Estimates for Electric Generating Stations," *Journal of the Air Pollution Control Association, 20*(3), 1970.

Turner, D. B. *Workbook of Atmospheric Dispersion Estimates*. Washington, DC: U.S. Environmental Protection Agency, 1970.

U.S. Environmental Protection Agency. *User's Guide for the Industrial Source Complex (ISC 3) Dispersion Models*. Vol. II, *Description of Model Algorithms*. EPA-454/B-95-003b. Research Triangle Park, NC, 1995.

Vatavuk, W. M. "Hoods, Ductwork, and Stacks," Chapter 10 in *OAQPS Control Cost Manual* (5th ed.). EPA 453/B-96-001. Research Triangle Park, NC, February, 1996.

Vatavuk, W. M., and Neveril, R. B. "Part VIII: Estimating Cost of Exhaust Stacks," *Chemical Engineering*, June 15, 1981.

Wark, K., and Warner, C. F. *Air Pollution, Its Origin and Control* (2nd ed.). New York: Harper & Row, 1981.

Williamson, S. J. *Fundamentals of Air Pollution*. Reading, MA: Addison-Wesley, 1973.

CHAPTER 21

INDOOR AIR QUALITY AND CONTROL

*Will we see a time when nonindustrial IAQ specifications are writ-
ten into codes or regulations? . . . upper limits for key chemical species or total VOC in indoor air will someday likely be in effect. It
would be wise to prepare for that time by learning as much as we
can about the health effects of low-level VOC mixtures and about
how to improve indoor air quality.*

Franklin D. Aldrich, M.D., Ph.D., 1992

♦ 21.1 INTRODUCTION ♦

Up to this point, we have focused on ambient (outdoor) air quality.
However, in the twenty-first century, a very large percentage of the
population (especially in developed countries) lives and works indoors:
at their homes, in factories, in commercial office buildings, in stores,
in airports, in hotels, or inside vehicles going to or from those places.
All things considered, many people may be indoors more than 20
hours per day on average! The quality of the indoor air they breathe
can have a direct impact on their short-term health and long-term
well being.

Indoor air quality (IAQ) is the term that describes the focus of
many environmental engineers and industrial hygienists; it is a professional area that has developed in recent years in response to the
emerging recognition of the importance of the indoor environment.
The possibility of exposure to poor quality air inside buildings is not
limited only to industrial workers. Indeed, there have been a number
of instances of office buildings or hotels with *sick building syndrome*
(SBS) in the news in recent years, resulting in a variety of lawsuits
against building owners and operators. The U.S. EPA has estimated

that poor IAQ costs the United States over $1 billion each year through direct medical costs, and another $60 billion in lost productivity (Hays, Gobbel, and Ganick 1995).

A huge amount of energy is used each year for home and office heating, ventilation, and air conditioning (HVAC); water heating; operating computers, office equipment, and home appliances; and for other needs in commercial and residential buildings. This energy use accounts for about 40% of total U.S. energy consumption, an even greater portion than transportation (Kay, Keller, and Miller 1991). As a result of the U.S. energy crisis of 1973–1974, significant efforts have been made to "tighten" buildings and make them more energy efficient. To tighten a building means to reduce the inflow (and thus the discharge) of air. This saves energy by not having to heat or cool a large volume of air that, in effect, simply has passed through the building on its way to being expelled to the outside atmosphere. While such tightening has been quite effective at conserving energy, it has not been good for indoor air quality. With much less outside air flowing through the building, concentrations of pollutants have increased indoors, sometimes with serious health effects to occupants.

Indoor air is considered different from ambient or outdoor air for two important reasons. The first is legal—indoor air generally is considered private property, and therefore is not subject to the same federal regulations as ambient air. The second is technical—although the "same" air is exchanged between indoors and outdoors, the pollutants, their sources, and their dispersion behavior can be quite different indoors than in the outdoor atmosphere. In addition, there may be removal or significant generation of pollutants inside many buildings. So, in fact, for some pollutants, the concentrations indoors may be lower than outdoors, while for others, the concentrations indoors may be significantly higher than in the ambient air.

Common indoor air pollutants include the familiar criteria pollutants (PM-2.5, ozone, NO_2, SO_2, and CO), as well as a wide variety of VOCs. Other important examples of indoor air pollution include gases and dusts given off during industrial processing, oils and smoke from cooking, tobacco smoke, radon, formaldehyde, molds, pollens, and animal dander. VOCs are evaporated from cleaners, solvents, paints, perfumes, soaps, and hair sprays; they are released from office equipment (such as copiers and fax machines) and from materials (such as desks, chairs, drapes, and carpets). VOCs are produced as respiration byproducts of molds, and can even be produced by people as part of their normal metabolic functions.

The problems associated with industrial work environments can be quite different from those in office buildings or homes. The concentrations of pollutants to which people are exposed can vary dramatically based on types and sources of pollutants, emission rates, specific locations of sources, age of the building (note that newer isn't always

better), design and construction of the building, design and maintenance of the ventilation system, and whether air pollution control
techniques are being used.

This chapter presents information on several of the most significant indoor air pollutants—where they come from, how they can accumulate in indoor air, and how we can predict and control IAQ. For the
most part, this chapter focuses on homes and commercial buildings
rather than industrial factories, since industrial workplace air is
already regulated, monitored, and controlled.

◆ 21.2 SOME POLLUTANTS OF CONCERN ◆

When dealing with indoor air contaminants, the National Ambient
Air Quality Standards do not automatically apply, since they were
developed for outdoor exposures of the general public. The National
Institute for Occupational Safety and Health (NIOSH) has established
permissible exposure limits (PELs) for a number of indoor contaminants; PELs specify the maximum recommended concentrations for
workplace exposure (eight hours per day, five days per week). Similarly, the American Congress of Government and Industrial Hygienists (ACGIH) has developed *time weighted average threshold limit
values* (TWA-TLVs) that list guideline concentrations for the workplace. Often the PELs and the TWA-TLVs are similar, and usually
they are greater than the NAAQSs (see Table 21.1). This is because
the standards apply to the workplace, and traditionally it has been
assumed that workplace air is breathed only by healthy adult workers
who can tolerate higher concentrations. In modern society, however,
we are beginning to realize that indoor air quality in retail and commercial buildings applies to the entire population, and so the PELs
and TWA-TLVs may be just starting points.

There are a number of pollutants found in indoor air. Generally,
they can be classified into five broad groups—VOCs, inorganic gases,
particles, radon, and biological contaminants. While more than 900
individual contaminants have been found (Hays et al. 1995), a few are
much more prevalent and/or pose a much greater risk than the others.
In this section, we will briefly review a few of the most important
indoor air pollutants. Table 21.2 presents some examples of indoor air
pollutants and lists common sources.

Table 21.1   Comparison of Selected PELs and NAAQSs for Three
Criteria Pollutants

Pollutant	8-hr PEL	NAAQS
O_3	0.1 ppm	0.12 ppm (1-hr), 0.08 ppm (8-hr)
NO_2	5.0 ppm	0.053 ppm (annual)
CO	50 ppm	35 ppm (1-hr), 9 ppm (8-hr)

VOCs (INCLUDING FORMALDEHYDE)

Numerous VOCs have been detected indoors. These pollutants have a multitude of sources (see Table 21.3). Human beings themselves are directly and indirectly responsible for a variety of VOCs (from clothing, deodorants, hair spray, perfumes, biological functions, etc.), as shown in Table 21.4.

Formaldehyde (HCHO) is a VOC and is one of the most common and yet most serious pollutants found inside buildings. Formaldehyde is emitted from a variety of laminates and glues, some types of foam

Table 21.2   Examples of Indoor Air Pollutants and Their Sources

Pollutant	Sources
VOCs	Paints, thinners, perfumes, hair sprays, furniture polish, cleaning solvents, carpet dyes, glues, dry cleaned clothing, air fresheners, candles, soaps, bath oils, molds, tobacco smoke
Formaldehyde	Particle board, plywood, veneers, glues, carpets, insulation, fuel combustion, tobacco smoke
CO, CO_2, NO_x	Gas-fired stoves and ovens, candles, fireplaces, woodstoves, kerosene space heaters, human respiration, tobacco smoke
Particles	Fireplaces, woodstoves, candles, tobacco smoke
Radon	Release from underlying soil, release from water use
Biological contaminants	Pets, house plants, insects, molds (mildews and fungi), humans, pillows, bedding, wet materials, HVAC systems, humidifiers

Table 21.3   Examples of VOCs and Their Sources Indoors

Pollutant	Sources
Formaldehyde	Carpets, plywood, particle board, laminated furniture, insulation, glue, fuel combustion, tobacco smoke
Mono-, di-, and tri-chloromethane	Solvents, aerosol sprays, hot showers[1]
Carbon tetrachloride	Ink pens, photo equipment, tapes, solvents, rubber products
Styrene	Photocopiers, plastics, synthetic rubber products, resins
Trimethylbenzene	Wallpaper, needle felt, adhesives, floor varnish
Hexane	Fuel, aerosol propellants, perfume, cleaners, paint, solvents
Methanol	Window cleaner, paints and thinners, cosmetics, adhesives, human breath

[1] Chlorine treatment of public water supplies often results in the delayed formation of chlorinated hydrocarbons within the water supply pipelines.

insulation, and from newly installed materials such as carpets, plywood paneling, and particleboard shelves and furniture. Because it is emitted from so many building materials and furnishings, it is more often a problem in new office buildings or newly constructed homes (especially manufactured or "mobile" homes). Table 21.5 displays some emission rates for HCHO.

Formaldehyde also is a relatively stable product of incomplete combustion of any carbonaceous fuel, so sources of formaldehyde include woodstoves and fireplaces, natural gas stoves and water heaters, or oil or coal furnaces with slight leaks. Cigarettes also emit formaldehyde.

The principal human health effect of formaldehyde in the concentrations that may be found indoors is eye irritation, which occurs at concentrations of 0.01 to 2.0 ppm. If the concentrations build up to 5 to

Table 21.4   Indoor Air Pollutants from Human Beings[1]

Chemical	Concentration in Air[2] ppb	Estimated Emission Rate mg/hour per person
Acetone	20.6	2.1
Acetaldehyde	4.2	0.26
Acetic acid	9.9	0.83
Amyl alcohol	7.6	0.91
Butyric acid	15.1	1.9
Diethyl ketone	5.7	0.87
Methanol	54.8	3.1
Phenol	4.6	0.40
Ammonia		1.3
Carbon monoxide		48,400
Carbon dioxide		642,000

[1] Adapted from Wang (1975) as presented in Hines et al. (1993).
[2] Concentrations measured in a large lecture hall with 389 people in class.

Table 21.5   Emission Rates of Formaldehyde

Material or Activity	Emission Rate, $\mu g/m^2$-day (or as stated)
Fiberboard	17,600–55,000
Hardwood plywood paneling	1500–36,000
Particleboard	2000–25,000
Urea formaldehyde foam insulation	1200–19,200
Softwood plywood	240–720
Paper products	260–680
Carpeting	0–65
Smoking	1,300 μg/cigarette (sidestream smoke)
Portable kerosene heater	660–4600 μg/hr
Gas stove or oven	8000–28,000 μg/hr

Adapted from Hines et al., 1993.

30 ppm, irritation of the lower airway and lungs will occur. Formaldehyde concentrations inside homes and offices range from 0.02 to 0.3 ppm, although in some new mobile homes, HCHO has been reported as high as 1 to 2 ppm (Godish 1985). Furthermore, formaldehyde concentrations have been shown to increase substantially with increasing indoor air temperature, and/or increasing inside air humidity (most likely because these parameters affect the off-gassing rate for HCHO from furniture, carpets, wallboard, and other items).

INORGANIC GASES—COMBUSTION PRODUCTS

Two criteria pollutants that come from combustion sources are CO and NO_x. High concentrations can be reached indoors if there is little or no ventilation for combustion products. Common sources of CO and NO_x include gas stoves, woodstoves, kerosene heaters, and fireplaces. Some of the highest "indoor" CO concentrations are routinely encountered in automobiles, trucks, and buses stuck in rush-hour traffic. Commercial vehicle drivers, for example, can be exposed to CO concentrations that significantly exceed the 8-hr NAAQS. Also, if the outside air intake for a commercial building happens to be located at street level or next to a parking garage, CO and NO_x can increase markedly inside the building.

The human health effects of CO were discussed in Chapter 1, but are briefly reviewed here. Since CO is readily absorbed by hemoglobin in preference to oxygen, we include mention of the level of carboxyhemoglobin (COHb) that results from CO exposure. Normal metabolic processes produce a COHb level in the blood of 0.5 to 1.0%. Continuous exposure to 20 ppm CO in air leads to a COHb level of about 3% within 4 to 5 hours; this level of COHb is about the lowest level at which effects become statistically significant in healthy persons. At concentrations of 5–17% COHb, effects begin to manifest themselves as decreases in visual perception and manual and mental dexterity (such as decreased performance in school or in operating a vehicle). It should be noted that some people die each year from CO poisoning when CO accumulates indoors due to the improper operation of a kerosene space heater, faulty furnace, a car left running in a closed garage, or even a charcoal grill inside a poorly ventilated room or other space. The CO exposure must exceed several thousand ppm to cause death.

The data available on health effects of NO and NO_2 are somewhat limited, but NO can act like CO in interfering with oxygen uptake by the blood. It has been stated that exposure to NO at 3 ppm is comparable to exposure to CO at 10–15 ppm (Hays et al. 1995). Nitrogen dioxide is corrosive and is an irritant to the deep lungs; it hydrolyzes to nitric acid when absorbed in water. Even at concentrations as low as 0.5 ppm, there are effects on people with asthma (Hays et al. 1995).

RESPIRABLE PARTICLES (INCLUDING TOBACCO SMOKE)

"Smoke" is a visible indicator of incomplete combustion, and generally includes fine particulate matter (PM-2.5) as well as a variety of gases. (Keep in mind that even if a combustion source does not emit visible smoke, many different pollutants may still be present in the exhaust gases.) Smoke from cigarettes, cigars, and pipes contains a number of pollutants including CO, ammonia, hydrogen cyanide, and formaldehyde (all gases), and tars, nicotine, benzo-a-pyrene, certain heavy metals (such as cadmium), phenols, and fluoranthene (all as PM-2.5). Many of these substances are known or suspected carcinogens. Smoking has been blamed for more than 10 million deaths in the last 30 years. While smokers voluntarily choose to pollute their lungs, non-smokers who share the same indoor space with smokers have no choice. Smoke, in addition to circulating throughout the air within the home, restaurant, office, or shopping area, can deposit on surfaces (such as curtains), and the smell can linger for a long time. Fortunately, the U.S. government's public relations campaign against smoking appears to be successful, and the percentage of smokers in the population is declining.

Combustion sources such as candles, woodstoves, and fireplaces also emit a variety of pollutants, many of them hazardous. Wood combustion often releases polycyclic aromatic hydrocarbons (PAHs) and trace metals (depending on the source of the wood burned). The term PAHs covers a large number of compounds that have two or more benzene rings and are carcinogenic. Other PM found indoors may include dusts and pollens that are brought in with ambient air, asbestos fibers, animal dander, mold spores, mites, cotton fibers, insect parts, hair, house dust, pesticides that come from the drying and dusting of indoor applications of bug sprays, and lead from lead-based paints. Problems associated with this noncombustive particulate matter are very site specific; furthermore, the effects can vary widely depending on the sensitivity of the people exposed.

RADON

Radon gas is a radioactive decay product of radium (which is itself a decay product of uranium) and is found in rocks and soils in many parts of the country. Although radon is not chemically active, its decay products (polonium, lead, and bismuth) are, and these elements can lodge in the lungs and emit alpha particles over time. Radon gas migrates upward through porous soils, and relatively large amounts of the gas can enter homes and other buildings if they are in its migration pathway. The gas enters through paths of least resistance (such as cracks in the concrete slab under a house, through semi-porous basement walls or flooring, or loose joints between floors and walls). The U.S. EPA maintains a map of the United States showing areas of high radon potential—go to the web address: www.epa.gov/iaq/radon and click on the link labeled "radon map."

The concentration of radon is measured with a Geiger counter and is reported as pico-Curies per cubic meter (pCi/m^3). A Curie (Ci) is much too large a unit of radioactivity to be useful in IAQ work, so we use the pCi (10^{-12} Ci). The soil flux of radon varies widely depending on location in the United States; it has been reported between 0.1 and 100 pCi/m^2-s, with 1.0 being most typical.

The principal health effect of radon is lung cancer caused by the radioactivity of radon itself or its daughter products (which when adsorbed onto small particles and settled in the lungs can emit radiation in a very vulnerable part of the body). Many homes have indoor air concentrations in the range of 1500 pCi/m^3 or 1.5 pCi/L. At this level, the risk of lung cancer is about equal to that obtained by receiving 75 chest X-rays per year. The U.S. EPA recommends remedial action to reduce radon levels in homes if the radon concentration exceeds 4 pCi/L. It has been estimated that radon is responsible for about 7,000 to 30,000 lung cancer deaths per year (about 5–20% of all lung cancer deaths) in the United States (U.S. EPA, 1997).

BIOLOGICAL CONTAMINANTS
(INCLUDING MOLDS, MILDEWS, AND ALLERGENS)

Biological-based indoor air pollutants include molds, mildews, allergens, viruses, bacteria, protozoa, plant pollens, pet dander, low molecular weight proteins from pet urine, and tiny body parts and feces from insects (Brooks and Davis 1992). In buildings with central air conditioning, the inside walls of the ductwork (especially in the first ten feet or so where the dew point is reached and the humidity is close to 100%) can become a breeding ground for mold, mildew, and bacteria. This problem can be especially bad in warm, humid climates, and has been exacerbated by energy-saving trends in building design and construction over the last thirty years. This statement is explained further in the next paragraph.

During the cooling of humid air, water is condensed. In old methods of air conditioning, the air was cooled much more than was needed for comfort, thus removing most of the water. The air was then reheated to the desired temperature. In modern practice, both the excess cooling and the reheating steps are gone, thus saving much energy. However, the cooled air that is sent into the building contains more moisture than in the past, and often moistens the inside of the duct walls, thus providing a perfect breeding ground for mold, fungi, bacteria, and other microorganisms.

Another energy-saving change employed in modern HVAC practice is to use a higher percentage of recirculated air (and a lower percentage of outside air). This change, along with modern cooling methods, has created IAQ problems and/or has made existing problems worse. In numerous investigations of "sick buildings," inspections of HVAC systems have found millions of fungal spores per square inch on the inside walls of ducts leading from the air conditioners in homes and commercial buildings. People may be allergic to molds, their spores, and/or their gaseous metabolic products.

One of the most infamous cases of IAQ problems occurred during the American Legion convention in Philadelphia in 1976, where 182 people became sick and 29 died from a pneumonia-like illness. The cause was a bacterium that was discovered to be living in the cooling-tower water of the air conditioning system for the hotel in which many of the visitors were staying. The bacterium later was named after the affected group, *Legionella pneumophila,* and the illness became known as Legionnaire's disease. The *Legionella pneumophila* bacterium is a major cause of respiratory illness, being responsible for 1–13% of all pneumonia cases seen in hospitals (Hays et al. 1995). These bacteria can survive outside the body in water for up to one year, and sources of *legionella* in indoor air have been found in hot tubs, humidifiers and vaporizers, hot water heaters, fire sprinkler systems, water fountains, and the cooling towers of many buildings.

Allergens is the name given to the large group of diverse substances that cause allergic reactions in some people. An *allergic reaction* is an immune response that is greatly exaggerated or inappropriate, and that can cause physical harm to the host. Allergens found indoors include the viable group: bacteria, fungi, mold spores, pollen, and algae, and the nonviable group: house dust, insect body parts, animal dander, dead fungal and bacterial cells, and dust mite and cockroach feces. Illnesses caused by airborne allergens include, among others, allergic rhinitis (acute inflammation of the nasal membranes, itching and watering eyes, and sneezing), allergic asthma (wheezing, shortness of breath, sneezing), and hypersensitivity pneumonitis (effects similar to those of influenza).

◆ 21.3 SOURCE CONTROL AND VENTILATION ◆

According to the U.S. EPA (1997), controlling indoor air quality involves integrating three main strategies:

1. Control the sources of pollutants within the indoor space,
2. dilute pollutants and remove them from the building through ventilation, or
3. clean the air by capturing or destroying pollutants in the recirculated air and/or the make-up air.

SOURCE CONTROL

Taking steps to reduce emission rates within the indoor space is the first strategy to consider for improving IAQ. Controlling indoor pollutants includes removing polluting materials from the building, reducing the activities that generate the pollutants, or isolating the sources from people. Such isolation can be accomplished through physical barriers, by hoods, air curtains, or other air pressurization techniques, or by controlling the timing of the source activities (such as cleaning during the night shift). Obviously, it is important to consider source control in the design and construction stage, but source control is crucial in the remediation of a sick building. Selecting appropriate building materials, reducing sources of indoor combustion, and removing sources of allergens are all appropriate and effective methods for reducing concentrations of indoor air pollution. Cleaning ducts, removing contaminated wallpaper, carpets, and so on, or controlling the growth of microorganisms in cooling towers can be extremely important in gaining control of a sick building. However, source control alone may not be enough to ensure good IAQ.

VENTILATION

Ventilation is defined as outside air coming in to replace inside air that is being exhausted to the outside. This process is called simply *air exchange*, and is one of the most important factors in indoor air quality. The American Society of Heating, Refrigeration and Air Conditioning Engineers (ASHRAE) publishes guidelines for building designers as to the minimum ventilation rates recommended for human comfort in different types of indoor space (see ASHRAE 1999). Some examples are 15 cfm per person (cfmpp) for barbershops or coin-operated laundries, 20 cfmpp for fast-food restaurants or office space, and 30 cfmpp for bars and gambling casinos. These recommended ventilation rates have changed over the years. For example, one guideline was 10 cfmpp in 1950, dropped to 5 cfmpp in 1975, and was raised to 15 cfmpp in 1989 (Hines et al. 1993). Increasing the ventilation rate is usually a recommended solution for SBS, and can be effective in diluting the concentrations of pollutants that are being generated indoors. However, increased ventilation could actually increase the indoor concentrations of certain pollutants if their ambient air concentrations are higher than they are indoors (one example might be CO if the building is adjacent to a parking garage).

Air exchange occurs in buildings in three ways: forced ventilation, natural ventilation, and infiltration. **Forced ventilation** uses fans or blowers to forcibly exchange the air, while **natural ventilation** permits natural air exchange through open windows or doors. **Infiltration** refers to the air exchange that occurs even when all windows and doors are closed. Air can leak into buildings through numerous gaps and openings in the building envelope such as the cracks around doors and windows, the gaps around pipes and electrical conduits, kitchen and bathroom vent pipes, floor-wall joints, mortar joints, and others. Ventilation and infiltration are depicted schematically in Figure 21.1.

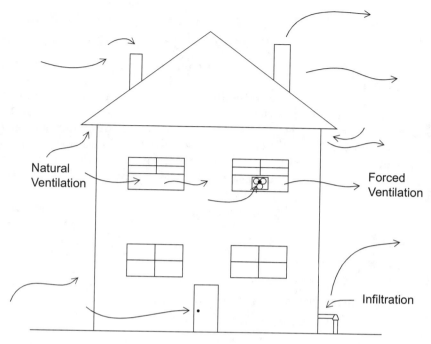

Figure 21.1
Ventilation and infiltration.

Outside air that enters a building must be heated or cooled to keep the inside conditions comfortable, and this energy use costs money. In the past thirty years, much effort has been devoted to reducing infiltration, and thus reducing energy costs. Even with tighter building designs, infiltration is still large and accounts for significant "excess" energy use. By some estimates, infiltration accounts for up to 10% of total U.S. energy use and costs about $10 billion per year (Masters 1998).

AIR CLEANING

Because of increasing energy costs, the trend toward tighter buildings continues. At the same time, increasing liability costs associated with litigation over SBS have prompted building owners to place more emphasis on indoor air pollution control. Air cleaning can be effective both for outside air (for example, if there is a lot of pollen in the air), and for recirculated air (to remove pollutants generated indoors). The major air-cleaning strategies reflect principles already presented in this book: the most popular approaches for PM control are filtration or electrostatic precipitation, and for VOC control, adsorption (activated carbon, molecular sieves, or silica gel) or catalytic oxidation are used.

◆ 21.4 MATERIAL BALANCE MODELS ◆ FOR INDOOR AIR QUALITY

The indoor concentrations of air contaminants can be predicted by simple mathematical models; the key variables are emission rates and ventilation rates. In some buildings, the airflow is very simple, and we can assume the whole building acts like a single, well-mixed room. For other situations, we might have to model the building as many such rooms connected in series and parallel. For the simple model of one well-mixed room (a box with air entering and leaving), the material balance equation is called a *box model*, and is represented as follows:

$$\frac{V dC_i}{dt} = QC_o + S - QC_i - kC_i V \qquad (21.1)$$

where

V = volume of the room, m^3

C_i = indoor concentration of the pollutant, $\mu g/m^3$

C_o = concentration of the pollutant in the outside air, $\mu g/m^3$

Q = ventilation rate, m^3/hr

S = source emission rate inside the room, $\mu g/hr$

k = removal reaction rate constant (here assumed to be first order), hr^{-1}

Equation (21.1) can be rewritten as

$$\frac{dC_i}{dt} + \left(\frac{Q}{V} + k\right)C_i = \frac{Q}{V}C_o + \frac{S}{V} \qquad (21.2)$$

The quantity $(Q/V + k)^{-1}$ is the characteristic time or time constant for this system, and is given the symbol τ. The general solution to Eq. (21.2) is:

$$C_i = \left\{C_0 - \tau\left(\frac{Q}{V}C_o + \frac{S}{V}\right)\right\}\{\exp[-t/\tau]\} + \tau\left(\frac{Q}{V}C_o + \frac{S}{V}\right) \qquad (21.3)$$

where C_0 = initial concentration in the room at time zero

The quantity Q/V is called the air exchange rate (air exchange was mentioned earlier—this quantifies it), and has units of hr^{-1}, but can be thought of as the number of room volumes of air exchanged per hour. It is a common measure of the degree of ventilation for a building or room. The solution to the steady-state case is just the last term of Eq. (21.3):

$$C_{i\,ss} = \tau\left(\frac{Q}{V}C_o + \frac{S}{V}\right) \qquad (21.4)$$

which can be written as:

$$C_{i\,ss} = \frac{\left(\dfrac{Q}{V} C_o + \dfrac{S}{V} \right)}{\dfrac{Q}{V} + k}$$
(21.5)

or

$$C_{i\,ss} = \left(\frac{AC_o + S/V}{A + k} \right)$$
(21.6)

where $A = Q/V$, air changes per hour

By inspecting Eq. (21.6), it can be seen that, as the air exchange rate increases, the indoor concentration approaches the outdoor concentration, regardless of the source emissions rate or the destruction rate by chemical reaction.

For buildings with forced ventilation, which include most modern office buildings and hotels, the make-up air is large compared with the infiltrated air. The make-up airflow rate is selected by architects and engineers based on ASHRAE guidelines or comparable standards during the design of the building. Once the make-up airflow rate is known, the make-up (outside) air fans can be sized, and the air conditioners and duct system can be designed. The ASHRAE guidelines (which often are adopted as building codes or standards) provide estimated occupancy rates for various types of rooms and the functions being held in them, and provide the minimum make-up air required. For example, the outdoor air required for (1) a restaurant dining area is 20 cfmpp, (2) a cocktail lounge is 30 cfmpp, and (3) a smoking lounge is 60 cfmpp (ASHRAE 1999).

For buildings without forced ventilation systems, the air exchange is entirely by natural ventilation and infiltration. This method of air exchange applies to buildings without central air conditioning, which includes many homes in northern climates. Of course, during the winter, these homeowners close the windows, so air exchange is greatly reduced. Table 21.6 presents some guidelines for estimating air infiltration rates into homes with windows closed.

Table 21.6   Air Infiltration Rates into Homes with Windows Closed

Layout of room	Air exchange rate, ach
No windows or exterior doors	0.5
Windows or exterior door on one wall	1.0
Windows or exterior doors on two walls	1.5
Windows or exterior doors on three walls	2.0

Adapted from de Nevers, 1995.

▶ ▶ ▶ ▶ ▶ ▶ ▶ ▶ ▶

Example 21.1

A posh country club has just had its sitting room paneled with new hardwood plywood paneling. The paneling emits formaldehyde at an emission rate of 20,000 $\mu g/m^2$-day; and 900 ft^2 of wall space is covered. Formaldehyde decays to carbon dioxide with a first-order rate constant of 0.40 per hour. The room measures 25 ft long by 20 ft wide by 10 ft high. The average ventilation rate is 1.5 air changes per hour, and the outdoor concentration is zero.

a. Assuming that the club opens the room to members immediately after the paneling is installed, what is the maximum concentration of formaldehyde to which people are exposed?

b. Assuming the initial HCHO concentration in the room is zero, how long does it take to reach 95% of the steady-state (maximum) concentration?

c. What do you suggest be done?

Solution

(a) At steady state, Eq. (21.6) applies, namely:

$$C_{i\,ss} = \left(\frac{AC_o + S/V}{A + k} \right)$$

The emissions rate is:

$S = 900 \; ft^2 \times 20.0 \; mg/day\text{-}m^2 \times 1 \; m^2/10.76 \; ft^2 \times 1 \; day/24 \; hr$
$= 69.7 \; mg/hr$ of formaldehyde emissions

The room volume is 5000 ft^3, and since C_o is zero, then

$$C_{i\,ss} = \frac{S/V}{A+k}$$

$$= \frac{69.7/5000}{1.5+0.4}$$

$$= 0.00734 \; mg/ft^3 \times 35.32 \; ft^3/m^3$$

$$= 0.259 \; mg/m^3$$

This concentration converts to 0.21 ppm, more than enough to cause serious eye and respiratory system irritation to the club members (and they thought it was just all those cigars they were smoking!).

(b) To estimate the time to reach 95% of the steady-state concentration, we must know how the emission rate function behaves. A simple approach is to assume that the emissions jump up immediately to a fixed level (a so-called step increase). For $C_0 = C_o$ = zero, Eq. (21.3) reduces to:

$$C_i = \frac{S/V}{A+k}(1-\exp[-t/\tau])$$
$$= C_{i\,ss}(1-\exp[-t/\tau])$$

Recall that

$$\tau = 1/(1.5+0.4) = 0.526 \text{ hr}$$

Now, set $C_i/C_{i\,ss} = 0.95$, and solve for t.

$$C_i/C_{i\,ss} = 0.95 = 1-\exp(-t/0.526)$$
$$\ln(1-0.95) = -t/0.526$$
$$t = 0.526 \times 2.996$$
$$t = 1.58 \text{ hours}$$

(c) The simplest and perhaps best thing to do in this case is to close off the room from the rest of the building and ventilate it with plenty of outside air until the new paneling "cures"—that is, until the emission rate of formaldehyde drops to a much lower level.

As pointed out by de Nevers (1995), an indoor air quality model can be made to be more realistic (and more complicated) than that described in Eq. (21.1). Consider the diagram in Figure 21.2 that schematically shows a building that experiences infiltration, forced ventilation with treatment, and recirculated air with separate treatment of that airstream. It is assumed that the air within the building is completely mixed. This model is still simplified compared to a real building in which there may be many separate rooms, with air flowing simultaneously from some rooms and into others, and with different emission rates and different concentrations in many of the rooms.

The material balance equation describing the processes shown in Figure 21.2 is

$$V\frac{dC_i}{dt} = Q_1 C_o + Q_3 C_o(1-\eta_1) + Q_4 C_i(1-\eta_2) - (Q_2 + Q_4 + Q_5)C_i + S - R$$

(21.7)

where all terms were defined earlier or are identified in Figure 21.2.

Note that η_1 and η_2 in Eq. (21.7) are the removal efficiencies of the two "filters" shown in that figure. Furthermore, note that if the term "filter" is used broadly, it can refer not only to a particulate filter but also to a carbon adsorber or catalytic oxidizer, so that a variety of pollutants can be removed by these "filters."

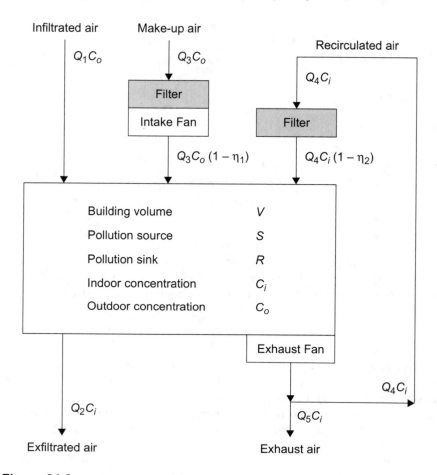

Figure 21.2
Schematic diagram for indoor air quality model.
(after de Nevers, 1995).

Numerical methods are often the only practical way to solve complicated equations such as 21.7. With the widespread availability of personal computers, engineers are expected to be able to solve such problems relatively quickly using a simple approach in a spreadsheet. Consider the next example problem.

▶ ▶ ▶ ▶ ▶ ▶ ▶ ▶ ▶

Example 21.2

A homeowner operates a gas stove 45 minutes each day; the stove consumes 12.0 cubic feet of natural gas per hour while operating (each cubic foot of gas has an energy content of about 1000 kJ). The emission factor for NO_2 for this kind of stove is 15 µg/kJ. The kitchen has an infiltration rate of 4000 ft³/hr, and the ambient air has a concentration of NO_2 of 40 µg/m³. The kitchen has dimensions of 15 ft by 20 ft by 8 ft tall. The house is not air conditioned, so there is no recirculation of air. The stove has a separate hood and vent fan with an exhaust rate of 500 ft³/hr (when the homeowner remembers to turn it on), and it can remove 80% of the emissions from the stove before those gases mix into the rest of the air. Note that when the fan is on, the infiltration rate increases to a total of 4500 cubic feet/hour. The reaction rate of NO_2 is effectively zero, and there are no other sources or sinks of NO_2 in the room.

Prepare a graph of the NO_2 concentration in the kitchen versus time. Follow the concentration for three hours starting from when the stove is first turned on. Plot two lines—one with and one without the vent fan running. Solve this problem using a spreadsheet.

Solution

First, calculate the NO_2 emission rate with the stove operating:

$$\frac{12.0 \text{ ft}^3 \text{ gas}}{\text{hr}} \times \frac{1 \text{ hr}}{60 \text{ min}} \times \frac{1000 \text{ kJ}}{\text{ft}^3} \times \frac{15 \text{ µg}}{\text{kJ}} = 3000 \frac{\text{µg NO}_2}{\text{min}}$$

To start the numerical solution, we start by rewriting the material balance (Equation 21.7) as a finite difference equation, eliminating those terms that do not apply:

$$V \frac{\Delta C_i}{\Delta t} = Q_1 C_o + S - Q_2 C_i - Q_5 C_i$$

Defining Δt as 1 minute, and setting the initial (time = 0) concentration inside the room to the ambient concentration ($C_{i0} = C_o$), we set up a spreadsheet as shown in Figure 21.3 on page 666. Then simply copy down any number of rows and watch the results until the calculated C_i approaches a constant value or until the 3-hour limit is reached. Remember to turn off the stove after 45 minutes (S becomes zero at time = 45). Also, for the case with the fan on, we model the source term as being reduced to 600 µg/m³ (20% of its unventilated rate gets into the kitchen). See Figure 21.4 on page 667 for the graph of NO_2 concentration versus time.

Rows		Columns		
3	**B**	**C**	**D**	**E**

4 Example Problem 21.2 Spreadsheet solution for gas stove—with and without fan

5

6	**Define initial conditions**			Czero =	40 µg/m^3
7	**and convert units**			t zero =	0 minutes
8		infiltr	4000 cf/hr	$Q1$ =	1.888 cu m/min
9		fan	500 cf/hr	$Q5$ =	0.236 cu m/min
10		emissions		S =	3000 µg/min
11		room	2400 cu ft	V =	67.97 cu m
12	Define simulation params			deltime =	1 minutes

13

14 **Equation to be solved** $V \times \mathrm{del}C/\mathrm{del}\,t = Q_1 \times C_o + S - Q_2 \times C_i - Q_5 \times C_i$

15 $\mathrm{del}C/\mathrm{del}\,t = Q_1/V \times C_o + S/V - (Q_2 + Q_5)/V \times C_i$

16 **Method**   Start w/ C0; calc delC/del t from Eqn; mult by del t & add to Czero to get Cnew

17   Copy Cnew into next row cell for Cold and repeat

18

19 **Results for case with vent fan off**   **Results for case with vent fan on**

20	Time	C old	dCdt	C new	Time	C old	dCdt	C new
21	0	40.00	44.138	84.14	0	40.00	8.828	48.83
22	1	84.14	42.911	127.05	1	48.83	8.552	57.38
23	2	127.05	41.719	168.77	2	57.38	8.284	65.66
24	3	168.77	40.561	209.33	3	65.66	8.026	73.69
25	4	209.33	39.434	248.76	4	73.69	7.775	81.46
26	5	248.76	38.339	287.10	5	81.46	7.532	89.00
27	6	287.10	37.274	324.38	6	89.00	7.296	96.29
28	7	324.38	36.238	360.61	7	96.29	7.068	103.36
29	8	360.61	35.232	395.84	8	103.36	6.847	110.21
30	9	395.84	34.253	430.10	9	110.21	6.634	116.84
31	10	430.10	33.301	463.40	10	116.84	6.426	123.27
32	11	463.40	32.376	495.78	11	123.27	6.225	129.49
33	12	495.78	31.477	527.25	12	129.49	6.031	135.52
34	13	527.25	30.603	557.86	13	135.52	5.842	141.37
35	14	557.86	29.753	587.61	14	141.37	5.660	147.03
36	15	587.61	28.926	616.53	15	147.03	5.483	152.51
37	16	616.53	28.123	644.66	16	152.51	5.312	157.82
38	17	644.66	27.341	672.00	17	157.82	5.146	162.97
39	18	672.00	26.582	698.58	18	162.97	4.985	167.95
40	19	698.58	25.844	724.42	19	167.95	4.829	172.78
41	20	724.42	25.126	749.55	20	172.78	4.678	177.46
42	21	749.55	24.428	773.98	21	177.46	4.532	181.99
43	22	773.98	23.749	797.73	22	181.99	4.390	186.38
44	23	797.73	23.090	820.82	23	186.38	4.253	190.63
45	24	820.82	22.448	843.26	24	190.63	4.120	194.75
46	25	843.26	21.825	865.09	25	194.75	3.991	198.75

Figure 21.3
Spreadsheet solution to Example 21.2 (first 25 minutes).

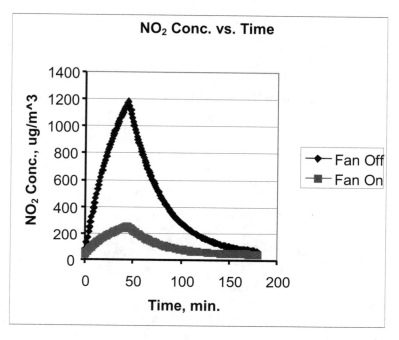

Figure 21.4
Graph of NO_2 concentration versus time for Example 21.2.

◀ ◀ ◀ ◀ ◀ ◀ ◀ ◀ ◀

21.5 PRACTICAL SOLUTIONS TO IAQ PROBLEMS

Equation (21.7) provides mathematical confirmation of earlier statements that the three approaches to solving IAQ problems are to reduce the emission rate, increase the ventilation rate, or install air cleaning devices. In ordinary cases, the solution most often used in the past has been to increase the ventilation rate. Source control was mentioned in Section 21.3, and is used when applicable and cost effective. That the use of treatment devices is not widespread is mainly due to their high cost; however, in some instances, cost is not a concern and control devices are essential—consider the space shuttle or a submarine!

Even though it is widely practiced, increasing the ventilation rate causes increased energy costs on a continuing basis, and has been discouraged as a long-term solution (after all, it is only a "dilution" method). However, it is often the method used because it is so easy to do. The costs associated with increased ventilation are easy to estimate. The appropriate equations were presented in Chapters 4 and 8, and are summarized here in a more convenient form.

$$w_a = 0.0001174 \, Q \, \Delta P \tag{21.8}$$

where

w_a = rate of work (power) put into the air, kW

Q = airflow rate, cfm

ΔP = pressure drop experienced by the air being moved, in. H_2O

$$C_m = \frac{w_a t C_e}{\eta} \tag{21.9}$$

where

C_m = annual cost of moving the air, \$/year

t = time of operation of the air handling, hours/year

C_e = unit cost of electricity, \$/kWh

η = fan efficiency (typically between 0.65 and 0.8)

The cost C_m is only for moving the air; cooling (or heating) the air often is a larger cost. The total cost for heating or air conditioning a building depends quite strongly on the difference between the ambient and inside temperatures, radiant (sun) energy entering the building, internal heat load, building insulation, and the efficiency of the cooling or heating equipment. Architects and HVAC contractors have short-cut methods by which to estimate heat load and air conditioning costs for systems ranging in size from small home applications to large commercial buildings. However, from thermodynamic principles, we can estimate the incremental costs associated with heating or cooling just the make-up air. The cost for heating or cooling a stream of air is estimated from Eq. (21.10).

$$C_{h/c} = \frac{k \, C_p \, \rho \, Q \, |\Delta T| \, C_e}{\eta} \tag{21.10}$$

where

$C_{h/c}$ = cost of heating or cooling the make-up air, \$/min

k = units conversion factor, 0.000293 kWh/Btu

C_p = specific heat of air, Btu/lb-°F

ρ = density of air, lb/ft^3

Q = airflow rate, cfm

$|\Delta T|$ = absolute value of $T_i - T_o$, degrees F

C_e = unit cost of electricity, \$/kWh

η = equipment efficiency, dimensionless

If we (1) assume a typical efficiency for an electric air conditioner/heat pump, (2) assume average values for the specific heat and density of air, and (3) convert minutes to hours, then Eq.(21.10) can be simplified to the following:

$$C_{h/c} = 4.5(10)^{-4}\ Q|\Delta T|C_e \qquad \textbf{(21.11)}$$

where $C_{h/c}$ = cost of heating or cooling the make-up air, $/hr

Keep in mind the approximate nature of Eq. (21.11).

If we know exactly where the emissions are originating, and if we are smart about how and where to ventilate, localized venting within a building can be very effective. For example, to control cooking emissions in homes or commercial establishments, hoods and vent fans are installed over the stoves. In restaurants, it is essential to isolate the kitchen from the dining area. The vent fan exhausts only a small volume of air compared with the whole building, but that fan may capture and remove 90% or more of the emissions of certain pollutants. In the electroplating industry, hoods and vents located adjacent to the plating tanks serve to keep the workplace relatively free from various contaminants. In yet another example, if the problem is radon emissions entering through the basement of a house, the basement could be isolated, and then ventilated continuously (see Figure 21.5).

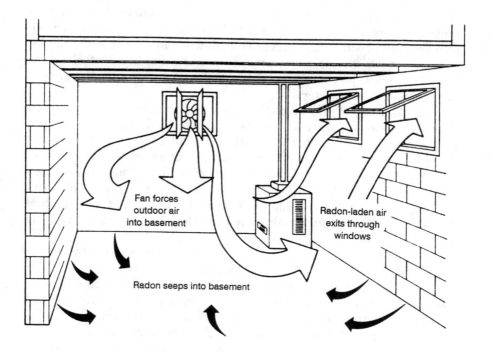

Figure 21.5

One method of controlling radon entry into the living space of a house. (From Cooper, Dietz, and Reinhart, 2000.)

For other problems, such as molds and mildews from serious water damage, a campaign to isolate, replace, repair, clean, and disinfect may be required (see the case studies in the next section). In recent years, IAQ problems related to smoking have been eased by bans on the smoking of tobacco within the confines of buildings. Each IAQ problem is unique and requires careful investigation, analysis, design, and implementation to solve the problem in a cost effective manner.

♦ 21.6 CASE STUDIES ♦

CASE I : A WATER DAMAGED HOTEL*

In late September of 1998, powerful Hurricane Georges struck the Florida Keys. A five-story hotel located adjacent to the ocean received extensive structural damage and water intrusion from the tidal surge and the wind-driven rain. Owing to the warm climate, microbial activity within the hotel rooms was evident within a few days after the hurricane, and became pervasive after 10 days. Given the nature and extent of the microbial activity, an environmental consultant was retained by the hotel to provide assistance.

Assessment of the hotel indicated that design deficiencies and deferred maintenance, coupled with the obvious effects of hurricane-force winds and rain, were responsible for the gross water intrusion into the hotel rooms. The excessive moisture within the hotel rooms, combined with the lack of temperature and humidity control due to loss of electrical power, provided an ideal environment for the growth of fungi, mold, and bacteria.

Remediation of the extensive microbial contamination required that the excess moisture be removed from the indoor space, and that proper humidity be maintained within the hotel rooms. In the case of the side of the hotel facing the ocean, large sections of the exterior wall were entirely removed and rebuilt. Desiccant (drying agent) was used and refrigerant dehumidification was conducted within the entire building. After a relative humidity below 40% within all the individual rooms was achieved and maintained for three consecutive days, the "deep drying" process was considered complete.

After deep drying was complete, a microbial remediation contractor was hired to remove all the sources of microbial growth and to sanitize the hotel rooms. Personal protection equipment and limited engineering controls were utilized to reduce employee exposure to airborne and surface microorganisms. After source removal was completed, the rooms were treated with anti-microbial chemical compounds.

A combination of airborne microbial sampling (both for viable organisms and for nonviable particles), surface microbial sampling,

*Adapted from Hewitt, 1999.

and visual observation was used to determine the completeness of the microbial cleaning process. For a variety of reasons, visual observations and surface sampling were the most effective indicators of the success of the cleaning process. The results of this project suggest the importance of thoroughly drying a facility after catastrophic water intrusion. Extensive microbial growth and contamination can be a consequence of an inadequate drying process.

CASE 2: ANOTHER CASE OF WATER DAMAGE*

A hotel in the normally dry southwestern region of the United States was on the verge of having to close its doors because of guest complaints of itchy eyes, allergic rhinitis, shortness of breath, and at least one diagnosis of reactive airway disease. Six months earlier there had been flash floods in the area, and the hotel had sustained water damage. The damaged areas were undergoing repair. However, the hotel was still renting rooms to guests during this time.

An investigation was conducted to determine whether the problems were caused by microbial contamination, and whether the water damage was "directly caused" by the floodwaters, aging of the building, or mechanical malfunctions. This determination of cause was of utmost importance to the insurance companies and the hotel—it would determine who would ultimately pay for the repairs!

During the site investigation, a variety of sporulating molds, worms, mites, and insects were found behind the wallpaper and under the carpets in several rooms; bed frames, doorways, expansion joints, and ceilings then were examined for microbial infestation as well. Potentially toxigenic molds as well as other molds at high (indoor) concentrations were identified along with free-living water nematodes.

Ultimately, the infestation was deemed to be extensive and pervasive. The hotel was closed for a period of time and many of the rooms were torn down and completely rebuilt. Although it was impossible to determine for sure what portion of the microbial growth and damage was preexisting and what portion was the direct result of water damage caused by the flash flood, a compromise was reached and the hotel repair costs were shared among the owners and insurance companies.

*Adapted from Vance, 1999.

◆ Problems ◆

21.1 Assume that 300 people are on a transatlantic flight, and 60 of them are smoking (an average of 2 cigarettes per hour). The inside volume of the airplane is 5000 ft^3. Calculate the make-up air ventilation rate required to prevent the formaldehyde concentration from exceeding 0.03 ppm.

21.2 Based on the methanol data in Table 21.4, estimate the ventilation rate (in cfm) in the lecture hall from which the data were taken. Does the classroom meet ASHRAE guidelines?

21.3 A restaurant seats 150 people and allows smoking. Estimate the required ventilation rate (make-up air only) for this restaurant. The restaurant changes hands and the new owner decides to switch to nonsmoking. What is the new flow rate of make-up air required?

21.4 If the estimated system pressure drop is 3.5 inches of water for the restaurant of Problem 21.3, estimate the weekly monetary savings realized by this restaurant due to switching to non-smoking. Base your estimate on the reduced airflow volume alone (not counting air conditioning costs). Assume the restaurant operates from 9:00 A.M. to 11:00 P.M., 7 days a week, and that electricity costs 8 cents/kWh.

21.5 Estimate the weekly savings in air conditioning costs for the restaurant of the previous problem. Assume that the temperature difference inside and outside averages 15 degrees, and that electricity costs 8 cents/kWh.

21.6 Consider a large dinner-show themed restaurant in Orlando, Florida. This nonsmoking restaurant seats 500 people. However, during the show, flaming torches are juggled, so the forced ventilation airflow is 150% of the recommended guideline, plus infiltration amounts to 15% of the total. There is one sold-out dinner show per night, and the facility is open from 4:00 P.M. to 10:00 P.M., 7 nights per week.

   a. Calculate the make-up air ventilation rate for this building, in cfm.

   b. Estimate the annual cost just for moving this air through the restaurant (assume a pressure drop of 4 inches of water).

21.7 The outside temperature in July in Orlando in late afternoon and early evening averages about 90 F. Assume that the make-up air for the restaurant of Problem 21.6 must be cooled to 60 F before being distributed into the room. Estimate the cost for cooling just the make-up air for the month of July.

21.8 An older home was built in an area that has a high radon flux (9.5 pCi/m^2-s). The concrete slab did not have a plastic vapor barrier installed, so assume that 90% of the radon flux passes through the slab and into the house. The house has a floor area of 2500 ft^2, an enclosed volume of 20,000 ft^3, and averages 1.5 ach. The decay coefficient for radon is 7.6 (10)$^{-3}$ hr^{-1}. Assuming the ambient concentration is negligible, estimate the steady-state indoor concentration of radon.

21.9 A person with no knowledge of the potential for carbon monoxide poisoning (and no common sense) brings his charcoal grill into his small (6000 ft^3) apartment when a sudden rainstorm begins. All the windows are closed because of the rain, so the ventilation rate is only 0.75 ach. The ambient CO concentration and the initial indoor concentration of CO are both zero. After one hour, the CO concentration is 190 mg/m^3. Calculate the emission rate of CO from the charcoal grill in mg/minute. What would the concentration be after two hours?

21.10 Using the same apartment volume and ventilation rate as in the preceding problem, and assuming that the emission rate of CO is 4.0 grams per minute, create a spreadsheet to solve for the indoor CO concentration as a function of time. Produce a figure to show how the CO concentration in the apartment varies with time. How long will it take before the CO concentration reaches the NAAQS of 40 mg/m^3? How long before it reaches the potentially fatal level of 800 mg/m^3?

REFERENCES

ASHRAE STANDARD. *Ventilation for Acceptable Indoor Air Quality,* ASHRAE 62-1999, American Society of Heating, Refrigeration and Air Conditioning Engineers, Atlanta, GA, 1999.

Brooks, B. O., and Davis, W. F. *Understanding Indoor Air Quality.* Boca Raton, FL: CRC Press, 1992.

Cooper, C. D., Dietz, J. D., and Reinhart, D. R. *Foundations of Environmental Engineering.* Prospect Heights, IL: Waveland Press, 2000.

de Nevers, N. *Air Pollution Control Engineering.* New York: McGraw-Hill, 1995.

Godish, T. *Air Quality.* Chelsea, MI: Lewis Publishers, 1985.

Hays, S. M., Gobbel, R. V., and Ganick, N. R. *Indoor Air Quality—Solutions and Strategies.* New York: McGraw Hill, 1995.

Hewitt, J. M. *Moisture and Microbial Remediation in a Hurricane-Damaged Hotel in Key West, FL,* a paper presented at the Florida Section AWMA Annual Conference, Orlando, FL, September 1999.

Hines, A. L., Ghosh, T. K., Loyalka, S. K., and Warder, Jr., R. C. *Indoor Air Quality and Control.* Englewood Cliffs, NJ: Prentice Hall, 1993.

Kay, J. G., Keller, G. E., and Miller, J. F. *Indoor Air Pollution—Radon, Bioaerosols, and VOC's.* Chelsea, MI: Lewis Publishers, 1991.

Masters, G. M. *Introduction to Environmental Engineering and Science* (2nd ed.). Englewood Cliffs, NJ: Prentice Hall, 1998.

U.S. Environmental Protection Agency. *An Office Building Occupant's Guide to Indoor Air Quality.* EPA-402-K-97-003 (available on the web: www.epa.gov/iaq/pubs/occupgd), Office of Air and Radiation, October 1997.

Vance, P. H. *Is Your Building Bugged?,* a paper presented at the Florida Section AWMA Annual Conference, Orlando, FL, September 1999.

Appendixes

Knowledge is of two kinds. We know a subject ourselves, or we know where we can find information upon it.

Samuel Johnson, 1775

♦ APPENDIX A: CONVERSION FACTORS ♦

How to use these tables:

> 1 unit from column A = **table entry** units from row B
> Examples: 1 lb_m = 453.6 g; 1 kg = 2.205 lb_m

Table A.1 Mass

			Row B			
Column A	**lb_m**	**g**	**gr**	**kg**	**ton**	**tonne**
lb_m	1.0	453.6	7,000	0.4536	0.00050	0.000454
g	0.002205	1.0	15.43	0.001	$1.10\,(10)^{-6}$	$1.0\,(10)^{-6}$
gr	0.000143	0.0648	1.0	$6.48\,(10)^{-5}$	$7.14\,(10)^{-8}$	$6.48\,(10)^{-8}$
kg	2.205	1,000	$1.54\,(10)^4$	1.0	0.0011	0.001
ton	2,000	$9.07\,(10)^5$	$1.40\,(10)^7$	907	1.0	0.907
tonne (metric ton)	2,205	$(10)^6$	$1.54\,(10)^7$	1,000	1.102	1.0

Table A.2 Length

			Row B			
Column A	**m**	**ft**	**in.**	**μm**	**km**	**miles**
m	1.0	3.281	39.37	10^6	0.001	$6.21\,(10)^{-4}$
ft	0.3048	1.0	12	$3.05\,(10)^5$	$3.05(10)^{-4}$	$1.894\,(10)^{-4}$
in.	0.0254	0.0833	1.0	$2.54\,(10)^4$	$2.54\,(10)^{-5}$	$1.578\,(10)^{-5}$
μm	10^{-6}	$3.28\,(10)^{-6}$	$3.94\,(10)^{-5}$	1.0	$1.0\,(10)^{-9}$	$6.22\,(10)^{-10}$
km	1,000	3,281	$3.94\,(10)^4$	$1.0\,(10)^9$	1.0	0.6215
miles	1,609	5,280	$6.336\,(10)^4$	$1.61\,(10)^9$	1.609	1.0

Table A.3 Volume

		Row B		
Column A	**ft^3**	**L**	**gal**	**m^3**
ft^3	1.0	28.32	7.481	0.02832
L	0.03531	1.0	0.2642	0.001
gal	0.1337	3.785	1.0	0.003785
m^3	35.31	1,000	264.2	1.0

Table A.4 Force

Column A	Row B N	lb$_f$	kg-m/s^2	lb$_m$-ft/s^2
N	1.0	0.2248	1.0	7.232
lb$_f$	4.448	1.0	4.448	32.17
kg-m/s^2	1.0	0.2248	1.0	7.232
lb$_m$-ft/s^2	0.1383	0.03108	0.1383	1.0

NOTE: 1 dyne = 10.0 μN

Table A.5 Pressure

Column A	Row B atm	psi	mm Hg	in. H$_2$O	mbar	Pa (N/m^2)
atm	1.0	14.70	760	406.8	1,013	101,300
psi	0.068	1.0	51.7	27.67	68.9	6,891
mm Hg	1.316 $(10)^{-3}$	0.0193	1.0	0.535	1.333	133.3
in. H$_2$0	0.002458	0.03614	1.868	1.0	2.49	249
mbar	9.87 $(10)^{-4}$	0.0145	0.750	0.4016	1.0	100
Pa	9.87 $(10)^{-6}$	1.45 $(10)^{-4}$	0.0075	0.00402	0.01	1.0

NOTE: 1 Pa = 1 N/m^2

Table A.6 Energy

Column A	Row B Btu	kJ	cal	ft-lb$_f$	kWh	liter-atm
Btu	1.0	1.055	252	778	2.93 $(10)^{-4}$	10.41
kJ	0.948	1.0	239	737.5	2.778 $(10)^{-4}$	98.62
cal	0.00397	0.004184	1.0	3.087	1.163 $(10)^{-6}$	0.0413
ft-lb$_f$	0.001285	0.001356	0.3239	1.0	3.766 $(10)^{-7}$	0.01338
kWh	3,412	3,600	8.60 $(10)^5$	2.66 $(10)^6$	1.0	3.55 $(10)^4$
liter-atm	0.0961	0.01014	24.22	74.74	2.82 $(10)^{-5}$	1.0

NOTE: 1 J = 1 N-m

Table A.7 Power

Column A	Row B				
	W	kW	ft-lb$_f$/s	hp	Btu/hr
W	1.0	0.001	0.737	0.00134	3.412
kW	1,000	1.0	737.6	1.341	3,412
ft-lb$_f$/s	1.356	0.001356	1.0	0.001818	4.63
hp	745.5	0.7455	550	1.0	2,545
Btu/hr	0.293	$2.93 (10)^{-4}$	0.216	$3.93 (10)^{-4}$	1.0

NOTE: 1 W = 1 J/s

Table A.8 Speed

Column A	Row B			
	ft/s	m/s	mi/hr	ft/min
ft/s	1.0	0.3048	0.6818	60.0
m/s	3.281	1.0	2.237	196.8
mi/hr	1.467	0.447	1.0	88.0
ft/min	0.01667	0.00508	0.01136	1.0

Table A.9 Viscosity

Column A	Row B			
	cp	g/cm-s	lb$_m$/ft-hr	kg/m-hr
cp	1.0	0.01	2.42	3.61
g/cm-s	100	1.0	242	361
lb$_m$/ft-hr	0.413	0.00413	1.0	1.492
kg/m-hr	0.277	0.00277	0.670	1.0

◆ APPENDIX B: PROPERTIES OF AIR ◆ AND OTHER MATERIALS

Table B.1 Concentrations of Gases Comprising Modern Air (dry basis)

Gas	Concentration (units as listed[a])	
Nitrogen	78.09	%
Oxygen	20.94	%
Argon	0.93	%
Carbon Dioxide	370	ppm
Neon	18	ppm
Helium	5.2	ppm
Methane	1.7	ppm
Krypton	1.0	ppm
Hydrogen	500	ppb
Nitrous Oxide	300	ppb
Xenon	80	ppb
Criteria Pollutants[b]		
CO	100	ppb
O_3	20	ppb
NO_2	1	ppb
SO_2	200	ppt
Others[b]		
Ammonia	10	ppb
Freon (CFC-11)	230	ppt
Hydrogen Sulfide	200	ppt
***		

[a] Units are *percent, parts per million, parts per billion, and parts per trillion*—all on a volume basis.

[b] Values given here represent (approximately) natural "rural background" concentrations; ambient concentrations in urban areas are often considerably higher.

*** A variety of other relatively stable organic and inorganic gases may exist at various concentrations, especially in urban areas.

Values in this table were obtained from various sources.

Table B.2 Some Properties of Air

Temp. °F	Specific Heat at Constant Pressure (C_p) Btu/lb$_m$ °F	Absolute Viscosity (μ) lb$_m$/hr-ft	Thermal Conductivity (k) Btu/hr-ft-°F	Prandtl No. ($C\mu/k$) (Dimensionless)	Density (ρ)[a] lb$_m$/ft^3
0	0.240	0.040	0.0124	0.77	0.0863
20	0.240	0.041	0.0128	0.77	0.0827
40	0.240	0.042	0.0132	0.77	0.0794
60	0.240	0.043	0.0136	0.76	0.0763
80	0.240	0.045	0.0140	0.77	0.0734
100	0.240	0.047	0.0145	0.76	0.0708
120	0.240	0.047	0.0149	0.76	0.0684
140	0.240	0.048	0.0153	0.76	0.0662
160	0.240	0.050	0.0158	0.76	0.0639
180	0.240	0.051	0.0162	0.76	0.0619
200	0.240	0.052	0.0166	0.76	0.0601
250	0.241	0.055	0.0174	0.76	0.0558
300	0.241	0.058	0.0182	0.76	0.0521
350	0.241	0.060	0.0191	0.76	0.0489
400	0.241	0.063	0.0200	0.76	0.0460
450	0.242	0.065	0.0207	0.76	0.0435
500	0.242	0.067	0.0214	0.76	0.0412
600	0.242	0.072	0.0229	0.76	0.0373
700	0.243	0.076	0.0243	0.76	0.0341
800	0.244	0.080	0.0257	0.76	0.0314
900	0.245	0.085	0.0270	0.77	0.0295
1,000	0.246	0.089	0.0283	0.77	0.0275
1,200	0.248	0.097	0.0308	0.78	0.0238
1,400	0.251	0.105	0.0328	0.80	0.0212
1,600	0.254	0.112	0.0346	0.82	0.0192
1,800	0.257	0.120	0.0360	0.85	0.0175
2,000	0.260	0.127	0.0370	0.83	0.0161

[a] ρ taken at pressure of 29.92 in. of mercury.

Adapted from *Air Pollution Engineering Manual* (2nd ed.), AP-40, J. A. Danielson, Ed. Washington, DC: U.S. Environmental Protection Agency, 1973.

Table B.3   Henry's Law Constants for Selected Gases in Water

where $\overline{P}_a = H_a x_a$

$\overline{P}_a$ = partial pressure of the solute a in the gas phase, atm

x_a = mole fraction of solute a in the liquid phase, mole fraction

H_a = Henry's law constant, atm/mole fraction

Temp.	$H_a \times 10^{-4}$, atm/mole fraction									
°C	Air	CO_2	CO	C_2H_6	H_2	H_2S	CH_4	NO	N_2	O_2
0	4.32	0.0728	3.52	1.26	5.79	0.0268	2.24	1.69	5.29	2.55
10	5.49	0.104	4.42	1.89	6.36	0.0367	2.97	2.18	6.68	3.27
20	6.64	0.142	5.36	2.63	6.83	0.0483	3.76	2.64	8.04	4.01
30	7.71	0.186	6.20	3.42	7.29	0.0609	4.49	3.10	9.24	4.75
40	8.70	0.233	6.96	4.23	7.51	0.0745	5.20	3.52	10.4	5.35
50	9.46	0.283	7.61	5.00	7.65	0.0884	5.77	3.90	11.3	5.88
60	10.1	0.341	8.21	5.65	7.65	0.103	6.26	4.18	12.0	6.29
70	10.5		8.45	6.23	7.61	0.119	6.66	4.38	12.5	6.63
80	10.7		8.45	6.61	7.55	0.135	6.82	4.48	12.6	6.87
90	10.8		8.46	6.87	7.51	0.144	6.92	4.52	12.6	6.99
100	10.7		8.46	6.92	7.45	0.148	7.01	4.54	12.6	7.01

NOTE: To use this table, extract the table entry, then multiply by 10^4 to get the H_a. For example, the value of H_{H_2S} at 20 °C is 483 atm/mole fraction.

Adapted from Foust, Wenzel, Clump, Maus, and Anderson, *Principles of Unit Operations*, John Wiley & Sons, New York, 1960.

Table B.4   Solubility Data for NH_3 and SO_2 in Water

Mass NH_3 per 100 Masses H_2O	(a) Ammonia Partial Pressure NH_3, mm Hg						
	0 C	10 C	20 C	30 C	40 C	50 C	60 C
100	947						
90	785						
80	636	987					
70	500	780					
60	380	600	945				
50	275	439	686				
40	190	301	470	719			
30	119	190	298	454	692		
25	89.5	144	227	352	534	825	
20	64	103.5	166	260	395	596	834
15	42.7	70.1	114	179	273	405	583
10	25.1	41.8	69.6	110	167	247	361
7.5	17.7	29.9	50.0	79.7	120	179	261
5	11.2	19.1	31.7	51.0	76.5	115	165
4		16.1	24.9	40.1	60.8	91.1	129.2
3		11.3	18.2	29.6	45.0	67.1	94.3
2			12.0	19.3	30.0	44.5	61.0
1					15.4	22.2	30.2

Mass SO_2 per 100 Masses H_2O	(b) Sulfur Dioxide Partial Pressure of SO_2, mm Hg							
	0 C	7 C	10 C	15 C	20 C	30 C	40 C	50 C
20	646	657						
15	474	637	726					
10	308	417	474	567	698			
7.5	228	307	349	419	517	688		
5.0	148	198	226	270	336	452	665	
2.5	69	92	105	127	161	216	322	458
1.5	38	51	59	71	92	125	186	266
1.0	23.3	31	37	44	59	79	121	172
0.7	15.2	20.6	23.6	28.0	39.0	52	87	116
0.5	9.9	13.5	15.6	19.3	26.0	36	57	82
0.3	5.1	6.9	7.9	10.0	14.1	19.7		
0.1	1.2	1.5	1.75	2.2	3.2	4.7	7.5	12.0
0.05	0.6	0.7	0.75	0.8	1.2	1.7	2.8	4.7
0.02	0.25	0.3	0.3	0.3	0.5	0.6	0.8	1.3

Adapted from Foust, Wenzel, Clump, Maus, and Anderson, *Principles of Unit Operations*, John Wiley & Sons, New York, 1960.

Table B.5   Mass Diffusivities of Selected Substances in Air and in Water

(a) Gases in air (at 25 C, 1 atm)		
Substance	$\mathscr{D}$, sq cm/s	$(\mu/\rho\mathscr{D})*$
Ammonia	0.229	0.67
Carbon dioxide	0.164	0.94
Hydrogen	0.410	0.22
Oxygen	0.206	0.75
Water	0.256	0.60
Carbon disulfide	0.107	1.45
Ethyl ether	0.093	1.66
Methanol	0.159	0.97
Ethyl alcohol	0.119	1.30
Propyl alcohol	0.100	1.55
Butyl alcohol	0.090	1.72
Arnyl alcohol	0.070	2.21
Hexyl alcohol	0.059	2.60
Formic acid	0.159	0.97
Acetic Acid	0.133	1.16
Propionic acid	0.099	1.56
i-Butyric acid	0.081	1.91
Valeric acid	0.067	2.31
i-Caproic acid	0.060	2.58
Diethyl amine	0.105	1.47
Butyl amine	0.101	1.53
Aniline	0.072	2.14
Chlorobenzene	0.073	2.12
Chlorotoluene	0.065	2.38
Propyl bromide	0.105	1.47
Propyl iodide	0.096	1.61
Benzene	0.088	1.76
Toluene	0.084	1.84
Xylene	0.071	2.18
Ethylbenzene	0.077	2.01
Propylbenzene	0.059	2.62
Diphenyl	0.068	2.28
n-Octane	0.060	2.58
Mesitylene	0.067	2.31

* The group $(\mu/\rho\mathscr{D})$ in the table above is evaluated for mixtures composed largely of air.
Adapted from Foust, Wenzel, Clump, Maus, and Anderson, *Principles of Unit Operations*, John Wiley & Sons, New York, 1960.

Table B.5 *(continued)*

	(b) Liquids at 20 C, dilute solutions		
Solute	Solvent	$\mathscr{D} \times 10^5$ (sq cm/s) $\times 10^5$	$\left(\dfrac{\mu}{\rho \mathscr{D}}\right)^*$
O_2	Water	1.80	558
CO_2	Water	1.77	670
N_2O	Water	1.51	665
NH_3	Water	1.76	570
Cl_2	Water	1.22	824
Br_2	Water	1.2	840
H_2	Water	5.13	196
N_2	Water	1.64	613
HCl	Water	2.64	381
H_2S	Water	1.41	712
H_2SO_4	Water	1.73	580
HNO_3	Water	2.6	390
Acetylene	Water	1.56	645
Acetic acid	Water	0.88	1,140
Methanol	Water	1.28	785
Ethanol	Water	1.00	1,005
Propanol	Water	0.87	1,150
Butanol	Water	0.77	1,310
Allyl alcohol	Water	0.93	1,080
Phenol	Water	0.84	1,200
Glycerol	Water	0.72	1,400
Pyrogallol	Water	0.70	1,440
Hydroquinone	Water	0.77	1,300
Urea	Water	1.06	946
Resorcinol	Water	0.80	1,260
Urethane	Water	0.92	1,090
Lactose	Water	0.43	2,340
Maltose	Water	0.43	2,340
Glucose	Water	0.60	. . .
Mannitol	Water	0.58	1,730
Raffinose	Water	0.37	2,720
Sucrose	Water	0.45	2,230
Sodium chloride	Water	1.35	745
Sodium hydroxide	Water	1.51	665

*Based on μ/ρ = 0.01005 cm^2/s for water, 0.00737 for benzene, and 0.01511 for ethanol, all at 20 C. Applies only for dilute solutions.

Adapted from Foust, Wenzel, Clump, Maus, and Anderson, *Principles of Unit Operations*, John Wiley & Sons, New York, 1960.

Table B.6 Standard Heats of Combustion for Various Organic Compounds in the Gaseous State (products are $H_2O(g)$ and $CO_2(g)$ at 25 C)

Compound	Formula	MW	ΔH_c, kJ/kg
n-Alkanes			
Methane	CH_4	16.0	50,150
Ethane	C_2H_6	30.1	47,440
Propane	C_3H_8	44.1	46,350
n-Butane	C_4H_{10}	58.1	45,730
n-Pentane	C_5H_{12}	72.2	45,320
n-Hexane	C_6H_{14}	86.2	45,090
n-alkanes above C_6	C_nH_{2n+2}	MW	$\dfrac{[3.8868(10)^6 + 6.147(10)^5(n-6)]}{MW}$
1-Alkenes			
Ethylene	C_2H_4	28.1	47,080
Propylene	C_3H_6	42.1	45,760
1-Butene	C_4H_8	56.1	45,300
1-Pentene	C_5H_{10}	70.1	45,020
1-Hexene	C_6H_{12}	84.2	44,780
1-alkenes above C_6	C_nH_{2n}	MW	$\dfrac{[3.7704(10)^6 + 6.147(10)^5(n-6)]}{MW}$
Miscellaneous			
Acetylene	C_2H_2	26.0	48,290
Benzene	C_6H_6	78.1	40,580
1,3-Butadiene	C_4H_6	54.1	44,540
Cyclohexane	C_6H_{12}	84.2	43,810
Ethylbenzene	C_8H_{10}	106.2	41,310
Methylcyclohexane	C_7H_{14}	98.2	43,710
Styrene	C_8H_8	104.2	40,910
Toluene	C_7H_8	92.1	40,950
Carbon monoxide	CO	28.0	10,110
Hydrogen	H_2	2.016	120,900
Ethanol	C_2H_5OH	46.1	25,960
Methanol	CH_3OH	32.0	18,790
Water (liq. to gas)	H_2O	18.0	2,445

Adapted from *Introduction to Chemical Engineering Thermodynamics* (2nd ed.) by J. M. Smith and H. C. Van Ness, McGraw-Hill, New York, 1959.

Table B.7   Specific Enthalpies of Some Gases, All in Btu/lb$_m$ of Gas (datum temperature is 60 F)

Temp. °F	CO$_2$	N$_2$	H$_2$O[a]	O$_2$	Air
100	5.8	6.4	17.8	8.8	9.6
150	17.6	20.6	40.3	19.8	21.6
200	29.3	34.8	62.7	30.9	33.6
250	40.3	47.7	85.5	42.1	45.7
300	51.3	59.8	108.2	53.4	57.8
350	63.1	73.3	131.3	64.8	70.0
400	74.9	84.9	154.3	76.2	82.1
450	87.0	97.5	177.7	87.8	94.4
500	99.1	110.1	201.0	99.5	106.7
550	111.8	122.9	224.8	111.3	119.2
600	124.5	135.6	248.7	123.2	131.6
700	150.2	161.4	297.1	147.2	156.7
800	176.8	187.4	346.4	171.7	182.2
900	204.1	213.8	396.7	196.5	211.4
1,000	231.9	240.5	447.7	221.6	234.1
1,100	260.2	267.5	499.7	247.0	260.5
1,200	289.0	294.9	552.9	272.7	287.2
1,300	318.0	326.1	606.8	298.5	314.2
1,400	347.6	350.5	661.3	324.6	341.5
1,500	377.6	378.7	717.6	350.8	369.0
1,600	407.8	407.3	774.2	377.3	396.8
1,700	438.2	435.9	831.4	403.7	424.6
1,800	469.1	464.8	889.8	430.4	452.9
1,900	500.1	493.7	948.7	457.3	481.2
2,000	531.4	523.0	1,003.1	484.5	509.5
2,100	562.8	552.7	1,069.2	511.4	538.1
2,200	594.3	582.0	1,130.3	538.6	567.1
2,300	626.2	612.3	1,192.6	566.1	596.1
2,400	658.2	642.3	1,256.8	593.5	625.0
2,500	690.2	672.3	1,318.1	621.0	654.3
3,000	852.3	823.8	1,640.2	760.1	802.3
3,500	1,017.4	978.0	1,975.4	901.7	950.3

[a] The enthalpies tabulated for H$_2$O represent a gaseous system, and the enthalpies do *not* include the latent heat of vaporization. It is recommended that the latent heat of vaporization at 60 F (1,059.1 Btu/lb$_m$) be used where necessary.

Adapted from *Air Pollution Engineering Manual* (2nd ed.), AP-40, J. A. Danielson, Ed., U.S. Environmental Protection Agency, Washington, DC, 1973.

Table B.8 Enthalpies of Saturated Steam and Water

T, °F	P, atm	Enthalpy, Btu/lb$_m$ Sat. Liq.	ΔH_v	Enthalpy, Btu/lb$_m$ Sat. Vapor	T, °F	P, atm	Enthalpy, Btu/lb$_m$ Sat. Liq.	ΔH_v	Enthalpy, Btu/lb$_m$ Sat. Vapor
32	0.0060	0	1,075.1	1,075.1	116		83.92	1,027.5	1,111.4
34		2.01	1,074.0	1,076.0	118		85.92	1,026.4	1,112.3
36		4.03	1,072.9	1,076.9	120	0.115	87.91	1,025.3	1,113.2
38		6.04	1,071.7	1,077.7	122		89.91	1,024.1	1,114.0
40	0.0083	8.05	1,070.5	1,078.6	124		91.90	1,023.0	1,114.9
42		10.06	1,069.3	1,079.4	126		93.90	1,021.8	1,115.7
44		12.06	1,068.2	1,080.3	128		95.90	1,020.7	1,116.6
46		14.07	1,067.1	1,081.2	130	0.151	97.89	1,019.5	1,117.4
48		16.07	1,065.9	1,082.0	132		99.89	1,018.3	1,118.2
50	0.0121	18.07	1,064.8	1,082.9	134		101.89	1,017.2	1,119.1
52		20.07	1,063.6	1,083.7	136		103.88	1,016.0	1,119.9
54		22.07	1,062.5	1,084.6	138		105.88	1,014.9	1,120.8
56		24.07	1,061.4	1,085.1	140	0.196	107.88	1,013.7	1,121.6
58		26.07	1,060.2	1,086.3	142		109.88	1,012.5	1,122.4
60	0.0174	28.07	1,059.1	1,087.2	144		111.88	1,011.3	1,123.2
62		30.06	1,057.9	1,088.0	146		113.88	1,010.2	1,124.1
64		32.06	1,056.8	1,088.9	148		115.87	1,009.0	1,124.9
66		34.06	1,055.7	1,089.8	150	0.253	117.87	1,007.8	1,125.7
68		36.05	1,054.5	1,090.6	152		119.87	1,006.7	1,126.6
70	0.0247	38.05	1,053.4	1,091.5	154		121.87	1,005.5	1,127.4
72		40.04	1,052.3	1,092.3	156		123.87	1,004.4	1,128.3
74		42.04	1,051.2	1,093.2	158		125.87	1,003.2	1,129.1
76		44.03	1,050.1	1,094.1	160	0.322	127.87	1,002.0	1,129.9
78		46.03	1,048.9	1,094.9	162		129.88	1,000.8	1,130.7
80	0.0345	48.02	1,047.8	1,095.8	164		131.88	999.7	1,131.6
82		50.02	1,046.6	1,096.6	166		133.88	998.5	1,132.4
84		52.01	1,045.5	1,097.5	168		135.88	997.3	1,133.2
86		54.01	1,044.4	1,098.4	170	0.408	137.89	996.1	1,134.0
88		56.00	1,043.2	1,099.2	172		139.89	995.0	1,134.9
90	0.0475	58.00	1,042.1	1,100.1	174		141.89	993.8	1,135.7
92		59.99	1,040.9	1,100.9	176		143.90	992.6	1,136.5
94		61.98	1,039.8	1,101.8	178		145.90	991.4	1,137.3
96		63.98	1,038.7	1,102.7	180	0.511	147.91	990.2	1,138.1
98		65.98	1,037.5	1,103.5	182		149.92	989.0	1,138.9
100	0.0646	67.97	1,036.4	1,104.4	184		151.92	987.8	1,139.7
102		69.96	1,035.2	1,105.2	186		153.93	986.6	1,140.5
104		71.96	1,034.1	1,106.1	188		155.94	985.3	1,141.3
106		73.95	1,033.0	1,107.0	190	0.635	157.95	984.1	1,142.1
108		75.94	1,032.0	1,107.9	192		159.95	982.8	1,142.8
110	0.0867	77.94	1,030.9	1,108.8	194		161.96	981.5	1,143.5
112		79.93	1,029.7	1,109.6	196		163.97	980.3	1,144.3
114		81.93	1,028.6	1,110.5	198		165.98	979.0	1,145.0

Table B.8   *(continued)*

T, °F	P, atm	Enthalpy, Btu/lb$_m$			T, °F	P, atm	Enthalpy, Btu/lb$_m$		
		Sat. Liq.	ΔH_v	Sat. Vapor			Sat. Liq.	ΔH_v	Sat. Vapor
200	0.784	167.99	977.8	1,145.8	245		213.41	948.7	1,162.1
202		170.01	976.6	1,146.6	250	2.029	218.48	945.3	1,163.8
204		172.02	975.3	1,147.3	255		223.56	942.0	1,165.6
206		174.03	974.1	1,148.1	260	2.411	228.65	938.6	1,167.3
208		176.04	972.8	1,148.8	265		233.74	935.3	1,169.0
210	0.961	178.06	971.5	1,149.6	270	2.848	238.84	931.8	1,170.6
212	1.000	180.07	970.3	1,150.4	275		243.94	928.2	1,172.1
215		183.10	968.3	1,151.4	280	3.348	249.06	924.6	1,173.7
220	1.170	188.14	965.2	1,153.3	285		254.18	921.0	1,175.2
225		193.18	961.9	1,155.1	290	3.916	259.31	917.4	1,176.7
230	1.414	198.22	958.7	1,156.9	295		264.45	913.7	1,178.2
235		203.28	955.3	1,158.6	300	4.560	269.60	910.1	1,179.7
240	1.699	208.34	952.1	1,160.4					

Table B.9   Selected Properties of Liquid Water

Temperature °F	Density lb_m/ft^3	Viscosity cp	Vapor Pressure psi	Surface Tension dynes/cm
40	62.43	1.55	0.122	75.0
50	62.42	1.31	0.178	74.2
60	62.37	1.13	0.256	73.5
70	62.30	0.98	0.363	72.0
80	62.22	0.86	0.507	71.7
90	62.11	0.76	0.698	70.8
100	62.00	0.68	0.949	69.9
120	61.71	0.56	1.69	68.1
200	60.13	0.30	11.53	60.1

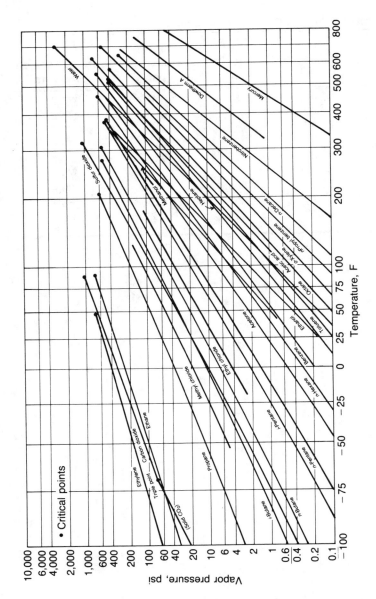

Figure B.1

Vapor pressures of some common liquids.

(Adapted from Foust, Wenzel, Clump, Maus, and Anderson, *Principles of Unit Operations*, John Wiley & Sons, New York, 1960.)

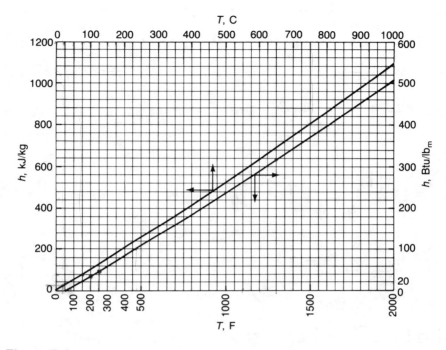

Figure B.2
Specific enthalpy of air. Datum temperatures are 60 F for curve in English
units or 0 C for curve in SI units.

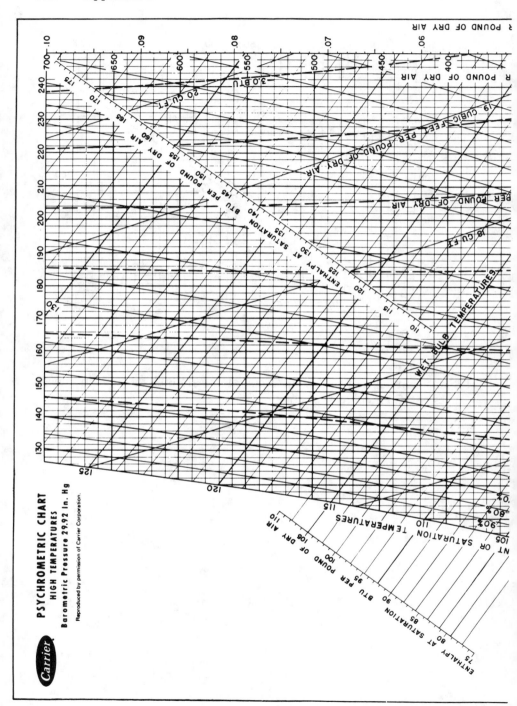

Figure B.3

Psychrometric chart for high temperature in I-P units.

(Courtesy of the Carrier Corporation.)

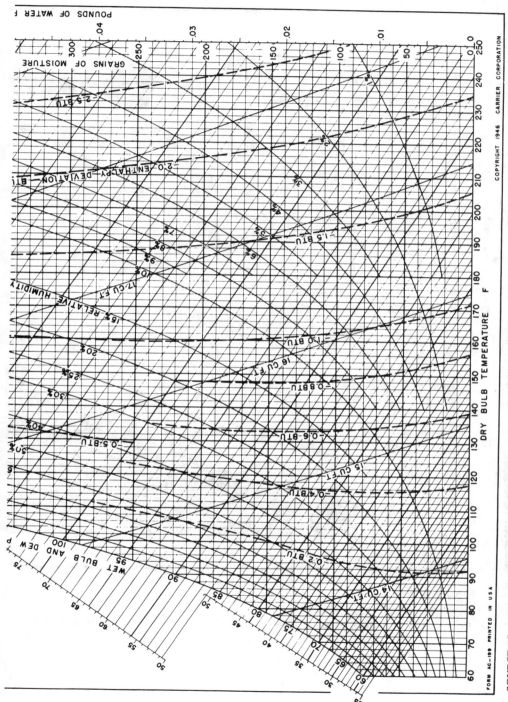

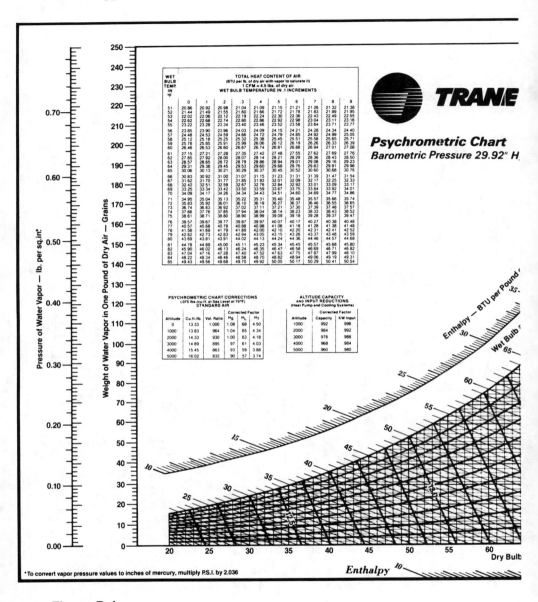

Figure B.4

Psychrometric chart.

(Courtesy of The Trane Company, Dealer Products Group, a Division of American Standard, Inc.)

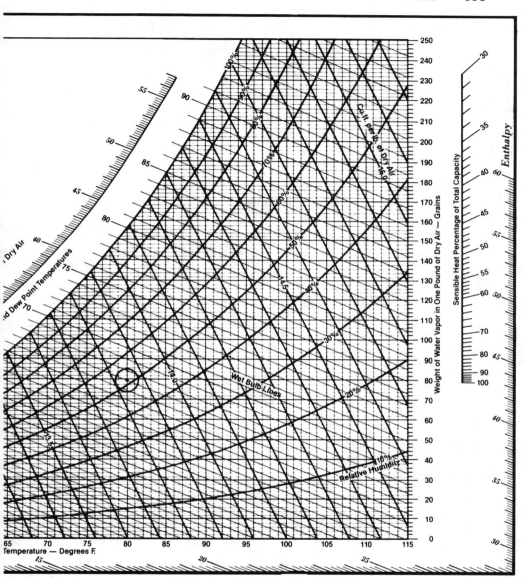

APPENDIX C: SOME PROPERTIES OF A GAUSSIAN DISTRIBUTION

A probability density function (pdf) of a continuous random variable, x, can be defined as a function, $g(x)$, that satisfies two conditions:*

$$(1) \quad g(x) \geq 0 \text{ for} - \infty < x < + \infty$$

and

$$(2) \quad \int_{-\infty}^{+\infty} g(x)\,dx = 1$$

A continuous random variable, x, is said to have a normal or Gaussian, distribution if its pdf is of the form:

$$g(x) = \frac{1}{\sqrt{2\pi}\sigma} \exp\left(-\frac{1}{2}\left(\frac{x-\mu}{\sigma}\right)^2\right) \tag{C.1}$$

where

μ = the mean of x

σ = the standard deviation of x

Note that if we define a variable $B = (x - \mu)/\sigma$, then the normalized or standard Gaussian distribution can be represented by

$$g(B) = \frac{1}{\sqrt{2\pi}} \exp\left(\frac{-B^2}{2}\right) \tag{C.2}$$

This "standard normal distribution function" has a mean of 0 and a standard deviation of 1. The standard Gaussian distribution is shown in Figure C.1.

The Gaussian distribution is the so-called "bell-shaped curve" with unit area. The value of the ordinate peaks at 0.4 at $B = 0$ ($x = \mu$), and it approaches 0 as $|B|$ approaches infinity. The curve is symmetrical about a vertical line through $B = 0$.

The area under the curve from $-\infty$ up to any value B_1, is given by

$$G(B_1) = \frac{1}{\sqrt{2\pi}} \int_{-\infty}^{B_1} \exp\left(\frac{-B^2}{2}\right) dB \tag{C.3}$$

This area is equal to the probability that the random variable B lies between $-\infty$ and B_1. Because the total area is equal to 1.0, the probability that B will be greater than B_1 is just $1 - G(B_1)$. Also, because of symmetry, we can say that

$$G(-B_1) = 1 - G(B_1) \tag{C.4}$$

---

*Meyer, P. L. *Introductory Probability and Statistical Applications*. Reading, MA: Addison Wesley, 1965.

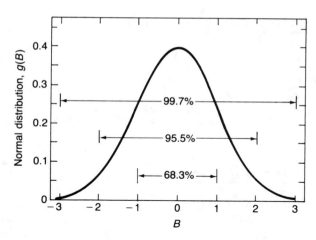

Figure C.1
The normal (Gaussian) distribution.

Furthermore, we can show that the probability that B lies between two values (say B_1 and B_2, where $B_2 > B_1$) is simply $G(B_2) - G(B_1)$. For a normal distribution, the areas contained between $-B$ and $+B$, where B is 1.0, 2.0, and 3.0, are 0.683, 0.955, and 0.997, respectively.

The pollutants in a plume from a point source are normally distributed in the y direction about the plume centerline at any downwind distance x. The plume "edge" could possibly be defined as the distance from the centerline at which the concentration is one-tenth of the centerline concentration [that is, $g(y) = 0.1g(0)$ at the same x]. At this ordinate value, B_1 and B_2 would be −2.15 and +2.15, and the plume would contain 96.84% of the total mass of pollutants. However, if we defined the plume edge as being three standard deviations away from the centerline (B_1 and $B_2 = -3.0$ and $+3.0$), then the plume would contain 99.74% of the pollutants, and the edge concentration would be only 1.1% of the centerline concentration. Values of the Gaussian distribution function, G, are given in Table C.1.

Table C.1   Values of the Gaussian Distribution Function

$$G(B) = \frac{1}{\sqrt{2\pi}} \int_{-\infty}^{B} \exp\left(\frac{-B^2}{2}\right) dB$$

where $B = \dfrac{(x - \mu)}{\sigma}$

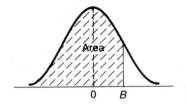

B	0	1	2	3	4	5	6	7	8	9
−3.0	0.0013	0.0010	0.0007	0.0005	0.0003	0.0002	0.0002	0.0001	0.0001	0.0000
−2.9	0.0019	0.0018	0.0017	0.0017	0.0016	0.0016	0.0015	0.0015	0.0014	0.0014
−2.8	0.0026	0.0025	0.0024	0.0023	0.0023	0.0022	0.0021	0.0021	0.0020	0.0019
−2.7	0.0035	0.0034	0.0033	0.0032	0.0031	0.0030	0.0029	0.0028	0.0027	0.0026
−2.6	0.0047	0.0045	0.0044	0.0043	0.0041	0.0040	0.0039	0.0038	0.0037	0.0036
−2.5	0.0062	0.0060	0.0059	0.0057	0.0055	0.0054	0.0052	0.0051	0.0049	0.0048
−2.4	0.0082	0.0080	0.0078	0.0075	0.0073	0.0071	0.0069	0.0068	0.0066	0.0064
−2.3	0.0107	0.0104	0.0102	0.0099	0.0096	0.0094	0.0091	0.0089	0.0087	0.0084
−2.2	0.0139	0.0136	0.0132	0.0129	0.0126	0.0122	0.0119	0.0116	0.0113	0.0110
−2.1	0.0179	0.0174	0.0170	0.0166	0.0162	0.0158	0.0154	0.0150	0.0146	0.0143
−2.0	0.0228	0.0222	0.0217	0.0212	0.0207	0.0202	0.0197	0.0192	0.0188	0.0183
−1.9	0.0287	0.0281	0.0274	0.0268	0.0262	0.0256	0.0250	0.0244	0.0238	0.0233
−1.8	0.0359	0.0352	0.0344	0.0336	0.0329	0.0322	0.0314	0.0307	0.0300	0.0294
−1.7	0.0446	0.0436	0.0427	0.0418	0.0409	0.0401	0.0392	0.0384	0.0375	0.0367
−1.6	0.0548	0.0537	0.0526	0.0516	0.0505	0.0495	0.0485	0.0475	0.0465	0.0455
−1.5	0.0668	0.0655	0.0643	0.0630	0.0618	0.0606	0.0594	0.0582	0.0570	0.0559
−1.4	0.0808	0.0793	0.0778	0.0764	0.0749	0.0735	0.0722	0.0708	0.0694	0.0681
−1.3	0.0968	0.0951	0.0934	0.0918	0.0901	0.0885	0.0869	0.0853	0.0838	0.0823
−1.2	0.1151	0.1131	0.1112	0.1093	0.1075	0.1056	0.1038	0.1020	0.1003	0.0985
−1.1	0.1357	0.1335	0.1314	0.1292	0.1271	0.1251	0.1230	0.1210	0.1190	0.1170
−1.0	0.1587	0.1562	0.1539	0.1515	0.1492	0.1469	0.1446	0.1423	0.1401	0.1379
−0.9	0.1841	0.1814	0.1788	0.1762	0.1736	0.1711	0.1685	0.1660	0.1635	0.1611
−0.8	0.2119	0.2090	0.2061	0.2033	0.2005	0.1977	0.1949	0.1922	0.1894	0.1867
−0.7	0.2420	0.2389	0.2358	0.2327	0.2297	0.2266	0.2236	0.2206	0.2177	0.2148
−0.6	0.2743	0.2709	0.2676	0.2643	0.2611	0.2578	0.2546	0.2514	0.2483	0.2451
−0.5	0.3085	0.3050	0.3015	0.2981	0.2946	0.2912	0.2877	0.2843	0.2810	0.2776
−0.4	0.3446	0.3409	0.3372	0.3336	0.3300	0.3264	0.3228	0.3192	0.3156	0.3121
−0.3	0.3821	0.3783	0.3745	0.3707	0.3669	0.3632	0.3594	0.3557	0.3520	0.3483
−0.2	0.4207	0.4168	0.4129	0.4090	0.4052	0.4013	0.3974	0.3936	0.3897	0.3859
−0.1	0.4602	0.4562	0.4522	0.4483	0.4443	0.4404	0.4364	0.4325	0.4286	0.4247
−0.0	0.5000	0.4960	0.4920	0.4880	0.4840	0.4801	0.4761	0.4721	0.4681	0.4641
0.0	0.5000	0.5040	0.5080	0.5120	0.5160	0.5199	0.5239	0.5279	0.5319	0.5359
0.1	0.5398	0.5438	0.5478	0.5517	0.5557	0.5596	0.5636	0.5675	0.5714	0.5753
0.2	0.5793	0.5832	0.5871	0.5910	0.5948	0.5987	0.6026	0.6064	0.6103	0.6141
0.3	0.6179	0.6217	0.6255	0.6293	0.6331	0.6368	0.6406	0.6443	0.6480	0.6517
0.4	0.6554	0.6591	0.6628	0.6664	0.6700	0.6736	0.6772	0.6808	0.6844	0.6879
0.5	0.6915	0.6950	0.6985	0.7019	0.7054	0.7088	0.7123	0.7157	0.7190	0.7224
0.6	0.7257	0.7291	0.7324	0.7357	0.7389	0.7422	0.7454	0.7486	0.7517	0.7549
0.7	0.7580	0.7611	0.7642	0.7673	0.7703	0.7734	0.7764	0.7794	0.7823	0.7852
0.8	0.7881	0.7910	0.7939	0.7967	0.7995	0.8023	0.8051	0.8078	0.8106	0.8133
0.9	0.8159	0.8186	0.8212	0.8238	0.8264	0.8289	0.8315	0.8340	0.8365	0.8389

Table C.1   *(continued)*

B	0	1	2	3	4	5	6	7	8	9
1.0	0.8413	0.8438	0.8461	0.8485	0.8508	0.8531	0.8554	0.8577	0.8599	0.8621
1.1	0.8643	0.8665	0.8686	0.8708	0.8729	0.8749	0.8770	0.8790	0.8810	0.8830
1.2	0.8849	0.8869	0.8888	0.8907	0.8925	0.8944	0.8962	0.8980	0.8997	0.9015
1.3	0.9032	0.9049	0.9066	0.9082	0.9099	0.9115	0.9131	0.9147	0.9162	0.9177
1.4	0.9192	0.9207	0.9222	0.9236	0.9251	0.9265	0.9278	0.9292	0.9306	0.9319
1.5	0.9332	0.9345	0.9357	0.9370	0.9382	0.9394	0.9406	0.9418	0.9430	0.9441
1.6	0.9452	0.9463	0.9474	0.9484	0.9495	0.9505	0.9515	0.9525	0.9535	0.9545
1.7	0.9554	0.9564	0.9573	0.9582	0.9591	0.9599	0.9608	0.9616	0.9625	0.9633
1.8	0.9641	0.9648	0.9656	0.9664	0.9671	0.9678	0.9686	0.9693	0.9700	0.9706
1.9	0.9713	0.9719	0.9726	0.9732	0.9738	0.9744	0.9750	0.9756	0.9762	0.9767
2.0	0.9772	0.9778	0.9783	0.9788	0.9793	0.9798	0.9803	0.9808	0.9812	0.9817
2.1	0.9821	0.9826	0.9830	0.9834	0.9838	0.9842	0.9846	0.9850	0.9854	0.9857
2.2	0.9861	0.9864	0.9868	0.9871	0.9874	0.9878	0.9881	0.9884	0.9887	0.9890
2.3	0.9893	0.9896	0.9898	0.9901	0.9904	0.9906	0.9909	0.9911	0.9913	0.9916
2.4	0.9918	0.9920	0.9922	0.9925	0.9927	0.9929	0.9931	0.9932	0.9934	0.9936
2.5	0.9938	0.9940	0.9941	0.9943	0.9945	0.9946	0.9948	0.9949	0.9951	0.9952
2.6	0.9953	0.9955	0.9956	0.9957	0.9959	0.9960	0.9961	0.9962	0.9963	0.9964
2.7	0.9965	0.9966	0.9967	0.9968	0.9969	0.9970	0.9971	0.9972	0.9973	0.9974
2.8	0.9974	0.9975	0.9976	0.9977	0.9977	0.9978	0.9979	0.9979	0.9980	0.9981
2.9	0.9981	0.9982	0.9982	0.9983	0.9984	0.9984	0.9985	0.9985	0.9986	0.9986
3.0	0.9987	0.9990	0.9993	0.9995	0.9997	0.9998	0.9998	0.9999	0.9999	1.0000

Adapted from Meyer, P.L. *Introductory Probability and Statistical Applications.* Reading, MA: Addison-Wesley, 1965.

◆ Appendix D: Computer Programs ◆

A number of air quality models are available to the public from the EPA Support Center for Regulatory Air Models (SCRAM) Web site—www.epa.gov/scram001. This Web site lists a number of models in several categories. The *preferred / recommended* models are those typically used (and accepted automatically by EPA) in modeling for regulatory purposes (permits, impact assessments, and other actions taken to legally comply with state or federal regulations). *Screening* models are used for quick evaluations to estimate the maximum impacts that may occur from a facility. *Alternative* models are used for special cases, such as a spill and escape of a toxic gas. Table D.1 lists some examples of the models (with brief descriptions). The SCRAM Web site also has a wealth of meteorological data that can be used in the models, official EPA guidance on using some of these models, and access to many User's Guides.

Table D.2 presents a simple Fortran model for finding the maximum downwind concentration for emissions from a single stack; this shows the simplicity of calculations contained in Gaussian-based models. Obviously, for a single calculation or even for several hundred calculations, a spreadsheet would suffice. The main advantage of the EPA computer models is their ability to deal with historical meteorological data (e.g., wind speed, wind direction, ambient temperature, and atmospheric stability class for 8760 hours per year), process millions of calculations (solving the Gaussian equation repeatedly for multiple emission points and multiple receptors), aggregate the results, and then sort the output into organized summary tables.

Table D. 1   EPA Models and Brief Descriptions[1]

Preferred/Recommended Models (used for regulatory modeling)

ISC3 (Industrial Source Complex Model)   A steady-state Gaussian plume model which can be used to assess pollutant concentrations from a wide variety of sources associated with an industrial complex. This model can account for the following: settling and dry deposition of particles; downwash; point, area, line, and volume sources; plume rise as a function of downwind distance; separation of point sources; and limited terrain adjustment. ISC3 operates in both long-term and short-term modes.

CALINE3   A steady-state Gaussian dispersion model, CALINE3 is designed to determine air pollution concentrations at receptor locations downwind of "at-grade," "fill," "bridge," and "cut section" highways located in relatively uncomplicated terrain.

CDM2 (Climatological Dispersion Model)   This is a climatological steady-state Gaussian plume model for determining long-term (seasonal or annual) arithmetic average pollutant concentrations at any ground-level receptor in an urban area.

CTDMPLUS (Complex Terrain Dispersion Model Plus Algorithms for Unstable Situations)   This model is a refined point source Gaussian air quality model for use in all stability conditions for complex terrain. The model contains, in its entirety, the technology of CTDM for stable and neutral conditions.

Table D. 1   *(continued)*

UAM-IV (Urban Airshed Model IV)   This is an urban-scale, three-dimensional, grid-type numerical simulation model. The model incorporates a condensed photochemical kinetics mechanism for urban atmospheres. UAM-IV is designed for computing ozone (O_3) concentrations under short-term, episodic conditions lasting one or two days resulting from emissions of oxides of nitrogen (NO_x), volatile organic compounds (VOCs), and carbon monoxide (CO). The model treats VOC emissions as their carbon-bond surrogates.

Screening Models (very useful for screening studies; often accepted by agencies)

CAL3QHC/CAL3QHCR (CALINE3 with queuing and hot spot calculations)   This screening model is a CALINE3-based CO model with a traffic model to calculate delays and queues that occur at signalized intersections; CAL3QHCR requires local meteorological data.

SCREEN3   A single-source Gaussian plume model which provides maximum ground-level concentrations for point, area, flare, and volume sources, as well as concentrations in the cavity zone, and concentrations due to inversion break-up and shoreline fumigation. SCREEN3 is a screening version of the ISC3 model.

VALLEY   A steady-state, complex-terrain, univariate Gaussian plume dispersion algorithm designed for estimating either 24-hour or annual concentrations resulting from emissions from up to 50 (total) point and area sources.

Alternative Models (used and accepted by regulatory agencies on a case-by-case basis)

AFTOX (Air Force Toxics Model)   This is a Gaussian dispersion model that will handle continuous or instantaneous liquid or gas elevated or surface releases from point or area sources. Output consists of concentration contour plots, concentration at a specified location, and maximum concentration at a given elevation and time.

DEGADIS (Dense Gas Dispersion Model)   This model simulates the atmospheric dispersion at ground-level, area source dense gas (or aerosol) clouds released with zero momentum into the atmospheric boundary layer over flat, level terrain. The model describes the dispersion processes which accompany the ensuing gravity-driven flow and entrainment of the gas into the boundary layer.

EKMA   An empirical, city-specific model, EKMA is used to fill the gap between more sophisticated photochemical dispersion models and proportional (rollback) modeling techniques.

MESOPUFF II   A short-term, regional-scale puff model designed to calculate concentrations of up to 5 pollutant species (SO_2, SO_4, NO_x, HNO_3, NO_3). Transport, puff growth, chemical transformation, and wet and dry deposition are accounted for in the model.

UAM-V (The UAM-V Photochemical Modeling System)   The Urban Airshed Model (UAM-V) is a three-dimensional photochemical grid model that calculates concentrations of pollutants by simulating the physical and chemical processes in the atmosphere. The updated version of the UAM-V modeling system (version 1.3) includes process-analysis capabilities, an enhanced chemical mechanism (enhanced treatment of hydrocarbon and toxic species), updated deposition and nested-grid algorithms, a flexible coordinate system (including Lambert conformal), and user-selection of a "standard" or "fast" solver.

---

[1] These descriptions obtained from the SCRAM web site. For a complete listing of these and other models, see the EPA Web site: www.epa.gov/scram001.

Table D.2 A Gaussian Point Source Program

```
C       AIR—AN AIR POLLUTION DISPERSION MODEL
C
C       BASED ON GAUSSIAN EQUATION FOR CONT. POINT SOURCE EMITTING
C         FROM AN ELEVATED STACK INTO A CONSTANT WIND
C
C       USES CORREL. FROM MARTIN (JAPCA, 26, 2, 1976) FOR SIG Y & SIG
C         Z
C
C       USES GOLDEN SECTION SEARCH METHOD FOR FINDING MAX OF A
C         FUNCTION
C
C       Q MUST BE IN G/SEC, X IN KM, H IN M, U IN M/S, KLS (CLASS) IN
C         A NUMERIC CODE (1 =A, 2 =B, 3 =C, ETC), Y IN M, DEL, Al, B1
C         ALL IN KM
C
        COMMON Q, H, U, KLS
        READ (15, 10) NRUNS
10      FORMAT (I2)
        DO 99 ICOUNT = 1, NRUNS
        READ (15, *) Q, H, U, KLS, OPT
C
C       OPT IS AN OPTION PARM. IF OPT=1, JUST CALC. CONCS. FOR X, Y
C         PTS.
C       IF OPT=2, THEN IT IS DESIRED TO FIND THE MAX DOWNWIND
C         CONC.
20      FORMAT (1H1////)
21      FORMAT (8X, "INPUT DATA":
     1  "Q(G/S)  H(M)  U(M/S)  CLASS")
22      FORMAT (20X, F7.3, 5X, F5.1, 6X, F4.1, 8X, I1//)
23      FORMAT (11X, F7.3, 9X, F5.0, 15X, F8.1)
24      FORMAT (10X, "FOR THE MAXIMIZATION OPTION, Y IS SET TO
          ZERO"//)
25      FORMAT (5X, "THE DIST. TO THE MAX CONC IS", F7.3, 5X,
     1  "THE MAX GROUND LEVEL CONC IS", F 8.1)
26      FORMAT (////)
27      FORMAT (13X,"X(KM)",10X,"Y(M)",15X,CONC(UG/M**3)")
        IF (OPT. GT. 1.5) GO TO 50
        WRITE (16, 20)
        WRITE (16, 21)
        WRITE (16, 22) Q, H, U, KLS
        WRITE (16, 27)
        READ (15, 10) NPTS
        DO 30 I=1, NPTS
        READ ( 15, *) X, Y
        CALL EVAL (X, Y, C)
        WRITE (16, 23) X,Y,C
30      CONTINUE
        WRITE (16, 26)
        GO TO 99
```

Table D.2   *(continued)*

```
50      WRITE (16, 20)
        WRITE (16, 24)
        READ (15,*) DEL, A1, B1
        CALL GOLDN (DEL, A1, B1, X, C)
        WRITE (16, 25) X,C
99      CONTINUE
        END

        SUBROUTINE EVAL (X, Y, C)
        COMMON Q, H, U, KLS
        CALL SIGMA (X, SY, SZ)
        C = 1000000.*Q/(6.2832*U*SY*SZ)
        C = C*EXP(-Y*Y/(2*SY*SY)
        C = C*2.*EXP(-H*H/(2*SZ*SZ))
        RETURN
        END
        SUBROUTINE SIGMA (X, SY, SZ)
        COMMON Q, H, U, KLS
        B = 0.894
        IF (X .GT. 1.0) GO TO 30
        IF (KLS .GT. 1.5) GO TO 22
        A = 213.
        C = 440.8
        D = 1.941
        F = 9.27
        GO TO 50
22      IF (KLS .GT. 2.5) GO TO 23
        A = 156.
        C = 106.6
        D = 1.149
        F = 3.3
        GO TO 50
23      IF (KLS . GT. 3. 5) GO TO 24
        A = 104.
        C = 61.0
        D = 0.911
        F = 0.0
        GO TO 50
24      IF (KLS .GT. 4.5) GO TO 25
        A = 68.
        C = 33.2
        D = 0.725
        F = -1.7
        GO TO 50
25      IF (KLS .GT. 5.5) GO TO 26
        A = 50.5
        C = 22.8
        D = 0.678
        F = -1.3
        GO TO 50
```

Table D.2 *(continued)*

```
26      A = 34.
        C = 14.35
        D = 0.74
        F = –0.35
        GO TO 50
30      IF (KLS .GT. 1.5) GO TO 32
        A = 213.
        C = 459.7
        D = 2.094
        F = –9.6
        GO TO 50
32      IF (KLS .GT. 2.5) GO TO 33
        A = 156.
        C = 108.2
        D = 1.098
        F = 2.0
        GO TO 50
33      IF (KLS .GT. 3.5) GO TO 34
        A = 104.
        C = 61.0
        D = 0.911
        F = 0.0
        GO TO 50
34      IF (KLS .GT. 4.5) GO TO 35
        A = 68.
        C = 44.5
        D = 0.516
        F = –13.0
        GO TO 50
35      IF (KLS .GT. 5.5) GO TO 36
        A = 50.5
        C = 55.4
        D = 0.305
        F = –34.0
        GO TO 50
36      A = 34.
        C = 62.6
        D = 0.18
        F = –48.6
50      SY = A*X**B
        SZ = C*X**D+F
        IF (SZ .LT. 0.1) THEN SZ = 0.1
        RETURN
        END

        SUBROUTINE GOLDN (DEL, A1, B1, X, C)
        COMMON Q, H, U, KLS
        DIMENSION A (99), B (99), D (99), E (99)
```

Table D.2 *(continued)*

```
C
C       GOLDEN SECTION SEARCH ALGORITHM
C
C       THIS VERSION FINDS THE MAX OF A CONCAVE FUNCTION
C       TO FIND THE MIN OF A CONVEX FUNCTION JUST REVERSE THE SIGN
C          OF THE FUNCTION OR REVERSE THE SIGN OF THE INEQUALITY IN
C          THE STATEMENT RIGHT AFTER LINE 10
C
C       DEFINITIONS: A(K)=LEFT SIDE OF INTERVAL, B(K)=RIGHT SIDE
C       D(K)=LAMBDA(K)=NEW LEFT SIDE, E(K)=MU(K)=NEW RIGHT
C          SIDE
C       C1=FUNCTION EVALUATED AT D(K), C2=FUNCTION AT E(K)
C
        FR = 0.618
        Y = 0.0
        A(1) = A1
        B(1) = B1
        D(1) = A(1) + (1. –FR)*(B(1) –A(1) )
        E(1) = A(1) + FR*(B(1) –A(1) )
        CALL EVAL (D(1),Y, C1)
        CALL EVAL (E(1), Y, C2)
        K = 1
10      IF ( (B (K) – A (K) ) .LT. DEL) GO TO 80
        K1 = K + 1
        IF (C1 .GT. C2) GO TO 30
        C1 = C2
        A(K1) = D(K)
        B(K1) = B(K)
        D(K1) = E(K)
        E(K1) = A(K1) + FR*(B(K1) – A(K1) )
        CALL EVAL (E (K1), Y, C2)
        GO TO 50
30      C2 = C1
        A(K1) = A(K)
        B (K1) = E(K)
        E(K1) = D(K)
        D(K1) = A (K1) + (1. –FR)*(B (K1) – A (K1) )
        CALL EVAL (D (K1) , Y, C1)
50      K = K + 1
        GO TO 10
80      X = (B(K) + A(K) ) /2.
        CALL EVAL (X, Y, C)
        WRITE (16, 81)
        WRITE (16, 82) K,D(K),E(K),C1,C2
81      FORMAT (2X////2X, "NO. ITERATIONS", 2X, "LEFT SIDE",
      1    2X, "RIGHT SIDE", 12X, "CONC. 1", 5X, "CONC 2"//)
82      FORMAT (10X, I2, 7X, F8.3, 3X, F8.3, 2X, F8.1, 2X, F8.1///)
        RETURN
        END
```

APPENDIX E: PRACTICE PROBLEMS (WITH SOLUTIONS) IN AIR QUALITY FOR THE P.E. EXAMINATION IN ENVIRONMENTAL ENGINEERING

The problems in this Appendix have been prepared as a study aid for the Air Quality portion of the *Principles and Practice of Engineering Examination in Environmental Engineering*. These problems are not actual problems that have appeared on previous P.E. examinations, but are presented in a similar format to help the student prepare for this important exam. In addition, various tips for the P.E. Exam candidate are offered.

Given

$$PV = nRT = \frac{M}{MW} RT$$

where

P = absolute pressure

V = volume of gas

n = moles of gas

M = mass of gas

MW = molecular weight of gas

T = absolute temperature

R = universal gas constant

For use in the above equation, two values of R (choose the most convenient one) are:

$$R = 0.73 \frac{(\text{atm})(\text{ft}^3)}{(\text{lbmole})(^\circ \text{R})}$$

$$R = 0.0821 \frac{(\text{atm})(\text{L})}{(\text{gmole}) (^\circ \text{K})}$$

Tip No. 1: The ideal gas law is the basis for a large number of air quality calculations.

1. The volume occupied by one gmole of an ideal gas at 25 °C and 1 atm is most nearly:

    a. 2.05 L

    b. 22.4 L

    c. 24.5 L

    d. 28.9 L

2. The ozone concentration at a monitoring site is measured as 0.11 ppm, at 25 °C and 1 atm. What is the concentration (in $\mu g/m^3$) at 25 °C and 1 atm? (MW of ozone is 48)

   a. 155 $\mu g/m^3$

   b. 190 $\mu g/m^3$

   c. 215 $\mu g/m^3$

   d. 260 $\mu g/m^3$

3. The SO_2 concentration in a stack is 500 ppm, the stack diameter is 12 ft, and the stack gas velocity is 65 ft/sec. The gas temperature and pressure are 390 °F and 1 atm. Calculate the approximate SO_2 mass emission rate in lb/day. (The atomic weight of S = 32 and the atomic weight of O = 16)

   a. 12,300 lb/day

   b. 16,400 lb/day

   c. 24,600 lb/day

   d. 32,800 lb/day

4. A contaminated airstream flows through a paint-drying oven in an automobile factory at 500 m³/min. The temperature is 50 °C and the pressure is 1 atm. It has been estimated that paint solvent evaporates in the oven at an average rate of 0.05 kg/min. Air has an average MW of 29 and the solvent has an average MW of 90. The solvent concentration in the oven exhaust, in ppm, is most nearly:

   a. 0.0001 ppm

   b. 0.1 ppm

   c. 2.4 ppm

   d. 29 ppm

Tip No. 2: Specific questions about laws or regulations generally do not appear on the P.E. exam because of frequent changes in the regulations and/or differences among states.

5. A stack (T = 550 F and P = 750 mm Hg) was sampled using EPA Method 5. The total gas volume that flowed through the dry gas meter was 2.785 cubic meters (at T = 60 F and P = 800 mm Hg). The mass of particles collected was 2.50 g. Also, 72.0 g of H_2O was collected in impingers. Calculate the approximate concentration of PM in the stack ($\mu g/m^3$) at stack conditions.

   a. $2.5(10)^5$ $\mu g/m^3$

   b. $4.2(10)^5$ $\mu g/m^3$

   c. $5.9(10)^5$ $\mu g/m^3$

   d. $7.3(10)^5$ $\mu g/m^3$

6. The concentration of PM inside a stack is 100,000 µg/m^3. The stack gas exits at an average velocity of 20 m/s, and the stack has an inside diameter of 2 meters. The emission rate of PM, in kg/day, is approximately:

   a. 1.5 kg/day

   b. 27.2 kg/day

   c. 138 kg/day

   d. 543 kg/day

Tip No. 3: The P.E. exam concentrates on stationary source emissions control, with much less emphasis on ambient air quality.

7. A power plant burns 4000 tons/day of coal with a heat content of 12,000 Btu/lb. Given a regulatory limit that is 0.03 lb of PM/million Btu of heat input, calculate the allowable daily emissions of PM, in lb/day.

   a. 2880 lb/day

   b. 3600 lb/day

   c. 8760 lb/day

   d. 14,400 lb/day

Tip No. 4: Some familiarity with air pollution control equipment, units, and definitions of terms (e.g., ESP, cyclone, baghouse, grains/ft^3, collection efficiency, etc.) is assumed.

8. Dusty air at a fertilizer plant flows through a 73% efficient cyclone and then through an ESP. The inlet air to the cyclone has a dust loading of 80 grains/cubic foot. In order to meet a standard of 98.5% collection efficiency for the fertilizer plant as a whole, what is the allowable concentration of dust (in grains/cubic foot) in the air that exits from the ESP?

   a. 0.15 gr/ft^3

   b. 1.2 gr/ft^3

   c. 21.6 gr/ft^3

   d. 58.4 gr/ft^3

9. Calculate the required collection efficiency for just the ESP in the above problem.

   a. 91.5%

   b. 94.5%

   c. 97.5%

   d. 98.5%

10. An industrial plant is currently operating a large baghouse for air pollution control. The air flows at a rate of 100,000 acfm. The gas

temperature is 250 °F and the inlet dust loading is 12 grains/scf. The emission regulation for the plant as a whole is 0.05 grains/scf. The required collection efficiency for the baghouse is about:

a. 90%

b. 96%

c. 99%

d. 99.6%

Tip No. 5: On the P.E. exam, the air-to-cloth ratio for the type of dust and baghouse would be supplied.

11. A fabric filter is treating 30,000 acfm with an average air-to-cloth ratio of 2 ft/min. The bags are 8 inches in diameter and 12 feet long. Estimate the number of bags in the baghouse.

a. 80 bags

b. 200 bags

c. 600 bags

d. 1200 bags

12. The following table provides the PM size distribution by mass and the collection efficiency for a proposed control device as a function of particle size.

Particle Size Range, μm	Mass fraction	Efficiency, %
0–6	0.10	20
6–16	0.30	50
16–30	0.40	83
30+	0.20	100

The overall collection efficiency of the control device on this PM is most nearly:

a. 50%

b. 60%

c. 70%

d. 80%

13. An ESP must treat 800,000 acfm with 99.1% efficiency. Assuming an effective drift velocity of 14 ft/min, calculate the required collection area.

a. 152,000 ft^2

b. 269,000 ft^2

c. 347,000 ft^2

d. 462,000 ft^2

Tip No. 6: The design method of Lapple has remained effective for many years, and thus is acceptable for use on a P.E. exam problem.

14. Consider a Lapple conventional cyclone with an inlet height of 1 m and width of 0.5 m. Standard air (viscosity = 0.0011 kg/m-min) flows into the cyclone at 600 m³/min. The PM in the airstream has a density of 1100 kg/m³. The cyclone cut diameter, in μm, is most nearly:

    a. 10 μm

    b. 13 μm

    c. 16 μm

    d. 20 μm

Tip No. 7: Always check to be sure you are using variables with the proper units in your equations. Some equations need only consistent units; others require specific units.

15. An ESP is treating the flue gas from a coal combustion unit and achieving a 98% PM removal efficiency. They start burning a new coal and the efficiency drops to 93%, but the gas flow rate has remained exactly the same. Estimate the ratio of the new effective drift velocity (w_{new}) to the old one (w_{old}).

    a. 0.50

    b. 0.68

    c. 0.77

    d. 0.84

16. A coal-burning power plant burns coal at a rate of 4000 tons/day. The ash content of the coal is 5%, and the sulfur content is 1.5% by weight. The rate of $CaCO_3$ needed to capture 90% of the SO_2 generated is about: (Atomic weights are: C = 12, O = 16, S = 32, Ca = 40)

    a. 54 tons/day

    b. 85 tons/day

    c. 108 tons/day

    d. 169 tons/day

Tip No. 8: Always draw a simple diagram when starting a mass balance type problem. It helps you understand the problem better and directs you to the correct answer quickly.

17. A coal-burning power plant burns coal at a rate of 4000 tons/day. The ash content of the coal is 5%, and the sulfur content is 1.5% by weight. Assume 30% of the ash drops out in the furnace and 70%

becomes fly ash. Calculate the efficiency required of a final control device to meet a PM emissions limit of 2.0 tons/day.

a. 98.6%

b. 99.1%

c. 99.5%

d. 99.9%

Tip No. 9: When calculating the temperature change of a gas that is being heated or cooled (or when estimating the heat to be added or removed to achieve a desired temperature change), use Appendix B (either Table B.7 or Figure B.2).

18. An afterburner is being designed to control hexane emissions. A contaminated airstream from the source flows at 4000 acfm at 200 °F and 1 atm. What is the approximate rate of heat addition needed to raise the temperature to 1400 °F?

a. 47,700 Btu/min.

b. 62,300 Btu/min.

c. 73,900 Btu/min.

d. 93,400 Btu/min.

19. One lb of fuel is being burned with 30 lb of air in a perfectly insulated chamber. The fuel has a heating value of 15,000 Btu/lb. The fuel and the air are coming into the chamber at 75 °F. The temperature of the exhaust gases is most nearly:

a. 1500 °F

b. 1900 °F

c. 2200 °F

d. 2500 °F

20. An exhaust gas flows from an afterburner at 12,000 acfm (at 1400 °F and 1 atm). If the exhaust is cooled to 400 °F by heat exchange in a waste heat boiler, how much steam can be generated? Assume that about 1100 Btu is needed to generate 1 lb of steam.

a. 3600 lb/hr

b. 9500 lb/hr

c. 12,300 lb/hr

d. 16,800 lb/hr

21. If 10,000 acfm of exhaust gases is cooled from 1400 °F to 400 °F using dilution air at 70 °F, determine the volumetric flow of the cooled mixed stream.

a. 18,600 cfm

b. 25,200 cfm

c. 31,800 cfm

d. 43,400 cfm

22. If 10,000 acfm of exhaust gases is cooled from 1400 °F to 400 °F by spraying 70 °F water into the airstream, estimate the water addition rate needed, in gal/min.

   a. 5.3 gpm

   b. 8.5 gpm

   c. 18.5 gpm

   d. 43.3 gpm

23. An airstream contaminated with 500 ppm benzene flows at 2000 m^3/min at T = 30 °C and 1 atm. The process operates 24 hours/day. A carbon adsorption unit is to be used to control the emissions. The working capacity of the carbon is 0.2 g benzene/g of carbon. Estimate the carbon required in both beds of a 2-bed adsorber if each bed is run for 3 hours (then regenerated while the other bed is on-line). The molecular weight of benzene is 78.

   a. 2830 kg

   b. 4150 kg

   c. 5650 kg

   d. 7290 kg

Tip No. 10: Superficial gas velocity is simply the gas volumetric flow rate divided by the total face area (Q/A). Although not the true velocity at any point in the bed, it is used in calculations involving fabric filters, carbon adsorbers, and many other devices.

24. Determine the cross-sectional area of a carbon bed to treat 20,000 cfm if the maximum acceptable superficial velocity through the adsorber is 70 ft/min.

   a. 3.5 ft²

   b. 14 ft²

   c. 72 ft²

   d. 286 ft²

25. Assume that the carbon bed in question #24 is 2 feet deep and consists of carbon that is sized between 4 and 10 mesh. Estimate the pressure drop in this bed.

   a. 3 in. H_2O

   b. 6 in. H_2O

   c. 9 in. H_2O

   d. 12 in. H_2O

Tip No. 11: You can never escape from having to know how to balance chemical reactions!

26. The volume of air needed for complete combustion of 1 tank of propane (C_3H_8) in a backyard gas grill is most nearly: (One tank of propane contains about 45 lbs, which is about 400 scf as a gas.)

    a. 400 scf

    b. 2000 scf

    c. 9500 scf

    d. 12,500 scf

27. A test of a pilot-scale catalytic oxidizer provides a removal efficiency of 95%. The catalyst bed was 2.0 inches in length. The required length of catalyst in a full-scale catalytic oxidizer that is designed to remove 99.8% of the pollutant is closest to:

    a. 3.0 inches

    b. 4.2 inches

    c. 5.6 inches

    d. 7.1 inches

28. The uncontrolled NO_x exhaust emission rate on a new U.S. passenger car is about 3.0 g/mile. The conversion efficiency of a catalytic converter to meet an emission standard of 0.4 g/mile is most nearly:

    a. 87%

    b. 90%

    c. 98%

    d. 99%

Tip No. 12: Mobile source control is not addressed in much detail in the P.E. exam.

29. Determine stoichiometric air required for the combustion of octene (C_8H_{16}) in pounds of air per pound of fuel.

    a. 12.5 lb/lb

    b. 13.6 lb/lb

    c. 14.7 lb/lb

    d. 15.8 lb/lb

30. Estimate the downwind, centerline, ground-level concentration, at a downwind distance of 5000 meters from the stack, of a nonreactive pollutant which is emitted from a 90-m-tall stack at the rate of 40 g/s. The atmospheric stability is Class D, and the wind speed

at stack height is 5.0 m/s. The plume rise is 40 m. Give your answer in $\mu g/m^3$.

a. 12 $\mu g/m^3$

b. 34 $\mu g/m^3$

c. 142 $\mu g/m^3$

d. 198 $\mu g/m^3$

Tip No. 13: The Gaussian equation is the standard for simple dispersion calculations, but the box model is widely used as well.

31. A fireplace is releasing NO_x into a room at the rate of 5 mg/min. The room has dimensions of 2 m × 5 m × 6 m, and is ventilated at the rate of 1 air change per hour with fresh air ($NO_x = 0$). Estimate the steady-state concentration of NO_x in the room. Ignore any NO_x reactions that might occur within the room.

a. 25 $\mu g/m^3$

b. 180 $\mu g/m^3$

c. 600 $\mu g/m^3$

d. 5000 $\mu g/m^3$

SOLUTIONS

1. **The correct answer is (c)**

   From the ideal gas law: $\dfrac{V}{n} = \dfrac{RT}{P}$; and with $P = 1$ atm, $T = 25$ °C

   $$\frac{V}{n} = \frac{0.0821\left(\dfrac{atm\ L}{gmol°K}\right) \times (25 + 273)°K}{1\ atm}$$

   $$= 24.5\ L/gmole$$

2. **The correct answer is (c)**

   The answer can be derived from the ideal gas law as shown in Ch. 1, or obtained directly from Eq. (1.9):

   $$C = \frac{1000(0.11)48}{24.5}$$

   $$C = 215\ \mu g/m^3$$

3. **The correct answer is (d)**

   MW of $SO_2 = 32 + 2(16) = 64$

   Volume flow rate of "pure" $SO_2 = 0.0005 \times 65$ ft/sec

   $$\times \pi \frac{(12\ ft)^2}{4} \frac{86,400\ sec}{day} = 3.176\ (10)^5\ ft^3/day$$

From the ideal gas law: $\dot{M} = \dfrac{VPMW}{RT}$

$$\dot{M} = \frac{3.176\ (10)^5\ \text{ft}^3/\text{day} \times 1\ \text{atm} \times 64}{0.73 \left(\dfrac{\text{atm ft}^3}{\text{lb mol}^\circ\text{R}} \right)(460 + 390)^\circ\text{R}} = 32{,}800\ \text{lb/day}$$

4. **The correct answer is (d)**

   (Remember, ppm for gases is almost always by volume or moles.)

   $$n_{\text{air}} = \frac{PV}{RT} = \frac{1\ \text{atm}\ 500\ \text{m}^3 \times \dfrac{1000\ \text{L}}{\text{m}^3}}{0.082(50 + 273)}$$

   $n_{\text{air}} = 18{,}880\ \text{gmol/min}$

   $$n_{\text{solvent}} = \frac{0.05\ \text{kg/min} \times 1000\ \text{g/kg}}{90\ \text{g/gmol}} = 0.555\ \text{gmol/min}$$

   mole fraction of solvent $= \dfrac{0.555}{18{,}880} \times 10^6 = 29\ \text{ppm}$

5. **The correct answer is (b)**

   note 60 °F = 15.6 °C = 289 °K; also 550 °F = 561 °K

   $$n_{\text{dry}} = \frac{\left(\frac{800}{760}\right)\text{atm}\ 2785\ \text{L}}{0.082\dfrac{\text{L} \cdot \text{atm}}{\text{gmol K}}289\text{K}} = 123.7\ \text{gmol}$$

   $$n_{\text{H}_2\text{O}} = \frac{72.0\text{g}}{18(\text{g/gmol})} = 4.0\ \text{gmol}$$

   $n_T(\text{at stack}) = 123.7 + 4.0 = 127.7\ \text{gmol}$

   $$V\ (\text{at stack}) = \frac{127.7 \times 0.082 \times 561}{(750/760)}$$

   $$= 5950\ \text{L}$$

   $$C_{\text{part}} = \frac{2.50\text{g}}{5950\ \text{L}} \times \frac{1000\ \text{L}}{\text{m}^3} \times \frac{10^6\,\mu\text{g}}{1\ \text{g}}$$

   $$= 4.20(10)^5\ \mu\text{g/m}^3$$

6. **The correct answer is (d)**

   $\dot{M} = QC$

   $$= 20\frac{\text{m}}{\text{s}} \times \pi \frac{(2\ \text{m})^2}{4} \times 100{,}000\ \frac{\mu\text{g}}{\text{m}^3} \times \frac{1\ \text{kg}}{10^9\mu\text{g}} \times 86{,}400\ \frac{\text{s}}{\text{day}} = 543\ \text{kg/day}$$

7. **The correct answer is (a)**

$$PM = 4000 \frac{ton}{day} \times \frac{2000 \text{ lb}}{ton} \times \frac{12,000 \text{ Btu}}{lb} \times \frac{0.03 \text{ lb PM}}{10^6 \text{Btu}} = 2880 \frac{lb}{day}$$

8. **The correct answer is (b)**
   Overall penetration = 80 (1 − .985) = 1.2 gr/ft^3

9. **The correct answer is (b)**
   Cyclone penetration = 80(1 − 0.73) = 21.6 gr/ft^3

   $$\text{ESP penetration} = \frac{1.2 \text{ gr/ft}^3}{21.6 \text{ gr/ft}^3} = 0.055$$

   ESP effic = 1 − 0.55
   $$= 0.945 = 94.5\%$$

10. **The correct answer is (d)**

    $$\eta = \frac{12 - 0.05}{12} = 0.996 = 99.6\%$$

11. **The correct answer is (c)**

    Since air to cloth ratio = filtering velocity = $V = \dfrac{Q}{A}$

    then, $A = \dfrac{Q}{V} = \dfrac{30,000 \text{ acfm}}{2 \text{ ft /min}} = 15,000 \text{ ft}^2$

    The area of one bag $= \pi DL = \pi \left(\dfrac{8}{12}\right)12 = 25.1 \text{ ft}^2$

    Therefore: Number of Bags $= \dfrac{15,000}{25.1} = 597 \approx 600$

12. **The correct answer is (c)**

Size Range, μm	Mass fract.	Efficiency %	% Collected
0–6	0.10	20	2
6–16	0.30	50	15
16–30	0.40	83	33
30+	0.20	100	20
		Total	**70%**

13. The correct answer is (b)

Use Eq. (5.7):

$$\eta = 1 - e^{-Aw/Q}$$

$$e^{-Aw/Q} = 1 - \eta$$

$$A = \frac{-Q}{w} \ln (1 - \eta)$$

$$A = -\frac{800,000}{14} \ln (1 - .991)$$

$$A = 269,000 \text{ ft}^2$$

14. The correct answer is (a)

Use Eq. (4.6) and ignore the density of gas:

$$V_i = \frac{600 \text{ m}^3/\text{min}}{1 \text{ m} \times 0.5 \text{ m}} = 1200 \text{ m/min}$$

$$d_{pc} = \left[\frac{9\mu W}{2\pi N_e V_i (\rho_P - \rho_g)} \right]^{1/2}$$

$$= \left(\frac{9 \times 0.0011 \times 0.5}{2\pi \times 6 \times 1200 \times 1100} \right)^{1/2}$$

$$= 9.97 (10)^{-6} \text{ m} = 10 \text{ } \mu\text{m}$$

15. The correct answer is (b)

Use Eq. (5.7) twice

$$-\frac{Aw_{new}}{Q} = \ln (1 - \eta_{new}) \quad (1)$$

$$-\frac{Aw_{old}}{Q} = \ln (1 - \eta_{old}) \quad (2)$$

Divide (1) by (2)

$$\frac{w_{new}}{w_{old}} = \frac{\ln (1 - .93)}{\ln (1 - .98)}$$

$$= 0.68$$

16. The correct answer is (d)

$$S + O_2 \rightarrow SO_2 \qquad \text{Atomic weight of CaCO}_3 = 40 + 12 + 3(16) = 100$$
$$32 \quad 32 \qquad 64$$

$$SO_2 + CaCO_3 + \tfrac{1}{2}O_2 \rightarrow CaSO_4 + CO_2(g)$$
$$ 64 100$$

$$4000 \, \frac{\text{tons coal}}{\text{day}} \times \frac{0.015 \text{ ton S}}{\text{ton coal}} \times \frac{64 \text{ ton SO}_2}{32 \text{ ton S}} \times (0.90)$$

$$\times \frac{100 \text{ ton CaCO}_3}{64 \text{ ton SO}_2} = 169 \text{ tons/day}$$

17. The correct answer is (a)

Ash into furnace $= 4000$ tons/day coal $\times \dfrac{0.05 \text{ ton ash}}{\text{ton coal}} = 200$ ton/day

Fly ash out of furnace $= 200$ ton/day $\times (1 - 0.3) = 140$ ton/day

$$\eta = \frac{140 - 2}{140} \times 100\% = 98.6\%$$

18. The correct answer is (c)

(Refer to Appendix B, Table B.7)

$$H_{\text{air, 200}} = 33.6 \text{ Btu/lb} \qquad H_{\text{air, 1400}} = 341.5 \text{ Btu/lb}$$

From Table B.2, $\rho_{\text{air, 200}} = 0.060 \text{ lb/ft}^3$

$$\Delta H = 4000 \text{ ft}^3/\text{min} \times 0.060 \text{ lb/ft}^3 \times (341.5 - 33.6) \text{Btu/lb}$$
$$= 73{,}900 \text{ Btu/min}$$

19. The correct answer is (b)

Assume incoming fuel and air have similar enthalpies, about 5 Btu/lb.

Use mass balance to get exhaust flow:
1 lb + 30 lb = 31 lb

Use enthalpy balance to get temperature of exhaust:
$1 \times (5 + 15{,}000) + 30 \times (5) = 31 \times (h_E)$

$$h_E = \frac{15{,}155 \text{ Btu}}{31 \text{ lb}} = 488.9 \text{ Btu/lb}$$

Interpolating, from Appendix B, table B.7, T = 1910 °F

20. The correct answer is (a)

$$\Delta H = 12{,}000 \text{ acfm} \times 0.0212 \text{ lb/ft}^3 \times (341.5 - 82.1) \text{ Btu/lb}$$
$$= 65{,}991 \text{ Btu/min}$$

steam generated $= 65{,}991$ Btu/min $\times \dfrac{1 \text{ lb steam}}{1100 \text{ Btu}} \times \dfrac{60 \text{ min}}{\text{hr}} = 3600$ lb/hr

21. The correct answer is (a)

From Eq. (8.21)

$$Q_d = Q_e \left(\frac{T_e - T_f}{T_f - T_d}\right)\left|\frac{T_d}{T_e}\right|$$

$$Q_d = 10,000\left(\frac{1400 - 400}{400 - 70}\right) \times \left(\frac{70 + 460}{1400 + 460}\right) = 8635 \text{ cfm}$$

Total gas volume at 400 °F $= 10,000\left(\frac{400 + 460}{1400 + 460}\right) + 8635\left(\frac{400 + 460}{70 + 460}\right)$

$$= 18,635 \text{ acfm} = 18,600 \text{ acfm (rounded)}$$

22. The correct answer is (a)

Use Eq. (8.24)

$$\dot{M}_a C_{p(\text{avg})}\left(T_a - T_f\right) = \dot{M}_w\left[\Delta H_v + C_{pwv}\left(T_f - T_w\right)\right]$$

First, from the ideal gas law

$$\dot{M}_a = \left(\frac{10,000 \text{ cfm}}{0.73} \frac{1 \text{ atm}}{(1400 + 460)}\right)29 \text{ lb/lb mol} = 213.6 \text{ lb/min}$$

$$C_{pwv} \cong 0.45$$

$$C_{p(\text{air})} = 0.25 \text{ Btu/lb-°F}, \quad \Delta H_v = 1060 \text{ Btu/lb},$$

$$213.6 \times 0.25 \times (1400 - 400) = \dot{M}_w[1060 + 0.45 \times (400 - 70)]$$

Solving for $\dot{M}_w$,

$$\dot{M}_w = 44.2 \text{ lb/min}$$

The density of water is 8.33 lb/gal

$$Q_w = \frac{44.2}{8.33} = 5.3 \text{ gal/min}$$

23. The correct answer is (c)

The mass flow rate of benzene is

$$\dot{M}_B = \frac{2000 \text{ m}^3/\text{min (1 atm)}}{0.082 \frac{\text{L atm}}{\text{g mol K}} \; 303\text{K}} \times \frac{500}{10^6} \times \frac{78 \text{ g}}{\text{g mol}} \times \frac{1000 \text{ L}}{\text{m}^3} = 3140 \text{ g/min}$$

Amount adsorbed in 3 hours in one bed = 3140 g/min × 60 min/hour × 3 hours = 565,000 g benzene

Carbon in one bed $= \dfrac{565 \text{ kg benzene}}{0.2 \; \frac{\text{kg benzene}}{\text{kg C}}} = 2826 \text{ kg of carbon}$

Total Carbon = 5652 kg

24. The correct answer is (d)

$$A = \frac{20,000 \text{ ft}^3/\text{min}}{70 \text{ ft/min}} = 286 \text{ ft}^2$$

25. The correct answer is (c)

From Figure 12.6, $\Delta P = \dfrac{4.5 \text{ in. H}_2\text{O}}{\text{ft. of bed}}$

$\Delta P = 4.5 \text{ in. H}_2\text{O/ft} \times 2.0 \text{ ft} = 9 \text{ in. H}_2\text{O}$

26. The correct answer is (c)

$$C_3H_8 + 5O_2 \rightarrow 3CO_2 + 4H_2O$$

1 mole of propane requires 5 moles of O_2, thus 1 scf propane needs 5 scf O_2

But, $O_2 = 21\%$ of air

$$\text{air needed} = \frac{5 \text{ scf } O_2/\text{scf propane}}{0.21 \text{ scf } O_2/\text{scf air}} = \frac{23.8 \text{ scf of air}}{\text{scf of propane}}$$

400 scf propane $\times$ 23.8 = 9524 scf air

27. The correct answer is (b)

From Eq. (11.29),

$$\frac{C_L}{C_O} = e^{-\frac{L}{L_m}}$$

For 95% efficiency, $C_L = 1 - .95 = 0.05$

$$0.05 = e^{\frac{-2.0}{L_m}}$$

$$\ln(.05) = \frac{-2.0}{L_m}$$

$$L_m = \frac{-2.0}{-2.996}$$

$L_m = 0.6676$ inches

To achieve 99.8% efficiency,

$$0.002 = e^{\frac{-L}{0.6676}}$$

$$\ln(.002) = \frac{-L}{0.6676}$$

$L = 4.15$ inches

28. The correct answer is (a)

$$\eta = \frac{3.0 - 0.4}{3.0} \cdot 100 = 86.7\%$$

29. The correct answer is (c)

The raw equation is:

$$C_8H_{16} + X\ O_2 + 3.76(X)N_2 \rightarrow 8CO_2 + 8H_2O + 3.76(X)N_2$$

The balanced equation is:

$$C_8H_{16} + 12\ O_2 + 45.1\ N_2 \rightarrow 8CO_2 + 8H_2O + 45.1N_2$$

$$\text{lb air/lb fuel} = \frac{12(32)\ +\ 45.1(28)}{8(12) + 16(1)} = 14.7$$

30. The correct answer is (b)

$$C = \frac{Q}{2\pi\ u\ \sigma_y \sigma_z} \times 2\ \exp\left(-\frac{H^2}{2\sigma_z^2}\right)$$

For Class D, at x = 5000 m, σ_y = 287 m and σ_z = 89 m

$$C = \frac{40\ (10)^6}{2\pi\ 5\ (287)(89)} \times 2\ \exp\left(-\frac{(90 + 40)^2}{2(89)^2}\right)$$

$$C = 34\ \mu g/m^3$$

31. The correct answer is (d)

$V_{room} = 2 \times 5 \times 6 = 60\ m^3$
So, 1 air change per hour means $Q = 60\ m^3/hr$
A steady-state mass balance is:

$$0 = \frac{60\ m^3}{hr} \times 0\ \mu g/m^3\ +\ \frac{5\ mg}{min} \times \frac{60\ min}{hr} \times \frac{1000\ \mu g}{1\ mg}\ -\ \frac{60\ m^3}{hr} C$$

$$C = 5000\ \mu g/m^3$$

◆ APPENDIX F: ANSWERS TO ◆
ODD-NUMBERED PROBLEMS

1.1   1989: 32.1% from transport; 27.6% from utilities
        1998: 31.7% from transport; 24.9% from utilities

1.3 $NO_x = 61.5$ T/D; $SO_x = 61.5$ T/D; PM = 3.1 T/D

1.5   The oil-fired plant emits more SO_x

1.7   (a) $1.18(10)^5$ µg/m^3; (b) 2870 kg/day

1.9   (b) $Q = 2.52(\Delta P)^{1/2} + 0.44$; (c) 5.79 ft^3/min; (d) 246 g/min

1.11   40 mg/m^3

1.13   1.03 g/liter

1.15   737 ppm

1.17   (a) 8.9 km; (b) 1.6 km

1.19   10% HbCO is 27% of the saturation level

1.21   Los Angeles—95—Moderate
         Houston—136—Unhealthful
         Honolulu—57—Moderate

1.23   22,000 kg/day of SO_2

1.25   0.080 ppm

1.27   (a) $2.52(10)^5$ µg/m^3; (b) 8550 kg/day; (c) $4.96(10)^5$ µg/m^3

1.29   0.01476 mg/cm^3

2.1   (a) 2.4%; (b) $9.28(10)^4$ mg/m^3

2.3   250 lb$_m$/day

2.7   \$253,000

2.9   choose water injection

2.11   (a) 11.48 mol/hr; (b) 0.0196 mol fract

3.1   (a) 2.5 gr/ft^3; (b) 13,300 kg/day

3.3   Yes, it is log-normal. $\sigma_g = 2.0$; $d_{50} = 10.8$ µm

3.5   96.2%

3.7   95%

3.9   5.5% (by mass) less than 2.0 µm

3.11   7.2

3.13   (a) $3.66(10)^{-3}$ m/s; (b) $1.0(10)^{-3}$ m/s

3.15   98.5%

3.17   93%

3.19   Ratio (mass) = 21.9

3.21   (a) 0.75 gr/ft^3; (b) 96.7%

3.23   50%

3.25   15%

4.1   81%

4.3   (a) 87%; (b) 89%

4.5   3.30 kPa or 13.2 in. H_2O

4.7　2.0 kPa or 7.9 in. H_2O

4.9　57.1 ft/sec; Yes

4.11　$\dfrac{\Delta P_{\text{High Efficiency}}}{\Delta P_{\text{High Throughput}}} = 14.1$

4.13　$14,300

4.15　(a) $8,350/year; (b) $16,100/year

4.17　91.6%

4.19　1.51 kPa; 2.52 kW

4.21　Lapple: 67% and 10.2 in. H_2O
　　　Swift: 65% and 10.2 in. H_2O

5.1　(a) 5433 m^2 (required), 5760 m^2 (actual); (b) 51 plates

5.3　(a) 96,000 ft^2; (b) 244 plates

5.5　99.1%

5.7　$A_{99}/A_{90} = 2.0$

5.9　DEC (1998 $) = $7,250,000; TIC (1998 $) = $16,100,000
　　　TIC (2002 $) = $17,400,000

5.11　DEC = $1,172,000; TIC = $2,600,000

5.13　97.6%

5.15　(a) 3430 m^2; (b) 56 plates

5.17　(a) ratio of w (new)/w (old) = 0.72; (b) w (new) = 4.3 m/min

6.1　K_s = 15.9 Pa-min-m/g; K_e = 490 Pa-min/m

6.3　ΔP increases with temperature

6.5　(a) 3 compartments; (b) 2.5 ft/min; (c) 4022 ft^2; (d) 384 bags;
　　　(e) 5.3 in. H_2O

6.7　135 bags

6.9　$2.69(10)^{-11}$ ft^2

6.11　(a) 905 bags; (b) 222 kW; (c) 316 kW

6.13　$893,000

6.15　(a) 9.6 in. H_2O; (b) 819 bags

6.17　V_{des} = 2.25 fpm; N = 5; V_{N-1} = 2.25 fpm; V_N = 1.80 fpm

6.19　2.7 in. H_2O

6.21　(a) 80 bags; (b) 33.8 kW; (c) 20 kW

6.23　(a) 5.67 ft; (b) 65 hp; (c) $104/day; (d) $114/day

7.1　80%

7.3　65%

7.5　(a) 53 in. H_2O; (b) 3.7 kWh/1000 m^3

7.7　(a) 99.9%; (b) 94%

7.9　For Q_L/Q_G = 5, η = 66% for 1 μm and 88% for 6 μm
　　　For Q_L/Q_G = 10, η = 92.9% for 1 μm and 99.7% for 6 μm

7.11　62.7%

7.13　0.86 gpm

8.1　(a) 2000 rpm; (b) 9.4 in. H_2O

8.3 $D_{new}/D_{old} = 1.26$
8.5   0.0001575
8.7   80.7%
8.9   6.3 in. H_2O
8.11   (a) 26 inches × 26 inches; (b) 48 inches × 16 inches
8.13   (a) 28,280 cfm; (b) 71,000 cfm; (c) 290,000 cfm
8.15   14,000 ft^2
8.17 $68,000
8.19   21 inches
8.21   (a) 60 inches with heat exchanger; 80 inches without;
       (b) $350,000
8.23   11.7 gpm

10.1 $\overline{P}_{NO} = 0.0264$ atm; $y_{NO} = 0.0264$
10.3 $\overline{P} = 0.5$ psi; $C = 34,000$ ppm
10.5   12 C
10.7   (a) $2.2(10)^{-6}$; (b) $8.4(10)^{-5}$
10.9   300,000 liters
10.11 $E = 26.2$ kcal/mole; $A = 1.05(10)^7$ sec^{-1}
10.13 0.30 psia; 13,200 ppm
10.15 (a) 37,000 L; (b) 135 kg/day

11.1   (a) 1800 F; (b) 1670 F
11.3   (a) 9.2 kg/min; (b) 14 m^3/min
11.5   1270 F
11.7 $L = 16$ ft; $D = 6.9$ ft; methane = 5.5 kg/min
11.9   (a) fuel gas can be reduced by 1.9 lb_m/min; (b) preheated waste
       gas $T = 1150$ F
11.11 (a) Vol = 12.0 ft^3; (b) Length = 3.0 ft
11.13 $465,000
11.15 17.1:1; 34.2:1; 0.10
11.17 Autoignition: 1120 F; Lee et al.: 1230 F; Cooper et al.: 1310 F
11.19 (a) 2.16 kg; (b) 3.30 m^3
11.21 4.9 ppm

12.1   22.4 lb_m hexane/100 lb_m carbon
12.3   Either method: 10 lb_m CS_2/100 lb_m carbon
12.5   100 hp
12.7 $347,000
12.9   1385 kg/bed
12.11 9.0 in. H_2O
12.13 $\Delta P = 1.5, 4.25,$ and 8.9 in. H_2O, respectively
12.15 (a) 3000 lb; (b) 5 ft × 10 ft × 1 ft deep

13.1   (a) 0.0415; (b) 0.0083; (c) 4.3 lbmol/(hr-ft^3-Δy); (d) 44 ft;
       (e) 8.4; (f) column is already operating above flooding point

13.3   31 ft
13.5   23,000 lb_m/hr
13.7   96 gpm
13.9   8.25 ft

14.1 H_D = 19.7; ethane is not good
14.3   97.6%
14.5   (a) 1.7 gpm; (b) 83 °F, 1965 m^3/min; (c) yes, provide soaker hoses
14.7   Filter 1: 3.67 g/m^3-hr; Filters 2 & 3: 3.14 g/m^3-hr
14.9   Range is 0.9 to 2.6 million dollars
14.11   (a) 98.8%; (b) 97.6%
14.13 D = 3 m, Z = 7.5 m

15.1   1476 tons/day
15.3   1211 lb/hr
15.5 $9.85(10)^7$ Btu/hr
15.7   (a) 3564 ppm; (b) 285 ppm; (c) 25,200 lb/hr; (d) 1800 hp

16.1   823 ppm
16.3   1730 ppm
16.5   (a) 408 ppm; (b) 5.79 kg/hr
16.7   3380 kg/day
16.9   13.7 kg air/kg fuel
16.11   (a) NO = 10,600 ppm; N_2 = 0.725; O_2 = 0.0447; (b) 0.215 mol/m^3-s; (c) 776 ppm
16.13   9.1 ppm

18.1 VOC_{1990} = 90% VOC_{1975}; $NO_{x_{1990}}$ = 149% $NO_{x_{1975}}$
18.3 VOC_{1990} = 105% VOC_{1975}; $NO_{x_{1990}}$ = 150% $NO_{x_{1975}}$
18.5   France = 4.5 mmt; Thailand = 0.42 mmt; U.S. = 41.0 mmt
         France/U.S. = 0.11; Thailand/U.S. = 0.01; Thailand/France = 0.09
18.7   U.S. = $5.0(10)^{11}$ kg/yr or 33%
         Rest of OECD: $6.0(10)^{11}$ kg/yr or 40%
         Non-OECD: $4.0(10)^{11}$ kg/yr or 26%
18.9   14.5
18.11   CO = 15.1 g/mile; NO = 1.89 g/mile
18.13   (a) 6.44; (b) 8.95; (c) 17.2; (d) 34.3
18.15   (a) 13.7; (b) 16.7; (c) 19.7
18.17   (a) 8.95; (b) 0.151 scfm

19.3   17,350 ft
19.5   3125 ft
19.7   345 million gallons
19.9   Southeast

20.1   (a) 526 $\mu g/m^3$; (b) 66 $\mu g/m^3$
20.3   (a) $5.5(10)^4$ $\mu g/m^3$; (b) $1.7(10)^4$ $\mu g/m^3$
20.5   6.4 $\mu g/m^3$
20.7   53 $\mu g/m^3$
20.9   547 $\mu g/m^3$
20.11 $C_{max} = 300$ $\mu g/m^3$; $x_{max} = 5.6$ km
20.13   (a) 44 m; (b) 97 m; (c) 126 m
20.15 $x_{max} = 0.707\,H/c$; $(Cu/Q)_{max} = 0.234\,c/(aH^2)$
20.17   1.2 m
20.19 $3.4 million
20.21 $h = 140$ m; $D = 4.34$ m
20.23   (a) $u_c = 1.7$ m/s; (b) $u_c = 2.2$ m/s
20.25   47 $\mu g/m^3$

21.1   2460 cfm
21.3   old—4500 cfm; new—3000 cfm
21.5 $79/week
21.7 $3013
21.9   765 mg/min; 280 mg/m^3

INDEX